AF530786

MAGNETODYNAMIC PHENOMENA IN THE SOLAR ATMOSPHERE

PROTOTYPES OF STELLAR MAGNETIC ACTIVITY

MAGNETODYNAMIC PHENOMENA IN THE SOLAR ATMOSPHERE

Prototypes of Stellar Magnetic Activity

PROCEEDINGS OF THE 153RD COLLOQUIUM OF THE
INTERNATIONAL ASTRONOMICAL UNION
HELD IN MAKUHARI, NEAR TOKYO, MAY 22–27, 1995

EDITED BY

YUTAKA UCHIDA
Science University of Tokyo, Japan

TAKEO KOSUGI
National Astronomical Observatory, Tokyo, Japan

and

HUGH S. HUDSON
University of Hawaii, Honolulu, HI, U.S.A.

KLUWER ACADEMIC PUBLISHERS
DORDRECHT / BOSTON / LONDON

Library of Congress Cataloging-in-Publication Data

International Astronomical Union. Colloquium (153rd : 1996 : Makuhari, Japan)
Magnetodynamic phenomena in the solar atmosphere : prototypes of stellar magnetic activity : proceedings of the 153rd Colloquium of the International Astronomical Union, held in Makuhari, near Tokyo, May 22-27, 1995 / edited by Yutaka Uchida, Takeo Kosugi, and Hugh S. Hudson.
p. cm.
At head of title: International Astronomical Union; Union astronomique internationale.
ISBN 0-7923-4176-7 (alk. paper)
1. Solar magnetic fields--Congresses. 2. Stars--Magnetic fields--Congresses. I. Uchida, Y. (Yutaka) II. Kosugi, Takeo. III. Hudson, Hugh S. IV. International Astronomical Union. V. Title.
QB539.M23I58 1996
523.7'2--dc20 96-30322

ISBN 0-7923-4176-7

Published by Kluwer Academic Publishers,
P.O. Box 17, 3300 AA Dordrecht, The Netherlands.

Kluwer Academic Publishers incorporates
the publishing programmes of
D. Reidel, Martinus Nijhoff, Dr W. Junk and MTP Press.

Sold and distributed in the U.S.A. and Canada
by Kluwer Academic Publishers,
101 Philip Drive, Norwell, MA 02061, U.S.A.

In all other countries, sold and distributed
by Kluwer Academic Publishers Group,
P.O. Box 322, 3300 AH Dordrecht, The Netherlands.

Printed on acid-free paper

Printed in the Netherlands

This volume is dedicated
to our dear friend Prof. Yoshiaki Ogawara
without whose selfless contributions
the satellite *Yohkoh* would not have come into existence

Memorial: **Prof. Keizo Kai** (1934-1991)

Keizo Kai, Professor of Astronomy and Director of the Nobeyama Solar Radio Observatory, played a major role in the development of solar physics in Japan. His research work always centered on high energy solar physics, especially in the joint interpretation of radio/X-ray bursts. His career began in the early 1960's at the Tokyo Astronomical Observatory (later the National Astronomical Observatory). He participated in the early development of the Nobeyama Radio Observatory and made two extended visits to the Culgoora Solar Observatory in Australia. These visits produced many fascinating discoveries, and these led to his great motivation for the development of the Nobeyama Solar Radio Observatory, and especially for the creation of the recently-completed 17 GHz radioheliograph. He served as Director of the Nobeyama Solar Radio Observatory until the end. He also strongly encouraged the development of *Yohkoh* and served as the Principal Investigator of its Hard X-ray Telescope. Many of the results presented at the conference represent his observational dreams.

Table of Contents

I.2. High Temperature Outer Atmospheres of Other Stars

II. Wind and Mass-Loss from the Sun and Stars

II.1. Wind and Mass-Loss from the Sun

II.2. Wind and Mass-Loss from Stars

III. Production of Superhot Plasma and High-Energy Particles in the Sun and Stars

III.1. Solar Flares

III.2. Stellar Flares

III.3. Models of Flares

IV. Magnetic Behavior of the Sun and Stars

IV.1. Active Zones and Coronal Holes of the Sun and Their Cycle Variation

IV.2. Observed Domain of Activity on Stars

V. Governing Factors of Solar/Stellar Activity

Poster Papers

Posters for Session I

Posters for Session II

Posters for Session III

Posters for Session IV

Preface

These are the Proceedings of Colloquium No. 153 of the International Astronomical Union, held at Makuhari near Tokyo on May 22 - 26, 1995, and hosted by the National Astronomical Observatory.

This meeting was intended to be an interdisciplinary meeting between researchers of solar and stellar activity, in order for them to exchange the newest information in each field. While each of these areas has seen remarkable advances in recent years, and while the researchers in each field have felt that information from the other's domain would be extremely useful in their own work, there have not been very many opportunities for intensive exchanges of information between these closely related fields. We therefore expected much from this meeting in providing stellar researchers with new results of research on the counterparts of their targets of research, spatially and temporarily resolved, as observed on the Sun. Likewise we hoped to provide solar researchers with new results on gigantic versions of their targets of research under the very different physical circumstances on other active stars. It was our greatest pleasure that we had wide attendance of experts and active researchers of both research fields from all over the world. This led to extremely interesting talks and very lively discussions, thereby stimulating the exchange of ideas across the fields.

The Scientific Organizing Committee consisted of Drs. L. W. Acton, Ai Guoxian, J. L. Culhane, V. Gaizauskas, T. Hirayama, J. Linsky, M. Pick, E. Priest, M. Rodono, R. Rosner, J. Schmitt, A. V. Stepanov, and Y. Uchida (chair). We thank them and colleagues such as Drs. R. Pallavicini, B. Byrne, and others, for very useful suggestions and input towards making the program of the meeting scientifically more interesting. The Local Organizing Committee consisted of Drs. S. Enome, H. Hudson, M. Kojima, T. Kosugi (chair), H. Kurokawa, Y. Ogawara, T. Sakurai, K. Shibata, S. Tsuneta, T. Watanabe, and A. Yamaguchi, and we are grateful to them for their efficient efforts in organizing the meeting together with the assistance of junior staff and graduate students. We express our appreciation to Miss M. Kuroiwa for her efficient help in editing the Proceedings. The scientific preparation of this meeting in Japan was supported by the Ministry of Education, Science, Culture and Sports (Grants-in-aid for Cooperative Research Nos. 04302012 and 06352002; PI: Y. Uchida).

Finally, we acknowledge the sponsorship of the meeting by the International Astronomical Union, the National Astronomical Observatory, and the Astronomical Society of Japan. The meeting was held with financial support from the Ministry of Education, Science, Culture and Sports for the promotion of international symposia. Financial support also came from IBM Japan, the Inamori Foundation, and others. These helped us to assist in the attendance of some key speakers and active young research workers.

Tokyo, March 1996 Y.Uchida, T.Kosugi, and H.S.Hudson

I. Magnetic Atmospheres of the Sun and Stars

I.1. The Solar Corona, the Magnetic Atmosphere of the Sun

CORONAL STRUCTURES, LOCAL AND GLOBAL

L. W. ACTON
Montana State University
Bozeman, Montana 59717 USA

Abstract. The purpose of this paper is to present an overview of the magnetic topology and physical properties of the sun's corona, as revealed primarily by X-ray imagery.

1. Introduction

Stellar observations in the visible, as well as theoretical work on stellar evolution and energy generation, gave no hint that stars might possess atmospheres hot enough to produce X-rays. However, two independent lines of research caused early workers to postulate that the outer envelope of the sun had unexpected energetic properties. Firstly, eclipse observations revealed a corona of much greater extent than expected for hydrostatic equilibrium at the photospheric temperature. Also, very broad coronal emission lines suggested that the sun is surrounded by a million degree outermost atmosphere. Secondly, the presence of ionized layers high in the atmosphere of the earth, which reflected radio waves, required an ionizing source. Regarding the latter E. O. Hulbert (1938) wrote, '. . . (layer) E is caused by a moderately penetrating solar radiation which approximately travels in straight lines into the terrestrial atmosphere and is absorbed exponentially. The radiation might be ultraviolet light, X-rays or particles of zero average charge." Finally, Edlén (1952) identified the coronal emision lines as transitions in highly ionized atoms produced by a very hot corona.

Now, thanks to observations from space, it is known that X-ray emission is a common property of rotating stars with convective envelopes. Such stars are magnetically active. This is seen as the cause of the coronal activity and emission. The form of the corona will reflect the form of the coronal magnetic fields – as modified by evolution, reconnection, and stellar winds.

The sun provides the opportunity to observe the magnetic topology of a corona in detail. While the coronas of other stars might be quite differ-

Y. Uchida et al. (eds.),
Magnetodynamic Phenomena in the Solar Atmosphere – Prototypes of Stellar Magnetic Activity, 3–11.

ent in ways that are difficult to predict, they will undoubtedly have many characteristics in common with the solar example. Thus, it is reasonable to organize a symposium such as this one, which considers the sun as a prototype system for the broader population of magneto-active stars.

2. X-ray Imagery

The X-ray observability of coronal magnetic structures depends both upon the properties of the plasma and the telescope. Density and temperature, which reflect heating, trapping and energy loss processes, are the dominant physical factors. For the case of the SXT, image brightness responds primarily to increasing plasma temperature up to about 3 MK beyond which the square of the electron density is dominant.

The appearance of the X-ray corona is determined by the configuration of the coronal magnetic fields. The coronal fields are, in turn, rooted in the photosphere where flows, field eruptions, and subsidence provide a source surface of constant change and turmoil. Within the corona the magnetic field topology is modified by coronal processes such as relaxation, reconnection and eruption. In rare cases the plasma beta is sufficiently high that coronal gas pressure may play a detectable role in shaping coronal structures.

Notably, the solar magnetic cycle requires that all of this structural complexity be processed cyclically, ever seeking but never reaching stable equilibrium. The appearance and radiative output of the X-ray corona directly reflects the cyclical variation of the photospheric magnetic activity. However, on a moment to moment basis we have not successfully related coronal effects to photospheric field evolution. It seems analogous to a clock which is driven by the mainspring but the ringing of the alarm is controlled by other factors.

3. Geometrical Properties

The X-ray corona comprises:

- bright local structures associated with active regions,
- larger magnetic loops of intermediate brightness which connect strong-field centers of activity,
- a more diffuse component within large closed-field structures underlying coronal streamers,
- coronal holes – faint regions within which the field lines are predominantly open to space or sufficiently extended as to be effectively open as far as plasma trapping and containment are concerned.

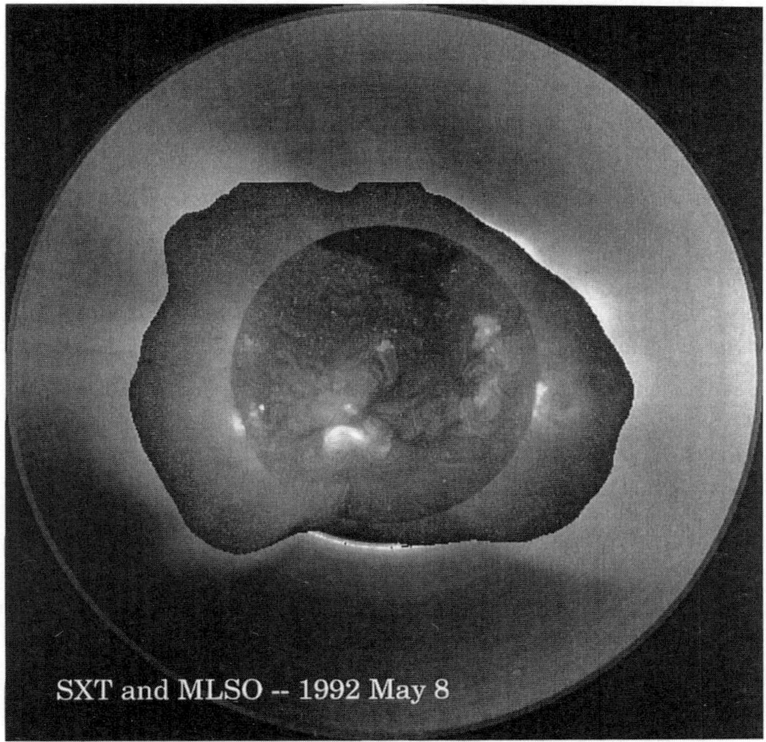

Figure 1. Composite of an X-ray image from SXT and a white light coronagram from the HAO Mk III coronagraph on Mauna Loa for 1992 May 8. North is at the top and east to the left. The black region at the solar south pole is the occulting disk of the coronagraph. The enhanced brightness near the outer limits of the coronagram, most prominent in the northwest, is an artifact.

The spatial relationship between the X-ray (thermal emission) and white-light (electron scattered) corona is illustrated in Figure 1. The X-ray image has an intensity range of greater than 10^5 so is displayed with a logarithmic intensity scale. Note the good correspondence between extended X-ray corona and white light streamers.

One of the most striking properties of the corona, immediately evident in the earliest X-ray images, was the non-uniformity of coronal brightness. As telescopes improved it became obvious that X-ray emission was largely confined to magnetic "loops" whose length was considerably longer than their cross-sectional diameter (Viana, Krieger and Timothy, 1973).

Some details of X-ray structures are shown in figure 2. The following types of features are illustrated:

A. Large helmet arcade (1992 Jan 25 08:15).
B. Loop with cusp at the limb (1991 Oct 25 23:12).
C. Eruptive feature, 30 km/sec (1991 Nov 5 21:24).
D. Small, symmetrical flaring loop (1992 Jan 15 10:44).
E. Two cusped loops (1991 Oct 24 22:28).
F. Active region loops and twisted X-ray jet, 200 km/sec (1991 Oct 3 05:20).
G. Sinuous loops between active regions (1992 Jan 16 09:30).

The discovery by SXT of X-ray loops with a cusp at the top is taken as evidence for magnetic reconnection at higher levels. Helmet arcade structures as illustrated in panel A form following coronal mass ejections. The beautiful sinuous structure of the inter-region loops in panel G demonstrate the helicity of the coronal magnetic fields.

4. Physical Properties

The X-ray emissivity of the corona is strongly non-uniform. This is illustrated in Figure 3. The north polar coronal hole is very faint – barely detectable with SXT. At the 4 o'clock position about 0.8 of the way to the limb is a long thin loop with its eastern end anchored in a faint X-ray bright point. Strong, et al. (1992) have shown that it is common for X-ray bright points to inject energetic particles into adjacent long loops. However, the process is quite discrete, as shown here, and does not result in the heating of very much corona.

The large loop system near the equator at the east limb is expanding. The physics of loop expansion is not well understood theoretically although it is a common feature of SXT movies.

The lower panel of the figure quantifies the extreme non-uniformity of coronal radiance. The 3 levels of gray and white delineate areas wherein the soft X-ray intensity falls within specified limits as given in Table 1. The white circle indicates the limb of the sun. Black contours separating the gray areas are included for clarity. Each contour reports a change in intensity of approximately a factor of 15. The total area corresponds to the detectible corona, i.e., excluding the black background.

The conversion of SXT signal to solar X-ray flux requires specifying the temperature of the source and the spectral passband of interest. Analysis of SXT filter ratios gives 3 MK as a characteristic temperature of the corona in 1992. Once a temperature is specified the choice of passband is somewhat arbitrary as the flux is derived by integrating over the synthetic spectrum of Mewe et al. (1985, 1986). The passband of the thinnest SXT filter is roughly 3 - 40 Å. However, to better illustrate the total X-ray radiance of the sun the intensity levels of Table 1 and the radiance curve of Figure 4

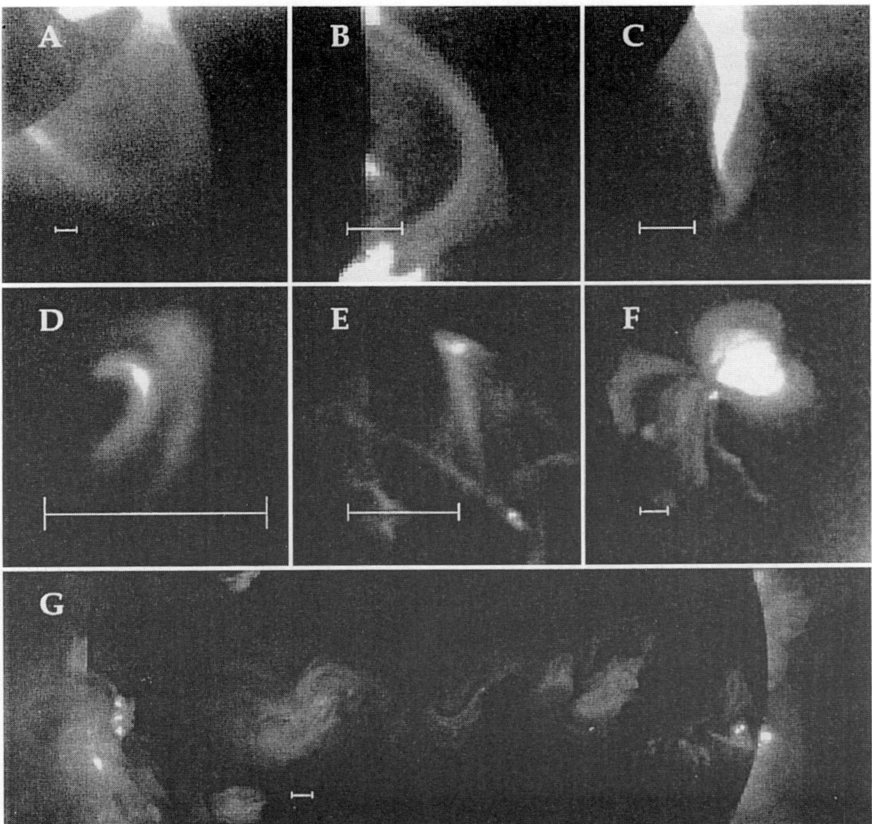

Figure 2. Examples of coronal X-ray structures observed by the SXT. The horizontal line segment in each panel indicate an angle of one minute of arc, equal to about 43,500 km on the sun.

are expressed in terms of the integrated X-ray flux below 300 Å. Differential emission measure model calculations indicate that these assumptions provide X-ray radiance values which are better than a factor of two except in extreme cases.

Table 1 Radiation from the 92 May 8 Corona

Level	Lower Intensity Threshold ($erg/cm^2/ster$)	Temperature (MK)	Fractional Area	Fractional Radiance
Dark gray	20	2.0	0.75	0.095
Med gray	4.7×10^2	2.2	0.23	0.34
Light gray	7.0×10^3	2.8	0.017	0.43
White	1.0×10^5	5.6	0.0008	0.13

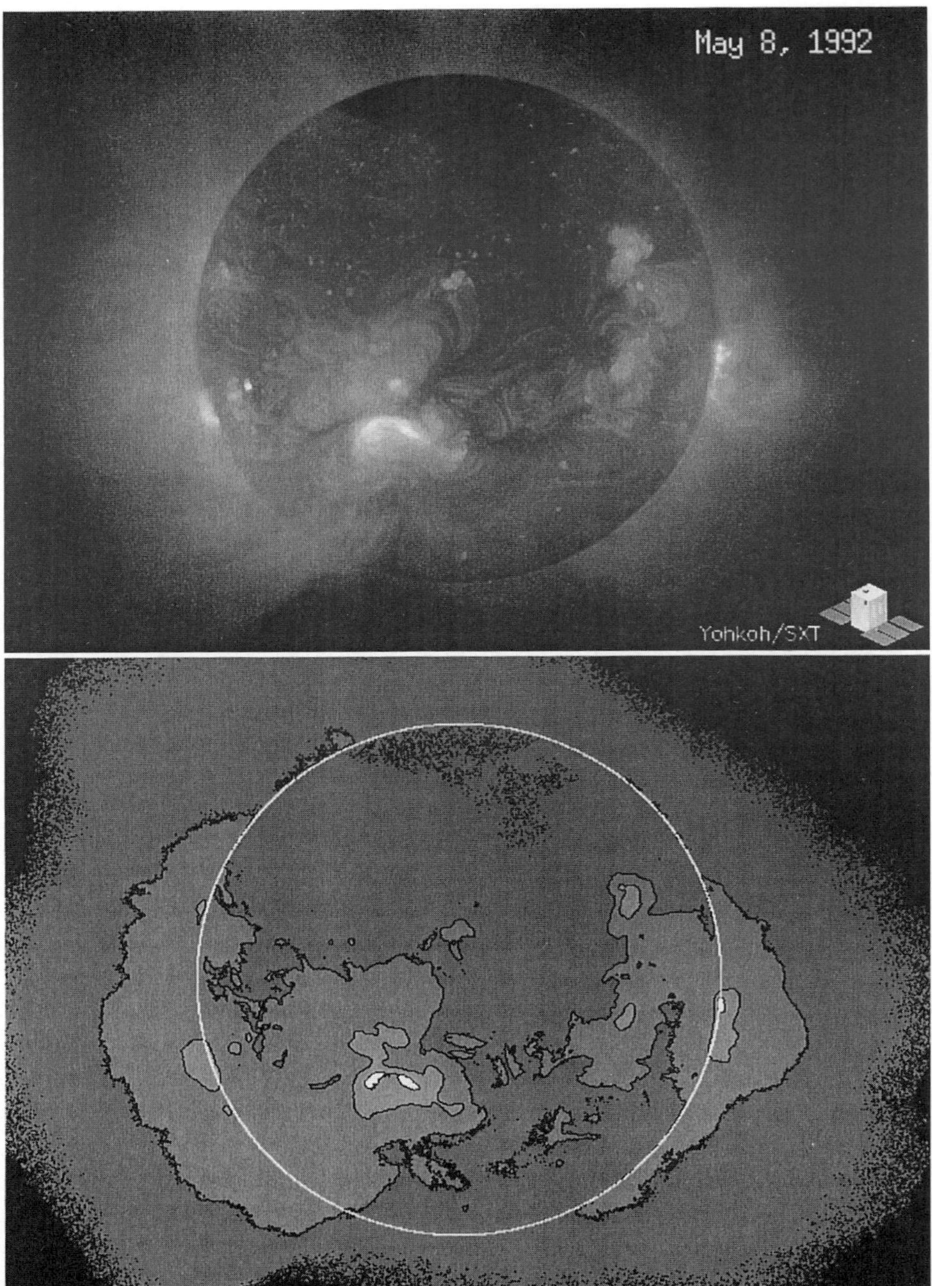

Figure 3. (Top) An enhanced SXT image of the X-ray corona on 1992 May 8 using a logarithmic scaling. The intensity range of the original image is 300,000. North is up and east is to the left. (Bottom) The distribution of image intensity as a contour diagram with four logarithmically spaced intensity intervals.

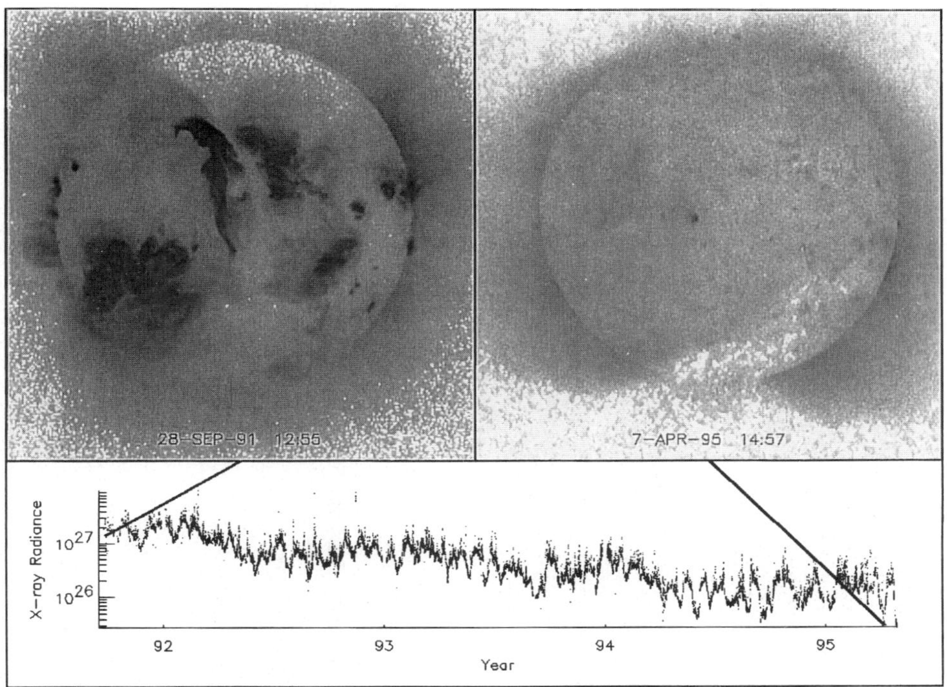

Figure 4. Change in X-ray appearance and radiance of the sun over a 3.5 year period. Units of radiance are erg/sec in the 0-300 Å band.

There is the following rough correspondence between these intensity regimes and coronal features.

Dark gray	Diffuse corona and coronal holes.
Medium gray	Inter-region loops and coronal arcades
Light gray	Active regions.
White	Hot cores of active regions.

Note that less than 2 percent of the area is responsible for over half of the X-ray signal. Accounting for the 3-dimensional structure of the corona, the concentration of most of the emission into a tiny fraction of the coronal volume would be even more emphasized. The temperature given for each of the intensity bands is an average derived from the first year of the *Yohkoh* mission (1991 November to 1992 November).

5. Temporal Properties

It has been established from the earliest satellite measurements (Acton et al. 1963) down to GRO that the sun's X-ray emission is highly variable

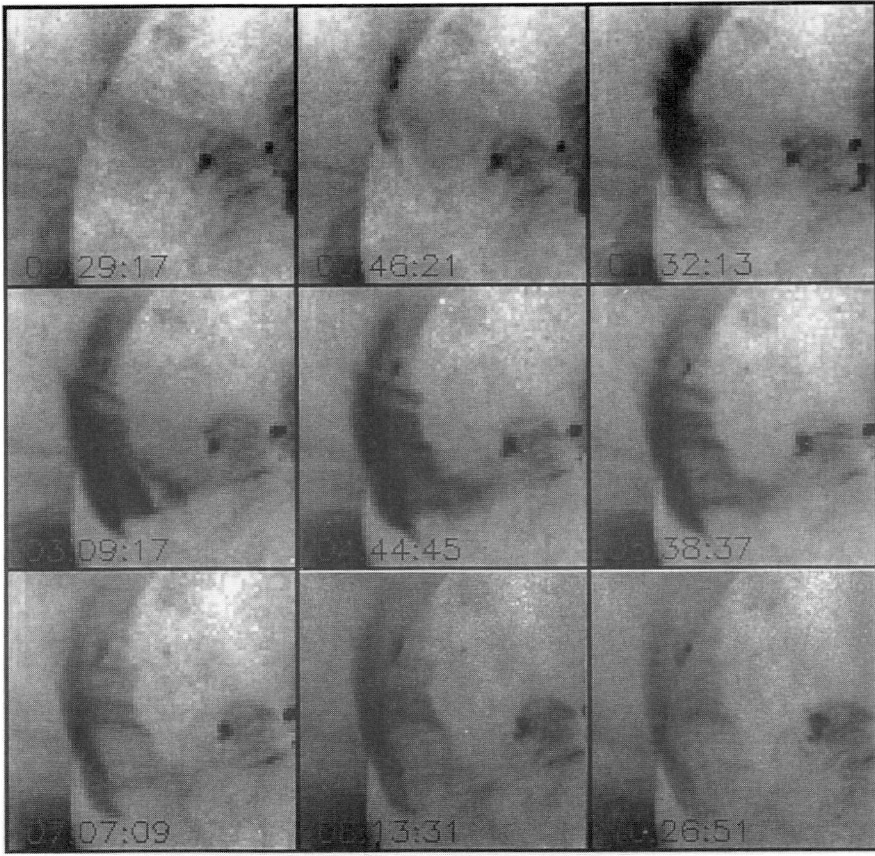

Figure 5. Gowth of an arcade of X-ray loops following a prominence eruption on the east limb on 1992 July 31. The images have been displayed with a reversed color table. Each image is 11.5 arcmin square. All but the last 2 images were taken with 10"x10" pixels.

on all time scales – down to the limit of measurement. *Yohkoh* has offered the best opportunity yet to relate this variability to magnetic structures in the atmosphere of the sun. Figure 4 illustrates the remarkable change in the appearance of the X-ray corona, and the factor of 100 peak to peak decrease in solar X-ray radiance, over a 3.5 year period. Each data point in the radiance curve has been derived from a single SXT image. Flares have been excluded. The 0-300 Åband has been chosen for convenience to include most of the coronal emission. The absolute value is expected to be correct to about a factor of 2 – based upon modelling of the effect of using an isothermal assumption (3 MK) for the conversion of signal to radiance (to be published).

Magnetic instabilities leading to large scale eruptions appear to play an important role in the evolution of coronal structure. Figure 5 shows the growth of a long arcade of X-ray loops following the eruption of a prominence.which occurred near 1992 July 31 00:00 UT. On a full-disk X-ray monitor this appeared as a weak gradual flare. The prominence eruption was well observed by the Nobeyama radioheliograph at 17 Ghz. (Hanaoka, et al. 1994) It was well on its way by 00:13. At that time a dimming of the high X-ray corona was observed by SXT above this position angle. It is too subtle to be seen in this figure. That dimming was caused by the sweeping out of the corona by the coronal mass ejection which preceded the cool prominence material. The X-ray arcade is clearly a result of magnetic reconnection behind the erupting prominence. These observations are broadly consistent with the eruption-reconnection scenario treated by Carmichael, Sturrock, Hirayama, Kopp and Pneuman (the CSHKP model) and now considered the "standard" model of such eruption-flare events.

Acknowledgements I am deeply indebted to the ISAS, NASA and the *Yohkoh* Team for the work that has made the SXT such a great success. The U.S. Yohkoh effort is supported by NASA contract NAS8-37334 with the Lockheed Martin Palo Alto Research Laboratory. I thank M. Kuroiwa for doing the LateX version of this paper.

References

Acton, L.W., Chubb, T.A., Kreplin, R.W., and Meekins, J.F. (1963) Observations of solar X-ray emission in the 8 to 20 A band, *J. Geophys. Res.*, **68**, pp.3335-3344.

Edlén, Bengt (1952) Observations on the HeI isoelectronic sequence, *Arkiv för Fysik* **4**, pp.441-454.

Hanaoka, Y., Kurokawa, H., Enome, S., Nakajima, H., Shibasaki, K., Nishio, M., Takano, T., Torii, C., Sekiguchi, H., and Kawashima, S. (1994) Simultaneous observations of a prominence eruption followed by a coronal arcade formation in radio, soft X-rays and H(alpha),.*Publ. Astron. Soc. Japan* **46**, pp. 205-216.

Hulburt, E.O. (1938) Photoelectric ionization in the ionosphere, *Phys. Rev.* **53**, pp.344-352.

Mewe, R., Gronenschild, E. H. B. M., and van den Oord, G. H. J. (1985) Calculated X-radiation from optically thin plasmas. *V, Astron. Astrphys. Suppl. Series* **62** pp. 197-254.

Mewe, R., Lemen, J. R., and van den Oord, G. H. J. (1986) Calculated X-radiation from optically thin plasmas. *VI, Astron.Astrophys. Suppl. Series* **63**, pp. 511-536.

Strong, K.T., Harvey, K.L., Hirayama, T., Nitta, N., Shimizu, T., and Tsuneta, S. (1992) Observations of the variability of coronal bright points with the Soft X-ray Telescope, *Publ. Astron. Soc. Japan 44*, pp.L161-L166.

Vaiana, G.S., Krieger, A.S., and Timothy, A.G. (1973) Identification and analysis of structures in the corona from X-ray photography, *Solar Physics*, **32**, pp.81-116.

DYNAMICAL PROCESSES IN THE SOLAR CORONA

– X-ray Jets and X-ray Plasma Ejections from Impulsive Flares –

KAZUNARI SHIBATA
National Astronomical Observatory
Mitaka, Tokyo 181, Japan

Abstract: *Yohkoh*/SXT discovered *X-ray jets* ejected from microflares and *X-ray plasma ejections* (loop-like or blob-like ejections; the latter are often referred to as *plasmoids*) from impulsive compact loop flares. The fundamental properties of these newly discovered hot plasma ejections are reviewed. A unified scheme based on the magnetic reconnection hypothesis is presented to understand these mass ejections.

1. Introduction

Yohkoh/SXT (Tsuneta et al. 1991) has revealed that the solar corona is much more dynamic than had been thought; i.e., the corona is full of transient loop brightening (microflares), jets, various mass ejections, and so on. They are important not only for understanding coronal dynamics but also for clarifying coronal heating mechanism and solar wind acceleration mechanism.

In this article, I would like to review the fundamental properties of the newly discovered mass ejections in the corona, i.e., (1) X-ray jets ejected from microflares, and (2) X-ray plasma ejections from impulsive compact loop flares. On the basis of these observations, I will propose that these mass ejections and various flares (microflares, impulsive flares, and LDE flares) can be understood in a unified scheme in which the magnetic reconnection associated with plasmoid ejections plays a key role (Shibata 1996).

2. X-ray Jets

X-ray jets have been discovered as transitory X-ray enhancements with apparent collimated motion (Shibata et al., 1992b, 1994a,b, 1996a, Strong et al. 1992, Shimojo et al. 1996a,b). Almost all jets are associated with

Y. Uchida et al. (eds.),
Magnetodynamic Phenomena in the Solar Atmosphere – Prototypes of Stellar Magnetic Activity, 13–20.

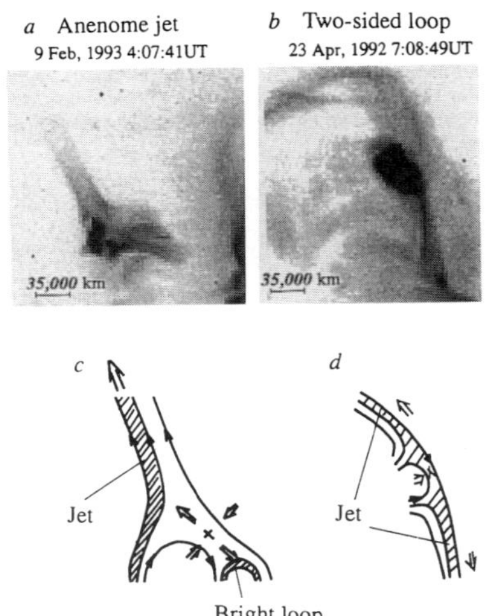

Figure 1. SXT images of two types of interaction of emerging flux with overlying coronal magnetic fields. (a) *Anemone-Jet* type observed on 9 Feb. 1993. (b) *Two-sided-loop* type observed on 23 Apr. 1992. Schematic illustration of (c) *Anemone-Jet* type and (d) *Two-sided-loop* type (from Yokoyama and Shibata 1995).

microflares or subflares. The length and the apparent velocity of the jets are $10^3 - 4\times10^5$ km and $10 - 1000$ km/s. The number of jets decreases as the length, the width, or the velocity increases. This trend is more clearly seen in the distribution of the lifetimes of the jets; the number of jets decreases as the lifetime increases from 100 sec to 10 hours, and shows a power-law distribution.

According to the preliminary analysis by Shimojo et al. (1996b), the temperature of some small jets in active regions is $4 - 6$ MK and is comparable to that of microflares at the footpoints of the jets. The estimated electron density of the jets ranges from 3×10^8 cm^{-3} (for larger jets) to 5×10^9 cm^{-3} (for smaller jets). The kinetic energy is estimated to be of the order of $10^{25} - 10^{28}$ erg. Shimojo et al. (1996a) found that in many jets the width of the jets is nearly *constant* or decreases with height (i.e., *converging* shape), which is similar to the shape of Hα surges observed in emerging flux regions (Kurokawa and Kawai 1993).

Many jets are ejected from emerging flux regions (EFRs) or unresolved X-ray bright points. Shibata et al. (1994b) examined how jets are ejected from well resolved emerging flux region, and found that jets are formed as a result of an interaction between emerging flux and pre-existing coronal magnetic field. There are basically two types of interaction, depending on the place where emerging flux appeared (Fig. 1). One is the case where emerging flux appears in coronal holes and is called the *anemone-jet* type. In this case, a jet is ejected in a vertical direction along open (vertical) coronal field. The footpoint of the jet is usually seen as an anemone-type active region. [1] The other is the case where emerging flux appears in quiet regions and is called *two-sided-loop/jet* type. In this case, two loop brightenings (or jets) occur in the horizontal direction at both sides of the emerging flux.

2.1. RELATION TO MICROFLARES

Shibata et al. (1996b) studied the relation between microflares and jets in some emerging flux regions. Figure 2 shows time variation of the total soft X-ray intensity of the active region NOAA 7176 from its birth on 19 May 1992 to 24 May 1992. This region appeared in a coronal hole and showed characteristics of an *anemone-jet* type. The short vertical lines in the figure show the time when the jets are ejected. It is remarkable that more than 20 jets were produced in the 5 days after the birth of the region. It is interesting to note that the emerging flux region began to decay just after ejecting several jets on 21 May, suggesting that the jets might play a role in active region decay. Detailed analysis of SXT images of this region show that all jets during these 5 days were associated with microflares (see sharp spikes in the intensity curve of Fig. 2). On the other hand, the fraction of microflares associated with jets is about 60 percent. Since the probability of detection of jets is not 100 percent because of the limited time cadence of observations, this would suggest that more microflares may be associated with jets.

It should be noted here that these microflares are seen as *transient loop brightenings* in SXT images (Shimizu et al. 1992, Shimizu 1995). From comparison of NSO/Kitt Peak magnetogram with SXT images of many jets, Shimojo et al. (1996b) found that more than 70 percent of X-ray jets are ejected from the mixed polarity regions such as satellite spots embedded in opposite polarity regions. They also found that if the magnetic field is closed, only transient loop brightenings occur even if the magnetic polarity is mixed. This suggests that the transient loop brightenings may be physically the same as X-ray jets as illustrated in Fig. 1.

2.2. RELATION TO Hα SURGES AND TYPE III BURSTS

Although there are many Hα surges which are not associated with X-ray jets (e.g., Schmieder et al. 1995), Shibata et al. (1992b), Canfield et al. (1996a,b), Okubo et al. (1996) found some examples showing both X-ray

[1] According to Sheeley (personal communication 1995), the *anemone type active region* was already discovered by Skylab (Sheeley et al. 1975). It was called *fountain regions*.

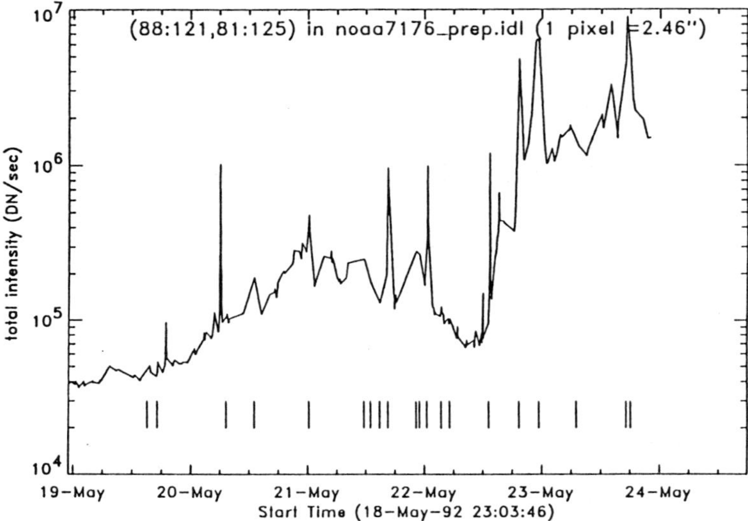

Figure 2. Time variation of the total soft X-ray intensity of the active region 7176 from its birth on 19 May 1992 to 24 May 1992 (Shibata et al. 1996b). This region appeared in coronal hole and and showed characteristics of *anemone-jet* type. The short vertical lines in the figure show the time when the jets are ejected.

jets and Hα surges in the same direction. Canfield et al. (1996a,b) found several new observational signatures of magnetic reconnection in surges, that is, *converging footpoints* and *moving blue shifts.*

Kundu et al. (1995) found that a Type III burst was associated with an X-ray jet on Aug. 16 1992, and that the density derived from the X-ray jet is consistent with that derived from the Type III burst. The discovery of association with Type III bursts implies the existence of high energy electrons in these small flares and jets, and supports the view that the generation mechanism of X-ray jets and microflares may be physically similar to that for larger flares.

3. X-ray Plasma Ejections from Impulsive Compact Loop Flares

Masuda et al. (1994, 1995) discovered hard X-ray sources well above the soft X-ray loop in some impulsive flares observed near the limb, and suggested that magnetic reconnection is occurring above the soft X-ray loop. If the reconnection hypothesis similar to the CSHKP model is correct, the plasmoid ejection would be found high above the soft X-ray loop (Hirayama 1991) as illustrated in Figure 4. Shibata et al. (1995) searched for such plasmoid ejections in the Masuda flare on 13 Jan. 1992, and indeed discovered X-ray plasma ejections high (around $\sim 10^5$ km) above the hard X-ray source (Figure 3).

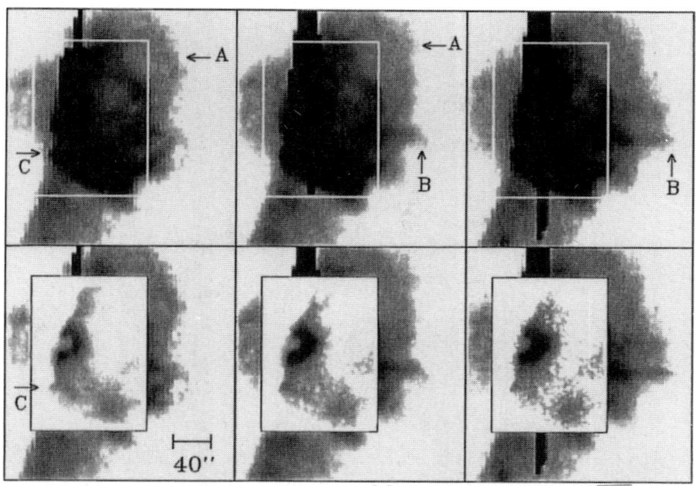

Figure 3. Soft X-ray (negative) images of X-ray plasma ejections found in the Masuda flare on 13 Jan. 1992 (Shibata et al. 1995). The upper three shows long-exposure images at 5 arcsec spatial resolution, and the bottom shows short-exposure images at 2.5 arcsec resolution (at nearly the same time) composited on the long-exposure images. The time in the brackets denotes the exact time when the short-exposure images are taken. 40 arcsec correspond to about 29000 km. A careful examination of the long-exposure images revealed at least two very faint erupting features high ($4 \times 10^4 - 8 \times 10^4$ km) above the soft X-ray loop. They are indicated by the arrows A and B; the ejection A seems to be loop-like, and the ejection B looks more like a jet. The velocity of these ejections is about 100 – 150 km/s. The onset of both ejections are nearly simultaneous with the hard X-ray impulsive peak.

Shibata et al. (1995) further surveyed such ejections in 8 impulsive limb flares which were selected in an unbiased manner by Masuda (1995) with the following two selection criteria: (1) The peak count rate in the HXT M2-band (33 – 53 keV) exceeds 10 cts/s/subcollimator. (2) The heliocentric longitude exceeds 80 degrees. It is remarkable that plasma ejections were found in all 8 impulsive limb flares. The ejections seen were loop-like, blob-like, or jet-like. It was further found that the range of velocity of the ejections is 50 – 400 km/s. Interestingly, flares with hard X-ray sources well above (5 – 10 arcsec) the loop top show systematically higher ejection velocities. The size of the ejections is typically $(4 - 10) \times 10^4$ km. The soft X-ray intensity of the ejections is $10^{-4} - 10^{-2}$ of the peak soft X-ray intensity in the bright soft X-ray loop. The strong acceleration of the ejections occurs nearly simultaneously with the hard X-ray impulsive peaks (Ohyama and Shibata 1996).

Ohyama and Shibata (1996) analyzed the temperature and emission measure distribution of X-ray plasma ejections in some cases showing bright

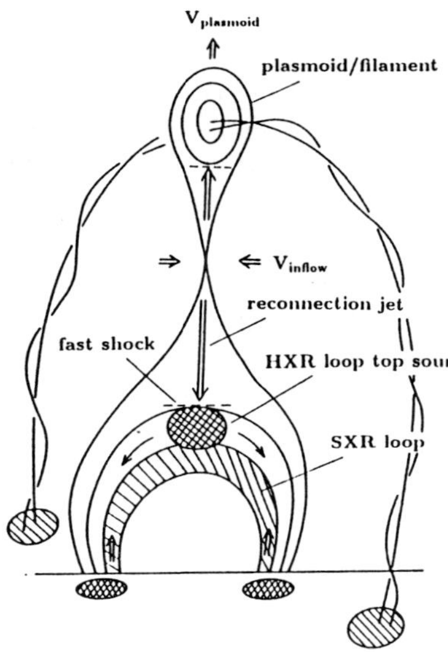

Figure 4. A reconnection – plasmoid ejection model for compact loop flares. This is an extension of the CSHKP (Carmichael-Sturrock-Hirayama- Kopp-Pneuman) model of flares (Hirayama 1991, Moore and Roumeliotis 1991). Note that plasmas confined by a closed field (in two dimensions) or by a helically twisted flux tube (in three dimensions) are called *plasmoids*. The cool ($\sim 10^4$ K) plasmas associated with the twisted flux tube is the *filament* or *prominence*. Hot ($> 10^6$ K) plasma ejections are expected to be associated with the twisted tube or expanding loop high above the reconnected (soft X-ray) loop. The cross-hatched region at the footpoints of the soft X-ray loop shows the bright hard X-ray/soft X-ray double sources. The hatched region at the footpoints of the expanding (helical) loop penetrating the plasmoid shows predicted hard X-ray/soft X-ray distant sources.

blob-like ejections (often referred to as *plasmoids*), and found the following. The temperature of the plasmoids is $\sim 6 - 13$ MK, the electron density is $\leq 1.5 \times 10^{10}$ cm^{-3}. The temperature of the plasmoids is less than that of the flare loop. The thermal and kinetic energy of the plasmoid is an order of magnitude smaller than the thermal energy of the soft X-ray flare loop.

4. Modeling and Interpretation

Yokoyama and Shibata (1995, 1996a,b, Yokoyama 1995) successfully modeled the two types of reconnection of emerging flux with overlying preexisting magnetic fields (Fig. 1) and the associated formation of X-ray jets and Hα surges in each case, by extending the previous numerical simulations (Shibata et al. 1992a). In these models, the reconnection produces not only hot/cool jets but also hot loops (Fig. 1). The latter would correspond to *transient loop brightenings* (Shimizu et al. 1992) and hence a part of the *microflare* process.

These numerical simulations showed that the reconnection proceeds with the formation and ejection of magnetic islands (plasmoids). In the Yokoyama and Shibata model, cool gas confined by magnetic islands eventually become a cool jet (Hα surge), and hot gas surrounding an island become a hot jet (X-ray jet). This may be in some sense similar to the X-ray plasma ejection observed in impulsive flares (see Fig. 4 for a model). In fact, Ohyama and Shibata (1996) found that the temperature of the X-ray plasma ejection (plasmoid) is cooler inside than outside. (It should be noted here that the loop-like X-ray plasma ejection such as in Fig. 3 would correspond to a side view of the plasmoid ejection as shown in Fig. 4.) A model for impulsive flares accompanied by the plasmoid ejections has been developed by Magara et al. (1995) (see also Ugai 1989).

Traditionally, the emerging flux model and the CSHKP model have been developed independently to explain *compact loop flares* and *eruptive flares* separately (e.g., Priest 1981). However, the new observations have revealed various common phenomena such as ejection of hot plasmas in various flares (Table I). It is now not easy to classify flares into two classes. In fact, statistical studies show no essential differences between larger flares and microflares (e.g., Shimizu 1995). On the other hand, recent numerical simulations showed also some common physical processes, such as an ejection of plasmoids, in both the emerging flux model and the CSHKP model. Hence, I would like to propose that the flare phenomenon should be understood in a unified scheme in which the ejection of plasmoids plays a key role in triggering fast reconnection.

TABLE 1. Unified View of Various Mass Ejections from Flares

Hot Mass Ejections	Cool Mass Ejections	Flares
X-ray jets	Hα surges	microflares
X-ray plasma ejections	sprays or Hα filament eruptions	impulsive flares
CME	Hα filament eruptions	LDE flares or giant arcades

5. Conclusions

Conclusions are summarized in the following.

(1) Both *X-ray jets* and *transient loop brightenings* are part of microflares. In open field regions, X-ray jets and loop brighenings are formed, whereas in closed field regions, only loop brightenings occur.

(2) Various observational signatures of reconnection have been found in X-ray jets and Hα surges. These are successfully modeled by Yokoyama (1995) and Yokoyama and Shibata (1995, 1996a,b).

(3) Discovery of X-ray plasma ejections in impulsive flares gives us further evidence of reconnection (in addition to the Masuda's loop top HXR source) and leads to a unified view of mass ejections in flares (Table I).

Acknowledgement

The author would like to thank M. Shimojo, M. Ohyama, T. Yokoyama, and other *Yohkoh* team members for their various help and discussion.

References

Canfield, R. C. et al., 1996a, *Ap. J.*, in press.
Canfield, R. C. et al., 1996b, in these proceedings.
Hirayama, T. 1991, in *Flare Physics in Solar Activity Maximum 22*, (eds. Y. Uchida et al.) (Lecture Note in Physics, No. 387, Springer Verlag) p. 197.
Kundu, M. R. et al., 1995, *Ap. J. Lett.*, 447, L135.
Kurokawa, H. and Kawai, G., 1993, *Proc. The Magnetic and Velocity Fields of Solar Active Regions, ASP Conf. Series 46*, ed. H. Zirin et al., p. 507.
Magara, T., Mineshige, S., Yokoyama, T., and Shibata, K., 1995, *Ap. J.*, submitted.
Masuda, S., 1995, Ph. D. Thesis, Univ. Tokyo.
Masuda, S., Kosugi, T., Hara, H., Tsuneta, S., and Ogawara, Y. 1994, *Nature*, **371**, 495.
Masuda, S., et al., 1995, *Publ. Astr. Soc. Japan*, **47**, 677.
Moore, R. L. and Roumeliotis, G. 1991, in *Eruptive Solar Flares*, (eds. Z. Svestka et al.) (Lecture Note in Physics, No. 399, Springer Verlag, 1992) p. 69.
Ohyama, M., and Shibata, K., 1996, in these proceedings.
Priest, E. R., 1981, *Solar Magnetohydrodynamics*, Reidel.
Schmieder, B., K. Shibata, L. van Driel-Gesztelyi, and S. Freeland, 1995, *Solar Phys.*, 156, 245.
Sheeley, N., et al. 1975, *Solar Phys.*, 40, 103.
Shibata, K., 1996, *Adv. Space Res.*, **17**, (4/5)9.
Shibata, K., Nozawa, S., and Matsumoto, R., 1992a, *Publ. Astr. Soc. Japan*, **44**, 265.
Shibata, K., et al., 1992b, *Publ. Astr. Soc. Japan*, **44**, L173.
Shibata, K., et al., 1994a, *Ap. J. Lett.*, **431**, L51.
Shibata, K., et al., 1994b, *in Proc. "X-ray Solar Physics from Yohkoh", eds. Y. Uchida, H. Hudson, T. Watanabe, and K. Shibata, Univ. Academy Press*, p. 29.
Shibata, K., et al., 1995, *Ap. J. Lett.*, 451, L83.
Shibata, K., Yokoyama, T., and Shimojo, M., 1996a, *Adv. Space Res.*, **17**, (4/5)197.
Shibata, K., et al., 1996b, in preparation.
Shimizu, T., et al., 1992, *Publ. Astr. Soc. Japan*, 44, L147.
Shimizu, T., 1995, *Publ. Astr. Soc. Japan*, 47, 251.
Shimojo, M., Hashimoto, S., Shibata, K., Hirayama, T., Hudson, H., and Acton, L., 1996a, *Publ. Astr. Soc. Japan*, in press.
Shimojo, M., et al., 1996b, in these proceedings.
Strong, K. T., et al., 1992, *Publ. Astr. Soc. Japan*, 44, L161.
Tsuneta, S. *et al.*, 1991, *Solar Phys.*, 136, 37.
Ugai, M., 1989, *Phys. Fluids B*, 1, 942.
Yokoyama, T., 1995, Ph. D. Thesis, National Astronomical Observatory.
Yokoyama, T., and Shibata, K., 1995, *Nature*, **375**, 42.
Yokoyama, T., and Shibata, K., 1996a, in these proceedings.
Yokoyama, T., and Shibata, K., 1996b, *Publ. Astron. Soc. Japan*, in press.

HEATING MECHANISMS OF THE SOLAR CORONA

T. SAKURAI
National Astronomical Observatory
Mitaka, Tokyo, Japan

Abstract. Heating mechanisms of the solar corona are briefly addressed. Magnetic vs. non-magnetic, and DC vs. AC mechanisms are compared. AC or wave heating mechanisms are discussed, and future directions of research which is important in clarifying the contribution of wave heating are suggested.

1. Introduction

The magnetic fields on the sun and solar-type stars are created by an MHD dynamo process which is driven by the combination of differential rotation and surface convection. The magnetic fields thus created leads to various kinds of magnetic activity such as sunspots, flares, and the heating of the upper atmosphere.

Decades ago the solar corona was thought to be heated (uniformly) by the dissipation of sound waves generated by the turbulent convection in the photosphere. However, the energy flux of upward-propagating sound waves was measured to be insufficient in heating the corona (e.g. Mein and Schmieder 1981). X-ray observations from space showed magnetic structuring of the solar corona; the corona is made of coronal magnetic loops (Vaiana and Rosner 1978). Therefore the heating is inherently connected to the magnetic field.

2. Magnetic vs. Non-magnetic Heating

At the photospheric level the magnetic field is inhomogeneous and is bundled into small-scale flux tubes (Stenflo 1976). The field strength in the flux tubes is about 1000 G, and the magnetic pressure inside the tube is of the same order of the gas pressure of the surrounding medium. The concentra-

Y. Uchida et al. (eds.),
Magnetodynamic Phenomena in the Solar Atmosphere – Prototypes of Stellar Magnetic Activity, 21–27.

tion of magnetic flux into small-scale flux tubes is a natural consequence of the interaction between the magnetic field and the convective motion (Galloway et al. 1977).

The magnetic field lines from the flux tubes fan out and fill the upper atmosphere above a height of 700–1500 km (the merging height) (Spruit 1981). The presence of very cool (< 4000K) material in the solar atmosphere indicates that the non-magnetic part of the atmosphere is not significantly heated (e.g. Solanki et al. 1994). Therefore, the current picture is that the solar atmosphere is made of two components, a magnetically-heated flux tube atmosphere which dissolves into the corona, and a weakly heated non-magnetic atmosphere which terminates at the merging height.

Observations show that the mean magnetic field strengths on various stars scale as $fB \sim \Omega_{\rm rot}\tau_{\rm conv}$, where f stands for the fraction of stellar surface covered with magnetic field of strength B, $\Omega_{\rm rot}$ is the rotation frequency, and $\tau_{\rm conv}$ is the convective turn-over time (Noyes et al. 1984). The X-ray flux from stars scales as $F_{\rm X} \sim \Omega_{\rm rot}^2$ (Pallavicini et al 1981). However, in the limit of small $\Omega_{\rm rot}$ (i.e. small B), one still detects chromospheric and transition region activity, and these may reflect the existence of a non-magnetically heated atmosphere (basal heating) (Rutten et al. 1991; Schrjiver 1992).

3. AC vs. DC Heating Mechanisms

The heating mechanisms of the coronae of solar-type stars can be divided into two classes according to the time scale of the driving motions $\tau_{\rm dr}$ and the dynamical response time scale of the coronal loops (Alfvén time scale, $\tau_{\rm A}$). If the time scale of the driver is shorter than the Alfvén time scale ($\tau_{\rm dr} < \tau_{\rm A}$), waves are excited in the coronal loops. In the opposite case ($\tau_{\rm dr} > \tau_{\rm A}$), the coronal loops are subject to quasi-static deformation due to footpoint motions. The former and latter cases correspond to AC and DC heating mechanisms. The DC mechanisms are combinations of slow energy build-up followed by bursty energy release. Therefore the DC mechanisms are essentially the same as the mechanisms for energy release in flares; the heating of the corona is attributed to miniature flares (micro-flares; Parker 1988). The name 'DC' is misleading because it only implies the initial half of the process. Heating mechanisms associated with mass flows (spicules, transition region explosive events, jets) might also be classified as DC if the central mechanism is magnetic reconnection.

The DC mechanisms are reviewed by G.Vekstein in these proceedings, and this review focuses on AC (wave) heating mechanisms. From statistics of microflares observed with Yohkoh (Shimizu 1995) and from a differential emission measure analysis of flares (Watanabe et al. 1995), microflares are responsible for the high temperature (5 MK or hotter) component of the corona. This component is a transient component and is not the major energy container of the corona. The majority of coronal thermal energy resides in the stationary, 2MK component. This might be due to some wave heating mechanism.

4. Physical Conditions in the Solar Corona

The solar corona is made of a plasma of temperature $T = 2 \times 10^6$K and density $n_e = 10^{8-9}$ cm^{-3}. Required heat input per unit area is derived from observations as (Withbroe and Noyes 1977)

$$3 \times 10^5 \text{erg cm}^{-2}\text{s}^{-1} \quad \text{in quiet regions,}$$

and

$$1 \times 10^7 \text{erg cm}^{-2}\text{s}^{-1} \quad \text{in active regions.}$$

By using an energy balance model one can estimate the volume heating rate ϵ_H as (Rosner et al. 1978)

$$\epsilon_H = 4 \times 10^{-3} \left(\frac{T_{max}}{10^6\text{K}}\right)^{7/2} \left(\frac{l}{10^9\text{cm}}\right)^{-2} \text{erg cm}^{-3}\text{s}^{-1},$$

where l is the length of the loop, and T_{max} is the temperature at the loop apex.

Some of the plasma parameters are summarized in TABLE 1. The plasma of the solar corona is collisional (collision times $\ll$ characteristic time scale), and a fluid (MHD) description is applicable (mean-free-path $\ll$ system size). The magnetic force dominates pressure and gravity forces, therefore it is a so-called low-β plasma, where $\beta = 8\pi p/B^2$. Since $\tau_e \Omega_e \gg 1$ and $\tau_p \Omega_p \gg 1$, the thermal conductivity and viscosity are anisotropic, and cross-field thermal conduction and viscous stress are orders of magnitude smaller than those along the field.

5. Wave Heating

5.1. ACOUSTIC WAVES

According to the classical acoustic heating theory, the power of acoustic noise generated by turbulent convection is proportional to the 8-th power of the Mach number of the convective motions, and therefore is hard to estimate precisely. Recent, more elaborate analyses so far have been unable to provide large-enough acoustic wave flux theoretically (Narain and Ulmschneider 1990) and observationally (Fontenla et al. 1993). Even if this

TABLE 1. Plasma Parameters in the Solar Corona

collision times
$\tau_e = 1.4 \times 10^{-2} \left(\frac{\ln \Lambda}{20}\right)^{-1} \left(\frac{T_e}{10^6\ \mathrm{K}}\right)^{3/2} \left(\frac{n_e}{10^9\ \mathrm{cm}^{-3}}\right)^{-1} \mathrm{s} \simeq 10^{-2}\ \mathrm{s}$
$\tau_p = 8 \times 10^{-1} \left(\frac{\ln \Lambda}{20}\right)^{-1} \left(\frac{T_p}{10^6\ \mathrm{K}}\right)^{3/2} \left(\frac{n_p}{10^9\ \mathrm{cm}^{-3}}\right)^{-1} \mathrm{s} \simeq 1\ \mathrm{s}$
gyro-frequencies
$\Omega_e = 2 \times 10^7 B\ \mathrm{s}^{-1}$
$\Omega_p = 1 \times 10^4 B\ \mathrm{s}^{-1}$
mean-free-paths
$\lambda_{e,p} = 10^7 \left(\frac{\ln \Lambda}{20}\right)^{-1} \left(\frac{T}{10^6\ \mathrm{K}}\right)^2 \left(\frac{n}{10^9\ \mathrm{cm}^{-3}}\right)^{-1} \mathrm{cm} \simeq 100\ \mathrm{km}$

difficulty is resolved, another problem is how to selectively heat the magnetic part of the atmosphere. If the acoustic waves in the non-magnetic atmosphere excite other waves in magnetic flux tubes (Spruit 1981), the heating along the magnetic field lines may be explained.

5.2. MAGNETIC WAVES

Although waves in elementary flux tubes are not yet directly observed, waves localized in sunspots (umbral oscillations and running penumbral waves) have been observed. The absorption of acoustic waves by sunspots (Braun et al. 1988) also indicates a coupling of five minute acoustic oscillations and sunspot magnetic fields. The fate of the absorbed acoustic wave energy is not yet clarified. It may escape upward into the upper atmosphere, or escape downward into the convection zone, or it may be dissipated in sunspots.

Although Alfvén waves are observed in situ in the interplanetary space, evidence for Alfvén waves in the solar corona is inconclusive (Koutchmy et al. 1983; Rusin and Minarovjech 1991; Dermendjiev 1991). Turbulent line broadening of a few tens of kilometers per second observed in the transition region and in the corona may indicate unresolved waves (Hassler et al. 1990), but the wave modes have not been identified.

5.3. DISSIPATION OF WAVES

Compressive waves tend to steepen into shock waves and dissipate their energy (e.g. Kuperus et al. 1981). This is particularly true for slow-mode waves, and the observed decay of wave flux with height is presumably due to this effect.

Fast-mode waves have longer wavelengths and suffer less steepening.

However, the increase in Alfvén speed at the transition region deflects the fast-mode waves away from the corona. The observed decay of wave flux with height is for longitudinal wave motions and does not necessarily rule out the existence of the fast-mode waves in the corona. However, the fast-mode waves in the corona will propagate nearly isotropically, and it is hard to explain why the heating is localized in elongated loops.

Alfvén waves are incompressible and are able to carry energy into the corona. Contrary to compressive waves, the difficulty with the Alfvén waves is to identify effective dissipation mechanisms. Nonlinear wave-wave coupling (Uchida and Kaburaki 1974), resonance absorption (Davila 1991), and phase mixing (Sakurai 1986) are examples of possible mechanisms. The observed nonlinearity of Alfvén waves in the corona is small ($\delta v / V_{\rm A} \lesssim 1/20$), and so will be the nonlinear coupling. The resonance absorption and the phase mixing both come from a peculiar nature of the Alfvén waves, namely their continuous spectrum. Which of the two is more effective depends on the nature of the driving motions and on the magnitude of nonuniformity in the Alfvén speed in the corona.

The resonance phenomena have the general property that, if the growth of wave amplitudes is limited by a dissipation mechanism, the amount of dissipated energy does not depend on the details of the dissipation mechanism and is determined by the power of the driving motions. For a fixed amplitude of the motion which shakes a coronal loop at its feet, the wave amplitude excited in the loop is larger (smaller) if the resistivity or viscosity is smaller (larger), respectively, and the total dissipation is the same. The mechanism-independent dissipation rate determined by the driving source is generally sufficient for the heating of the corona. The expected wave amplitude does depend on the efficiency of the dissipation mechanism. The more efficient the mechanism is, the smaller the amplitude of the excited wave. The consistency of a theory with observations is examined by comparing the expected and observed wave amplitudes. Often the expected amplitude exceeds the observed wave amplitude or the magnitude of turbulent motions. It is not clear at present whether the discrepancy can be removed by refining the theoretical models.

On the other hand if the wave amplitude is not limited by dissipation itself but by another process (i.e. the escape of waves, Hollweg 1984), the wave amplitude is independent of the dissipation mechanism, and the amount of dissipated energy is larger for a system of high resistivity/viscosity. Generally the dissipation mechanism is not effective enough, so that the expected dissipation rate falls short of the required coronal heating rate.

5.4. FUTURE DIRECTIONS

In order for a wave-heating scenario of the corona to be confirmed observationally, it is of vital importance to identity the wave mode and to evaluate the wave energy flux. The energy flux has to be large enough at the coronal base, and has to decay at higher layers. Two possible ways for wave-mode identification are (1) to derive a relation between wavelengths and frequencies, and (2) to determine the direction (polarization) of the oscillations.

For longitudinal waves on the disk, the wave diagnostics have already been done by looking into phase relations of waves either in two spectral lines formed at different layers, or in two quantities of a single spectral line.

For transverse waves the observations have to be done near or off the limb. So far, only the turbulent line widths (v_t) have been observed. The quantity $\frac{1}{2}\rho v_t^2$ times the propagation speed is an upper bound for the wave flux. If the wave excited in a coronal loop is approximately a standing wave, the actual energy flux would be much smaller than this.

If slow-mode waves are generated in the corona by magnetic reconnection as suggested by Porter et al. (1994), most criticisms on the acoustic theory will be discarded. The propagation direction of slow mode waves thus generated will be random, and may be difficult to be mode-identified. Imaging observations of such phenomena and to carry out number statistics will be more adequate than spectroscopic studies.

References

Braun, D.C., Duvall, T.L., Jr., and LaBonte, B.J. (1988) The absorption of high-degree p-mode oscillations in and around sunspots, *Astrophys. J.* **335** 1015–1025.

Davila, J. (1991) Resonance Absorption Heating, in P. Ulmschneider, E.R. Priest, and R. Rosner (eds.), *Mechanisms of Chromospheric and Coronal Heating*, Springer Verlag, Berlin, pp. 464–479.

Dermendjiev, V.N. (1991) Height-dependent short-period oscillations in the Fe XIV (530.3nm) solar corona above a sunspot group crossing the limb, in P. Ulmschneider, E.R. Priest, and R. Rosner (eds.), *Mechanisms of Chromospheric and Coronal Heating*, Springer Verlag, Berlin, pp. 33–35.

Fontenla, J.M., Rabin, D., Hathaway, D.H., and Moore, R.L. (1993) Measurement of p-mode energy propagation in the quiet solar photosphere, *Astrophys. J.* **405**, 787–797.

Galloway, D.J., Proctor, M.R.E., and Weiss, N.O. (1977) Formation of intense magnetic fields near the surface of the sun, *Nature* **266**, 686–689.

Hassler, D.M., Rottman, G.J., Shoub, E.C., and Holzer, T.E. (1990) Line broadening of Mg X $\lambda\lambda$ 609 and 625 coronal emission lines observed above the solar limb, *Astrophys. J. Letters* **348**, L77–L80.

Hollweg, J.V. (1984) Alfvénic resonant Cavities in the solar atmosphere: simple aspects, *Solar Phys.* **91**, 269–288.

Koutchmy, S., Zhugzda, Y.D., Locans, V. (1983) Short period coronal oscillations: observation and interpretation, *Astron. Astrophys.* **120**, 185–191.

Kuperus, M., Ionson, J.A., and Spicer, D.S. (1981) On the theory of coronal heating mechanisms, *Ann. Rev. Astron. Astrophys.* **19**, 7–40.

Mein, N. and Schmieder, B. (1981) *Astron. Astrophys.* Mechanical flux in the solar chromosphere, **97**, 310–316.

Narain, U. and Ulmschneider, P. (1990) Chromospheric and coronal heating mechanisms, *Space Sci. Rev.* **54**, 377–445.

Noyes, R.W., Hartmann, L.W., Baliunas, S.L., Duncan, D.K., and Vaughan, A.H. (1984) Rotation, convection, and magnetic activity in lower main-sequence stars, *Astrophys. J.* **279**, 763–777.

Pallavicini, R., Golub, L., Rosner, R., Vaiana, G.S., Ayres, T., and Linsky, J.L. (1981) Relations among stellar X-ray emission observed from Einstein, stellar rotation and bolometric luminosity, *Astrophys. J.* **248**, 279–290.

Parker, E.N. (1988) Nanoflares and the solar X-ray corona, *Astrophys. J.* **330**, 474–479.

Porter, L.J., Klimchuk, J.A., and Sturrock, P.A. (1994) The possible role of MHD waves in heating the solar corona, *Astrophys. J.* **435**, 482–501.

Rosner, R., Tucker, H.W., and Vaiana, G.S. (1978) Dynamics of the quiescent solar corona, *Astrophys. J.* **220**, 643–665.

Rusin, V and Minarovjech, M. (1991) Short-term oscillations in green and red coronal lines, in P. Ulmschneider, E.R. Priest, and R. Rosner (eds.), *Mechanisms of Chromospheric and Coronal Heating*, Springer Verlag, Berlin, pp. 30–32.

Rutten, R.G.M., Schrjiver, C.J., Lemmens, A.F.P., and Zwaan, C. (1991) Magnetic structure in cool stars XVII. Minimum radiative losses from the outer atmosphere, *Astron. Astrophys.* **252**, 203–219.

Sakurai, T. (1986) Heating mechanisms of the solar corona, in Y.Osaki (ed.), *Hydrodynamic and Hydromagnetic Problems in the Sun and Stars*, University of Tokyo, pp. 17–36.

Schrjiver, C.J. (1992) The basal and strong-field components of the solar atmosphere, *Astron. Astrophys.* **258**, 507–520.

Shimizu, T. (1995) Energetics and occurrence rate of active-region transient brightenings and implications for the heating of the active-region corona, *Publ. Astron. Soc. Japan* **47**, 251–263.

Solanki, S.K., Livingston, W., and Ayres, T. (1994) New light on the heart of darkness of the solar chromosphere, *Science* **263**, 64–66.

Spruit, H.C. (1981) Magnetic flux tubes, in S. Jordan (ed.), *The Sun as a Star*, NASA SP-450, pp.385–412.

Stenflo, J.O. (1976) Small-scale solar magnetic fields, in V. Bumba and J. Kleczek (eds.), *Basic Mechanisms of Solar Activity*, D.Reidel Publishing Company, Dordrecht, pp. 69–99.

Uchida, Y. and Kaburaki, O. (1974) Excess heating of corona and chromosphere above magnetic regions by non-linear Alfvén waves, *Solar Phys.* **35**, 451–466.

Vaiana, G.S. and Rosner, R. (1978) Recent advances in coronal physics, *Ann. Rev. Astron. Astrophys.* **16**, 393–428.

Watanabe, T., Hara, H., Shimizu, T., Hiei, E., Bentley, R.D., Lang, J., Phillips, K.J.H., Pike, C.D., Fludra, A., Bromage, B.J.I., and Mariska, J.T. (1995) Temperature structure of active regions deduced from the helium-like sulfur lines, *Solar Phys.* **157**, 169–184.

Withbroe, G.L. and Noyes, R.W. (1977) Mass and energy flow in the solar chromosphere and corona, *Ann. Rev. Astron. Astrophys.* **15**, 363–387.

SOLAR CORONAL HEATING: MHD MODELS AND OBSERVATIONAL SIGNATURES

G. E. VEKSTEIN
UMIST, Manchester, M60 1QD, England, UK

1. Introduction

Although it is generally accepted that the mechanism for maintaining high coronal temperatures is magnetic, understanding of the process is still far from complete. It is likely that different heating mechanisms operate in various large-scale observable structures (loops, coronal holes, X-ray bright points, etc). All heating theories broadly divide into two classes: wave models, if the time-scale of the driving photospheric motions is fast compared with Alfven transit time, and quasi-static models, if the driving time is slow (Browning, 1991; Sakurai, 1996). In the latter case, more relevant for compact and strongly magnetized active regions, the coronal field evolves through a series of magnetostatic equilibria: photospheric motions generate electric currents (approximately field-aligned), and this is a source of excess magnetic energy available for heating. However, high electrical conductivity of the hot coronal plasma makes classical Ohmic dissipation on global length scales completely inefficient. Thus , for heating to be effective, the coronal magnetic field must possess fine-scale structure such as current sheets (Parker, 1972), within which even small resistivity can break the topology constraints of ideal MHD, allowing fast transition of coronal configuration to a state of lower magnetic energy via magnetic reconnection. Such a scenario of storage and release of magnetic energy is reminiscent of flares, and indeed the coronal heating process can be viewed as a superposition of numerous "nanoflare" events (Parker, 1988). This concept bridges two historically separate research areas, flares and coronal heating, considering nanoflares as the lower-energy population of a wide spectrum of flare-like events. Unfortunately, individual nanoflares cannot be resolved observationally, at present nor in the foreseeable future. Therefore it is

Y. Uchida et al. (eds.),
Magnetodynamic Phenomena in the Solar Atmosphere – Prototypes of Stellar Magnetic Activity, 29–36.

extremely important to reveal observational signatures of the nanoflare-heated corona. One is likely to be a fine structure in coronal plasma, with hot and cool loops insulated by the magnetic field (the so called filling factor). Though existing information about this is quite controversial, more reliable data are expected from the upcoming SOHO mission.

2. Current Sheets and Magnetic Reconnection

The present understanding of how current sheets may develop within an initially smooth coronal magnetic field can be summarised as follows. There are two basic causes of fine structure formation in the corona: the first is the turbulent character of photospheric motions, and the second is the complex topology of the coronal magnetic configuration. In the first case fine scales can develop even within a coronal field of simplest geometry, which may be represented by the model due to Parker (1972): an initially uniform magnetic field, $\mathbf{B} = B_0\hat{\mathbf{z}}$, anchored in two rigid planes (the photosphere) at $z = 0$ and $z = L$, with perfectly conducting tenuous coronal plasma occupying the space between. Then the magnetic field lines are tangled about each other by slowly displacing their footpoints by a prescribed photospheric motion with typical space scale l_{ph}. The question is whether magnetic structures with scale $\delta << L, l_{ph}$ can arise (typical values of L, the coronal field length, are of order of $10^9 - 10^{10}$cm, while $l_{ph} \sim 10^8$cm). A simple, purely kinematic mechanism for creating such fine structures was pointed out by Van Ballegooijen (1988). The idea is that photospheric velocities, while being a smooth function of space coordinates (x, y) at any given moment, vary randomly in time, with some correlation time τ_c, typically equal to several convective eddy turnover times. Therefore the displacement of photospheric footpoints, viewed as a continuous mapping of the photospheric plane to itself, becomes more and more structured, with typical scale l decreasing like $l \sim l_{ph} \exp\left(-\frac{t}{\tau_c}\right)$. Consequently, a force-free magnetic field, generated between the planes by such a mapping, will possess electric currents growing exponentially in time (until finite resistivity eventually intervenes, providing statistically steady state balance). An alternative scenario might be a dynamical development of fine scales as a result of current-driven MHD instability. In this case the quasistatic evolution of the coronal field becomes disrupted once field-aligned currents exceed an instability threshold. Then the subsequent evolution of the system is governed mainly not by photospheric motions but by nonlinear dynamics of instability (so the former can be even switched off). Such a behaviour has been observed in a recent numerical simulation of Parker's model based on the reduced resistive MHD equations (Longcope & Sudan, 1994). The appearance of fine scale is evident in Fig.1.

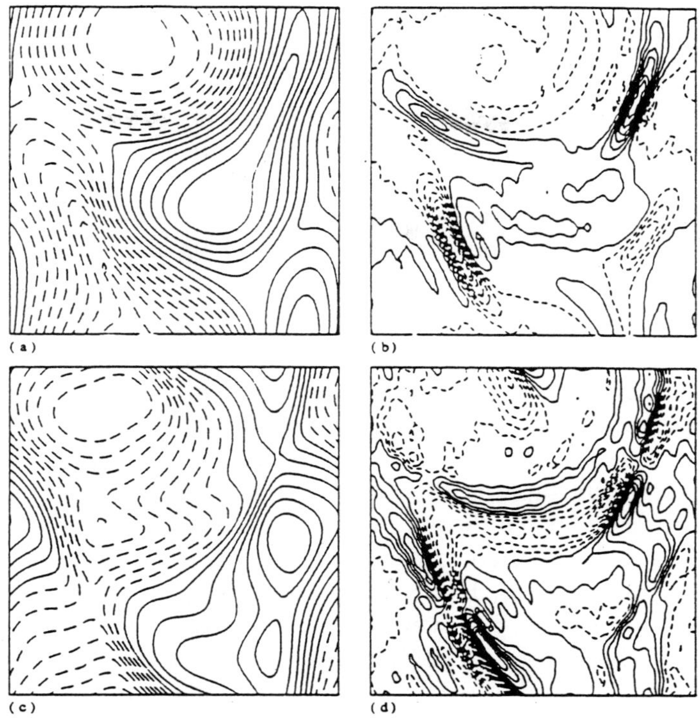

Figure 1. Contour plots of the magnetic field (a), electric current (b), fluid streamlines (c) and vorticity (d) at $z = \frac{L}{2}$ and $t = 28\tau_A$ (Longcope & Sudan, 1994)

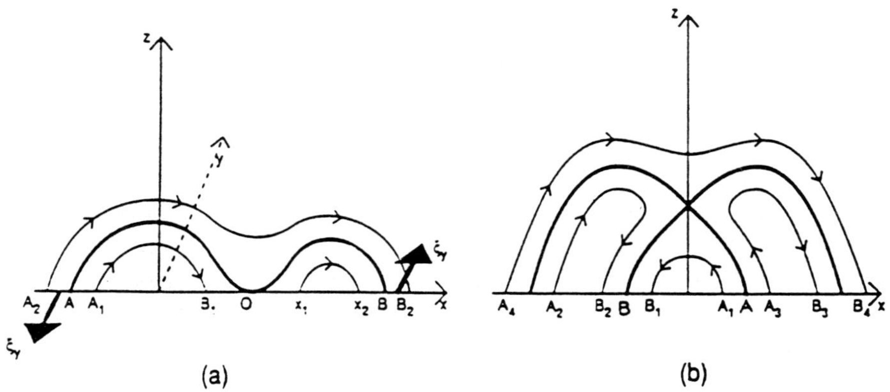

Figure 2. Poloidal magnetic configurations with separatrix lines

Current sheet formation caused by complexity of the coronal magnetic configuration is illustrated in Fig.2, where initially potential magnetic fields have separatrix lines, either without a neutral point (a) or with an X-point (b). The essential point is that these continuous magnetic fields have discontinuous connectivity of their footpoints on the photospheric plane $z = 0$. For example, field lines A_1B_1 and A_2B_2 on Fig.2(a) have their left-hand side footpoints A_1 and A_2 very close, while the right- hand side counterparts B_1 and B_2 are far apart. Thus even smooth shearing displacements ξ_y result in current sheet formation along the separatrix line AOB (Low & Wolfson, 1988; Vekstein et al, 1991).

Since both these factors, namely turbulent shuffling of photospheric footpoints and complex structure of magnetic field, are quite common features of the corona, the development of current sheets and their subsequent magnetic reconnection seem to be generic processes in the solar atmosphere. This view is strongly supported by Yohkoh SXT observations of continuous global restructuring in the corona (Tsuneta et al, 1992), as well as by high-resolution images of isolated and simpler coronal structures such as X-ray bright points (Strong et al, 1992). The latter may be considered as elementary reconnection events, following a scenario suggested recently by Priest et al (1994) . Their converging flux model involves an interplay between two opposite polarity photospheric magnetic fragments and an overlying background field (Fig.3). These fragments, while moving towards each other, remain magnetically unconnected until reaching the critical separation $2d$ (Fig.3b (ii)), when a null point at the photosphere appears. In a perfectly conducting corona the further approach of the fragments would result in formation of a vertical current sheet (Fig.3b (iii)). However, finite resistivity allows magnetic reconnection, so the fragments become partially connected to produce a lower-energy configuration shown in Fig.3b (iv). The released magnetic energy, typically of $10^{27} - 10^{28}$ ergs (microflare), heats coronal plasma, which is injected along the newly reconnected field lines, thus forming an observable X-ray bright point. This model provides a plausible explanation for the observed X-ray brightenings: hot plasma structures predicted from Kitt Peak magnetograms of the photosphere below the bright points show good agreement with simultaneous NIXT soft X-ray images (Parnell et al, 1994). A similar comparison has been reported recently (Van Driel-Gesztelyi et al, 1996) for X-ray bright points caused by magnetic flux emerging at the photosphere.

3. Coronal Heating as Driven Magnetic Relaxation

Once the hot corona is assumed to be an integral effect of multiple small-scale bursts, the statistical properties of these energy-release events become

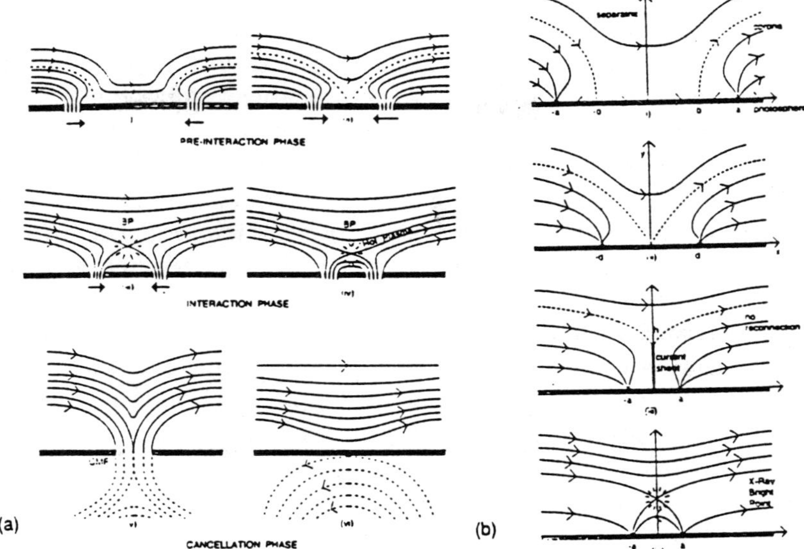

Figure 3. Different phases of the interaction of magnetic fragments (a), and the corresponding MHD scheme (b) (Priest et al, 1994)

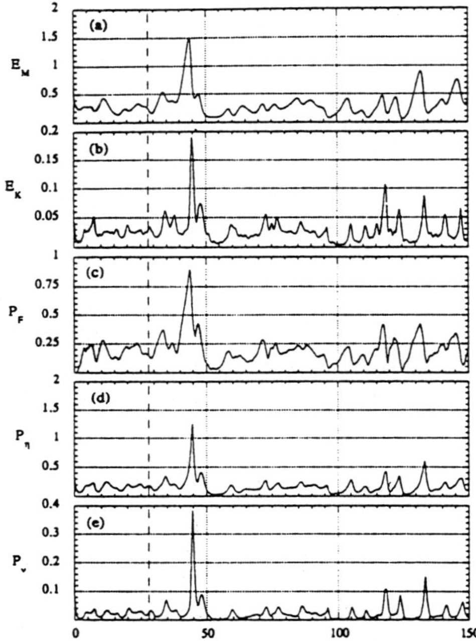

Figure 4. Evolution of the global coronal quantities (Longcope & Sudan, 1994)

an essential issue. The energy spectrum of microflares has been recently derived from the Yohkoh SXT data of active-region transient brightenings (Shimizu, 1995), and conclusion is that they add only slightly to the global energy balance. Thus the job is left to presumably more frequent but elusive events of lower energy scale (nanoflares?). An example of theoretical predictions on energetics and occurrence rate of coronal heating events is given in Fig.4, which shows results of numerical simulation for Parker's model (Longcope & Sudan, 1994). Here, the time history of global quantities in the corona, such as the free magnetic energy E_M, the plasma kinetic energy E_K, the photospheric power input P_F, and the Ohmic P_η and viscous P_ν dissipation rates is followed until a statistical steady state is reached, when energy input into the corona becomes balanced by dissipation. Though such simulations provide some insight into the problem, the crucial difficulty is that they can be performed only for quite moderate values of the magnetic Reynolds number (5×10^2 in this case), while its typical value in the solar corona is around 10^{12}. It is completely hopeless to get any realizable computations for such a large Reynolds number, while available numerical results cannot be extrapolated to real coronal parameters since the proper scaling is unknown. Therefore it might be more productive to tackle this problem starting not from the first principles, but applying more general phenomenological considerations. For example, viewing the solar corona as a complex turbulent system, the concept of self-organised criticality, originally introduced to describe a self-similar behaviour of avalanches, can be applied also to the coronal flare-like events (Lu & Hamilton, 1991). A dynamical basis for these "kinematic" avalanche-type models has been discussed recently by Chiuderi et al (1996). Another approach to the coronal heating process has been proposed by Heyvaerts & Priest (1984), who generalised a very successful idea from laboratory plasma physics due to Taylor (1974). This is to regard the effect of complex magnetic reconnection processes as being a relaxation of the system to a minimum magnetic energy state subject to the constraint imposed by a large value of the magnetic Reynolds number, namely of conservation of the total magnetic helicity. This allows derivation of the average heat flux into the corona in terms of the photospheric driver and coronal parameters as (Vekstein et al, 1993):

$$q \simeq 2\times10^{-2}\frac{B^2\langle v_{ph}^2\rangle}{L}\frac{\tau_{ph}^2}{\tau_R}$$

where τ_R , the relaxation time, is an imposed phenomenological parameter. For a typical active region with magnetic field $B\sim(1-3)\times10^2$G and the size $L\sim$ 10Mm, and photospheric motions with the mean square velocity $\langle v_{ph}^2\rangle\sim10^6-10^7$ m^2 s^{-2} and period $\tau_{ph}\sim10^3$s, this requires the relaxation time to be of the order of 10^4-10^5s to produce a "standard" active region maintaining an energy flux of 10^7 erg cm^{-2} s^{-1}. This may be considered as a reasonable value, especially from the point of view that observed global restructuring in the corona (Tsuneta et al, 1992) has a similar time scale of $10 - 20$ hours.

4. Coronal Filling Factors

Though the spatial scale of primary energy release events (say, nanoflares) is likely to be far beyond the observational capabilities, an integral effect of multiple nanoflares producing the hot corona may have a detectable signature such as filamentary structure of coronal plasma, when hot loops constitute only some fraction (known as the filling factor) of the whole volume. Then, since only these loops contribute to the observed X-ray emission, the magnitude of the filling factor, ϕ, can be estimated by comparing the average density in the corona (derived from the integral emission measure) with the actual density of hot plasma, which can be obtained from density-sensitive spectral lines ratios. Theoretical prediction of filling factors requires statistical analysis of an ensemble of impulsively heated coronal loops cooled by radiation and heat conduction (Cargill, 1993; Kopp & Poletto, 1993), but a crude estimate can be obtained quite simply. Assuming that the heat flux q, which maintains the corona hot ($q \sim 10^7$ ergs cm^{-2} s^{-1} for active region), is provided by elementary energy releases inside the corona, each of energy ΔW, the occurrence rate (per unit area) of the latter has to be equal to $\frac{q}{\Delta W}$. Thus, viewing coronal structure as a superposition of flux loops, each of the length L and cross-section S, it is on average every $\Delta t = S\frac{\Delta W}{q}$ sec that an energy burst will occur somewhere along a fixed loop. Being heated in such a way, the loop will remain hot for some time τ_h which is determined by radiative and conductive losses. Then the filling factor can be derived as (Cargill, 1993):

$$\phi = \frac{\tau_h}{\Delta t} \simeq 8.5 \times 10^{-5} \frac{qLS^{\frac{7}{6}}}{(\Delta W)^{\frac{7}{6}}}$$

This result shows a strong dependence on the loop cross-section S which is actually determined by the specific details of the energy release mechanism, and therefore cannot be imposed arbitrarily. A self-consistent derivation of S might be quite a complicated problem, since newly reconnected and heated field lines are convected away from the reconnection side. However, the minimal value of S can be simply estimated from the requirement for pressure balance between the hot loop and the surrounding magnetic field, yielding eventually prediction for ϕ which does not depend on the energy release "quantum" ΔW:

$$\phi \simeq 0.7\times 10^{-2}\frac{\tilde{q}}{\tilde{B}^{\frac{7}{3}}\tilde{L}^{\frac{1}{6}}}$$

Here $\tilde{q} = q/10^7$ ergs cm^{-2} s^{-1}, $\tilde{B} = B/10^2$ G, and $\tilde{L} = L/10$ Mm are dimensionless heat flux, magnetic field and loop length normalised by their typical values in the corona. Thus small values of the coronal filling factors are likely to be expected, and if so, this may affect the energy balance analysis of active regions (Shimizu, 1995) as well as that of scaling for coronal loops heating (Klimchuk & Porter, 1995).

References

Browning, P. K. 1991, *Plasma Phys. Contr. Fusion*, **33**, 539.
Cargill, P. J. 1993, *Solar Phys.*, **147**, 263.
Chiuderi, C. 1996, these Proceedings.
Heyvaerts, J. & Priest, E. R. 1984, *Astron. Astrophys.*, **137**, 63.
Klimchuuk J. A.& Porter, L. J. 1995, *Nature*, **377**, 131.
Kopp, R. A. & Poletto, G. 1993, *Astrophys. J..*, **418**, 496.
Longcope, D. W. & Sudan, R. 1994, *Astrophys. J.*, **437**, 491.
Low, B. C. & Wolfson, R. 1988, *Astrophys. J.*, **324**, 574.
Lu, E. T. & Hamilton, R. J. 1991, *Astrophys. J.*, **380**, L89.
Parker, E. N. 1972, *Astrophys. J.*, **174**, 499.
Parker, E. N. 1988, *Astrophys. J.*, **330**, 474.
Parnell, C. E., Priest, E. R. & Golub, L. 1994, *Solar Phys..*, **151**, 57.
Priest, E. R., Parnell, C. E. & Martin, S. F. 1994, *Astrophys. J.*,**427**, 459.
Sakurai, T. 1996, these Proceedings.
Shimizu, T. 1995,*Publ. Astron. Soc. Japan*, **47**, 251.
Strong, K. et al. 1992, *Publ. Astron. Soc. Japan*, **44**, L161.
Taylor, J. B. 1974, *Phys. Rev. Lett.*, **33**,1139.
Tsuneta, S. et al. 1992, *Publ. Astron. Soc. Japan*, **44**, L211.
Van Ballegooijen, A. A. 1988, *Geophys.Astrophys.Fluid Dyn.*,**41**, 181.
Van Driel - Gesztelyi, L. et al. 1996, these Proceedings.
Vekstein, G. E., Priest, E. R. &Amari, T. 1991, *Astron. Astrophys.* , **243**, 492.
Vekstein, G. E., Priest, E. R. &Steele, C. D. C. 1993, *Astrophys. J.*, **417**, 781.

HEATING OF ACTIVE REGION CORONA BY TRANSIENT BRIGHTENINGS (MICROFLARES)

— Yohkoh/SXT -'La Palma Observations —

T. SHIMIZU AND S. TSUNETA
Institute of Astronomy, The University of Tokyo
2-21-1 Osawa, Mitaka, Tokyo 181, Japan

AND

A. TITLE, T. TARBELL, R. SHINE AND Z. FRANK
Lockheed Palo Alto Research Laboratory
3251 Hanover Street, Palo Alto, CA 94304, U.S.A.

The *Yohkoh* Soft X-Ray Telescope (SXT) has been revealing the coronal characteristics of a microflaring phenomenon in X-rays: Microflares are observed as transient brightenings of compact coronal loops (Shimizu *et al.*, 1992); they appear mainly in form of multiple or single loops, although they are sometimes point-like (Shimizu *et al.*, 1994). They recur frequently in two localized areas of active regions, namely, in growing emerging magnetic flux regions and around the outer edge of the penumbra of well-developed large spots (Shimizu, 1993). The sum of transient brightenings does not make much contribution to the heating of the active-region corona, although they occur very numerously in active regions (Shimizu, 1995a).

Simultaneous observations made in soft X-rays and visible light are essential for understanding magnetic coupling between transient brightenings in the corona and magnetic activity at the photosphere. Using a coordinated set of observations by SXT and LPARL's narrow-band tunable filter at La Palma (Swedish Solar Observatory), we have studied the evolution of the photospheric magnetic field associated with point-like transient brightenings (Shimizu, 1995b). A small-scale emergence of magnetic flux is found to start 5 - 15 minutes prior to the onset of X-ray brightenings in at least 4 of the 16 events studied, although no evolutionary changes of the magnetic field are found in the other cases. Figure 1 is an example clearly showing a close connection between a transient brightening and a small-scale flux emergence. Results from the detailed comparison will be published elsewhere in the future.

Y. Uchida et al. (eds.),
Magnetodynamic Phenomena in the Solar Atmosphere – Prototypes of Stellar Magnetic Activity, 37–38.

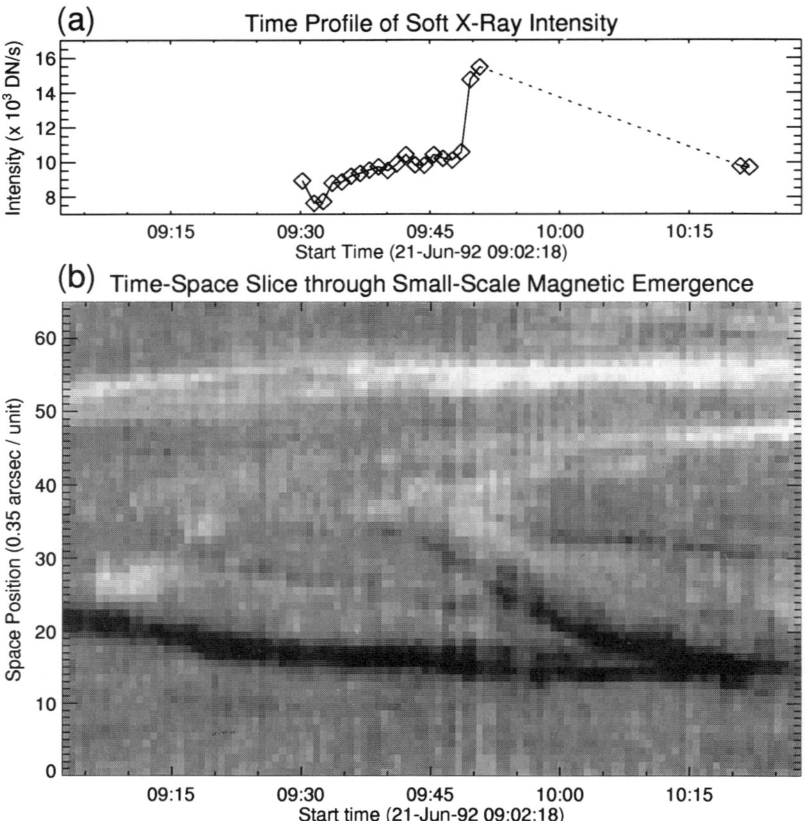

Figure 1. (a) Time profile of soft X-ray intensity of the 21-June-92 9:48 UT transient brightening; (b) Time-space slice of longitudinal magnetogram with time increasing horizontally along a line through the X-ray brightening centering at position 35 with the size of 5". The counterpart magnetic flux starts to emerge at about 9:38 UT.

References

Shimizu, T., Tsuneta, S., Acton, L. W., Lemen, J. R. and Uchida, Y. (1992) Transient Brightenings in Active Regions Observed by the Soft X-Ray Telescope on Yohkoh, *Publ. Astron. Soc. Japan,* **44**, L147-153.

Shimizu, T. (1993) Active-Region Transient Brightenings in NOAA 7260, in T. Sakurai et al. (eds.), *Proceedings of the 2nd Japan-China Seminar on Solar Physics,* National Astronomical Observatory of Japan, Tokyo, pp. 193-196.

Shimizu, T., Tsuneta, S., Acton, L. W., Lemen, J. R., Ogawara, Y. and Uchida, Y. (1994) Morphology of Active Region Transient Brightenings with the Yohkoh Soft X-Ray Telescope, *Astrophys. J.,* **422**, 906-911.

Shimizu, T. (1995a) Energetics and Occurrence Rate of Active-Region Transient Brightenings and Implications for the Heating of the Active Region Corona, *Publ. Astron. Soc. Japan,* **47**, 251-263.

Shimizu, T. (1995b) Studies of Transient Brightenings (Microflares) Discovered in Solar Active Regions, PhD thesis, University of Tokyo.

THE HEATING OF SOFT X-RAY CORONAL LOOPS

J. A. KLIMCHUK
E. O. Hulburt Center for Space Research
Naval Research Laboratory, Code 7675JAK
Washington, DC 20375, USA

AND

L. J. PORTER
Massachusetts Institute of Technology
24-204A, 77 Massachusetts Ave.
Cambridge, MA 02139, USA

1. Summary

The physical process which heats the multi-million degree coronae of late-type stars such as the Sun is one of the fundamental mysteries of astrophysics. A number of plausible theories have been proposed, but none of them has yet been demonstrated to be correct. We must therefore rely on new coronal observations, such as those from *Yohkoh*, to constrain and distinguish among the competing ideas.

The coronal magnetic field organizes the plasma into loops—the basic structural elements—and to understand coronal heating, we must understand the heating of these loops. It is important to remember that the magnetically-closed corona is completely filled with loops, and that most loops are not readily distinguishable from their neighbors. Klimchuk and Gary (1995) have recently studied the heating of the diffuse corona, which is comprised of indistinguishable loops, and we here consider the heating of individual loops that can be easily identified in *Yohkoh* Soft X-Ray Telescope (SXT) images. Details of our study can be found in Klimchuk and Porter (1995) and Porter and Klimchuk (1995), and related work has been reported by Kano and Tsuneta (1995) and Sturrock and Acton (1995).

We measured the temperatures, pressures, and lengths of 47 apparently steady loops observed by SXT. The median temperature is 5.7×10^6 K and the median pressure is 1.6×10^{16} cm^{-3} K. Using nonparametric statistical

Y. Uchida et al. (eds.),
***Magnetodynamic Phenomena in the Solar Atmosphere – Prototypes of Stellar Magnetic Activity*, 39–40.**

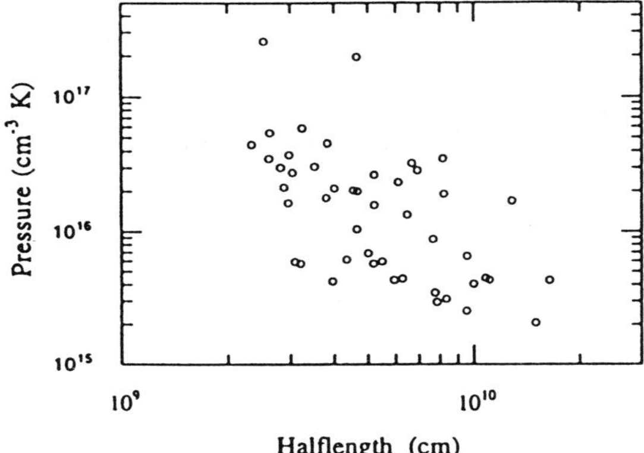

Figure 1. Plot of pressure versus halflength for the 47 loops.

methods, we determined that temperature is uncorrelated with length and that pressure varies inversely with length to approximately the first power (see Figure 1). Because the observed lifetimes of the loops are much longer than their cooling times, we then assumed quasistatic equilibrium and used scaling laws to infer that the volumetric heating rate varies inversely with length to approximately the second power.

This is an important result, because different coronal heating models predict different scalings. For example, the popular models for (1) the tangling of coronal magnetic fields by random shuffling motions at the footpoints, (2) the twisting of coronal flux tubes by vortical motions at the footpoints, and (3) the resonance absorption of Alfven waves are able to reproduce the observations only if the coronal magnetic field strength scales with loop length as $B \propto L^{\delta}$, where δ = -0.5, 0, and -2, respectively. The actual value of δ is not presently known, but it could be determined from coronal extrapolations of photospheric magnetograms. Once δ is determined, we will be able to draw definitive conclusions about the viability of these and other models for coronal heating.

References

Klimchuk, J. A. and Gary, D. E. (1995) A Comparison of Active Region Temperatures and Emission Measures Observed in Soft X-Rays and Microwaves and Implications for Coronal Heating, *Ap J* **448**, pp. 925–937

Klimchuk, J. A. and Porter, L. J. (1995) Scaling of Heating Rates in Solar Coronal Loops, *Nature* **377**, pp. 131–133

Porter, L. J. and Klimchuk, J. A. (1995) Soft X-Ray Loops and Coronal Heating, *Ap J*, in press (Nov. 20)

Kano, R. and Tsuneta, S. (1995) Temperature Distributions and Energy Scaling Law of Solar Coronal Loops Obtained with *Yohkoh*, *PASJ*, submitted

Sturrock, P. A. and Acton, L. W. (1995) in these proceedings

TEMPERATURE STRUCTURE OF SOLAR ACTIVE REGIONS

T. YOSHIDA AND S. TSUNETA
Institute of Astronomy, The University of Tokyo
2-21-1, Osawa, Mitaka, Tokyo, 181 JAPAN

Introduction

Yohkoh X-ray images of the Sun show ubiquitous bright loops. Most of the analyses of *Yohkoh* data so far made on the temperature structure of the coronal loops concentrated on the individual X-ray loops. We, however, need to remember that the high temperature corona apparently exists outside the distinct loops and in the diffuse region. The fundamental questions in the context of coronal heating are why specific magnetic structures are bright in X-rays and what the difference is between bright X-ray loops and faint diffuse regions. As a first step to answer these questions observationally, we obtain high-quality temperature maps of the entire active regions from X-ray maps summed over a certain time interval by sacrificing the time resolution.

Observations and Analysis

The temperatures of active regions are obtained using a pair of broad-band filters. We use the *Yohkoh* soft X-ray images taken from 1992 April 24 through July 5. During this period, the Soft X-Ray Telescope (SXT) aboard *Yohkoh* took images of several active regions every 1 minute with 2 different filters alternately. We sum the images for each filter in units of spacecraft orbit period (97 minutes). The number of frames taken in about 60 minutes of the spacecraft daytime is about 30 for each filter. The summation allows us to obtain high-quality temperature maps even outside the bright loops at the expense of time resolution. The estimated photon-statistics errors in temperature range from about three percent in bright X-ray loops to about ten percent in dark diffuse regions.

Temperature Structure of Active Regions

The soft X-ray maps and temperature maps appear quite different. The temperatures of the loops whose sizes are similar and whose X-ray intensities are similar are not always similar. There are various loops with tem-

Y. Uchida et al. (eds.),
Magnetodynamic Phenomena in the Solar Atmosphere – Prototypes of Stellar Magnetic Activity, 41–42.

perature from 3 MK to 10 MK. The time variability of the temperature structures suggest that there are two distinct time scales in the coronal temperature structures. Structures with temperatures higher than 5 MK are seen for several hours, and those with temperature less than 5 MK can be seen over half a day without significant change in shape and temperature. Transient structures with lifetimes shorter than several hours have the following features: (1) The temperatures are higher than 5 MK, and do not depend on the soft X-ray intensities. (2) The structures are associated with the cusp structures or multipleloop structures in most cases. Steady structures with lifetimes longer than half a day have the following features: (1) The temperatures of the loop tops are from 4 to 5 MK, and those of the footpoints are around 3 MK. (2) The difference in temperature between nearby loops are too small to recognize individual loop structures in the temperature maps.

Discussion

The conductive loss time scales of the structures with life time shorter than several hours are about 30 minutes ($T = 6$ MK , $n = 10^{9.5}$ cm^{-3} , $l = 10^5$ km), and the radiative loss time scales are about 10 hours. The observed time scales are close to these time scales. This indicates that these structures are heated transiently, since these are associated with the cusp structures or multipleloop structures in most cases. The transient heating thus may be due to magnetic reconnection.

The conductive loss time scales of the structures with life time longer than half a day are about 50 minutes ($T = 4.5$ MK , $n = 10^{9.5}$ cm^{-3} , $l = 10^5$ km), and the radiative loss time scales are about 6 hours. The observed time scales are much longer than these time scales. This indicates that the steady plasmas are continuously heated. It is sometimes hard to find the temperature structures corresponding to the steady loop structures seen in X-ray images. Thus, the steady coronal loops may be heated more or less in a loop-independent way.

This indicates that the coronal heating consists of both basal heating and transient heating. The basal heating mechanism maintains the plasma temperature around 3–5 MK, and the occasional transient heating, perhaps due to magnetic reconnection, produces higher temperature plasma with temperatures up to 10 MK.

References

T. Yoshida and S. Tsuneta, (1996), *ApJ*, 459, 342

TEMPERATURE AND HEATING DISTRIBUTIONS ALONG THE STEADY CORONAL LOOPS

R. KANO AND S. TSUNETA
Institute of Astronomy, The University of Tokyo, 2-21-1 Osawa, Mitaka, Tokyo 181, Japan

The temperature distributions along the steady coronal loops reflect their heating distributions. We selected 16 bright steady coronal loops from the *Yohkoh* SXT images from May 1992 through July 1992, and obtained the temperature distributions $T(s)$ along these loops. The steady loops generally have constant temperatures along the loop top regions (plateau) with decreasing temperatures toward both footpoints ("trapezoidal" temperature distribution, figure 1a). Some of the loops have sharp peaks in their temperature distributions ("triangular" temperature distribution, figure 1b), suggesting the concentration of the heat input at the peak.

We find a good correlation between total energy loss $\mathcal{L}_{\rm T}$ (erg/sec/cm^2) and gas pressure $p_{\rm g}$ (dyn/cm^2) for the steady loops (Kano & Tsuneta 1996);

$$(\log \mathcal{L}_{\rm T} - 8.0) = (0.99 \pm 0.12)(\log p_{\rm g} - 0.9), \tag{1}$$

or

$$\mathcal{L}_{\rm T} = 1.1 \times 10^7 p_{\rm g}^{0.99 \pm 0.12}. \tag{2}$$

The total energy loss is the radiative loss plus conductive flux estimated from the observed temperature and density distributions. The gas pressure is obtained from the observed temperature and emission measure at the top. This correlation fits well with the equation,

$$\mathcal{L}_{\rm T} = 1.0 \times 10^7 \sqrt{T_{\rm m}/(6 \times 10^6)} \cdot p_{\rm g}, \tag{3}$$

theoretically derived from two relations of the scaling law;

$$T_{\rm m} = 1.4 \times 10^3 (p_{\rm g} \cdot L)^{1/3} \quad \text{(Rosner, Tucker \& Vaiana 1978)}, \tag{4}$$

$$\mathcal{L}_{\rm T} = 1.5 \times 10^5 (p_{\rm g}^{7/6} \cdot L^{1/6}), \tag{5}$$

where L (cm) is the observed length of the loop, and $T_{\rm m}$ (K) is the maximum temperature ($T_{\rm m} \sim 6 \times 10^6$ (K) for all the loops analyzed here).

Y. Uchida et al. (eds.),
Magnetodynamic Phenomena in the Solar Atmosphere – Prototypes of Stellar Magnetic Activity, 43–44.

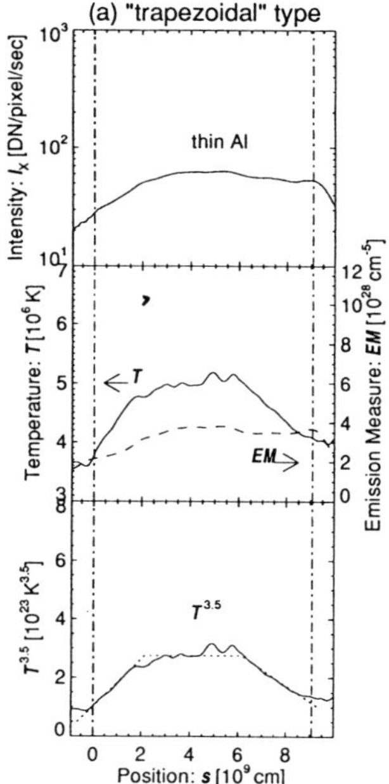

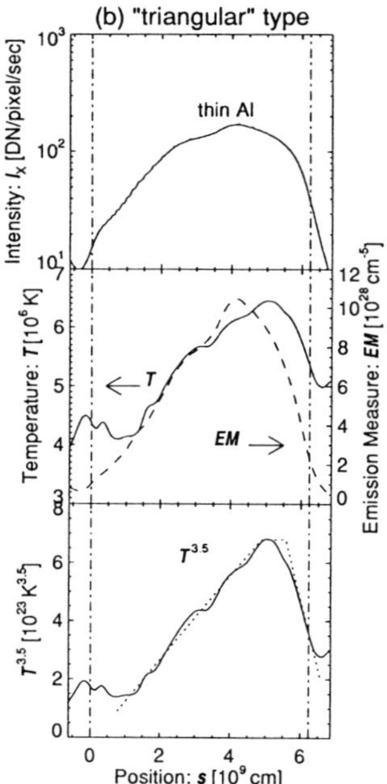

Figure 1. Two extreme temperature distributions along the loop; (a) "trapezoidal" type and (b) "triangular" type. The upper panels are the X-ray intensity taken with the thin aluminum filter. The middle panels are the temperature (solid line) and the emission measure (dashed line) distributions. The lower panels are the 3.5 power of the temperature distribution. We fit the dotted lines to the $T^{3.5}$ profiles to derive the conductive fluxes. The vertical dot-dashed lines in these panels show the positions of the apparent footpoints.

We also find that $T_{\rm m}$ is correlated with $(p_{\rm g} \cdot L)$ (Kano 1994, Kano & Tsuneta 1995);

$$T_{\rm m} = 3.8 \times 10^4 (p_{\rm g} \cdot L)^{1/(5.1 \pm 0.5)}. \tag{6}$$

Although the power-law index 1/5.1 is different from the index 1/3 of the theoretically derived scaling law, we conclude that this observed correlation is consistent with the theoretical scaling law within the systematic errors of our analysis.

References

Kano, R. (1994) Scaling Law and Heating Function of Coronal Loops Obtained by Yohkoh SXT, Master Thesis, The University of Tokyo

Kano, R. and Tsuneta, S. (1995) Scaling Law of Solar Coronal Loops Obtained with Yohkoh, *Astrophys. J.*, **454**, 934–944

Kano, R. and Tsuneta, S. (1996) Temperature Distribution and Energy Scaling Law of Solar Coronal Loops Obtained with Yohkoh, *Publ. Astron. Soc. Japan*, submitted

Rosner, R., Tucker, W., and Vaiana, G. (1978) Dynamics of the Quiescent Solar Corona, *Astrophys. J.*, **220**, 643–665

LONG TIME STATISTICS OF MAGNETICALLY DRIVEN MHD TURBULENCE, SOLAR FLARES AND CORONAL HEATING

C. CHIUDERI, M. VELLI
Dip. di Astronomia e Scienza dello Spazio 50125 Firenze, Italy
G. EINAUDI
Dipartimento di Fisica, 56100 Pisa, Italy
AND
A. POUQUET
Observatoire de la Côte d'Azur, 06304 Nice Cedex 04, France

The properties of the coronal turbulence, and in particular the way the energy can be stored in the magnetic field and then dissipated, represent a very difficult problem. In fact, due to the large values of the magnetic Reynolds number S $\sim 10^{13}$, the magnetic field contributing to the available free energy must be structured over spatial scales of the order of one meter or less. As a result the local release of magnetic energy occurs on the dynamical timescale and is concentrated inside current sheets which are continuously formed and dissipated throughout the system (Einaudi and Velli 1994).

Such a scenario results from our present knowledge of MHD turbulence. The 3-D simulations performed by Mikic et al. (1989), Strauss (1993) and Longcope and Sudan (1994) show that the presence of a mean large scale field allows the separation of the dynamical behaviour along that field from that perpendicular to it which turns out to be very similar to the 2-D case. Such simulations however have too low resolution and have been performed for too short times to give reliable information about the long term response of the corona to the stresses induced by the photospheric motions.

Einaudi et al. (1995) have performed numerical simulations of a 2D section of a coronal loop, subject to random magnetic forcing. These simulations last much longer than both the coronal Alfvén time and the time scale of the driver and it is therefore possible to follow in time the properties of the current sheets which continuously form and disrupt in the system. The forcing models the link between photospheric motions and en-

Y. Uchida et al. (eds.),
Magnetodynamic Phenomena in the Solar Atmosphere – Prototypes of Stellar Magnetic Activity, 45–46.

ergy injection in the corona. The boundary disturbances propagate along the vertical mean field B_0 with the associated Alfvèn velocity and give rise to a perpendicular magnetic field $b_\perp$. The appearance of $b_\perp$, and consequently of $v_\perp$, can be followed in time by solving the MHD equations in 2-D in which the terms coupling the perpendicular dynamics to the vertical one are treated as imposed forcing terms. The results show the highly intermittent spatial distribution of current concentration generated by the coupling between internal dynamics and external forcing. The total power dissipation is a rapidly varying function of time, with jumps of orders of magnitude even at low Reynolds numbers, and is due to the superposition of current dissipation in a number of localized current sheets. The histogram of dissipation events as a function of dissipated power follows a power-law behaviour.

The fact that both the spatial and temporal intermittency increases with the Reynolds number (resolution) suggests that the turbulent nature of the corona can physically motivate statistical theories of solar activity such as the self-organized criticality (avalanche) model (Lu and Hamilton, 1991; Vlahos et al. 1995). In this models the fundamental ingredient is the non-gaussian response to a gaussian forcing which results from the locality of the instability criterion and of the relaxation process. We have performed a statistical analysis of the data resulting from our simulations at 256x256 resolution. There is evidence of intermittent behaviour both in time and space, which means the existence of a non-gaussian tail in the distribution of the current around its mean (spatial or temporal) value. This spatial and temporal intermittency is not well understood, but is a natural tendency of the MHD turbulence and could be the physical constituent of the avalanche model for solar flares.

It is our feeling that the above preliminary results substantiate the idea that heating and flares could be considered as manifestations of the same basic physics,the difference being quantitative more than qualitative.

This work has been partially supported by EEC through the contract ERBCHRXCT930410.

References

Einaudi, G. and Velli, M.: 1994, *Advances in Solar Physics*, eds. G. Belvedere, M. Rodonò and G.M. Simnett, (Springer-Verlag), p.149.

Einaudi, G., Velli, M., Politano, H. and Pouquet, A.: 1995, *Astrophys. J. Letts.*, "Energy release in a turbulent corona", in press.

Longcope, L. and Sudan, L.: 1994, *Astrophys. J.*, **437**, 491.

Lu, E.T. and Hamilton, R.J.: 1991, *Astrophys. J.*, **380**, L89.

Mikic, Z., Schnack, D.D. and Van Hoven, G.: 1989, *Astrophys. J.*, **338**, 1148.

Strauss, H.: 1993, *Geophys. Res. Letts.*, **20**, 325.

Vlahos, L.,Georgoulis, M., Kluiving, R. and Paschos, P.: 1995, *Astron. Astrophys.*, **299**, 897.

MASS-FLOW EFFECTS ON WAVE HEATING BY RESONANT ABSORPTION

R. ERDÉLYI VON FÁY-SIEBENBÜRGEN
Center for Plasma-Astrophysics, K.U. Leuven
Celestijnenlaan 200 B, B-3001 Heverlee, Belgium

1. Introduction

To heat the solar atmosphere by waves one requires some efficient dissipation mechanisms. Resonant absorption, suggested by Ionson (1978), might be a candidate to explain the heating mechanism(s) in the solar corona. In the former studies of resonant absorption one has mostly assumed static equilibria. However, Hollweg et al. (1990) showed that resonant absorption can be severely influenced by the presence of velocity shears. Erdélyi et al. (1995) studied analytically the resonant absorption in linear compressible MHD for background equilibrium states with flow.

In this paper we show numerically the effect of background equilibrium flows on resonant absorption. Lou's (1990) cylindrical equilibrium model is extended with an equilibrium plasma flow, $\underline{\mathbf{v}}_0(r) = f \times \underline{\mathbf{v}}_{Alfv}(r)$, where f denotes the flow strength parameter.

2. Results

In Figure 1 we have plotted the absorption rate versus the flow strength parameter, f. Fig. 1 shows that the flow has a very determinant effect on the absorption rate. By increasing the flow strength parameter the absorption rate enhances very strongly. Increasing further the equilibrium flow strength, the absorption rate shows a strong oscillatory behaviour between ca. the value of total resonant absorption and the value of total reflection. Increasing even further the equilibrium flow parameter, f, the absorption rate smoothly decreases. Roberts et al. (1984) have shown that homogeneous, non-twisted magnetic flux tubes can have infinite eigenvalues. If one

Y. Uchida et al. (eds.),
Magnetodynamic Phenomena in the Solar Atmosphere – Prototypes of Stellar Magnetic Activity, 47–48.

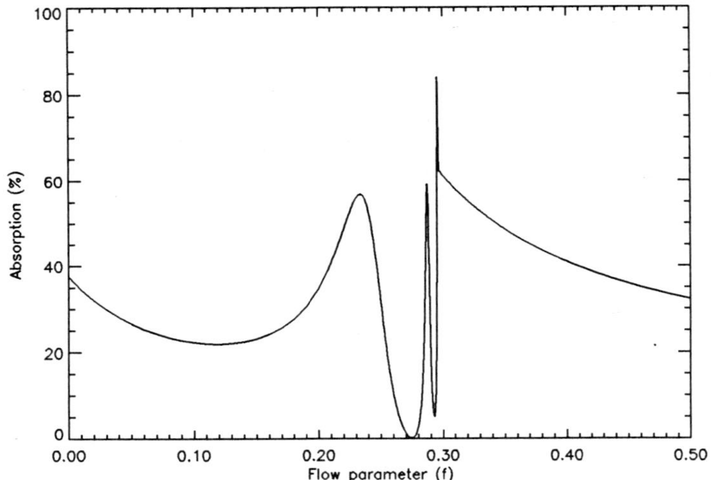

Figure 1. Absorption coefficient α versus the equilibrium flow strength parameter, f

drives a flux tube with a frequency close to one of its eigenfrequencies the driving wave will be strongly absorbed (Goossens and Hollweg, 1993). We guess this is the case when in Fig. 1 $f \in [0.25 - 0.31)$, the absorption rate shows so many maxima. We cannot prove this statement now because we have to solve the eigenvalue problem of our non-homogeneous driven problem, but this work is under investigation.

The presence of the equilibrium flow may therefore be very determinant for resonant absorption and may significantly affect the resonant absorption rate for acoustic oscillations in sunspots or for Alfvén waves in the solar corona.

Acknowledgements The author is grateful to M. Goossens and M. Ruderman for suggesting the work and for stimulating discussions.

References

Erdélyi, R., Goossens, M. and Ruderman, M. S. (1995), *Sol. Phys.*, in press

Goossens, M. and Hollweg, J. V. (1993), *Sol. Phys.*,**145**, pp. 19–44

Hollweg, J. V., Yang, G., Cadez, V. M. and Gakovic, B. (1990), *ApJ.*, **349**, pp. 335–344

Ionson, J. A. (1978), *ApJ.*,**226**, pp. 650–673

Lou, Y. Q. (1990), *ApJ.*, **350**, pp. 452–462

Roberts, B., Edwin, P. M. and Benz, A. O. (1984), *ApJ.*, **279**, pp. 857–865

Hα SURGES AND X-RAY JETS IN AR7260

RICHARD C. CANFIELD, KEVIN P. REARDON, K. D. LEKA
Institute for Astronomy, University of Hawaii
2680 Woodlawn Drive, Honolulu, HI 96822

K. SHIBATA, T. YOKOYAMA
National Astronomical Observatory of Japan
Mitaka, Tokyo 181, Japan

AND

M. SHIMOJO
Department of Physics, Tokai University
1117 Kitakaname, Hiratuka, Kanagawa, Japan

We have studied nine events, which we observed simultaneously from Yohkoh as jets in X-rays and from /MSO/ as surges in Hα, in AR7260 at approximately N16 W30 on 19-20 August, 1992. We attribute these events to the reconnection of moving magnetic bipoles with surrounding coronal fields. The X-ray jets share many features with those discovered by Yohkoh in active regions, EFRs, and X-ray bright points (Shibata *et al.* 1992). The Hα surges are adjacent to the X-ray jets. Typically the footpoints of two closed loops are observed near the base of the surge, and flare in association with the surge (Figure 1). At the surge bases we observe both blueshifts (initially) and redshifts (1–2 minutes later). All the observed surges spin in the same sense.

Two new phenomena have been discovered, which indicate the role of magnetic reconnection: footpoint convergence and moving-blueshift features. In footpoint-convergence events, bright Hα features are observed at both ends of Hα fibrils. These features converge during the early stages of the jet/surge event, indicating that reconnection involves successively lower-lying field lines in the moving-bipole flux system. In moving-blueshift events, Hα blueshifted features are observed to move along fibrils at about 10% of the Alfven speed, which we attribute to the whip-like response of newly-reconnected magnetic field lines previously in the moving-bipole flux system, as they adjust to their new topology.

Y. Uchida et al. (eds.),
Magnetodynamic Phenomena in the Solar Atmosphere – Prototypes of Stellar Magnetic Activity, 49–50.

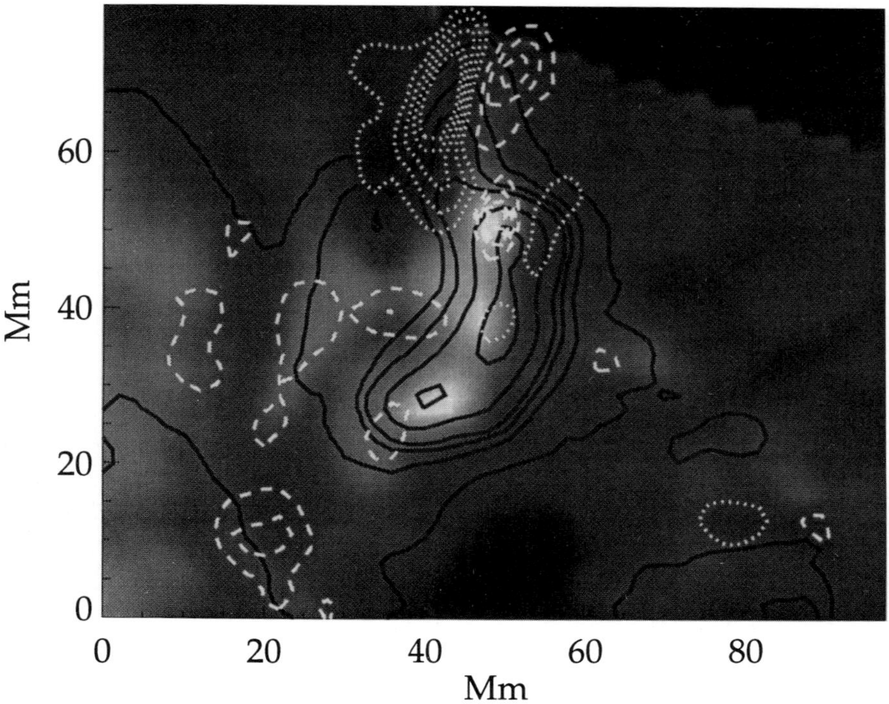

Figure 1. Contours of Hα velocity (white) and X-ray intensity (black) overlaid on an Hα image of the surge/jet event of 23:11 UT on 19 August, 1992. The SXT X-ray contours show the jet extending North (up) from the northern tip of two contiguous end-to-end flaring loops whose footpoints are seen as three bright patches in the MCCD line-center Hα spectroheliogram. The adjacent oval regions of Hα redshift (dashed white) and blueshift (dotted white) centered over the jet are due to the spinning motion of the surge. We find that the total angle through which the surge spins is comparable to the product of the measured force-free field parameter α and length of the reconnecting moving magnetic bipole, implying that release of stored twist is the mechanism for the spin. We associate the X-ray jet with the upward reconnection jet and the region of strong Hα redshift at its base with the downward reconnection jet.

We have submitted for publication a model of the morphology of the reconnection sequence and quantitative analyses of the new phenomena (Canfield *et al.* 1995).

References

Canfield, R. C., Reardon, K. P., Leka, K. D., Shibata, K., Yokoyama, T., and Shimojo, M. (1995) *Astrophys. J.*, submitted

Shibata, K. *et al.* (1992) *Publ. Astronomical Society Japan*, **44**, L173

SURFACE DRIVEN EVOLUTION AND ACTIVITY OF ATMOSPHERIC MAGNETIC STRUCTURES

G. VAN HOVEN, Y. MOK, AND D.L. HENDRIX
Department of Physics and Astronomy
University of California, Irvine, CA 92625, USA

Z. MIKIĆ
Science Applications International Corporation
San Diego, CA 92121, USA

The magnetic structure of the solar corona influences solar activity, including coronal heating and solar flares. Magnetic loops and arcades are observed to dominate the structure of the corona; they are formed by the interaction of high-β photospheric convection with low-β coronal magnetic fields. We show the results of 3-D MHD simulations of the response of coronal fields to convective and shear-flow driving from the photosphere. The principal examples include the micro-flare activity in the Parker loop-heating archetype (Schnack and Mikić 1994), and the formation and evolution of 3-D coronal loops (Van Hoven *et al.* 1995a).

- -

The question of the mechanism for *coronal heating* is a classic problem of solar and stellar physics. One main line of study of this question is devoted to the model proposed by Parker (1983), wherein the moderate-scale, slow, random, convective, surface twisting of the solar surface footpoints of a magnetic loop leads to fine-scale ohmic dissipation in the atmosphere. We have concentrated on a 3-D MHD representation of this model and have studied the contribution of magnetic reconnection (resistive tearing) to the dissipation (Van Hoven *et al.* 1995b).

We have developed the following understanding of this mechanism for coronal heating. The electric current spontaneously collapses into thin filaments in a process characterized by intermittency, magnetic reconnection and turbulence: 1) the current sheets driven by the footpoint motions eventually become thin enough to become unstable to magnetic tearing/reconnection; 2) this dynamic process further intensifies the current density, leading nonlinearly to an extended turbulent (apparently steeper than Kraichnan (1965)) spectrum; and 3) the ohmic heating rate is thereby enhanced significantly over what would arise from the more broadly distributed currents, allowing the dissipation to rise sufficiently to match the input Poynting flux. The nonlinear turbulent spectrum is shown in Fig. 1(a) (Hendrix & Van Hoven 1995); the filamentary current structure is shown in Fig. 1(b) (Hendrix *et al.* 1995); a magnified view of a current filament illustrating magnetic reconnection is shown in Fig. 1(c).

Coronal loops are prominent features of the *Yohkoh* X-ray observations (Klimchuk *et al.* 1992). Two limiting models for the formation of a simple, isolated, magnetic loop can be conceived. Either they erupt/emerge with substantial twist from below the convective solar photosphere, or they appear in a nearly unstressed

Y. Uchida et al. (eds.),
Magnetodynamic Phenomena in the Solar Atmosphere – Prototypes of Stellar Magnetic Activity, 51–52.

(current-free) state and are twisted by differential or vortical flows in the solar surface. We have demonstrated both of these formation mechanisms, but here describe the latter case (Van Hoven *et al.* 1995). Photospheric twist applied to a pre-existing current-free bipole propagates into the corona, causing current to flow along and confine a flux tube, as shown in Fig. 2(a). This S-shaped loop (McClymont & Mikić 1994) is in a force-free equilibrium state with a twist of 2.8π on the central field line. The loop field lines are surrounded by overlying field lines that remain relatively undistorted.

It is known that coronal loops can become unstable to kink instabilities if they are twisted sufficiently (Mikić *et al.* 1990). Furthermore, they may exhibit magnetic nonequilibrium that would make them erupt. When a loop with an overlying current-free arcade field is twisted beyond the amount described in the previous paragraph, it undergoes magnetic reconnection at the apex, as shown in Figure 2(b). Shibata (1996) has proposed that such reconnection could be a trigger for loop flares and their outflow jets, a topic that we are presently studying.

The strategic goal of such simulations is to use observable inputs to produce accurate models of the solar corona and inner heliosphere. We have made the first steps in being able to drive the evolution of empirical coronal fields by convective flows in the photosphere. Predictions from these models could provide the solar-influence component of space-weather forecasting.

This research was supported in part by NASA, NSF, and AFPL; computations were provided by NSF and UCI @ SDSC, and DOE @ NERSC.

Hendrix, D.L., & Van Hoven, G. 1995, "Magnetohydrodynamic Turbulence and Implications for Coronal Heating," *Ap. J.,* in draft.

Hendrix, D.L., Van Hoven, G., Schnack, D.D., Mikić Z. 1995, "On the Viability of Ohmic Dissipation as a Coronal Heating Source," *Ap. J.*, in draft.

Klimchuk, J.A., *et al.* 1992, *Publ. Astron. Soc. Japan*, **44**, L181.

Kraichnan, R.H., *Phys. Fluids*, **8**, 1965.

McClymont, A.N., & Mikić, Z. 1994, *Ap. J.*, **422**, 899.

Mikić, Z., Schnack, D.D., & Van Hoven, G. 1990 *Ap. J.,* **361**, 690.

Parker, E.N. 1983, *Ap. J.*, **264**, 642.

Schnack, D.D., and Mikić, Z. 1994, in *Solar Active Region Evolution: Comparing Models with Observations*, eds. K.S. Balasubramaniam and G.W. Simon, Astron. Soc. Pac., Conf. Series, **68**, 180.

Shibata, K. 1996, in Proceedings of IAU Colloquium 153 (this volume).

Van Hoven, G., Hendrix, D.L., and Schnack, D.D. 1995b, "The Diagnosis of General Magnetic Reconnection." *J. Geophys. Res.*, (in press).

Van Hoven, G., Mok, Y., and Mikić, Z. 1995a, *Ap. J. Letts.*, **440**, L105.

I. Magnetic Atmospheres of the Sun and Stars

I.2. High Temperature Outer Atmospheres of Other Stars

CHROMOSPHERES OF CORONAL STARS

JEFFREY L. LINSKY AND BRIAN E. WOOD
JILA, University of Colorado
Boulder CO 80309-0440 USA

Abstract. We summarize the main results obtained from the analysis of ultraviolet emission line profiles of coronal late-type stars observed with the Goddard High Resolution Spectrograph (GHRS) on the Hubble Space Telescope. The excellent GHRS spectra provide new information on magnetohydrodynamic phenomena in the chromospheres and transition regions of these stars. One exciting new result is the discovery of broad components in the transition region lines of active stars that we believe provide evidence for microflare heating in these stars.

1. Introduction

Solar and stellar coronae are now observed routinely in X-rays with the Yohkoh, ROSAT, and ASCA satellites and at radio wavelengths with the VLA and other radio telescopes. These data provide our main source of information concerning coronal structure, dynamics, and heating rates, which are controlled by locally strong magnetic fields. High-resolution ultraviolet spectra provide complementary information on the dynamics and energetics of plasmas at lower temperatures in a stellar chromosphere and transition region where the structure, dynamics, and heating are controlled by strong magnetic fields that are connected to the coronal fields. As a result, the combination of X-ray fluxes and high-resolution UV spectroscopy can provide a more complete picture of magnetodynamic phenomena in the atmospheres of the Sun and stars than is provided by the X-ray and radio data alone. In this paper we summarize the new results concerning magnetodynamic phenomena in late-type stars with coronae that have emerged from the analysis of spectra obtained with the Hubble Space Telescope (HST).

Since 1991 we have been observing late-type stars and RS CVn-type

Y. Uchida et al. (eds.),
Magnetodynamic Phenomena in the Solar Atmosphere – Prototypes of Stellar Magnetic Activity, 55–62.

binary systems with the Goddard High Resolution Spectrograph (GHRS) on HST. The GHRS can obtain UV spectra with low ($R = \lambda/\Delta\lambda = 2000$), moderate ($R = 20,000$), and high ($R = 90,000$) resolutions. (See Brandt *et al.* 1994 for a description of the GHRS.) Prior to the installation of the COSTAR optics to correct for the spherical aberration of the primary mirror in December 1993, the spectral resolutions of these modes were somewhat degraded when using the Large Science Aperture (LSA) but not appreciably degraded when using the Small Science Aperture (SSA). We will report here on the analysis of GHRS spectra of 9 stars, including two A-F stars (Altair = α Aql and Procyon = α CMi), three G-K stars (β Dra, α Cen A, and α Cen B), two M dwarfs (AU Mic and VB10), and two RS CVn-type binary systems (Capella = α Aur and HR 1099 = V711 Tau). The spectral types and references to the data analysis papers are listed in Table 1. This is not a complete list of all GHRS observations of late-type stars, but these 9 stars, which are known to have coronae, provide examples of important phenomena across the H-R diagram, and the analysis of the data from most of these stars is now either published or in press.

TABLE 1. Active Late-type Stars Observed with HST/GHRS

Star	Spectral Type	Comments	Reference
Altair	A7 IV-V	Hottest active star	Walter *et al.* (1995) Simon *et al.* (1994)
Procyon	F5 IV-V	Inactive F star	Wood *et al.* (1995)
β Dra	G2 Ib-II	Active supergiant	Obs. 1995 Apr 29
α Cen A	G2 V	Old star	Obs. 1995 May 1
α Cen B	K1 V	Old star	Obs. 1995 May 5
AU Mic	dM0e	Very active dMe star	Linsky & Wood (1994) Maran *et al.* (1994)
VB10	M8 Ve	End of main seq.	Linsky *et al.* (1995a)
Capella	G8 III + G1 III	Long Per. RS CVn	Linsky *et al.* (1995b)
HR 1099	K1 IV + G5 IV	RS CVn	Wood *et al.* (1995)

We have obtained data for these stars at all three spectral resolutions to address different objectives. The lowest-resolution mode (grating G140L) provides line fluxes in the 1150–1700 Å region suitable for emission measure analyses, estimates of the radiative loss rates from the chromosphere and transition region, and studies of whether photospheric or coronal abundances better characterize the plasma in the transition region. The medium-resolution modes (gratings G140M, G160M, G200M, and G270M) provide sufficient resolution (typically 15 km s^{-1}) to separate close blends, measure line shifts, and measure line profile shapes quite accurately. Most of the results described later in the paper are based upon the analysis of these data. Finally, the highest-resolution data (typically 3.5 km s^{-1}) are obtained with the two echelle gratings (echelle-A and echelle-B). Because the throughput is relatively low and the simultaneous spectral range small for the echelle modes, so far we have obtained spectra only for the Lyman-α and Mg II h and k lines. These data will be discussed elsewhere. We now describe 7 important new results obtained from GHRS data that are likely magnetodynamic in origin and thus relevant to the main thrust of this meeting. We start with phenomena that can be studied with low spectral resolution and proceed to phenomena that require moderate or high spectral resolution.

2. The Hottest Stars with Transition Region Plasmas

X-ray surveys with *Einstein* and ROSAT have identified Altair (A7 IV-V; $T_{\rm eff}$ = 8000 K) as the hottest star with a solar-type corona. The hot OB and chemically peculiar B-type stars are also bright X-ray sources, but shocks in their winds likely play a major role in heating their coronae, and these stars are not generally deemed to be solar-like. Why coronae are not present in the hotter A-type stars is not known, but it is commonly thought to be a result of their thin convective zones. Chromospheric Mg II and Lyman-α emissions have been detected from Altair, but the bright photospheric emission from this star in the UV has made it extremely difficult to observe transition-region lines against this bright background. The high S/N and low scattered-light background of the GHRS makes it feasible to search for transition-region emission lines in such stars. Simon *et al.* (1994) and Walter *et al.* (1995) detected emission of the C II 1334, 1335 Å doublet, the former by subtracting the spectrum of the less active A7 IV-V star 80 UMa and the latter by using spectral synthesis to determine and subtract the underlying photospheric spectrum. Thus transition regions and the nonradiative heating required to balance their radiative losses exist in stars at least as hot as $T_{\rm eff}$ = 8000 K (B–V = 0.22). The most important and unexpected result presented by Walter *et al.* (1995) is that the bulk of the C II emission cannot be explained by acoustic heating (by extrapolation of the basal flux from the early-F stars), but rather requires an additional heating source that must be magnetic in character.

3. Flaring on a Very Low Mass M Dwarf Star

Flares, the topic of many papers at this conference, are arguably the most intensely studied solar magnetodynamic phenomenon. Flares are also observed at radio, optical, UV, and X-ray wavelengths from a variety of late-type stars with luminosities that can exceed 10^{32} ergs s^{-1} (see review papers by Butler, Byrne, Haisch, and Houdebine in these Proceedings). GHRS spectra added a new component to this study when Linsky *et al.* (1995a) detected a flare on one of the lowest-mass stars known–VB10. This star, which is also called Gl 752B, has been classified as an M8 Ve star with $T_{\rm eff}$ = 2600 K and mass about 9% of solar. It thus lies at the very end of the hydrogen-burning main sequence and is very nearly a brown dwarf. Linsky *et al.* (1995a) observed VB10 with the low-dispersion G140L grating on the GHRS for about 1 hour and detected no emission, except for the last 5 minutes when bright emission lines of C II, Si IV, and C IV were clearly present. The peak enhancement of the C IV emission was at least a factor of 40 over quiescent, and the flare luminosity of the transition region and coronal gas was estimated to be about 4×10^{31} ergs s^{-1}. The existence of flares on very late M dwarfs, which have fully convective interior structures, requires a dynamo different from the $\alpha\Omega$-type dynamo thought to operate near the boundary of the radiative core and convective envelope of the Sun (e.g., Parker 1993). Concepts for alternative dynamos have been proposed (e.g., Durney *et al.* 1993), and the VB10 flare data should stimulate the development of quantitative models for such dynamos.

4. The Fe XXI 1354 Å Coronal Emission Line in dMe Stars

Maran *et al.* (1994) identified the Fe XXI 1354 Å line in their medium-resolution (grating G160M) spectrum of the very active dM0e star AU Mic. This line, formed in coronal plasma at 1×10^7 K, is often detected during solar flares, but it had not yet been detected reliably in other stars because of blending with a nearby C I line. To our knowledge no other coronal line has been detected in stellar UV spectra, although the lower-resolution EUVE spectra contain many coronal lines at wavelengths below 400 Å. The detection for the first time of a coronal emission line at moderate spectral resolution allows one to determine both random and systematic motions. The AU Mic line profile displayed no significant bulk motion or profile asymmetry, which indicates that the emission was primarily from static plasma, perhaps located in closed field regions rather than in expanding open field regions. Maran *et al.* (1994) derived an upper limit of 38 km s^{-1} for the turbulent motions of the 1×10^7 K plasma. The emission measure corresponding to the Fe XXI line flux is consistent with the stellar X-ray flux observed by the EXOSAT satellite.

5. Electron Densities in Stellar Transition Regions

The integrated fluxes of collisionally excited transition region lines can be used to infer the plasma emission measure $EM = \int_{\Delta T} n_e^2 dV$ near the temperature where an ion is formed, and the fluxes of lines formed over a broad temperature range can be used to determine the emission measure distribution EM(T). Applications of emission measure analysis to the transition regions of the Sun and late-type stars can be found in Jordan & Brown (1981) and Jordan *et al.* (1987), but there may be difficulties with the standard methods (cf. Judge *et al.* 1995). To proceed from the empirical emission measure distribution to an atmospheric model, which is the run of temperature with pressure or height, requires an independent measurement of electron densities. Fortunately, flux ratios of intersystem lines within multiplets or flux ratios of intersystem lines to permitted lines of the same ion or ions formed at the same temperature are often density dependent. This is because the upper levels of the transitions are depopulated to different extents by collisions and radiation. For such types of line ratios there will typically be a range of n_e for which one line is depopulated primarily by collisions and the other line primarily by radiation. In this range the line ratio is density sensitive, but outside of the range it is not because both lines are depopulated by either collisions (the high-density limit) or radiation (the low-density limit). Intersystem lines or multiplets that can provide useful density-sensitive line ratios include the C II] 2325 Å, O III] 1660 Å, and O IV] 1400 Å multiplets and the Si III] 1892 Å, C III] 1909 Å, N IV] 1486 Å, and O V] 1218 Å lines. For a review of this topic, see Mason & Monsignori Fossi (1994) and Brage, Judge, & Brekke (1995).

Intersystem lines, unfortunately, tend to be weak, and in F- and G-type stars the Si III] and C III] lines are swamped by the bright photospheric spectrum. Perhaps the most useful density-sensitive lines are the four lines that constitute the O IV] 1400 Å multiplet, because these lines are reasonably bright, line ratios within the multiplet are not sensitive to the abundances or ionization uncertainties, and Cook *et al.* (1995) have computed theoretical line ratios that are consistent with solar observations. We have now obtained values of n_e in the transition regions of five stars (β Dra, the G1 III star in Capella, the K1 IV star in HR 1099, Procyon, and α Cen B). We were surprised when our preliminary analysis showed that the electron densities in all of these stars are very similar, $n_e \approx 1 \times 10^{10}$ cm^{-3}, even though the stars differ greatly in activity and luminosity. Deeper observations of more stars are needed to confirm this preliminary result. For the Capella system, Linsky *et al.* (1995b) were able to use the different stellar radial velocities to show that the intersystem lines are formed mainly in the transition region of the G1 III star and thus are useful in determining the electron densities and atmospheric model for this star.

6. Redshifts of Chromospheric and Transition-Region Lines

The centroid velocities of spatially averaged solar UV emission lines are known to be redshifted, with redshifts in quiet regions increasing systematically from near 0 km s^{-1} for chromospheric lines to about 10 km s^{-1} for lines formed at $T \approx 1.35 \times 10^5$ K (e.g., Doschek *et al.* 1976). At higher temperatures the redshifts decrease rapidly, but this result is uncertain because of possible errors in the laboratory wavelengths (e.g., Achour *et al.* 1995). Ayres *et al.* (1988) and others have discovered a similar trend in the redshifts of active stars observed with IUE. Although there is no generally accepted physical explanation for these redshifts, the enhancement of the redshift velocities in solar active regions relative to quiet regions suggests that the spatially averaged emission is dominated by downflows in magnetic flux tubes.

With the GHRS one can obtain more precise values for redshifts in a broader range of stars. Wood *et al.* (1995), for example, find for the inactive star Procyon a steady increase in redshift with temperature, reaching 11 km s^{-1} in the O V] 1218 Å line formed at 2.5×10^5 K. This trend, which is very similar to that seen in the quiet Sun, is also observed in spectra of the inactive dwarfs α Cen A and B. The active G1 III star in the Capella system shows redshifts in the C IV lines of about 19 km s^{-1} (Linsky *et al.* 1995b). At comparable temperatures, the redshift velocities of the active G1 III star in Capella are approximately doubled compared to those of the quiet stars; this velocity difference is very similar to that between active and quiet regions on the Sun. This finding strengthens the case for an intimate connection of redshifts with magnetic fields. However, the very active dMe star AU Mic shows only a very small redshift (about 3 km s^{-1}) in the C IV lines (Linsky & Wood 1994) and the very active K1 IV star in HR 1099 also shows small redshifts (about 5 km s^{-1}) in the C IV lines (Wood *et al.* 1995). Observations of a broader range of stars are needed to identify trends with magnetic field strengths and coverage factors to better understand the physical processes responsible for the redshifts.

7. Excess Blue Emission in Procyon's Transition Region Lines

Wood *et al.* (1995) fitted Gaussians to Procyon's transition-region line profiles obtained with moderate spectral resolution and very high signal/noise (S/N). They were surprised to find that the C IV, Si IV, and O V] lines show excess emission on the blue side of the line profiles centered near -90 km s^{-1}. There is as yet no explanation for this phenomenon.

8. Broad Profiles: A New Diagnostic of Microflare Heating

Linsky & Wood (1994) used the GHRS to obtain the first moderate resolution profiles with high-S/N of transition-region lines in a dMe star when they observed the C IV and Si IV lines in AU Mic. For data obtained at times when the star was not obviously flaring, they could not fit the line profiles with single Gaussians. Instead, they obtained good fits to the line profiles by using two Gaussians–a narrow Gaussian with FWHM $\approx$ 29 km s^{-1} and a broad Gaussian with FWHM $\approx$ 173 km s^{-1}. The narrow Gaussians have similar widths to what is observed in solar quiet and active regions, whereas the broad Gaussians have similar widths to what is observed during solar transition-region explosive events (e.g., Dere *et al.* 1989). Cook (1991) argued that these explosive event profiles, which are observed in regions of complex and changing magnetic fields, are broadened by the plasma turbulence generated by microflares.

On the Sun, the broad components are responsible for about 5% of the total C IV and Si IV emission, but on AU Mic this contribution is 40%. A similarly large contribution is observed for the active star in Capella, and for the very active K1 IV star in HR 1099 the broad component contributes 60% of the total line flux. On the other hand, the inactive stars α Cen A

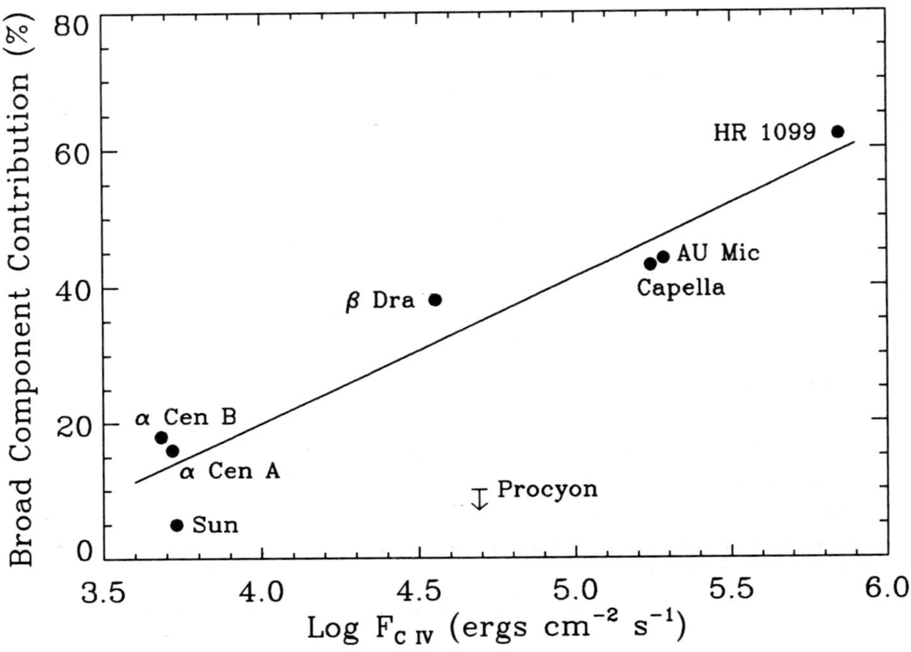

Figure 1. Percentage contribution of the broad component to the total flux in the C IV and Si IV lines is compared with the C IV and Si IV surface flux.

and B and Procyon show either small flux contributions or low upper limits for the broad components. We plot in Figure 1 our preliminary data on the fractional contribution of the broad component to the total C IV and Si IV flux vs. the C IV and Si IV surface flux, a good proxy for the transition-region heating rate. We find a better correlation when we plot the fractional contribution of the broad component vs. the X-ray surface flux, a good proxy for the coronal heating rate, because the Procyon data point then fits the correlation better. This preliminary analysis leads us to conclude that microflaring is the dominant heating mechanism in active stars, but that it plays only a minor role in quiescent stars like the Sun. We will publish our detailed analysis of this new data set elsewhere.

This work is supported by NASA Interagency Transfer S-56460-D to the National Institute of Standards and Technology.

References

Achour, H., Brekke, P., Kjeldseth-Moe, O., & Maltby (1995), *ApJ*, in press

Ayres, T.R., Jensen, E., & Engvold, O. (1988), *ApJS*, **66**, pp. 51–68

Brage, T., Judge, P., & Brekke, P. (1995), *ApJ*, in press

Brandt, J.C. *et al.* (1994), *PASP*, **106**, pp. 890–908

Cook, J.W. (1991), in Mechanisms of Chromospheric and Coronal Heating, ed. P. Ulmschneider, E.R. Priest, & R. Rosner (Berlin: Springer-Verlag), pp. 83–96

Cook, J.W. *et al.* (1995), *ApJ*, **444**, pp. 936–942

Dere, K.P., Bartoe, J.-D.F., & Brueckner, G.E. (1989) *Sol. Phys.*, **123**, pp. 41–68

Doschek, G.A., Feldman, U., & Bohlin, J.D. (1976) *ApJ*, **205**, pp. L177-180

Durney, B.R., De Young, D.S., & Roxburgh, I.W. (1993) *Sol. Phys.*, **145**, pp. 207–225

Jordan, C., Ayres, T.R., Brown, A., Linsky, J.L., & Simon, T. (1987), *MNRAS*, **225**, pp. 903–937

Jordan, C. & Brown, A (1981) in R.M. Bonnet & A.K. Dupree (eds.) *Solar Phenomena in Stars and Stellar Systems*, D. Reidel, Dordrecht, pp. 199–226

Judge, P.G., Woods, T.N., Brekke, P., & Rottman, G.J. (1995), submitted to *ApJ Letters*

Linsky, J.L., & Wood, B.E. (1994), *ApJ*, **430**, pp. 342–350

Linsky, J.L., Wood, B.E., Brown, A., Giampapa, M.S., & Ambruster, C. (1995a), *ApJ*, in press

Linsky, J.L., Wood, B.E., Judge, P., Brown, A., Andrulis, C., & Ayres, T.R. (1995b), *ApJ*, **442**, pp. 381–400

Maran, S.P. *et al.* (1994), *ApJ*, **421**, pp. 800–808

Mason, H.E., & Monsignori Fossi, B.C. (1994), *A&A Rev.*, **6**, pp. 123–179

Parker, E.N. (1993) *ApJ*, **408**, pp. 707–709

Simon, T., Landsman, W., & Gilliland, R. (1994), *ApJ*, **428**, pp. 319–323

Walter, F.M., Mathews, L.D., & Linsky, J.L. (1995), *ApJ*, **447**, pp. 353–363

Wood, B.E., Harper, G.M., Linsky, J.L., & Dempsey, R.C. (1995), *ApJ*, in press

EVIDENCE OF MAGNETIC CORONAE IN OTHER STARS

S. SERIO
Istituto e Osservatorio Astronomico
Palermo, Piazza del Parlamento 1, 90134 Italy

Abstract. X ray emission and activity in late type stars are generally believed to be caused by magnetic fields generated below the stellar photosphere. Coronal magnetic fields should induce structuring of the coronal plasma in loop like structures, as observed in the Sun. Do we have evidence of such structuring? I shall discuss how spectrally resolved X ray observations can evidence the existence of stationary stellar coronal loops. I shall also discuss how the lengths of stellar loops can be inferred by spectrally resolved observations during flare decay, and how the presence of coronal magnetic fields might influence the thermal stability of the transition region and induce characteristics signatures in line profiles.

1. Introduction

The evidence that the whole X-ray emitting solar corona is made out of magnetic loops has been derived from solar observations pioneered by the ATM mission, and now solidly established with the most recent *NIXT* and *YOHKOH* data.

The widespread observation of X ray emission from late type stars since the early results of the *Einstein* mission has been almost paradigmatically interpreted as evidence that stellar coronae should be similar to the sun, i.e. made up of loop like structures. However, the detailed interpretation and model fitting of stellar X ray spectral distribution and variability has only limitedly been based on loop models. For example, in most of the analyses of Imaging Proportional Counter, stellar spectra are usually presented in terms of one- or two-components isothermal models.

Y. Uchida et al. (eds.),
Magnetodynamic Phenomena in the Solar Atmosphere – Prototypes of Stellar Magnetic Activity, 63–70.

The reasons for this attitude - we firmly believe in loop coronae, but do not use this information for the interpretation of the data - is certainly to be found, but perhaps only in part, in the low count rate statistics and moderate spectral resolution generally available in stellar X ray observations.

2. Static Coronal Loops

2.1. THE DEGENERACY PROBLEM ...

One problem with the lack of popularity of loop modelling stays in a fundamental limitation. In fact, loop models are based on the two equations of hydrostatic equilibrium and energy conservation which, for a loop with constant cross section, take the form:

$$\frac{dp}{ds} = -nmg_s \tag{1}$$

$$E_H = n^2 P(T) - \frac{d}{ds}(\kappa T^{5/2}\frac{dT}{ds}) \quad , \tag{2}$$

where s is the coordinate along the field line and g_s the component of gravity along the same direction, p is the pressure, n the gas number density, m the average ion mass, E_H is the volumetric heating rate, T the temperature, $P(T)$ the energy loss function per unit emission measure, and κ the Spitzer coefficient of thermal conductivity. When the half-length L of the loop is much smaller than the pressure scale height $s_p = 2KT/mg$, where K is the Boltzmann constant, Rosner, Tucker and Vaiana (1978, RTV) scaling laws apply:

$$E_H = \frac{\kappa {T_{max}}^{7/2}}{L^2} = 10^5 p^{7/6} L^{-5/6} \quad , \tag{3}$$

$$T_{max} = 1.4 \cdot 10^3 (pL)^{1/3} \quad . \tag{4}$$

Using the RTV scaling laws, and $x = s/L$, it is easy to see that (2) becomes:

$$\frac{{T_{max}}^{1/2}}{3.1 \cdot 10^{19} \kappa K^2} P(T) - \frac{1}{{T_{max}}^{7/2}} \frac{d}{dx}(T^{5/2}\frac{dT}{dx}) = 1 \quad . \tag{5}$$

Hence, the distribution of temperature along the loop is independent on loop length and base pressure. If we assume that a stellar corona is dominated by a system of identical loops, then their dimensions cannot be determined by the spectral distribution of the observed photons. There is, essentially, a degeneracy in the loop model, that prevents extraction of the information one ideally would like to obtain, i.e. how a stellar corona looks like.

2.2. ...AND ITS CURES

This problem was recognized since the earliest attempts to fit stellar X ray data with loop models (see for example Schmitt *et al.*, 1985). Of course, the degeneracy is rigorous only within the framework in which (5) has been derived, and in particular, (i) when the loops are much shorter than the pressure scale height, (ii) when they have a constant cross section, and (iii) when Spitzer conductivity can be used with confidence.

2.2.1. *Gravity*

One contribution to breaking the degeneracy is gravity in (1), which becomes important when the height of the loops is comparable to the pressure scale height, or higher. When fitting a degenerate loop model to an observed proportional counter spectrum, the contours of constant χ^2 values in the p, L plane are open lines running parallel to the lines of constant effective temperature. Taking into account the pressure equation, (1), they become closed at lengths comparable to the scale height, while remaining open for high pressure and short lengths (see, however, Maggio and Peres, 1995, for loops longer than the star radius). So, while the fits are able to determine well the values of the effective temperature, they give typically high formal values of the loop length, although lower values are also compatible even with stringent confidence levels.

2.2.2. *Widening Loop Cross Section*

The degeneracy of loop models is also removed if the cross section of the loops increases with height. In this case (e.g. Vesecky, Antiochos and Underwood 1979; Serio *et al.*, 1981), the thermal stratification is the same as in a toroidal loop, although with slightly lower values of maximum temperature. The larger volume available to the hotter plasma near the top of the loop, however, results in larger values of effective temperature.

As an example, Fig. 1 shows the dependence of the *ROSAT PSPC* effective temperature on cross section expansion, assuming that it varies according to

$$\sigma = \sigma_o(1 + r/R_*)^\alpha \quad , \tag{6}$$

(of course, $\alpha = 2$ corresponds to radial expansion of the loops).

Although the data do suggest that solar coronal loops do not generally show any substantial variation with height of their cross section, successful fits have been obtained for stellar data, using model loops with variable cross section (for example Stern *et al.*, 1988, Lemen *et al.*, 1989).

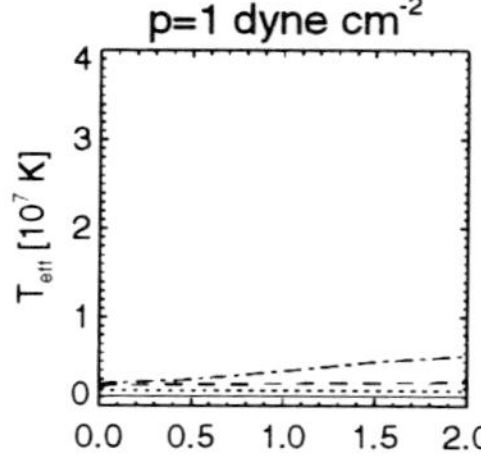

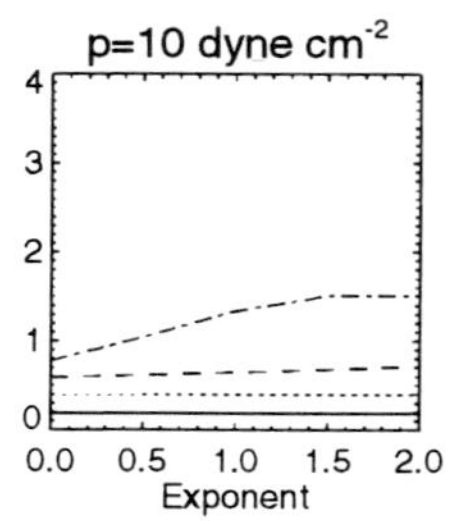

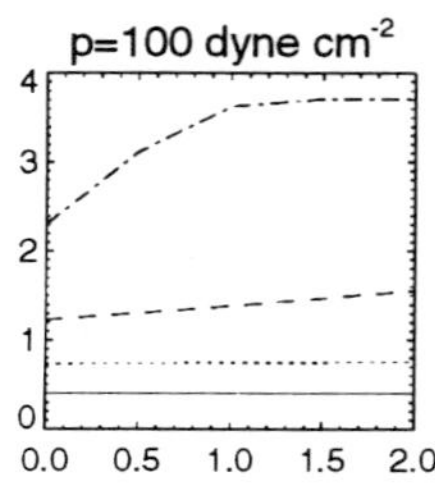

Figure 1. Dependence of the effective *ROSAT PSPC* temperature on the loop expansion parameter α. Solid lines: loops of semilength $L = 10^9\,cm$; dotted, $L = 10^{10}\,cm$; dashed, $L = 10^{11}\,cm$; dot-dashed, $L = 10^{12}\,cm$.

2.2.3. *Deviations from Spitzer Conductivity*

Deviations of heat conduction from the Spitzer regime are important when the ratio of electron mean free path to temperature scale height is greater than $\approx 10^{-3}$ (Gray and Kilkenny, 1980). As shown by Ciaravella *et al.* (1995), however, these deviations play only a minor role in determining the observed spectrum, since they are large enough only at heights much larger than the pressure scale height, and therefore where the emission measure is very low.

2.3. BUT ..., DO WE NEED CURES, AFTER ALL?

Independently of the effects described above, which contribute to remove the degeneracy in loop modelling of stellar coronae, a simple physical description of a loop dominated corona can be made, at the price of using appropriate variables.

In particular, I will show the equivalence of isothermal and loop models, for loops lower than the pressure scale height and of uniform cross section. The problem is only to interpret in terms of a loop model the parameters derived from the one-T fit, which we know to work reasonably well, at least to the first order: the temperature $T \propto (p_o L)^{1/3}$, and the luminosity

$$L_x \propto \frac{{p_o}^2 P(T)}{T^2} \, L f S_* \quad , \tag{7}$$

where S_* is the surface area of the star, and f the surface filling factor of loop like structures. Using the scaling law in (4), we easily see that $L_x \propto T P(T) f p_o S_*$. Therefore, a description of a degenerate one loop corona in terms of its maximum (or effective) temperature, and of the product of filling factor times base pressure, is essentially equivalent to a one-T model. The left panel of (2) shows a typical "banana" shaped fit of a loop

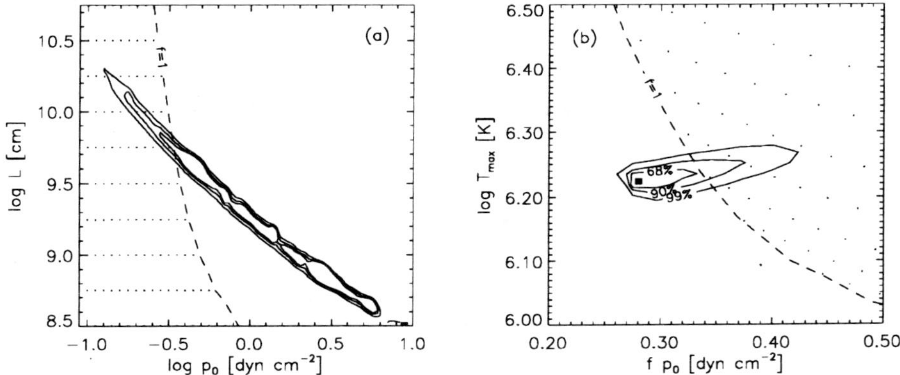

Figure 2. Analysis of the *ROSAT PSPC* spectrum of Procyon with hydrostatic loop models (adapted from Maggio *et al.*, 1995) Confidence regions at the 68%, 90% and 99% levels in the p_0 - L (a), and in the fp_oT (b) loop parameter space. The dashed line shows the locus where the surface filling factor $f = 1$, and the shading the unphysical region where $f > 1$.

model to *ROSAT* data in the p, L plane, dramatizing the uncertainty in the determination of the parameters of the fitting model (notice that the upper and lower limits in L and in p are accounted for by the fact that only a limited region of the parameter space has ben explored). Similar results are obtained in the LT and in the pT planes. In contrast, the right panel shows that the average coronal pressure fp_o and the maximum loop temperature T_{max} are reasonably well determined.

A further point of interest in this discussion is the possible presence of multiple systems of loops, which has not yet been investigated systematically. The discussion above can easily be extended, and two-temperature fits can be shown to be roughly equivalent to two-loops models.

The main question arising from these considerations, however, is the following: considering that one-loop models give generally χ^2 values better than one-T isothermal models and comparable to two- temperature models (Maggio and Peres, 1995), are not they more acceptable, since they are based on sound physical concepts, or should we continue to prefer isothermal models?

3. Evidence from Transients

It is well known that the size of a flaring region can be somewhat constrained by the decay time of the light curve. For instance, Reale *et al.* (1988) gave a picture of a coronal loop in Proxima Centauri obtained by the analysis of an *Einstein* X-ray flare by means of a hydrodynamical code.

The improved efficiency of *ROSAT* gives now many more flares accessible to such analysis. Since hydrodynamic modelling is time consuming, we have developed a technique to analyze flares without having to build a specialized model for each event. The technique is based on the finding that the entropy in a flaring loops undergoes a phase of linear decay, whose rate is proportional to the length of the loop. During this phase, temperature and density are related by a scaling law (Serio *et al.*, 1992):

$$T \propto n^{\xi} \quad , \tag{8}$$

where the exponent $\xi = 2$ for the temperature of the top of a solar loop, and assumes different values for the effective temperature measured in the pass band of different instruments, or when excess heat is deposited during the decay phase of the flare (Jakimiec *et al.*, 1992; Sylwester *et al.*, 1993; Reale *et al.*, 1993). Recently Cargill, Mariska and Antiochos (1995) have re-obtained the same scaling law on the basis of an analytical model, and have also discussed the dependence of ξ on abundancies.

It turns out that we can derive the length of loops with a reasonable precision with this method. For example Sciortino *et al* (1995), using *ROSAT PSPC* spectra, determine loop lengths for 3 flares observed in Pleiades K stars, and 1 in a field M star. The lengths derived are much smaller than the scale height at flare maximum, indicating confinement. Since we also measure peak emission measure (Σ) and temperature, we can estimate the gas pressure, given the flare volume. Assuming that confinement is obtained from a magnetic field B in a loop with ratio of semilength to cross section radius η, one can estimate $B\eta^{1/2} \approx 5.310^{-8} T^{1/2} (\Sigma/L^3)^{1/4}$.

Although we do not know the value of η, solar evidence and the results of Reale *et al.* (1988) for the Proxima Centauri loop indicate values of the order of .1. Using the values of temperature and loop length derived by Sciortino *et al.* (1995), the inferred magnetic fields are, therefore, $B > 170$ *Gauss* for Hz 892, $B > 35$ *Gauss* for Hz 1100, $B > 360$ *Gauss* for Hz 1516, and $B > 120$ *Gauss* for AD Leo.

It remains to investigate further along this line how these results depend on abundancies and on the assumption, implicit in (8), that the flare volume remains constant during the decay.

4. Redshifts in the Transition Region

Finally, I would like to suggest a novel approach to obtain evidence on the existence of magnetic fields in stellar coronae.

One puzzling result of Skylab and rocket solar UV observations was the discovery of $5 - 10 Km/s$ red shifts in solar transition region lines (e.g. Brekke, 1993, and references therein). More recently, similar results have been obtained on a variety of stars (Linsky, Wood and Andrulis, 1994; Linsky and Wood, 1995).

Reale, Peres and Serio (1995 a) have modelled in details the dynamic evolution of isobaric perturbations in a loop atmosphere. They have shown that condensations in the transition region can cause downdrafts driven by radiative cooling. As a condensation falls through the atmosphere, it tends to cool down and to increase its density, thus enhancing radiative losses. Therefore, falling perturbations emit more than the background atmosphere.

A thorough exploration of the parameter space shows that significant redshifted emission, at speeds of several km/s, can be produced both in active and quiet solar regions. The initial perturbations need to be moderately non linear (with density contrast ≈ 1), and with sizes typically larger than several tens of kilometers (Reale, Peres and Serio, 1995 b).

This mechanism might cause redshifts consistent with those observed, provided some constrains on the spectrum of the sizes of the perturbations and on their frequency are satisfied. As an example, figure 3 shows the distribution of time averaged flux enhancement in one of their models, representing an active region, as a function of velocity, in the hypothesis that the atmosphere is unmagnetized. The largest enhancements occur here for downward velocities of 4 to 7 km/s.

Even a small magnetic field, of the order of 1 *Gauss*, is strong enough to constrain these perturbations to fall along the field lines. When this happens, the falling perturbations can reach even higher values of velocity. In fact, suppression of motions and of thermal conduction in the directions perpendicular to the field lines conspire to make the condensations last longer and, therefore, to let them reach higher velocities.

While the origin of the perturbations has not been studied in detail, it appears unlikely that they could be due to purely hydrodynamic effects. Non linear MHD waves might have to be invoked in their generation. Doubtlessly, detailed measurements of transition region line profiles, together with dynamical modelling of the transition region, have the potential of furnishing some further diagnostic of the existence of magnetic fields in the coronae of stars other than the sun.

Acknowledgments

I thank A. Ciaravella, A. Maggio, G. Peres, F. Reale, and S. Sciortino for discussion. This research has been funded by the Italian Ministry of University (MURST) and by the Italian National Research Council (CNR).

References

P. Brekke, 1993, *Ap. J.*, **408**, 735.

A. Ciaravella, A. Maggio, G. Peres, and S. Serio, 1995, *A. & A.*, in press.

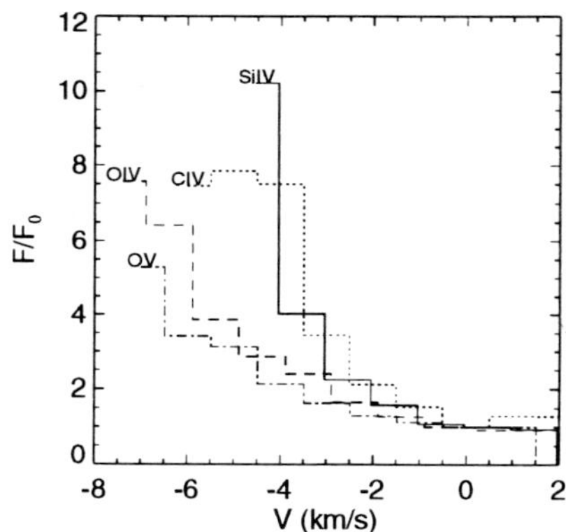

Figure 3. Time averaged flux enhancement in four UV lines (Si IV 1402.77 Å, C IV 1548.20 Å, O IV 1401.16 Å, O V 1371.30 Å), as a function of fluid velocity, during the evolution of an initially isobaric condensation (Radius 3 10^6 cm, initial density contrast 1) inside the transition region of an unmagnetized solar atmosphere with base pressure 3 dyne cm^{-2}.

P.J. Cargill, J.T. Mariska and S.K. Antiochos, 1995, *Ap. J.*, **439**, 1034.

D.R. Gray, and J.D. Kilkenny, 1980, *Plasma Phys.*, **22**, 81

J. Jakimiec, B. Sylwester, J. Sylwester, G. Peres, S. Serio, and F. Reale, 1992, *A. & A.*, **253**, 269.

J.R. Lemen, R. Mewe, C.J. Schrijver, A. Fludra, 1989, *Ap. J.*, **341**, 474.

J.L. Linsky, B.E. Wood and C. Andrulis, 1994, in J.P. Caillault (ed.), *Cool Stars, Stellar Systems and the Suns*, A.S.P. Conference series **64**, 59.

J.L. Linsky and B.E. Wood, 1995, *Ap. J.*, in press.

A. Maggio, and G. Peres, 1995, *A. & A.*, in press.

A. Maggio, G. Peres, J.P. Pye, S.T. Hodgkin, and J.E. Morley 1995, *in preparation*

F. Reale, G. Peres, S. Serio, R. Rosner, and J.H. Schmitt, 1988, *Ap. J.*, **328**, 256.

F. Reale, G. Peres, and S. Serio, 1993, *A. & A.*, **272**, 486.

F. Reale, G. Peres and S. Serio, 1995 a, *A. & A.*, in press.

F. Reale, G. Peres and S. Serio, 1995 b, in preparation.

R. Rosner, W.H. Tucker, and G.S. Vaiana, 1978, *Ap. J.*, **220**, 643.

J. Schmitt, F.R. Harnden, R. Rosner, G. Peres, and S. Serio, 1985, *Ap. J.*, **288**, 751.

S. Sciortino, G. Micela, F. Reale, V. Kashyap, R. Rosner, and F. R. Harnden, Jr, 1995, this volume.

S. Serio, G. Peres, G.S. Vaiana, L. Golub, and R. Rosner, 1981, *Ap. J.*, **243**, 288.

S. Serio, F. Reale, J. Jakimiec, B. Sylwester, and J. Sylwester, 1991, *A. & A.*, **241**, 197.

R.A. Stern, S.K. Antiochos, and F.R. Harnden Jr., 1986, *Ap. J.*, **305**, 417.

B. Sylwester, J. Sylwester, S. Serio, F. Reale, R.D. Bentley, and A. Fludra, 1993, *A. & A.*, **267**, 586.

J.F. Vesecky, S.K. Antiochos, and J.H. Underwood, 1979, *Ap. J.*, **233**, 987.

RECENT VLBI OBSERVATIONS OF STELLAR CORONAE

R. L. MUTEL
University of Iowa, Iowa City, IA 52242, USA

1. Introduction

This paper summarizes recent (1990-1995) VLBI observations of radio emission from active stars and how they have contributed to our understanding of stellar coronae. The observational results are first summarized and then applied to three broad questions concerning the physics of stellar coronae:

1. Which radiation mechanisms are responsible for the radio emission?
2. What is relationship between the X-ray emitting thermal plasma and the radio emitting regions?
3. What is the geometry of the radio corona?

I will not discuss exotic stars with collapsed companions, such as X-ray binaries or cataclysmic variables. For a more complete discussion of stellar radio emission see the recent reviews of Lang (1994, this conference) and Drake (1993).

2. Recent VLBI Observations: Summaries by Class

2.1. PRE-MAIN SEQUENCE STARS

Phillips *et al.* (1991) and Andre *et al.* (1992) have conducted VLBI surveys of the ρ Oph and Taurus-Auriga star formation regions, detecting a total of eleven young stellar objects (YSO's). Radio emission from YSO's is detected preferentially from the weak-lined T Tauri stars or young B-A stars without emission lines (Andre *et al.*, 1991). The emission is often mildly ($\sim$20%) circularly polarized with a nearly flat spectrum, similar to the well-studied RS CVn systems and consistent with a gyro-synchrotron emission process (e.g. Andre *et al.* 1992; Skinner 1993). The VLBI observations confirm the non-thermal nature of the emission, with typical brightness temperatures $T_B \sim 10^8$ K or higher. The corresponding linear sizes are 10-25 stellar

Y. Uchida et al. (eds.),
Magnetodynamic Phenomena in the Solar Atmosphere – Prototypes of Stellar Magnetic Activity, 71–80.

diameters, which is larger than the radio coronae of other stars, but appears to be consistent with the dead zone radius predicted by the model of Morris *et al.* (1990).

A particularly interesting case is the T Tauri star HD283447 (Feigelson *et al.*, 1994), in the Taurus-Auriga star formation region. It it the most luminous radio source in the region (O'Neal *et al.*, 1990), is highly variable in total flux, and has measurable circular and *linear* polarization (Feigelson *et al.*, 1994). VLBI observations at several epochs (Phillips *et al.*, 1991) show a variable spatial structure, varying from unresolved ($<$ 2 stellar diameters) at 5 GHz to a well sampled Gaussian brightness distribution of FWHM 2.4 mas (14 stellar diam.) at 1.6 GHz three days later. Unfortunately, simultaneous multi-frequency observations were not available, so it is unclear whether the size change was primarily due to temporal evolution or is a result of frequency dependent source size.

The well known protostar T Tau (S) was *not* detected using a sensitive VLBI array (Phillips *et al.*, 1993) in spite of strong evidence that the radio emission is non-thermal: It is variable, has moderate circular polarization, and a spectral index which switches from negative to positive during outbursts (Skinner and Brown, 1994). The source size lower limit derived from the VLBI non-detection is 1 a.u. (70 stellar radii) but this may be somewhat lower if the compact radio structure has opposite circular polarization than that used in the observations.

2.2. EARLY-TYPE MAIN SEQUENCE STARS

Phillips and Titus (1990) detected the supergiants Cyg OB2 Nr. 9 (O5If) and HD167971 (O8Ib) with brightness temperatures $T_B \sim 10^7$K and physical size 50-100 stellar radii. They interpret the radiation as gyro-synchrotron emission in which relativistic electrons are accelerated by shocks in strong stellar winds (White, 1985).

Felli and Massi (1991) detected the massive Wolf-Rayet binary system WR140 using the European VLBI Network (EVN) at 5 GHz, while Felli *et al.* (1991) reported the detection of the massive binary θ^1 Ori A (B0.5V+TTau?). Both were unresolved, indicating brightness temperatures $T_B > 10^7$K. This confirms the presence of a non-thermal emission mechanism since these stars lack hot coronae. In both cases, the size of the non-thermal emission is comparable with the binary system size. The relativistic electrons could be produced by the interaction of stellar winds.

2.3. CLOSE BINARIES

Active close binaries, especially the RS CVns, have been the most intensively studied stars using VLBI arrays. This is largely because several of the closest RS CVns have the highest average radio fluxes of any stellar source. Early VLBI studies of RS CVn and Algol binaries (e.g. Mutel *et al.*, 1985); Lestrade *et al.*, 1988; Massi *et al.*, 1988) found that the quiescent radio emission arises from an extended region comparable with the binary separation, whereas flare emission tends to be unresolved (less than a stellar diameter). Trigilio *et al.* (1994) detected HR1099 during the decay phase of a strong flare at 5 GHz using the EVN. They found that the angular size increased over a period of 3 hours as the flare decayed, reaching 4 mas (∼size of the binary system). The size evolution was based on a Gaussian fit to the projected baselines, since they did not have adequate spatial frequency coverage to produce a 2-dimensional map of the source.

Lestrade *et al.* (1993) used phase-referencing to a nearby extragalactic radio source to map the position of Algol's radio centroid with sub-mas accuracy. They observed during four epochs using a sensitive MkIII global VLB array. They established that the source's position in the binary orbit tracks the motion of the K subgiant rather than the B star, thus directly verifying the conjecture that the radio emission is associated with the more active K star.

In spite of considerable observational effort, the detailed structure and temporal evolution of the radio coronae of active binaries are still poorly understood. This is now beginning to change as VLBA maps of these systems become available. Bastian *et al.* (this conference) reports on dual-polarization 8.4 GHz maps of the RS CVN systems UX Ari and HR1099 at two epochs. During both quiescent and flaring states the source is extended over the binary system size. There may also be a displacement between the RCP and LCP centroid which might arise from oppositely polarized footprints of a large coronal loop (Beasley, *private comm.*).

Mutel and Scharringhausen (1996) have studied Algol at 8.4 and 15 GHz with the VLBA. Like Lestrade *et al.* they also used phase-referencing to provide sub-milliarcsecond positional information. Three 'snapshot' maps were made over eight hours referenced to a common phase center. The radio centroid was clearly seen moving in the same direction and at the same rate as the K star, thus confirming the result of Lestrade *et al.* in a single observing session. They also found significant changes in the size and morphology of the source, with an overall scale of several times the diameter of the K star.

2.4. DWARF M STARS

The first successful VLBI detection of a single M-dwarf was that of YZ CMi at 1.6 GHz (Benz and Alef, 1991) using the EVN. The visibility data was sparse, but consistent with an unresolved source whose size was less than

3.4 stellar diameters. The corresponding brightness temperature ($T_B > 2 \cdot 10^9$K) clearly indicated a nonthermal emission mechanism. Benz *et al.* (1995) reported VLBI detections of the the M-dwarfs EQ Peg B and AD Leo also using the EVN at 1.6 GHz during highly circularly polarized flares. AD Leo was also detected during a quiescent period. In all cases, the sources were unresolved, implying size upper limits of 1.8-1.9 stellar diameters for both stars. This is somewhat smaller than the expected size of a closed coronal structure predicted from pressure equilibrium between plasma and magnetic field, and is also considerably smaller than the radio coronae of several other stellar radio sources (e.g. RS CVns, WTT stars).

2.5. ASTROMETRIC VLBI: SEARCHING FOR PLANETS

Lestrade and colleagues (Lestrade *et al.* 1992; 1994; Jones *et al.* 1995) have undertaken a series of phase-referenced VLBI observations of several RS CVn systems over several years to search low mass companions. The observations are designed to search for small residual periodic motion after correction for parallax and proper motion. For the close binary σ CrB, they fitted 8 epochs over 5 years, finding a residual rms scatter of 0.22 mas (1 solar radius). This has two important implications. The first is that the system cannot have a Jupiter-size planets orbiting closer than ~4 a.u. Second, the radio emission cannot be centered on either star, since that would result in a 1.2 mas periodic displacement of the centroid at the orbital period, which is not seen.

While the spatial frequency coverage of the VLB array which has been used for astrometry is insufficient to allow mapping of the radio structure, this program of regular VLBI monitoring of the position should lead to interesting astrophysical results in addition to the primary goal of searching for extra-solar system planets.

3. Related Radio Observations

3.1. LONG-TERM MONITORING: EVIDENCE FOR PHASE DEPENDENCE

One of the most intriguing new discoveries in stellar radio observations over the past few years has been mounting evidence for an apparent phase dependence in the radio light curves of several close binary systems. Catalano (1990) showed a plot of 8 GHz radio flares on Algol versus orbital phase with several apparent peaks. Neidöfer *et al.* (1993) monitored the RS CVn system UX Arietis from December 1992 to May 1993 using the Bonn 100m radiotelescope at several frequencies from 1.4 to 32 GHz. They found that larger flux levels were seen preferentially near primary eclipse i.e., when the more active K star was in front. They also found tentative evidence for phase dependence of the degree of circular polarization, with the highest CP occurring near phase 0.5. Elias *et al.* (1994) monitored UX Arietis using the VLA daily for a period of three weeks at 5 GHz. They also found that the radio flux peaked near zero phase. They also observed UX Ari using broad band visual photometry during the same period and found that the radio and optical light curves were anti-correlated, i.e. the maximum radio flux occurred near minimum light, which corresponds to maximum starspot visibility. A radio flux-starspot correlation was also found by Lim *et al.* (1995) for the rapidly rotating single K dwarf AB Dor. The radio peaks occurred near quadrature, and were also correlated with the presence of a large starspot, as determined by simultaneous optical monitoring. Finally, LeFevre *et al.* (1994) monitored the radio light curves of six RS CVn and Algol binaries at 5 GHz using the VLA for a total of 70 hours over several years. While the temporal sampling was very non-uniform, there is some evidence for a recurrent phase dependent behavior.

Unfortunately, in all these cases, a proper analysis of the statistical significance of the flux density-phase correlation was not done, so it is difficult to judge whether the dependence is real. If it is real, it strongly suggests the presence of emission regions which are azimuthally asymmetric. For example, one can imagine the emission arising from one or more large coronal loops or possibly an interaction region between stars in a binary system. This is contrary to models which assume toroidal structures in which closed field lines trap radiating electrons (e.g., Morris *et al.* 1990, Linsky *et al.* 1992). With the enhanced sensitivity and dynamic range of the VLBA, it should be possible to directly measure the degree of asymmetry of the radio coronae by direct imaging.

3.2. MULTI-WAVELENGTH STUDIES

During the past several years, several multi-wavelength observations of active stars have been published which have included extensive radio observations. The observed wavelengths include radio (VLA, VLBI), optical (ground based, HST), UV (IUE, HST), X-ray (ROSAT,ASCA, EXOSAT), and gamma rays (GRO). The results have been largely disappointing: In the majority of cases, there appears to be little correlation between the radio light curve and activity at other wavelengths. A few examples include multi-wavelength studies of the YSO HD283447 (Feigelson *et al.*, 1994), UX Ari and Algol (Lang and Willson, 1988), several RS CVn systems (Fox *et al.*, 1994), and AR Lac (Mutel *et al.*, 1993). In some cases, there is a suggestion of correlation e.g., a strong radio flare precedes an X-ray flare in σ CrB by $\sim$ 30 min (Stern *et al.*, 1992). A quite interesting simultaneous observation of HR1099 at X-ray, EUV, UV, and radio is shown in this conference by Brown *et al.*, but the flares at different wavelengths were not coeval. In such cases, it is imperative to establish the frequency of flares at each wavelength over a long timescale in order to properly evaluate the statistical significance of a proposed correlations.

3.3. RADIO EMISSION MECHANISMS

3.3.1. *Quiescent Radio Emission*

The most common mechanism used to explain quiescent (non-flare) stellar radio emission is been gyro-synchrotron emission from mildly relativistic electrons in magnetic fields of 10-100 gauss. This mechanism has been repeatedly invoked for almost all stellar classes e.g., PMS stars (Andre, 1995), early-type chemically peculiar stars (Drake *et al.*, 1987), RS CVns (Chiuderi-Drago and Franciosini, 1993), Algols (Umana *et al.*, 1991), and late-type dwarfs (Güdel, 1994). This mechanism is consistent with the observed circular polarization, nearly flat spectral indices, radio luminosities, and brightness temperatures. Gyroresonant absorption from non-relativistic electrons may play role in some M-dwarfs (e.g. Linsky and Gary, 1983; Güdel and Benz 1989), but this requires substantially higher magnetic fields.

Some aspects of quiescent emission are still controversial. Most gyro-synchrotron models assume a power-law distribution of relativistic electrons, but Drake *et al.* (1992) argues that a 'super-thermal' tail of the X-ray emitting thermal distribution might be a plausible alternative. However, this would require very strong (B$\gtrsim$200G) magnetic field at large distances from the star, and would also require a very sharp cutoff ($S_\nu \propto \nu^{-8}$) in luminosity above the spectral peak, which has never been seen. If the electrons do have a power-law energy distribution, the exponent could in principle be measured by VLBI by determining the brightness temperature dependence on frequency, since $T_B(\nu) \propto \nu^{-0.9\delta}$ where δ is the energy power-law index of the electrons (Dulk, 1985).

White and Franciosini (1995) have suggested that the observed reversal of circular polarization with frequency (Morris *et al.*, 1990) is not due to mode-coupling as the emitting region becomes optically thick, as in the models of Morris *et al.* (1990) or Storey (1995). Rather, they suggest that the gyro-synchrotron component is only weakly polarized, and that the observed circular polarization arises from a highly polarized coherent component. They present VLA observations in which highly polarized, rapidly variable microflares are seen superposed on a slowly varying, unpolarized component. It is not clear how common this phenomenon is: Rucinski *et al.* (1993) monitored two RS CVn systems at the VLA for several hours and found no evidence for rapid flux variations on timescales of seconds to minutes.

A possible test of the White and Franciosini hypothesis would be to search for an offset between the centroid of the highly polarized unresolved ($T_B > 10^{12}$K) emission and the unpolarized, unresolved ($T_B \sim 10^{9-11}$K) emission using a dual circular polarization VLB array. A positional offset should be expected since the coherent emission region would be unlikely to be located exactly at the centroid of the extended radio corona.

3.3.2. *Radio Flares*

Unlike quiescent emission, stellar radio flares display a bewildering variety of characteristics. Many flares display telltale signs of coherent emission including impulsive, short duration temporal structure and very high circular polarization. Some examples include quasi-periodic, frequency drifting M-dwarf flares (Bastian *et al.*, 1990), the remarkable impulsive flares at 360 and 609 MHz on the RS CVn system II Peg (van den Oord and de Bruyn, 1994), and 'microflaring' on HR1099 (White and Franciosini, 1995). Proposed emission mechanisms include an electron-cyclotron maser (Bastian *et al.*, 1990), and plasma emission from double layers (van den Oord and de Bruyn, 1994) or from pulsed runaway electron acceleration (Kuijpers *et al.*, 1981). On the other hand, many flares last hours or days, have little or no polarization, and are very broadband. These flares are ascribed to synchrotron radiation from fully relativistic electrons.

In principle, VLBI observations can easily distinguish between coherent and incoherent processes by a simple measurement of brightness temperature, since coherent processes will have $T_B >> 10^{12}$K whereas incoherent processes cannot exceed the inverse-compton limit $T_B \sim 10^{12}$K. For weak sources, this can be problematical, since the maximum brightness temperature which can be resolved by a terrestrial baseline is

$$T_B(max) \sim 10^{11} S_{mJy} \text{ K}$$

independent of wavelength.

4. Implications for Coronal Models

4.1. WHAT IS THE RELATIONSHIP BETWEEN THE X-RAY AND RADIO EMITTING REGIONS?

Güdel and Benz (1993) and Benz and Güdel (1994) have found a surprising correlation between the radio and X-ray luminosities of a large number of magnetically active stars. The correlation ranges over ten orders of magnitude, from the active Sun to many classes of active single stars, binaries, and pre-main sequence objects. This correlation is especially surprising since it is clear that the thermal electrons responsible for the X-ray emission are not identical with the relativistic electron seen at radio wavelengths. The relationship is therefore statistical and seems to imply that the relativistic electrons are accelerated from the reservoir of thermal electrons. This implies that the energy sources for coronal heating and particle acceleration to relativistic energies are closely linked. The correlation is nearly linear, indicating that the mechanism maintains a roughly constant ratio of thermal to non-thermal electron densities. Linsky (1995) has recently suggested such a mechanism, in which electrons above the mean of the thermal distribution experience 'runaway' acceleration to relativistic energies by a Dreicer electric field, possibly generated by turbulent, time-varying magnetic fields.

Given the statistical relationship between radio and X-ray luminosities and the obvious familial coronal relationship, it is useful to compare the radio sizes found from VLBI observations with those derived from X-ray eclipses to try to understand the relationship between the two regions. A well documented case is the eclipsing RS CVn system Algol, for which there are both VLBA maps and X-ray eclipse observations from two satellites at several epochs. Ottman (1994, ROSAT) and Antunes *et al.* (1994, ASCA) have both observed minima near phase ∼0.45, with significant orbital modulation of the hot component (20MK) light curve outside eclipse. The overall light curve is remarkably similar in both observations (which were 1 year apart), suggesting a stable X-ray emitting region (loop?) on the K star's leading hemisphere with a scale height 0.8 times the K star radius.

By comparison, the 8.4 GHz VLBA map of Algol (Mutel and Scharringhausen, 1996) shows a radio FWHM size 2.8 times the radius of the K star. There is also some evidence for a phase dependent radio light curve on Algol (LeFevre *et al.*, 1994). Although we cannot register the X-ray and radio structures observationally, a plausible picture emerges in which the radio corona overlays a large hot thermal coronal loop from which the energetic electrons are accelerated by some undetermined process.

4.2. THE GEOMETRY OF THE RADIO EMITTING CORONAL GAS

Until very recently, stellar VLBI studies have largely been restricted to the most intense emitters and to rough estimates of angular structure at a single frequency, since the existing VLBI arrays had limited spatial frequency coverage and sensitivity. However, there has been precious little two dimensional structural information. Published models of radio coronae have largely used the observed spectra and polarization of quiescent radio emission as model constraints. The models often invoke trapped plasma in closed magnetic structures situated in a symmetric toroidal region centered on the magnetic equator of the active star. On the other hand, models based on recent X-ray eclipse observations of several late-type systems appear to show large-scale asymmetric structures (e.g. AR Lac, Siarkowski, 1992; TX Pyx, Pres *et al.*, 1995). In the latter case, it appears that large magnetic loop structures may connect the two stars and perhaps play an important role in particle acceleration. This is similar to V471 Tau (Lim *et al.* 1995) in which the evidence for radio emission between the K dwarf and the white dwarf is very strong. Future high fidelity VLBA images of these systems will play a key role in understanding the detailed geometry of the stellar radio coronae.

References

Andre, P. 1995, in *Radio Emission from the Sun and Stars*, ed. J.M. Paredes and A.R. Taylor (A.S.P. Conference Series), *in press.*

Andre, P., Deeney, B., Phillips, R., and Lestrade, J-F. 1992, *Ap. J.*, **401**, 667.

Andre, P., Phillips, R., Lestrade, J-F., and Klein, K. 1991, *Ap. J.*, **376**, 630.

Antunes, A., Nagase, F., and White, N. 1994, *Ap. J. (Letters)*, **436**, 83.

Bastian, T., Bookbinder, J., Dulk, G, and Davis, M. 1990, *Ap. J.*, **353**, 265.

Benz, A. and Alef, W. 1991, *Astr. Ap.*, **252**, L19.

Benz, A., Alef, W., and Güdel, M. 1995, *Astr. Ap.*, **298**, 187.

Benz, A., and Güdel, M. 1994, *Astr. Ap.*, **285**, 663.

Catalano, S. 1990, in in *Active Close Binaries*, ed. C. Ibanoglu, NATO ASI Series vol. 319 (Kluwer:Dordrecht), p.411.

Chiuderi-Drago, F. and Franciosini, E. 1993, *Ap. J.*, **410**, 301.

Drake, S. 1993, in *Physics of Solar and Stellar Coronae*, e.d J. Linsky and S. Serio (Dordrecht:Kluwer), p.393.

Drake, S., *et al.* 1987, *Ap. J.*, **322**, 902.

Drake, S., Simon, T., and Linsky, J. 1992, *Ap. J. Suppl.*, **82**, 311.

Dulk, G. 1985 *Ann. Rev. Astr. Ap.*, **23**, 169.

Elias, N. *et al.* 1994, *Ap. J.*, **439**, 983.

Feigelson, E., *et al.* 1994, *Ap. J.*, **432**, 373.

Felli, M., Massi, M., and Catarzi, M. 1991, *Astr. Ap.*, **248**, 453.

Felli, M. and Massi, M. 1991, *Astr. Ap.*, **246**, 503.

Fox, D. *et al.* 1994, *Astr. Ap.*, **284**, 91.

Güdel, M. 1994, *Ap. J. Suppl.*, **90**, 743.

Güdel, M. and Benz, A. 1989, *Astr. Ap.*, **211**, L5.

Güdel, M. and Benz, A. 1993, *Ap. J. (Letters)*, **405**, 63.

Jones, D., Lestrade, J-F., Preston, R., and Phillips, R. 1995, *Astr. Sp. Sci.*, **223**, 166,

Kuijpers, J., van der Post, P., and Slottje, C. 1981 *Astr. Ap.*, **103**, 331.

Lang, K. 1994, *Ap. J. Suppl.*, **90**, 753.

Lang, R. and Willson, R. 1988, *Ap. J.*, **328**, 610.

LeFevre, E., Klein, K-L., and Lestrade, J-F. 1994, *Astr. Ap.*, **283**, 483.

Lestrade, J-F., Mutel, R., Preston, R., and Phillips, R. 1988, *Ap. J.*, **328**, 232.

Lestrade, J-F., Preston, R., Phillips, R., and Gabuzda, D. 1992, *Astr. Ap.*, **258**, 112L.

Lestrade, J-F., Phillips, R., Hodges, M., and Preston, R. 1993, *Ap. J.*, **410**, 808.

Lestrade, J-F., Jones, D., Preston, R., and Phillips, R. 1994, *Astr. Sp. Sci.*, **212**, 251,

Lim, J., White, S., Nelson, G., and Benz, A. 1994, *Ap. J.*, **430**, 332.

Lim, J., White, S., and Cully, S. 1995, Ap.J. *in press.*

Linsky, J. 1995, in *Radio Emission from the Sun and Stars*, ed. J.M. Paredes and A.R. Taylor (A.S.P. Conference Series), *in press.*

Linsky, J. and Gary, D. 1983, *Ap. J.*, **274**, 776.

Linsky, J., Drake, S., and Bastian, T. 1992, *Ap. J.*, **393**, 341.

Massi, M. *et al.* 1988, *Astr. Ap.*, **197**, 200.
Morris, D., Mutel, R., and Su, B. 1990, *Ap. J.*, **362**, 299.
Mutel, R., Lestrade, J-F., Preston, R., and Phillips, R. 1985, *Ap. J.*, **289**, 262.
Mutel, R., Neff, J., Bookbinder, J, and Pagano, I. 1993, in *Physics of Solar and Stellar Coronae*, e.d J. Linsky and S. Serio (Dordrecht:Kluwer), p.409.
Mutel, R. and Scharringhausen, B. 1996, *in. prep.*
Neidöfer, J., Massi, M., and Chiuderi-Drago, F. 1993, *Astr. Ap.*, **278**, L51.
O'Neal, D., Feigelson, E., Mathieu, R., and Myers, P. 1990, *A. J.*, **100**, 1610
Ottman, R. 1994, *Astr. Ap.*, **286**, L27.
Phillips, R. and Titus, M. 1990, *Ap. J. (Letters)*, **359**, 15.
Phillips, R., Lonsdale, C., and Feigelson, E. 1991, *Ap. J.*, **382**, 261.
Phillips, R., Lonsdale, C., and Feigelson, E. 1993, *Ap. J. (Letters)*, **403**, 43.
Pres, P., Siarkowski, M., and Sylwester, J. 1995, *M. N. R. A. S.*, **275**, 43.
Rucinski, S., Krogulec, M., and Seaquist, E. 1993, *A. J.*, **105**, 2308.
Siarkowski, M. 1992, *M. N. R. A. S.*, **259**, 453.
Skinner, S. 1993, *Ap. J.*, **408**, 660.
Skinner, S. and Brown, A. 1994, *A. J.*, **107**, 1461.
Stern, R. *et al.* 1992, *Ap. J.*, **391**, 760.
Storey, M. 1995, in *Radio Emission from the Sun and Stars*, ed. J.M. Paredes and A.R. Taylor (A.S.P. Conference Series), *in press.*
Trigilio, C., Umana, G., and Migenes, V. 1994, *M. N. R. A. S.*, **260**, 903.
Umana, G., Catalano, S., and Rodono, M. 1991, *Astr. Ap.*, **249**, 217.
van den Oord, G. and de Bruyn, A. 1994, *Astr. Ap.*, **286**, 181.
White, R. 1985 *Ap. J.*, **289**, 698.
White, S. and Franciosini, E. 1995, *Ap. J.*, **444**, 342.

CORONAL STRUCTURE IN M DWARF STARS

M. S. GIAMPAPA
National Optical Astronomy Observatories, National Solar Observatory, POB 26732, Tucson, AZ 85726-6732 USA

R. ROSNER & V. KASHYAP
Department of Astronomy & Astrophysics, University of Chicago, Chicago, IL 60637 USA

T. A. FLEMING
Steward Observatory, University of Arizona, Tucson, AZ 85721 USA

J. H. M. M. SCHMITT
Max-Planck-Institut für Extraterrestrische Physik, 8046 Garching, Germany

AND

J. A. BOOKBINDER
Harvard-Smithsonian Center for Astrophysics, 60 Garden St., Cambridge, MA 02138 USA

1. Introduction

In view of the importance of these objects as observational probes for theories of stellar dynamos and coronal heating, as well as their potentially important role in the galactic environment, we implemented a program of pointed *ROSAT* PSPC observations of selected M dwarfs in the solar neighborhood. We derive key coronal plasma parameters based on the observed PSPC pulse-height distributions in order to investigate stellar coronal structure in more detail. In particular, we utilize temperatures and emission measures inferred for one or more distinct coronal components as constraints for the development of semi-empirical magnetic loop models as representations of the coronae of low mass stars (Giampapa et al. 1995).

Y. Uchida et al. (eds.),
Magnetodynamic Phenomena in the Solar Atmosphere – Prototypes of Stellar Magnetic Activity, 81–82.

2. Discussion

We find that the corona of low mass dwarfs consists of two distinct thermal components: a "soft" component with T $\sim$ 2-4 $\times$ 10^6 K and a "hard" component with T $\sim$ 10^7 K. The low-T emission component is characterized by loop scale lengths substantially smaller than the corresponding pressure scale height; thus these compact structures are consistently described by loop scaling laws (Rosner, Tucker & Vaiana 1978; RTV). By contrast, the high-T emission component is characterized by loop scale lengths of the order of, or larger than, the corresponding pressure scale height, unless the filling factor (f) is much smaller than unity. Thus, if $f \sim 1$ for this component, then these loops are *not* consistently described by the simple RTV scaling laws. Time-resolved spectroscopy of the X-ray data supports the idea that the low- and high-T emission components are physically distinct; the low-T component is relatively constant while the high-T component contributes preferentially to the observed count-rate variations. Hence the high-T component cannot be consistently modeled by quasi-static structures (RTV), and is more likely associated with a superposition of emission from flaring coronal structures. Our study therefore suggests that these stars do not have a large-scale magnetic field structure which is a feature of "classical" $(\alpha - \omega)$ dynamo theory. We therefore suggest that "classical" dynamo action does not occur in the cool, low mass dwarfs.

With regard to rotation, which has a direct bearing on dynamo action, we know from observations that the lowest-mass stars spin down (via magnetic braking) more slowly than the more nearly solar-type stars (Stauffer & Hartmann 1987). The compact loops we find for the low-temperature component suggests a natural explanation: The smaller moment arms implied by the high multipole-moments suggested by our loops models would lead to a reduced efficiency for magnetic braking. A preliminary model of the active G0 V star, π^1 UMa, shows that its hard-component can be described by relatively more extended structures, in contrast to that found for the M dwarfs. We therefore contend that it is the coronal field configuration which systematically modifies the wind and mass loss properties as a function of stellar mass that, in turn, leads to the observed mass-dependence of rotational evolution on the main sequence.

References

Giampapa, M. S., Rosner, R., Kashyap, V., Fleming, T. A., Schmitt, J.H.M.M., & Bookbinder, J. A. (1995) The Coronae of Low Mass Dwarf Stars, *ApJ*, submitted.

Rosner, R., Tucker, W. H., & Vaiana, G. S. (1978) Dynamics of the Quiescent Solar Corona, *ApJ*, **220**, 643–665.

Stauffer, J. & Hartmann, L. (1987) The Distribution of Rotational Velocities for Low-Mass Stars in the Pleiades, *ApJ*, **318**, 337–355.

X-RAY AND EUV OBSERVATIONS OF ACTIVE BINARY STELLAR CORONAE: RELATIONSHIPS TO SOLAR STUDIES OF FLARE PARAMETERS AND CORONAL ABUNDANCES

ROBERT A. STERN
Lockheed Martin Palo Alto Research Laboratory
O/91-30 Bldg. 252, 3251 Hanover Street
Palo Alto, CA 94304 USA

Extended Abstract

X-ray satellites have greatly increased our knowledge of physical parameters of coronae and flares in active binary systems. GINGA, in particular, observed large flares in UX Ari (Tsuru *et al.* 1989), V711 Tau (Stern *et al.* 1996), II Peg (Doyle *et al.* 1991), and Algol (Stern *et al.* 1992). These systems invariably show evidence of quiescent plasma temperatures of at least 10 MK, with thermal plasma temperatures exceeding 60 MK during flares. (A large flare apparently associated with the M dwarf system EQ1839.6+8002 reached temperatures near 100 MK; Pan *et al.* 1995). The total volume emission measures for the largest flares in active binaries can be as large as 10^{54}-10^{55} cm^{-3}. When plotted on a log EM–log T plot (developed by solar X-ray flare researchers), time histories of these flares may be studied in much the same manner as for solar flares (see Figure 1).

A related topic is the increasing evidence for reduced Fe abundances and abundance variations indicated in GINGA, ASCA and EUVE observations. Differences in solar photospheric and coronal abundances have, of course, been studied for well over a decade, and are thought to be related to the first ionization potential (FIP, e.g. Meyer 1985). In addition, abundance variations in solar active regions and flares are currently the subject of intense research (Saba and Strong 1992). Very recently, ASCA and EUVE observations are revealing curious apparent sub-solar Fe abundances in coronae of active binary systems (Drake *et al.* 1994, White *et al.* 1994, Antunes *et al.* 1994, Stern *et al.* 1995). Earlier flare observations of Algol by GINGA

Y. Uchida et al. (eds.),
Magnetodynamic Phenomena in the Solar Atmosphere – Prototypes of Stellar Magnetic Activity, 83–84.

suggest a sub-solar and time-varying Fe abundance, (Stern *et al.* 1992). Recently, Ottmann and Schmitt (1995) have interpreted ROSAT PSPC observations of another X-ray flare on Algol as showing abundance variations similar to the GINGA flare.

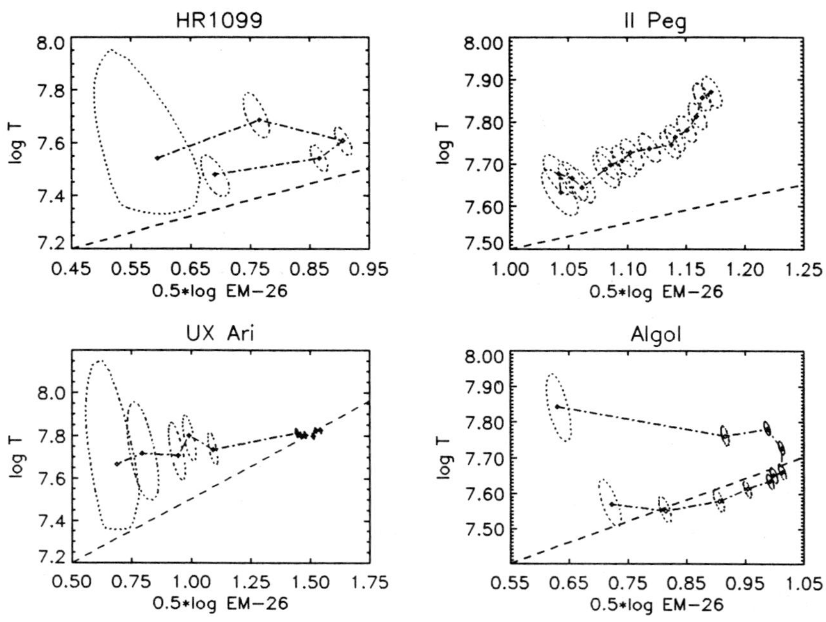

Figure 1. Evolution of stellar X-ray flares observed by GINGA in "log T − log ($EM^{1/2}$)" space. The abcissa is proportional to log n_e for a constant flare volume. The dashed lines show the predicted evolution for a "quasi-statically" evolving flare decay. The solid contours are 90% confidence regions for the derived parameters.

References

Antunes, A., Nagase, F., and White, N.E., 1994, ApJ, 436, L83.
Doyle, J.G., *et al.* , 1991, MNRAS, 248, 503.
Drake, S., Singh, K.P., White, N.E., and Simon, T., 1994, ApJ, 436, L87.
Meyer, J.P., 1985, ApJS, 57, 173
Ottmann, R., and Schmitt, J.H.M.M., 1995, A&A, in press.
Pan, H.C., Jordan, C., Makishima, K., Stern, R.A., Hayashida, K., and Inda-Koide, M., 1995, in *Proceedings of the IAU Colloquium No 151 "Flares and Flashes"* (Lecture Notes in Physics, Springer-Verlag).
Saba, J.L.R., and Strong, K.T., 1992, ESA SP-348 (Proc. First Ann. SOHO Workshop).
Stern, R.A., Uchida, Y., Tsuneta, S., and Nagase, F., 1992, ApJ, 400, 321.
Stern, R.A., Lemen, J.R., Schmitt, J.H.M.M., and Pye, J.P., 1995, ApJ, 444, L45.
Stern, R.A., Tsuneta. S., and Uchida, Y., 1996, in preparation.
White, N.E., *et al.* , 1994, PASJ, 46, L97.

ELEMENT ABUNDANCES IN STELLAR CORONAE

J.J. DRAKE
Center for EUV Astrophysics,
University of California, Berkeley CA 94720, USA

AND

J.M. LAMING AND K.G. WIDING
E.O. Hulburt Center for Space Research,
Naval Research Laboratory, Washington DC 20375, USA

About two years ago, the intrigue of the solar "FIP Effect"—that coronal abundances of elements with a low FIP ($\lesssim 10$ eV; e.g Fe, Mg, Si, Ca) are observed to be enhanced relative to those of high FIP elements ($\gtrsim 10$ eV; e.g. O, Ne, S; see e.g. the review by Meyer 1993)—prompted us to study stellar coronal abundances based on new EUV spectra obtained with the Extreme Ultraviolet Explorer (*EUVE*). Our results to-date indicate that some stars appear to show a solar-like FIP Effect, whereas others do not (Drake et al. 1995abc; Laming et al. 1995ab). These results are summarised in Fig. 1. Also in Fig. 1 is a column labelled "MAD"—metal abundance deficient. Recent analyses of both *ASCA* and *EUVE* spectra of RS CVn's, algol binaries and active single stars have found their coronae to be surprisingly metal deficient with respect to the Sun—by factors of 2-3 to as much as 10 (Singh, S.A.Drake & White 1995 and references therein; Stern et al. 1995; Schmitt et al. 1995). Are the coronae of these stars *depleted* in metals relative to their photospheres? Optical work indicates that the RS CVn photospheres are also metal poor (e.g. Randich et al. 1994 and references therein). However, the RS CVn's and Algols are in general relatively young ($\sim$ 2-3 Gyr and $<$ 1 Gyr, respectively), and not expected to be metal poor; are the photospheric lines in-filled by chromospheric activity? The photospheric abundances warrant further investigation.

If the MAD syndrome is a new coronal abundance anomaly shared by the very active stars, can both FIP and MAD be explained by the same processes? Recent diffusion models are appealing: von Steiger & Marsch (1994) show that a relatively simple chromospheric diffusion model yields a fractionation very similar to the observed solar FIP Effect. We notice that the middle regions of their model exhibit an *inverse* fractionation according

Y. Uchida et al. (eds.),
Magnetodynamic Phenomena in the Solar Atmosphere – Prototypes of Stellar Magnetic Activity, 85–86.

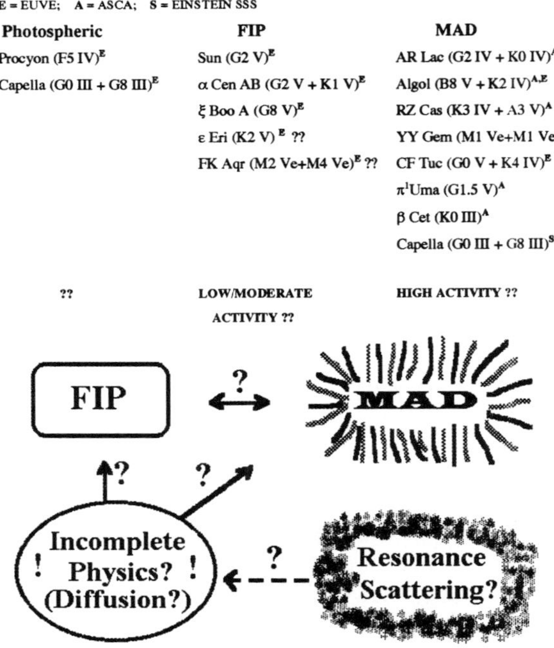

Figure 1. A summary (or confusion?!) of stellar coronal abundance results.

to FIP which is reminiscent of the MAD syndrome! We also note that a corona substantially enriched in He could mimic metal depletion through He's larger ($\sim Z^2$) contribution to the Bremsstrahlung continuum, lowering the metal line-to-continuum ratio. Such He enrichment is predicted by some solar diffusion models (e.g. Hansteen et al. 1993).

Future *EUVE*, *ASCA* and optical studies should produce some answers.

References

Drake, J.J., Laming, J.M., Widing, K.G., 1995a, ApJ, 443, 393

Drake, J.J., Laming, J.M., Widing, K.G., 1995b, in *Astrophysics in the EUV*, eds. S.Bowyer & R. Malina, Kluwer, in press.

Drake, J.J., Laming, J.M., Widing, K.G., 1995c, ApJ, submitted

Hansteen, V.H., Holzer, T.E., Leer, E., 1993, ApJ, 402, 334

Laming, J.M., Drake, J.J., Widing, K.G., 1995, 443, 416

Laming, J.M., Drake, J.J., Widing, K.G., 1995, ApJ, submitted

Meyer, J.-P., 1993, in *Origin and Evolution of the Elements*, eds. N. Prantzos, E. Vangioni-Flam & M. Cassé, CUP, 26.

Schmitt, J.H.M.M., Stern, R.A., Drake, J.J., Kürster, M., ApJ, 1995, submitted

Singh, K.P., Drake, S.A., White, N.E., 1995, ApJ, 445,

Stern, R.A., Lemen, J.R., Schmitt, J.H.M.M., Pye, J.P., 1995, ApJ, 44, L45

von Steiger, R., & Marsch, E., 1994, Space Sci. Rev., 70, 341

II. Wind and Mass-Loss from the Sun and Stars

II.1. Wind and Mass-Loss from the Sun

YOHKOH OBSERVATIONS OF CORONAL MASS EJECTIONS

H. HUDSON
Institute for Astronomy, University of Hawaii,
Honolulu, HI 96822, USA

Abstract. The soft X-ray telescope on *Yohkoh* has given us comprehensive views of the solar corona beginning in September 1991. A movie representation reveals numerous variations of configuration and structure, including material ejections of various kinds and the formation of arcades of loops, known from earlier observations to have a close association with coronal mass ejections (CMEs) observed with visible-light coronagraphs. *Yohkoh*/SXT observes solar mass loss closely related to the well-known CME phenomenon, both associated with solar flares and with large-scale arcade events outside the active latitudes. In addition there are other forms of mass loss, including the gradual ejection of loops from active regions and the impulsive formation of X-ray jets. These probably imply forms of stellar mass loss that had not been recognized prior to the *Yohkoh* observations.

1. Introduction

The Sun often launches mass outwards in discrete events. This mass, along with its frozen-in magnetic fields, may then become a part of the solar wind. Specifically, coronal mass ejections (CMEs) occur several times a day in active conditions. The defining observations of CMEs are those of white-light coronagraphs, but there are strong relationships with certain signatures obtained with *in situ* observations made by interplanetary spacecraft

This paper reviews a fundamentally new form of observation of coronal mass ejections, namely the view obtained from time series of soft X-ray images of the dynamic corona. The *Yohkoh* observations represent the first comprehensive data set of this type. It began in September, 1991, and continues through the time of writing of this paper. During this interval there have been no space-borne coronagraph observations, so that the precise

Y. Uchida et al. (eds.),
Magnetodynamic Phenomena in the Solar Atmosphere – Prototypes of Stellar Magnetic Activity, 89–96.

identification of the X-ray morphology with the familiar views of CMEs obtained in that manner is not possible. Klimchuk *et al.* (1994), however, showed that CME-like properties indeed could be easily seen in the *Yohkoh*/SXT limb observations. On the other hand, this has been an exciting interval for interplanetary observations, with the polar passages of the *Ulysses* spacecraft providing the first-ever views of the polar solar wind and the effects of coronal mass ejections at high latitudes.

2. The Nature of the SXT View of the Corona

Yohkoh/SXT responds to soft X-rays in ranges near 1 keV (10Å) which include the natural emission domain of plasmas at coronal temperatures. Tsuneta *et al.* (1991) give a full description of the instrument, which (in brief) uses a grazing-incidence mirror to feed a CCD sensor with a pixel array of 1024×1024 $2.45''$ pixels, subtending a total angular field of $42'$ square. The telemetry capacity of *Yohkoh* allows the transmission of about 20 whole-Sun images, with 2×2-pixel summations ($\sim 5''$ pixels), per 97-minute orbital period of the spacecraft. These images must be distributed among different broad-band filters and exposure times, so that the typical time cadence for large-scale coronal observations is on the order of ten minutes. However there are gaps in this coverage induced by various factors, including orbit night and flare occurrence (this causes an automatic mode switch that eliminates the full-Sun images).

There are several substantial differences between soft X-ray observations of the *Yohkoh* type, and white-light observations. In general the X-ray observations offer a better view near the Sun, and provide more diagnostic information (temperatures). However most of our knowledge of CMEs comes from the white-light data, so it is important to understand the differences in the data being reported here. The principal points to note are the following:

- Contrast. Because the X-ray emissivity depends strongly upon temperature, there is a much larger dynamic range in X-ray images of the solar corona than in white light.

- Field of view. The maximum height sampled by *Yohkoh*/SXT at the limb normally is below 1.5 R_o from Sun center, whereas white-light coronagraphs often do not look below this height.

- Disk observations. The soft X-rays show the entire visible hemisphere.

- Differential emission measure. The SXT response to coronal plasma emission is a (non-linear) function of temperature, as opposed (for example) to the flat response in the case of electron scattering (white-light coronagraph) or thermal free-free emission (dominant at radio wavelengths outside magnetic regions). This means that brightness does not transcribe so directly into mass or density.

Of these differences, perhaps the great contrast in soft X-rays is the most limiting. The CCD sensor views the entire Sun, including active regions as well as the faint corona. This means that long exposures appropriate for the corona may saturate. Of course this ability to detect active regions is also an advantage, because the action and consequences of CME launching can be studied in detail, even well away from the limb.

3. SXT Observations of Solar Mass Loss

Yohkoh/SXT provides a global view of the corona within its field of view, which is a square 0.70° across. The pointing axis of the telescope is almost always situated so that the entire disk is visible, except during special operations, and transmits full-Sun images at 2×2-pixel summation, *i.e.* with an effective pixel size of 4.91″. This view does not correspond well with that of a normal coronagraph, and this helps to explain the strong differences of morphology that are in fact observed. The *Yohkoh*/SXT data, in fact, show several forms of coronal mass loss, some of which differ strikingly from the classical CME phenomenon (*e.g.* Kahler, 1992). In addition to the eruptive events we associate with CMEs, which are described below in further detail, there are smaller-scale eruptions with no known counterparts in white light:

– *Expanding active-region loops.* One of the first discoveries in the new X-ray data was the tendency for some active-region loops to expand at intermediate speeds (10-50 km s^{-1}), rather than remain static (Uchida *et al.*, 1992). Perhaps this result should not have surprised us, because newly emerging flux is a characteristic property of solar active regions. This observation suggests that a magnetically-driven outward flow from active regions may contribute to the solar wind.

– *Soft X-ray jets.* The common occurrence of X-ray jets, invariably associated with flare properties at their feet, provide anothers possible channel of mass loss. *Yohkoh*/SXT has made the first observations of such jets (Shibata *et al.*, 1992; Strong *et al.*, 1992; Shimojo *et al.*, 1996) because of its greatly improved time coverage. The jets have recently been identified with Type III bursts (Aurass *et al.*, 1994; Kundu *et al.*, 1995; Raulin *et al.*, 1996). The X-ray jet consists of plasma flowing outwards, and the Type III burst depends upon non-thermal electrons that have been detected in the solar wind directly (*e.g.*, Lin 1985). This identification therefore means that – in some cases at least – the outward flow of the jet may continue into the outer corona and contribute to the solar wind.

The interpretation of these two newly-discovered phenomena as mechanisms for solar mass loss has not yet been placed on a quantitative footing, and much further research needs to be done.

4. The Coronal Mass Ejections Themselves

The classical CMEs, as identified in coronagraph images, have a broad range of characteristics but some simplifying patterns (Hundhausen, 1993). The structure consists of an outward-moving front, plus (ideally) a void following the front, and cooler material identified with an erupting filament emerging with the void. Standard interpretation (*e.g.* Low, 1994) relates the void to the flux rope thought to be necessary to support the filament before eruption. The mass driven outwards mainly comes from already-existing coronal material, so that there is no need to dredge it up in real time from lower altitudes. The solar counterparts of the mass ejections are eruptive flares of the type that often create post-flare Hα loops (the long-duration events, or LDEs). Similar but larger-scale phenomena without bright flare emissions may also accompany CME launches (Harvey, 1996). These often occur even at high latitudes, far from active regions. The filament itself seems not to be central to the CME launch, but the geometry in which a filament can sometimes form does seem to be universally required.

The *Yohkoh*/SXT data show parts of the corona disappearing, presumably during the formation of a CME (Hudson *et al.*, 1996; Lemen *et al.*, 1996). We illustrate this kind of observation with two examples in Figures 1 and 2, one of an LDE flare (13 Nov. 1994; Hudson *et al.*, 1996), and one of a high-latitude large arcade event (24 Jan. 1992; Hiei *et al.*, 1993). The photometry in these figures shows the time variation of integrated brightness in selected regions, which reflects the time scale of mass removal at the location chosen. More complete analyses could in principle determine the mass loss (by difference images) or the pattern of mass flow (by correlation tracking). We have made estimates of the mass, and find for the flare event that the mass of the discrete structure (shown in the Figure) is about 4×10^{14} g, a lower limit because of the difficulty of estimating the mass flow arising near the flare proper (not shown in the Figure). This difficulty results from the continuous nature of the outward flow during the interval of flare brightening, and from the contrast problem (the difficulty of detecting faint features near bright ones). For the 24 Jan. 1992 event, a rough estimate is about 10^{15} g. These estimates are within errors of the typical range of CME masses, although this means very little given the broad

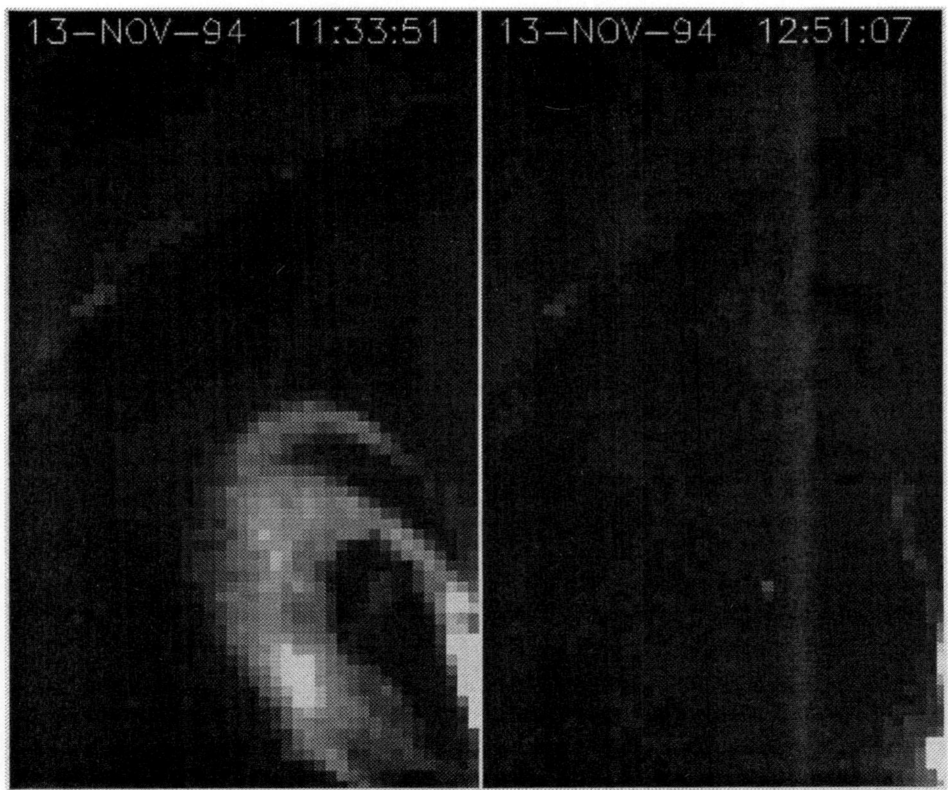

Figure 1. Example of mass loss associated with an LDE flare, 13 Nov. 1994 (Hudson *et al.*, 1996). The figure shows before (left) and after (right) images of a large structured cloud adjacent to the location of an M1.2 LDE flare. This flare exhibited multiple forms of mass ejection, but the one we believe most likely to be associated with a CME is the structured cloud shown here. This was observed to move slowly (projected velocity $\sim$100 km s^{-1}) outwards during the phase of increasing flare brightness and to disappear almost completely during a 63-minute data gap. This event was not observed as a true CME (*i.e.*, in a coronagraph) but would not have been easily observable because it occurred near disk center.

range of this property. A more explicit comparison would be interesting but would require simultaneous coronagraph and X-ray data.

5. Interplanetary Connections

A CME launched from the corona almost by definition produces effects in the solar wind, and sometimes striking results on the Earth and its magnetosphere. The advent of X-ray imaging with good time coverage makes the identification of the solar and interplanetary parts of this phenomenon much easier.

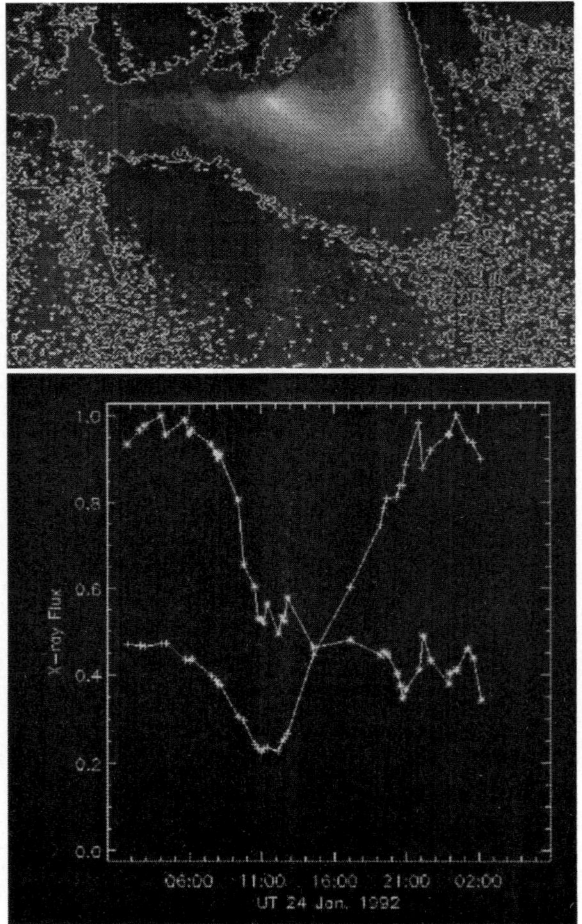

Figure 2. Example of mass loss associated with large arcade event, that reported by Hiei *et al.* (1993). The light curves at bottom show the soft X-ray fluxes observed in the two boxes shown on the difference image at the top. The contour shown for the latter is the zero contour; *i.e.*, the negative regions represent deficits between the times of the images (14:33 UT minus 06:01 UT). The subsequent rise in the light curve meaured directly *above* the arcade results from the increasing height and brightness of the large cusp (helmet streamer) following the event. The light curve from the area *adjacent* to the arcade just shows dimming. Note that these observations precisely identify the time of the mass ejection, which unfortunately occurred during night at the Mauna Loa Solar Observatory.

Lemen *et al.* (1996) studied a sample of interplanetary events detected during the *Ulysses* south polar passage. These events were in general detectable in *Yohkoh* soft X-ray images, either as a flare event (two examples out of seven), as a large-scale arcade event (four) or as a filament-related event. These categories are not mutually exclusive, of course; the main point is that the full-disk observations in soft X-rays allowed the probably identification of all seven events. Two of them occurred behind the limb, but portions of the soft X-ray sources rose into view.

The ionization temperatures of the interplanetary particles (Galvin *et al.*, 1996) showed no obvious correlation with the soft X-ray temperatures.

6. Conclusions: Geometry and Mechanisms

The *Yohkoh*/SXT data show several forms of mass ejection from the solar corona. As expected from the coronagraph data, events probably associated with "true" CMEs show clearly measurable dimmings of the X-ray corona near the site of the flare or large arcade brightening. Here we define a true CME as one detected in the usual manner by a white-light coronagraph. The analysis of such *Yohkoh* observations is in its infancy at the time of writing, but we expect many interesting results in the future. Much of this work will depend upon joint analysis of white light and X-ray data, for which we can look forward to the SOHO data. The correlation-tracking analysis of X-ray brightness fluctuations to determine the geometry of the CME motions near the Sun, using the *Yohkoh* data, is difficult and has not yet been seriously attempted. Since the images show the "dimming corona" so well, this may be possible in the future and in any case would be extremely desirable because of our ignorance of the geometry.

The dimming of the corona as seen in soft X-rays is a high-contrast phenomenon, as can be seen from the amplitude of the light curves shown in Figures 1 and 2. This helps to increase the signal-to-noise ratio, but it should be noted that still higher sensitivity than that of SXT would be better still. Stereoscopic observations would provide another great improvement in the directions that the *Yohkoh*/SXT data point.

Acknowledgements This work was supported by NASA under contract NAS 8–37334. *Yohkoh* is a mission of the Institute of Space and Astronautical Sciences.

References

Aurass, H., Klein, K.-L., & Martens, P. C. H., 1994, *Solar Phys.* **155**, 203
Galvin, A.T., *et al.*, 1996, *Solar Wind 8*, to be published
Kahler, S.W., 1992, *Annu. Revs. Astron. Astrophys.* **30**, 113
Harvey, K., 1996, these proceedings
Hiei, E., Hundhausen, A., & Sime, D. 1993, *Geophys. Res. Lett.* **20**, 24
Hudson, H.S., Acton, L.W., Alexander, D., Freeland, S.L., Lemen, J.R., & Harvey, K.L., 1996, *Solar Wind 8*, to be published
Hudson, H.S., Acton, L.W., & Freeland, S.F., 1996, *Ap. J.*, to be published
Hundhausen, A.J. 1993, J., *J. Geophys. Res.* **98** 13,177

Kahler, S., 1992, ARAA 30, 113

Klimchuk, J.A., Acton, L.W., Harvey, K.L., Hudson, H.S., Kluge, K.L., Sime, D.G., Strong, K.T., & Watanabe, Ta., 1993, in Y. Uchida *et al.* (eds.), 'X-ray Solar Physics from *Yohkoh*', Tokyo, Universal Academy Press, p. 181

Kundu, M.R., Raulin, J.P., Nitta, N., Hudson, H.S., Shimojo, J., Shibata, K., & Rault, A., 1995, *Ap. J. (Lett.)* **447**, L135

Lemen, J.R., Acton, L.W., Alexander, D.A., Galvin, A.B., Harvey, K., Hoeksema, T., Zhao, X., & Hudson, H.S., 1996, *Solar Wind 8*, to be published

Lin, R.P., 1985, *Solar Physics* **100**, 537

Low, B.C., 1994, *Phys. Plasmas* **1**, 1684

Raulin, J.P, Kundu, M.R., Hudson, H.S., Nitta, N., & Raoult, A., 1996, *Ap. J.* **306**, 299

Shibata, K., Ishido, Y., Acton, L.W., Strong, K.T., Hirayama, T., Uchida, Y., McAllister, A.H., Matsumoto, R., Tsuneta, S., Shimizu, T., Hara, H., Sakurai, T., Ichimoto, K., Nishino, Y., & Ogawara, Y. 1992, Publ. Astr. Soc. Japan **44**, L173–L180

Shimojo *et al.*, 1996, *Pub. Astr. Soc. Japan*, to be published

Strong, K.T., Harvey, K.L., Hirayama, T., Nitta, N., Shimizu, T., & Tsuneta, S. 1992, Publ. Astr. Soc. Japan **44**, L161–L166

Tsuneta, S., Acton, L, Bruner, M., Lemen, J., Brown, W., Caravalho, R., Catura, R., Freeland, S., Jurcevich, B. Morrison, M., Ogawara, Y., Hirayama, T., & Owens, J. (1991), *Solar Phys.*, **136**, 37

Uchida, Y., 1992, *Publ. Ast. Soc. Japan* **44**, L155

QUIET CORONA AND MASS LOSS

G. NOCI
Dipartimento di Astronomia e Scienza dello Spazio - Università di Firenze
Largo Fermi 5 - 50125 Firenze, Italy

1. Introduction

The quiet corona is a rather undefined entity, so that the argument of the title of this talk is rather wide. I have interpreted it as referring to the origin of the slow solar wind, although the 'slow' solar wind is also not terribly well defined.

The presence in the solar wind of high speed streams was recognized since the observations of Mariner 2 (Neugebauer and Snyder, 1966), which showed that several times, during the observational period (about four months), the solar wind speed changed abruptly from the 300 - 400 km/sec range to above 600 km/sec. These high speed streams have been studied extensively. It is outside the scope of this talk to discuss their characteristics in detail. It is necessary to recall, however, that they appeared, since the early observations, to differ from the slow wind for various physical parameters, beyond bulk speed, as the proton temperature and the density. Later works (see section 2.1) have shown that they differ also in the helium abundance.

To understand the fast wind/slow wind difference, and, in particular, the slow solar wind, one has to clarify a number of problems. The most interesting are, in my opinion, the following ones:

- which is the region of origin of the slow solar wind
- why the slow wind is so much slower than the fast one
- why the He abundance in the slow wind is lower than in the fast one.

I will discuss them in the rest of this talk.

Y. Uchida et al. (eds.),
Magnetodynamic Phenomena in the Solar Atmosphere – Prototypes of Stellar Magnetic Activity, 97–106.

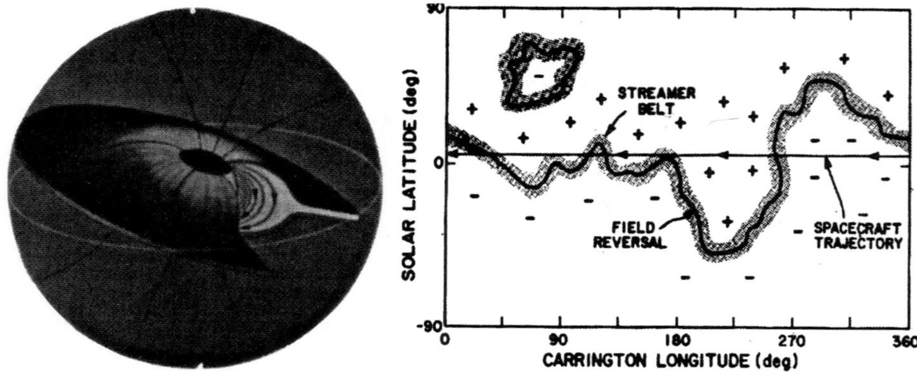

Figure 1. a) Artist's drawing of dipolar corona (from Hundhausen, 1977). b) Intersection of an irregularly shaped streamer belt with the 1 a.u. sphere: slow wind in the gray region around field reversal, fast wind outside it. (From Gosling et al., 1981.)

2. The region of origin of the slow solar wind

2.1. EARLY WORK

The structure of the heliosphere began to be understood when it was recognized that the recurrent streams of fast wind originate in coronal holes (Noci, 1973; Krieger et al. 1973; Bell and Noci, 1973, 1976; Neupert and Pizzo, 1974), coronal regions where the magnetic field is open towards the interplanetary space, and that the main holes occupy permanently the polar caps. Hole properties and heliospheric structure were studied extensively during the Skylab era (see Zirker, 1977, for a review).

A further clarification came from the study of Bame et al. (1977), who analyzed three years of IMP 6, 7, 8 data, and found that the plasma parameters, with the exception of the bulk speed, were much less variable in the fast than in the slow wind. Bame et al. (1977) gave particular attention to the α-particle concentration, for which they found the mean value 0.038, with 47% variability (standard deviation), in the slow wind, and 0.048, with 10% variability in the fast wind. Bame et al.'s conclusion was that the structureless solar wind is rather the fast than the slow wind, which is to be seen as a kind of perturbed version of the fast wind flowing from coronal holes.

In a further study based on the IMP data (Borrini et al., 1981; Gosling et al., 1981), the behaviour of various interplanetary plasma parameters (including helium abundance) was analysed, as a function of the distance from the boundary between two magnetic sectors. This analysis showed a minimum of the wind speed and also of the helium abundance at the sector boundary.

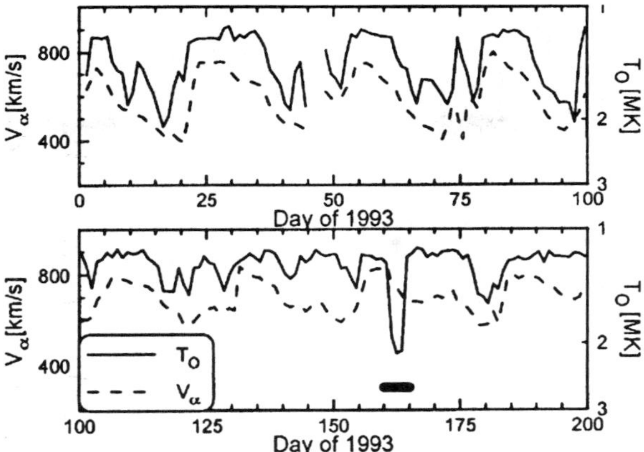

Figure 2. Plots of the solar wind speed (V_α, dashed) and freezeing-in O^{+7}/O^{+6} temperature (T_o) as a function of time, during recurring transitions of Ulysses between regions of slow solar wind and the fast solar wind originating in the southern polar hole. (From Geiss et al., 1995.)

As a conclusion of these early works the heliosphere came to fit into a rather simple overall pattern, having basically a dipolar structure, the streamer belt extending out into the interplanetary space around a current sheet (Figure 1a). In this picture the fast wind flows in both unipolar regions in which the interplanetary space is divided by the current sheet, the slow wind being confined to the borders of those regions. The slow wind, therefore, would come from the sides of the streamer belt, e.g. from the peripherical parts of the polar holes and of their low latitude extensions. This is shown in Figure 1b , which depicts a less idealized situation than that of Figure 1a. In this picture slow solar wind is observed only when the measuring spacecraft is close to the current sheet.

2.2. RESULTS FROM ULYSSES

Observations of different ionization stages present in the solar wind give information on the temperature of the coronal layers where the ions considered become frozen in the plasma. The coronal temperatures obtained in this way with the Solar Wind Ion Composition Spectrometer (SWIC-S) on board the Ulysses spacecraft (based on the O^{7+}/O^{6+} ratio) have been compared by Geiss et al. (1995) with the wind speed, and an inverse correlation was found (Figure 2). The association between high speed and low temperature is not surprising, since the source regions of the fast wind (coronal holes) are known to be the coolest coronal regions, but the found correlation also indicates that the slow wind comes from hotter regions, which is an important result.

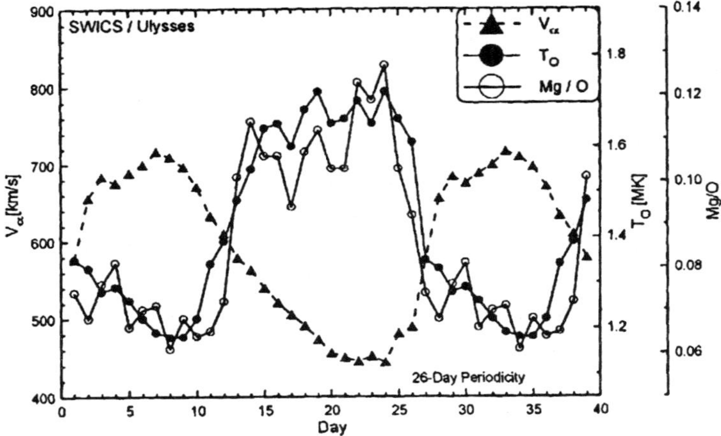

Figure 3. Superposed epoch plot showing the variations, during the solar rotation period, of wind speed, freezing-in O^{+7}/O^{+6} temperature and Mg/O ratio. (From Geiss et al., 1995.)

Another important information comes from the correlation between the coronal temperature, obtained as indicated above, and the abundance ratio Mg/O (Geiss et al., 1995), which is a good indicator of the FIP effect (the effect which makes overabundant, in the solar wind, the elements with potential of first ionization below $\sim$ 10 e.v.).

Figure 3 shows the coronal temperature, the Mg/O ratio and the wind speed, measured by the SWICS instrument, plotted, as a function of time, with the use of a superposed epoch tecnique to accumulate the effects of successive rotations. The anticorrelation between temperature and wind speed is apparent, as is the correlation between temperature and strength of the FIP effect. Figure 3 shows also that the borders between the region of origin of the slow wind and that of the fast wind are very narrow.

Since the FIP effect is believed to be indicative of the chromospheric conditions, this implies that the slow/fast difference originates in layers deep rooted in the solar atmosphere.

These Ulysses results fit in the picture described in the previous section, with the slow wind originating from the coronal hole borders/streamer flanks.

2.3. OTHER PROPOSED SLOW WIND SOURCES

Active regions have been proposed, recently, as possible sources of solar wind by Uchida et al. (1992), on the basis of the observations of the soft X-ray telescope on board the Yohkoh satellite, which suggest a continuous expansion of the coronal loops.

Another proposed possibility is a non-stationary solar wind, with a principally patchy nature, made up from plasmoids being pushed upwards by the magnetic pressure of the ambient corona. The evidence for such a scenario comes largely from the observation of explosive events near the solar limb; some additional evidence comes from the observation that the quiet solar wind is highly structured (Thieme et al., 1990, who found, however, that also the fast wind is structured, particularly inside 0.5 a.u.). See also, on this point, Neugebauer (1981). It is worth noting that this theory is in line with the greater variability of the physical parameters observed in the slow wind, compared to the fast one (section 2.1).

3. The fast/slow difference

It is well known that the steady models of solar wind are not able to reproduce the observations if 'ad hoc' energy deposition terms are not included in the equations, both for the slow and the fast solar wind. To attribute to differences in these terms the difference in outflow speed has, however, little meaning, since observational evidence of such an energy deposition is laking.

More significant have been studies on the effect of the geometry of the flow, in particular of the rapid expansion of the cross-sectional area of a flux tube. These studies have been initiated by Kopp and Holzer (1976), who found that an expansion of the flow such that the cross-sectional area increases with the heliocentric distance r more than proportionally to r^2, has an influence on the topology of the solutions $v(r)$ of the solar wind problem. The influence is small, and limited to the inner corona, if the expansion factor grows slowly, but becomes profound if it grows rapidly. (The expansion factor f is defined by comparing an infinitesimal cross section $d\sigma$ of a flux tube at r with the cross section $d\sigma_o$ of the same tube at $r = r_\odot$, $f(r) = d\sigma\, r_\odot^2 / d\sigma_o\, r^2$.) The effect of a rapidly growing expansion factor is that of decreasing the heliocentric distance of the critical point and increasing the flow speed both in the inner corona and in the interplanetary space. The observation of a rapid increase of the expansion factor in a polar hole (Munro and Jakson, 1977), therefore, presumably, in a fast wind flux tube, lent support to this theory.

The connection between wind speed and expansion factor in the low corona has been studied, statistically, by Wang and Sheeley in a series of papers (Wang and Sheeley, 1990; 1991; 1994; Sheeley et al., 1991). Wang and Sheeley (1990) analyse the expansion factor of different flux tubes within coronal holes, by calculating the coronal magnetic field with the source surface method, and notice that the individual expansion factors can be very different from the global one; it is improper, therefore, according to Wang and Sheeley, to use the Munro and Jackson result to infer an association between large expansion factors and fast solar wind. On the contrary, Wang and Sheeley find an inverse correlation between the expansion factor and the solar wind speed at 1 a.u., the former being calculated for the infinitesimal flux tube intersecting the source surface (located at $r_s = 2.5r_\odot$) at the Earth projected position.

To reconcile the results of Wang and Sheeley with those of the theoretical models indicated above, one needs to assume a dependence of the boundary conditions or of the free parameters of the models on the expansion factor. Indeed, Wang and Sheeley concluded (1991) that they could interpret their results assuming, at the coronal base, energy flux roughly independent of the expansion factor and mass flux increasing with it. To get high speed streams, the energy flux at the coronal base must be dominated by non-thermal inputs. If these are due to Alfvèn waves, the consequence of the Wang and Sheeley (1991) conditions is an inverse correlation between magnetic field strength and wave amplitude, which is rather surprising.

If one considers the individual magnetic flux tubes inside a magnetically open region, it appears that those which suffer the larger expansion are those at the border of the region, where the field lines bend to follow the slope of the adjacent streamer, or those from small isolated holes. On the contrary, according to Wang and Sheeley (1994), the flux tubes which suffer the smallest expansion are 'rooted along the facing edges of a polar hole and like polarity holes at lower latitudes', or rooted 'around gaps consisting of very weak or closed field within the polar hole itself'. This brought Wang and Sheeley to predict that the fastest solar wind originates from these regions, the wind from the center of the polar hole, although falling in the 'fast wind' category, being not so fast. However, when direct measurements of the solar wind velocity at high heliolatitudes were made with the instruments on board the Ulysses mission, the prediction of this theory, that the maximum speed would be encountered at intermediate latitudes, rather than above the poles, has not been confirmed.

In fact, when the Ulysses spacecraft began to move out of the ecliptic plane towards higher and higher heliolatitudes, in the southern hemisphere, the wind speed pattern began to change, with an increase of the speed of the wind encountered between two high speed streams, until the speed remained stationary at the fast stream value (Phillips et al., 1995; Figure 4). There is no indication, in the Ulysses data, of a lower speed at the southern pole.

Accordingly, the effect of the expansion factor variation on the wind speed is unclear, and the conclusions of Wang and Sheeley (1991) concerning the energy input at the coronal base, quoted above, are weakened, and

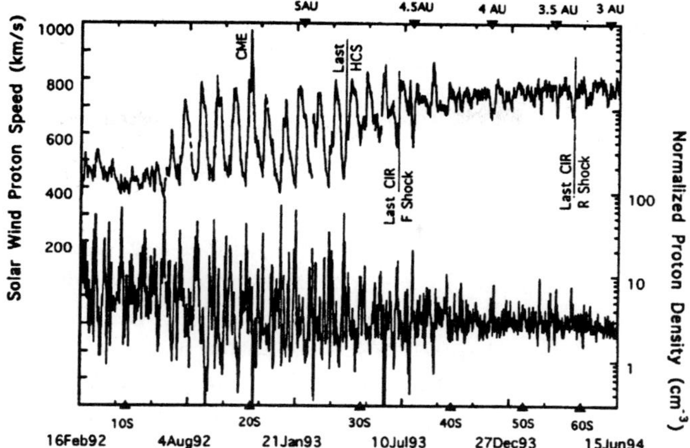

Figure 4. Solar wind speed and density observed by the Ulysses solar wind plasma experiment from 16 Feb. 1992 to 15 Jun. 1994 (from Phillips et al., 1995).

thus the problem of the cause of the speed differenece between slow and fast wind is completely open.

4. The helium concentration

The conclusion of Bame et al. (1977), that the helium content is smaller in the slow wind than in the fast one, described in section 2.1, has been confirmed by Schwenn (1983), as a result of an analysis of the HELIOS data taken between 1974 and 1982. The average values found for this quantity by Schwenn are, respectively, 2.5 for the slow wind, and 3.6 for the fast wind.

This helium deficiency in the slow wind is difficult to understand, since the theoretical analysis (Geiss et al., 1970) has shown that the He^{++} ions are dragged into the solar wind through Coulomb collisions with the protons, and we know that the proton flux has a small difference between fast and slow wind, and, in any case, is larger, on the average, in the slow wind (Schwenn, 1983).

A theory developped by Bürgi (1992) offers a rather simple explanation of this phenomenon. It assumes a large expansion factor at the interface between magnetically open and closed regions near the streamer cusp, as magnetic field calculations suggest. This makes the proton flux to decrease locally, which reduces the Coulomb drag, so that a much smaller number of α particles is carried in the solar wind close to the top of a streamer than in the adjacent coronal hole. Therefore the minimum of the helium abundance at the sector boundary in the interplanetary space.

This theory implies a very low helium concentration in the low corona, close to the stremer cusp, which should be observable.

5. Recent EUV Observations

At this point it is clear that the informations mainly needed to understand the slow solar wind, are those which concern the inner corona. For this part of the solar atmosphere we have UV and X-ray data (the only ones able to give information on the temperature and, possibly, on the outflow speed) at heliocentric distances lower than $\sim 1.1 r_{\odot}$, but very little of them at greater heights. Some observations in the UV, however, have been made recently with the coronagraph-spectrometer of the Spartan 201 mission, flown in April 1993 and in September 1994. The instrument has observed the Ly-α profile at several heights from $r = 1.5 r_{\odot}$ to $r = 3.5 r_{\odot}$, both in a streamer (Strachan et al., 1994) and in polar coronal holes (Kohl et al., 1994, 1995). Since the observed positions in the streamer correspond to heights large enough to be above the magnetically closed region, they should be characteristic of the slow solar wind. Hence, the Spartan 201 data concern both the slow and the fast solar wind. From the April 1993 data the authors report a not very variable temperature for the streamer (values ranging from 2.2 to $3.2 \times 10^6 K$, with uncertainty (1σ) of $0.25 \times 10^6 K$), while for the coronal holes they are not able to fit the observed profiles with a single Gaussian, the wings being too wide. They can fit the profiles with two Gaussian, very different in width, corresponding to two Maxwellian populations of neutral hydrogen atoms having quite different temperatures ($1.6 \times 10^6 K$ and 7.6 $\times 10^6 K$ at $r = 1.8 r_{\odot}$ in the south polar hole; 5.8 $\times 10^5 K$ and 3.0 $\times 10^6 K$ at $r = 2.13 r_{\odot}$ in the north polar hole, Figure 5).

The authors discuss some explanations for these non-Maxwellian profiles, each having considerable difficulties. It must be remarked that these kinetic temperatures concern the neutrals and, presumably, the protons, since the neutral/proton transition time is short enough in the solar corona. They are not, however, electron temperatures, so that they do not affect the ionization balance and, thus, the amount of neutrals. Electron temperatures can be obtained through the measurement of ion ratios in the solar wind (section 2.2). Such determinations, based on data from the SWICS instrument abroad Ulysses, and on a coronal model, are in progress (Cohen, 1995). Together with the proton temperatures from Spartan, they will establish a very important basis for solar wind models.

Concerning models based on very recent data, we mention Habbal et al. (1995) calculations, which use, as constraints, electron density data obtained from a white light coronagraph flown together with the UV instrument on Spartan 201, and from the Mauna Loa coronagraph. Supplement-

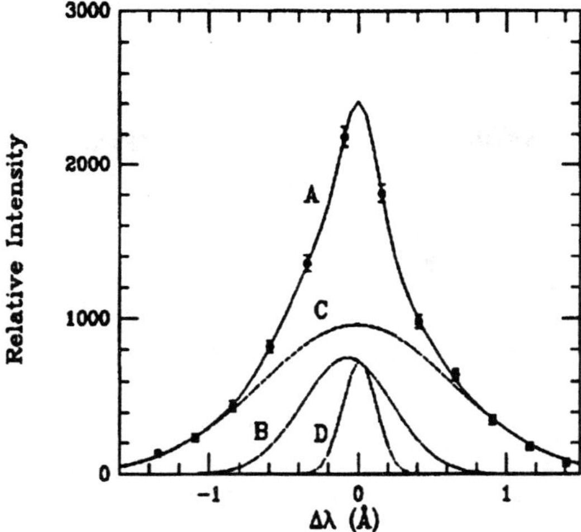

Figure 5. Profile of the HI Ly_α in the north polar hole at $r = 2.13 r_\odot$, from Spartan 201 (April 1993): curve A is a fit to the data, D is the geocoronal contribution, B and C are two Gaussian components such that A = B+C+D. (From Kohl et al., 1994.)

ing these density data with interplanetary measurements of bulk speed and mass flux, and assuming some unspecified source of energy deposition in the corona, Habbal et al. (1995) calculated a model able to fit the empirical constraints.

An interesting result of the model is a proton temperature twice as large as the electron temperature in the inner corona, reaching a peak of $2 \times 10^6 K$ at 2 $r_\odot$. Note that this work is not based on the Ly-α data discussed above. It is remarkable that similar results, for what concern the temperatures, had been obtained by Noci and Porri (1983), in their analysis of the Ly-α observations (quiet region) made with a rocket launch.

These models, although characterized by a proton temperature not as large as that indicated by the Spartan UV data, go however in the right direction.

For what concerns the problem of the solar wind source regions, as well as that of its acceleration, the Spartan observations are very important. Although not yet clearly interpreted, the difference, in the hydrogen velocity distribution, between streamers and coronal holes, and also the very high kinetic temperatures in the latter, are important constraint for acceleration theories.

In conclusion, the Spartan results are very promising: they reinforce the expectation that the yield of UV and visible data concerning the extended corona, and, in particular, the streamer/coronal hole interface, from the next solar space mission, SOHO, will permit a deeper insight in the problem of the origin of the solar wind.

References

Bame, S.J., Asbridge, J.R., Feldman, W.C., and Gosling, J.T.: 1977, *J. Geophys. Res.* **82**, 1487.

Bell, B., and Noci, G.: 1973, *Bull. AAS* **5**, 269.

Bell, B., and Noci, G.: 1976, *J. Geophys. Res.* **81**, 4508.

Borrini, G., Gosling, J.T., Bame, S.J., Feldman, W.C., and Wilcox, J.M.: 1981, *J. Geophys. Res.* **86**, 4565.

Bürgi, A.: 1992, in *'Solar Wind Seven'*, E. Marsch and R. Schwenn eds., Pergamon Press, p. 333.

Cohen, C.: 1995, private communication.

Geiss, J., Hirt, P., and Leutwyler, H.: 1970, *Solar Phys.* **12**, 458.

Geiss, J., Gloeckler, G., and von Steiger, R.: 1995, *Space Sci. Rev.* **72**, 49.

Gosling, J.T., Borrini, G., Asbridge, J.R., Bame, S.J., Feldman, W.C., and Hansen, R.T.: 1981, it J. Geophys. Res. **86**, 5438.

Habbal, S.R., Esser, R., Guhathakurta, M, and Fisher, R.R.: 1995, *Geophys. Res. Let.* **22**, 1465.

Hundhausen, A.J.: 1977, in *'Coronal Holes and High Speed Wind Streams'*, J.B. Zirker ed., Colorado Ass. Univ. Press, Boulder, p. 225.

Kohl, J.L., Gardner, L.D., Strachan, L., Hassler, D.M.: 1994, *Space Sci. Rev.* **70**, 253.

Kohl, J.L., Gardner, L.D., Strachan, L., Fisher, R., and Guhathakurta, M.: 1995, *Space Sci. Rev.* **72**, 29.

Kopp, R.A., and Holzer, T.E.: 1976, *Solar Phys.* **49**, 43.

Krieger, A.S., Timothy, A.F., and Roelof, E.C: 1973, *Solar Phys.* **29**, 505.

Munro, R.H., and Jackson, B.V.: 1977, *Astropys. J.* **213**, 874.

Neugebauer, M.: 1981, *Fund. Cosmic Phys.* **7**, 131.

Neugebauer, M., and Snyder, C.W.: 1966, *J. Geophys. Res.* **71**, 4469.

Neupert, W.M., and Pizzo, V.: 1974, *J. Geophys. Res.* **79**, 3701.

Noci, G.: 1973, *Solar Phys.* **28**, 403.

Noci, G., and Porri, A.: 1983, *IAGA 18th Gen. Ass., Hamburg*, paper 4L.04.

Phillips, J.L., Bame, S.J., Feldman, W.C., Gosling, J.T., Hammond, C.M., McComas, D.J., Goldstein, B.E., and Neugebauer, M.: 1995, *Adv. Space Res.* **16** (9), 85.

Schwenn, R.: 1983, in *'Solar Wind Five'* (NASA Conf. Publ 2280), M. Neugebauer ed., p. 489.

Sheeley, N.R. jr., Swanson, E.T., and Wang, Y.-M.: 1991, *J. Geophys. Res.* **96**, 13861.

Strachan, L, Gardner, L.D., Hassler, D.M., and Kohl, J.L.: 1994, *Space Sci. Rev.* **70**, 263.

Thieme, K.M., Marsch, E., and Schwenn, R.: 1990, *Ann. Geophysicae* **8** (11), 713.

Uchida, Y., McAllister, A., Strong, K.T., Ogawara, Y., Shimizu, T., Matsumoto, R., Hudson, H.S.: 1992, *Publ. Astron. Soc. Japan* **44**, L155.

Wang, Y.-M., and Sheeley, N.R. jr: 1990, *Astrophys. J.* **355**, 726.

Wang, Y.-M., and Sheeley, N.R. jr: 1991, *Astrophys. J.* **372**, L45.

Wang, Y.-M., and Sheeley, N.R. jr: 1994, *J. Geophys. Res.* **99**, 6597.

Zirker, J.B.: 1977, editor of *'Coronal Holes and High Speed Wind Streams'*, Colorado Ass. Univ. Press.

MECHANISMS OF SOLAR (AND STELLAR) MASS LOSS

R. ROSNER
Department of Astronomy and Astrophysics,
University of Chicago
Chicago, IL 60637, USA

1. Introduction

The fundamental theory for mass loss from late–type stars such as our Sun was developed by Gene Parker in the 1950's (Parker 1958), and has been largely understood since then; refinements since then have largely focused on physically more realistic fluid equations (which take into account, for example, the effects of multi–component fluids and corrections to energy transport by classical thermal conduction) and additional mechanisms for driving the wind (e.g., heating and momentum deposition by magnetohydrodynamic waves). Nevertheless, if one were to ask most professional solar wind theorists what the generally accepted model for the solar wind is, the answer is most likely a shrug of the shoulders: It has not been possible to develop a detailed theory for solar wind mass loss which takes into full account all of the observed complexities of this outflow; and the status of wind research for other late–type stars (meaning late–type giants and supergiants) is not in recognizably better shape. In this paper, I will attempt to focus on this embarassment, and will try to explain why it is that this problem has turned out to be so remarkably difficult. Along the way, I will provide a brief overview of the basic theory, and of the observational perspectives of a few years ago as contrasted with what is now known from the most recent data provided by the *Ulysses* spacecraft. Based on these discussions, I will then focus on a highly selective overview of what is to be explained, and the current status of theory in these regards.

Because of the brevity of available space, I will not be able to discuss a number of fascinating topics (which I will allude to only in passing, despite their importance): mass loss via transients (viz., CME events); elemental and isotopic "fractionation" (as measured by, for example, the He/H and

Y. Uchida et al. (eds.),
Magnetodynamic Phenomena in the Solar Atmosphere – Prototypes of Stellar Magnetic Activity, 107–114.

He^3/He^4 ratios); details of more complex physical models (which incorporate effects such as multi–component fluids and allow for corrections to the standard Navier–Stokes equations via, for example, higher–order moment equations). Instead, my starting point will be the by–now "classical" wind theory first developed by Gene Parker; and part of my intention is to "set the stage" for further discussions in the following papers.

2. The "Text Book" Picture of the Solar Wind

The classic mathematical and physical description of the solar wind was provided by Parker (1958), who showed that if one starts with the single–fluid conservation equations of mass and momentum for a fully–ionized hydrogen gas,

$$\frac{1}{r^2}\frac{d}{dr}\left(\rho u r^2\right) = 0\,, \tag{1}$$

$$u\frac{du}{dr} = \frac{1}{\rho}\frac{dp}{dr} - \frac{GM_\odot}{r^2}\,, \tag{2}$$

together with the very simplified energy equation $T = T_o =$ constant and the equation of state $p = 2nk_BT$ (where all quantities have their customary meaning, and we have assumed $n \sim n_p \sim n_e$), one then obtains with some algebra the classical wind equation

$$u\frac{du}{dr}\left(u^2 - \frac{2k_BT}{m}\right) = \frac{4k_BT}{mr} - \frac{GM_\odot}{r^2}\,. \tag{3}$$

The solutions to this equation have a number of important properties:

1. There is a critical point in the M–r solution plane, where the Mach number $M \equiv u/c_s$ equals unity ($c_s \equiv (2k_BT/m)^{1/2}$ is the isothermal sound speed); this critical point is located at $r = r_c \equiv GM_\odot m/4k_BT$. In this simple case, there is one transonic solution which starts at low wind speed at the solar surface; this solution's pressure variation is given by the analytical expression

$$p/p_0 = \exp\left(-\frac{GM_\odot m}{2k_BTR_\odot}\left(1 - \frac{R_\odot}{r}\right) - \frac{m}{4k_BT}\left(u^2 - u_o^2\right)\right)\,, \tag{4}$$

with the asymptotic value (as $r \to \infty$) $p/p_0 = 0$. (Quantities with subscript "o" refer to the base of the solar corona.)

2. The "breeze" solutions (which remain subsonic for all r) have instead the asymptotic pressure variation

$$p/p_0 = \exp\left(-\frac{GM_\odot m}{2k_BTR_\odot}\left(1 - \frac{R_\odot}{r}\right)\right)\,, \tag{5}$$

with the asymptotic value (as $r \to \infty$) $p/p_0 = \exp\left[-\frac{GM_\odot m}{2k_B T R_\odot}\right] = \text{constant}$.

3. Finally, if we adopt the above transonic branch as the only physically–meaningful solution, then one immediately discovers that the wind characteristics at 1 A.U. depend sensitively upon the physical conditions at the inner coronal base. For example, one has

$$nu|_{r=1\text{ A.U.}} \sim 0.4(n_0/10^8\text{cm}^{-3})(T_0/10^6\text{T})\exp\left[11.55\left[1-(T_0/10^6\text{K})^{-1}\right]\right] \tag{6}$$

One could ask if these results change significantly if additional physics is inserted (especially physics which is known *a priori* to be operative in the solar wind): For example, one could change the energy equation (3) so that both direct (*in situ*) mechanical heating and thermal conduction are accounted for. It turns out that increasing the sophistication of the models in this way does not change the basic fact that the mass flux at 1 A.U. is a sensitive function of coronal base conditions. Thus, it is very difficult to change the result that, for thermally–driven models, $u_E \sim 300$ km s^{-1} for $T_0 \sim 2\times10^6$ K.

One might next ask whether the above results of the "classical" wind theory are of any relevance, given the fact that magnetic fields are known to play an important role in solar wind dynamics; in particular, it has been long argued that Alfvén waves play an important role in determining solar wind acceleration (e.g., Hollweg 1978). In other words, why bother with such simple-minded models if we know that much more complex physical processes left out of this above description must play important roles? My answer is as follows: The wind equations form a highly non–linear system of partial differential equations, whose solutions depend in non–trivial ways on the parameters defining the solution space. In particular, it is essential to understand exactly what observational challenges one is trying to resolve by adding new physical effects to one's wind theory; and the only way to do this is to have a complete understanding of the limitations of the simpler models, i.e., an understanding of exactly what observations the simpler models cannot hope to describe adequately.

3. Will the "Real" Solar Wind Please Stand Up?

Give the above discussion, it is evident that a prerequisite for adequate wind modeling is a clear understanding of exactly what the physical conditions of the solar wind are, and how they relate to the coronal sources. Until very recently, however, it was not trivial to determine these conditions in a way that allows one to unequivocally associate them with locales in the inner solar corona which can be identified as the corresponding source regions. Instead, examination of the wind speed, wind density, and wind temperature at 1 A.U. within the ecliptic typically shows rapid variations over a large range of time scales, even in cases in which one can reasonably infer a connection to a clearly–identifiable feature in the inner corona, viz., a low–latitude coronal hole. For this reason, canonical mean reference values for a coronal hole–associated wind stream quoted by, for example, Hollweg (1986), $n_E = 7.8$ cm^{-3}; $T_E = 10^5$ K; $v_E = 536$ km s^{-1} (which were based on observations reported by Feldman et al. 1976) are not easily reconciled with observations such as were reported by Couzens & King (1986), whose data for a roughly month–long period in 1979 shows wide excursions in wind speed, density, and temperature about the above "canonical" values.

This state of affairs turns out to have been largely a consequence of where the measurements were taken, e.g., in the ecliptic, something that was widely suspected but could not be directly confirmed until one had explicit plasma measurements out of the ecliptic: With the new data from *Ulysses*, we now do have a much better understanding of the "real" solar wind conditions. In my estimation the most beautiful single illustration of the power of new data to clarify previous confusions is Figure 1 of the *Ulysses* Solar Wind Ion Composition Spectrometer (SWICS) team report (Geiss et al. 1995). Among other quantities, SWICS measured the speed of Helium ions (v_α) and the freeze–in temperature, T_O, for Oxygen (determined from measurements of the O^{7+}/O^{6+} ratio, under the assumption of local equilibrium between electron impact ionization and recombination). Geiss et al. show (see also Phillips et al. 1995):

(a) The "slow" solar wind is highly intermittent, with v_α generally in the range $4 - 500$ km s^{-1}; T_O in this domain fluctuates remarkably, but is generally in the range of $1.5 - 2 \times 10^6$ K. (Measurements were taken at radii ranging from $\sim$ 2 to 5 A.U.)

(b) The polar coronal hole–associated wind (i.e., the polar high speed wind stream) shows, in contrast, remarkably little variability; v_α is approximately 800 km s^{-1}, with fluctuation of order only 10%, and T_O is just above 10^6 K, with fluctuations of that are even smaller.

(c) The transition between the "slow" wind and the high speed solar wind stream shows extremely large variability, which is presumably associated with the complex spatial geometry of the interface region. Thus, as *Ulysses* descended from the ecliptic to solar latitudes below -10 degrees, it began to encounter the high speed wind stream; but because the outer stream boundary is not axially symmetric about the solar rotation axis, this initial encounter is modulated by the solar rotation. Indeed, the SWICS Helium wind speed measurements show a clear solar rotational modulation during this transition phase.

What one can conclude from all this, beyond the details of the measurements, is the crucial fact that high speed solar wind streams are the laboratory for exploring stellar wind theories: Contrary to what might have been thought, these wind streams are remarkably homogeneous structures, with physical conditions that show little evidence for fluctuations. As a consequence, the stream's mean physical properties can be meaningfully interpreted, something that cannot be said for the "slow" solar wind (whose mean properties may well be irrelevant for modeling purposes). For this reason, I will whenceforth focus solely on the high speed wind streams, and leave the explication of the "slow" solar wind for another time.

4. What Needs to be Explained I: The Solar Wind

The *Ulysses* observations, taken together with earlier observations, allow us to draw the following conclusions regarding the outflow in high speed solar wind streams:

Constraint 1. The simple (purely thermally–driven) models lead to predicted wind speeds far from the Sun which are much smaller than is observed: v_∞(observed) ~ 800 km s^{-1} >> v_∞("thermal") ~ 300 km s^{-1}.
Constraint 2. The inferred total wind mass loss is small; averaged over 4π steradians, the SWICS measurements imply that $\dot{M} \sim 10^{-14}\ M_\odot$ yr^{-1}.
Constraint 3. The mass flux is relatively constant, e.g., $\Delta(nv)/nv < 0.1$. The total mass loss rate is therefore apparently largely controlled by the total solid angle subtended by the high speed wind stream at large radii.

Constraint 4. The mass flux is very homogeneous. The absence of significant fluctuations in wind speed, density, or temperature argues strongly against impulsive wind acceleration processes, and strongly for processes associated with spatially fairly uniformly distributed momentum and energy deposition.

Constraint 5. There is considerable evidence for rapid flow acceleration very close to the solar surface in the polar solar wind, based both on observations using resonantly–scattered H I Lyman–α (e.g., Withbroe et al. 1982) and radio–scattering measurements (e.g., Grall et al. 1996); these observations suggest that most of the flow acceleration occurs well within $10R_\odot$ of the solar surface.

5. What Needs to be Explained II: Late-type Stars

Only two general classes of normal stars other than the Sun are known — from direct observations — to suffer substantial mass loss, namely OB stars and (evolved) red giants and supergiants; and only in the latter case, can one argue plausibly that theoretical ideas developed in the solar context should be applicable to these more exotic stars. To the extent that solar wind theoretical ideas ought to be applicable to these stars, observations of these stars will also serve to constrain wind theory. Cassinelli & MacGregor (1984) have published a thorough review of the observations for such stars; and Holzer & MacGregor's (1985) review nicely summarizes the relevant theory from the present perspective; I will therefore considerably abbreviate my discussion, and focus only on two key points:

Constraint 6. Purely thermally–driven models for winds from red giants fail totally in accounting for the observations. The fundamental problem is that wind base temperatures that are required by such models in order to drive the winds to the observed asymptotic wind speed, and also provide the observed mass loss rates, also lead to heating requirements (and implied radiative loss rates) directly contradicted by observations. Put differently: given the observed temperature structure of the base atmosphere of such stars, thermally–driven wind models cannot attain either the observed wind speeds, or the observed mass loss rates. What seems to be required is an additional mechanism for accelerating the wind without further heating the wind.

Constraint 7. The inferred total wind mass loss is relatively large; averaged over 4π steradians, observations imply that $\dot{M} \sim 10^{-8}\ M_{\odot}\ \mathrm{yr}^{-1}$.

6. Summary and Conclusions: The Current Theoretical Dilemma

The observations discussed above, when taken together with current theoretical solar wind models, strongly imply that some process — akin to the Alfvén wave acceleration process first discussed by Belcher (1971), Alazraki & Couturier (1971), and Hollweg (1973, 1978) — is required in order to account for the observed wind speeds. It is particularly striking that this conclusion follows, independently, from both solar wind high speed stream and red giant wind observations. Such models are attractive for a number of distinct reasons: First, one observes Alfvén waves directly in the solar wind; and although it has not been definitively established that the observed waves have their origin at the solar surface, it is nevertheless reassuring that the required wave modes in fact do exist. Second, Alfvén waves can deposit momentum without significant local plasma heating; this property is crucial for meeting the observational constraints 1. and 6., without violating the constraints on wind temperature. Third, it is possible to adjust the wave fluxes so that the observed mass loss rates for both the Sun and red giant winds are obtained.

What then are the remaining difficulties? What I am refering to here are difficulties that remain at the level of detail of the present description, and not the more sophisticated details such as the variation of elemental and isotopic abundance ratios with heliocentric distance, detailed electron, proton, Helium, and other temperature variations, etc. The first such conundrum, discussed in detail by Leer, Holzer, & Fla (1982), is the so-called "fine tuning problem": It turns out that the simple Alfvén wave-driven models which resolve Constraint 6 (e.g., Hartman & MacGregor 1980) require remarkable fine-tuning of the model parameters (in particular, tuning of the Alfvén wave damping length) in order to meet the observed result that v_∞(observed) $\sim v_{\rm escape}/2$ (in contrast to the solar case, for which v_∞(observed) $\sim (1-2) \times v_{\rm escape}$).

The second such puzzle is that for plausible Alfvén wave amplitudes, it is not possible to find solutions to the simple wave-driven wind equations that reach the asymptotic high speed wind velocity within $10R_\odot$; thus, models such as those of Leer & Sandbaek (1991) obtain flow speeds of order only 300 km s^{-1} at $10R_\odot$ for wave amplitudes $\delta v \sim 20$ km s^{-1}.

In order to address these two problems, it is important to note that most wave-driven wind models are based on the simple WKB approximation, which uses a momemtum conservation equation of the form

$$u\frac{d}{dr}u = -\frac{1}{\rho}\frac{dp}{dr} - \frac{GM_\odot}{r^2} + D\,, \tag{2'}$$

where $D = -\frac{1}{\rho}\frac{d}{dr}\left(\frac{\langle\delta B^2\rangle}{8\pi}\right)$ and $\langle\delta B^2\rangle \sim \exp(-r/L)$. (This formalism is directly applicable only to the propagation of linear torsional Alfvén waves; cf. Ferraro & Plumpton 1958.) The obvious question is: What about non-WKB waves? A number of authors, including An et al. (1990), Velli (1993, 1994), MacGregor & Charbonneau (1994), Lou & Rosner (1994), and Krogulec et al. (1994), have recently focused on this issue. The basic conclusion emerging from this work is, first, that in a certain well-defined sense, there may not be <u>any</u> WKB waves: That is, virtually all waves with periods longer than a few minutes show some reflection (due to gradients in the background Alfvén wave speed) somewhere in the flow. In particular, the notion of a single "damping length" for parametrizing the spatial variation of such waves is clearly far too simplistic; it is likely (though not as yet established) that the "fine tuning problem" can be avoided by taking more sophisticated account of the physics underlying Alfvén wave damping and reflection. A more speculative possibility, discussed in these Proceedings (Axford 1996), is that the inner solar wind is dominated by very high frequency waves, whose short damping lengths in the inner corona may lead to the required rapid flow acceleration. It would be very exciting if *SOHO*, which will start to return data on the solar corona and wind in 1996, can provide additional observational constraints to resolve these remaining problems.

Acknowledgements

Much of the work reported in this paper was supported by NASA through the Space Physics Theory Program. I would also like to thank my many collaborators in this work, including Y.-Q. Lou, R. Moore, Z. Musielak, S. Orlando, G. Peres, and S. Suess, for the many enjoyable discussions and arguments that have molded my views on this subject. I would also like to acknowledge a number of stimulating discussions with E. Leer and T. Holzer; and I would like to thank the organizers of this IAU Symposium for the opportunity to participate in the active scientific discussions at this meeting.

References

Alazraki, G., & Couturier, P. (1971) *A&A*, **13**, 380.
An, C.H., Suess, S.T., Moore, R.L., & Musielak, Z.E. (1990) *ApJ*, **350**, 309.
Axford, W.I. (1996), these proceedings.
Belcher, J.W. (1971) *ApJ*, **168**, 509.
Ferraro, V.C.A., & Plumpton, C. (1958) *ApJ*, **127**, 459.
Geiss, J., Gloeckler, G., von Steiger, R., Balsiger, H., Fisk, L.A., Galvin, A.B., Ipavich, F.M., Livi, S., McKenzie, J.F., Ogilvie, K.W., & Wilken, B. (1995) *Science*, **268**, 1033-1036.
Grall, R.R., Coles, W.A., Klinglesmith, M.T., Breen, A.R., Williams, P.J.S., Markkanen, J., & Esser, R. (1996) *Nature*, **379**, 429-432.
Hartman, L., & MacGregor, K.B. (1980) *ApJ*, **242**, 260.
Hollweg, J.V. (1978) *Solar Phys.*, **56**, 305.
Holzer, T.E., Fla, T., & Leer, E. (1983) *ApJ*, **275**, 808.
Holzer, T.E., & MacGregor, K.B. (1985) *Mass Loss from Red Giants*, ed. M. Morris & B. Zuckerman (Dordrecht: D. Reidel), 229-255.
Krogulec, M., Musielak, Z.E., Suess, S.T., Nerney, S.F., & Moore, R.L. (1994) *JGR*, **99**, 23489-23501.
Leer, E., Holzer, T.E., & Fla, T. (1982) *Space Sci. Rev.*, **30**, 161.
Leer, E., & Sandbaek, O. (1991) *Adv. Space Res.*, **11**(1), 197-211.
Lou, Y.-Q., & Rosner, R. (1994) *ApJ*, **424**, 429.
MacGregor, K.B., & Charbonneau, P. (1994) *ApJ*, **430**, 387-398.
Parker, E.N. (1958) *ApJ*, **128**, 664.
Phillips, J.L., Bame, S.J., Feldman, W.C., Goldstein, B.E., Gosling, J.T., Hammond, C.M., McComas, D.J., Neugebauer, M., Scime, E.E., & Suess, S.T. (1995) *Science*, **268**, 1030-1033.
Withbroe, G.L., Kohl, J.L., Weiser, H., Noci, G., & Munro, R.H. (1982) *ApJ*, **254**, 361.
Velli, M. (1993) *A&A*, **270**, 304.
Velli, M. (1994) *Adv. Space Res.*, **14**, 123.

ACCELERATION OF THE HIGH SPEED SOLAR WIND

W.I. AXFORD AND J.F. MCKENZIE
Max Planck Institut für Aeronomie,
37191 Katlenburg-Lindau, Germany

Abstract. A theory is developed for the high speed solar wind based on a simple dissipation length characterization of wave heating of the coronal plasma close to the Sun. It is shown that solutions with the correct particle and energy fluxes and with a realistic magnetic field, match the requirements on the density at the base of the corona provided the dissipation length is relatively small ($\sim$ 0.25 - 0.5 solar radii). The significant features of these solutions are that the acceleration is rapid, with the sonic point at about 2 solar radii, and the maximum proton temperatures are high, namely $8 - 10 \times 10^6$ K, in agreement with some recent observations. Such efficient dissipation requires any Alfvén waves responsible to have frequencies in the range 0.01 Hz - 10 kHz. This has implications for the nature of the plasma and energy source in the chromospheric network.

Key words: stars: coronal, winds-sun: solar wind - corona: Alfv́en waves

The high speed solar wind, which is directly associated with coronal holes, is the basic equilibrium form of the solar wind (Bame, et.al. 1977). As such it must be accounted for in any steady-state theory. In contrast, the low speed wind appears to be associated with transient openings of closed field regions in the corona which give rise to inherently unsteady streams which are not in equilibrium with the coronal base (Axford 1977).

The most important (average) properties of the high speed solar wind are (Schwenn, 1990; Mariani and Neubauer, 1990):

1. the asymptotic speed is $V_\infty \sim 750-800$km/sec with small fluctuations;
2. the particle flux at 1 a.u. is $\sim 2 \times 10^8/\mathrm{cm}^2\mathrm{sec}$;
3. the radial magnetic field strength at 1 a.u. is $\sim$ 2.8 nT with the field being unipolar in each stream;
4. the average proton and electron temperatures at 1 a.u. are $\mathrm{T}_p \sim$ 200,000 K and $\mathrm{T}_e \sim$ 100,000 K respectively;

Y. Uchida et al. (eds.),
Magnetodynamic Phenomena in the Solar Atmosphere – Prototypes of Stellar Magnetic Activity, 115–122.

5. the electron distribution function contains a field-aligned beam which evidently originates in a region close to the Sun where Te does not much exceed $\sim 10^6$ K;
6. the proton temperature perpendicular to the magnetic field is greater than that parallel to the field ($T_p^\perp > T_p^\|$);
7. magnetic field fluctuations with periods > 200 sec correspond to Alfvén waves propagating away from the Sun with energy fluxes (at ~ 0.3 a.u.) about 1% of the total solar wind energy flux;
8. minor species, including helium, have roughly the same temperature per atomic mass as the protons ($T_i \sim AT_p$) and move faster by approximately the Alfvén speed ($V_i \sim V + V_A$);
9. the composition appears to be constant and with helium abundance $\sim$ 5%;
10. in agreement with the behaviour of coronal holes, the high speed wind is dominant during periods of low solar activity and occupies the whole heliosphere at solar latitudes greater than about 20 deg;
11. the wind is fully developed at ~ 0.3 a.u. and, according to recent interplanetary scintillation (IPS) observations, is also close to its terminal speed at $\sim$20 and even ~ 8 R_s (R_s=solar radius) (Coles, private communication, 1995).

It is difficult to make measurements in coronal holes at the base of high speed streams as a result of contamination from denser regions of plasma in the fore- and background. However the following results appear to be sound:

12. the electron densities are low ($n_o \sim 2 \times 10^7 - 10^8/\mathrm{cm}^3$) in comparison with coronal streamers (Koutchmy, 1977);
13. the electron temperature does not appear to exceed about 10^6 K (Habbal et. al. 1993).

From these results we may deduce that pressure gradients associated with electrons and Alfvén waves do not play a major role in accelerating the wind because they act too slowly and cannot lead to rapid acceleration close to the sun. Furthermore, because the protons are hotter than the electrons and because the minor species are strongly favoured relative to protons, it appears that comparatively high frequency waves are involved, allowing discrimination in terms of Z/A (i.e. ion cyclotron waves). At 0.3 a.u., ion cyclotron waves have periods less than the Ferraro-Plumptom period at the base of the corona (i.e. $\tau_o = 2H/V_A \sim 30 - 100\,\mathrm{sec}$, where H is the scale height at the base of the corona). Waves with periods longer than τ_o do not propagate according to the WKB approximation and are inefficient transmitters of energy from the coronal base (Ferraro and Plumpton, 1958).

There appears to be only one possibility to account for the high speed solar wind, namely that it is mainly a consequence of heating close to the Sun by the dissipation of waves with periods considerably less than τ_o. These should preferentially heat protons and, especially, minor species including helium. Given that the magnetic field strength may be 5 - 10 gauss at the base of coronal holes, the frequency range of the waves in question is $\sim$ 0.01 Hz - 10 kHz.

In the absence of a complete and self-consistent theory for the origin and dissipation of the waves we adopt a simple approach to the heating and acceleration of the solar wind plasma assuming a single damping length L. The wave energy flux F_w is given by

$$F_w = p_w(3V + 2V_A), \tag{1}$$

where $p_w = \frac{1}{2}\rho < \delta V^2 >$ is the wave pressure, ρ and V the fluid density and speed, $< \delta V^2 >$ the mean square wave amplitude and $V_A = B/\sqrt{(8\pi\rho)}$ with B the magnetic field strength. The dissipation function Q is assumed to have the form

$$Q = Q_o \exp[-(s - s_o)/L] \tag{2}$$

where s is measured along the magnetic field and $s_o = r_o = 1$ R_s. The area factor A is such that magnetic flux is conserved:

$$BA = B(s_o)A(s_o) = \text{constant}. \tag{3}$$

To proceed further we need a suitable magetic field model: For solar minimum conditions the field is essentially that of an axisymmetric current sheet in the equatorial plane with the Sun's field being a dipole (Gleeson and Axford, 1976). Thus for the polar field line, for which $s = r$,

$$B(r)/M = s/r^3 + 1/a(a + r)^2 \tag{4}$$

where M is the magnetic dipole moment of the sun. We choose the parameter a such that the open field lines emerge from latitudes $\Theta > 60$ degrees; thus a $\approx$ 3.96 Rs. The pattern of field lines is shown in Fig.1. Note that the model does not allow for the region occupied by the slow solar wind ($\Theta < 20$ degrees), however it should be reasonably accurate close to the Sun (r $<$ 20 R_s). We compensate for this minor defect in estimating the field strength at the base of the coronal hole by mapping only the radial component of the field at large distances above 20 degrees latitude into the north and south polar coronal holes defined as $\Theta > 60$ degrees. Thus the average field strength in the holes is taken to be 5.5 gauss.

There are three integrals of the motion, the magnetic flux (see (3)), the particle flux J and the energy flux E:

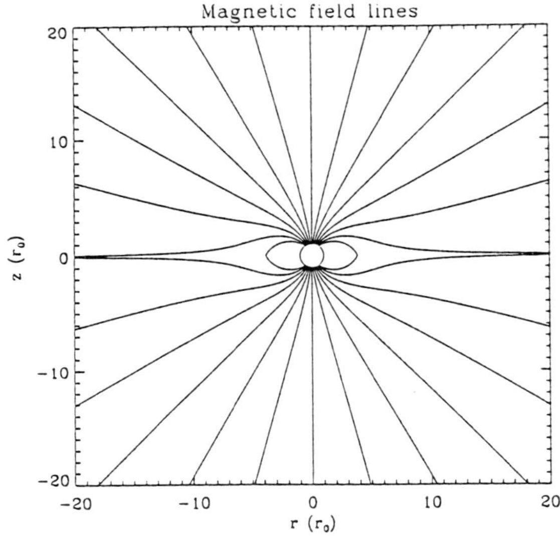

Figure 1. A model of the solar magnetic field appropriate to solar minimum conditions. The field lines emerging from latitudes above 60 degrees are open and form part of the high speed solar wind region. This model does not allow for the presence of an equatorial band of low speed solar wind and hence the relationship between the polar field strength and that at large distances must be modified accordingly.

There are three integrals of the motion, the magnetic flux (see (3)), the particle flux J and the energy flux E:

$$\rho V A = J \tag{5}$$

$$J[V^2/2 + 5p/2\rho - GM_o/r] + A(F_w + F_c) = E \tag{6}$$

where $p = p_i + p_e$ is the plasma pressure, GM_o/r the solar gravitational potential and $F_c = -\kappa_e dT_e/dr$ is the heat conduction flux. Proton heat conduction can be neglected since $\kappa_p$is presumed to be controlled by wave-particle interactions associated with the heating process and is small. However electron heat conduction controlled by Coulomb collisions may be the rule beyond the heating region; thus we assume

$$T_e = T_{oe} = 10^6 K \text{ for } r < 3R_s \tag{7a}$$

$$F_c = -\kappa_{oe} T_e^{\frac{5}{2}} dT_e/dr \text{ for } r > 3R_s, \tag{7b}$$

It can be shown that, if p_w/p is small at the base of the corona and the wave dissipation and plasma heating processes have the same length scales,

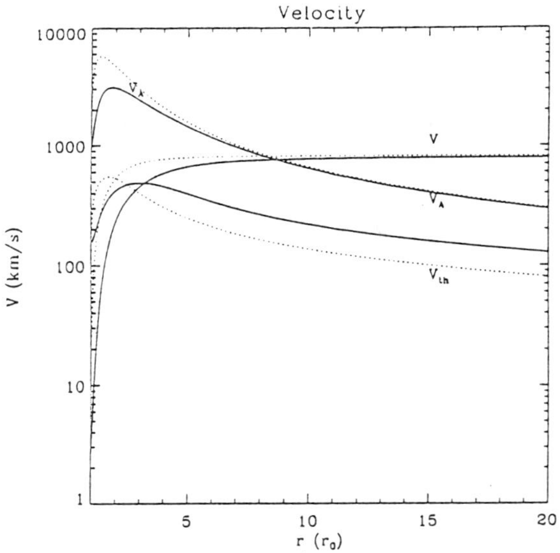

Figure 2. The variation of solar wind speed (V), Alfvén speed (VA) and thermal speed (V_{th}) in the range 1 - 20 R_s for L = 0,25 (dotted lines) and L = 0,5 R_s (full lines). The acceleration is rapid and the solar wind is essentially fully developed within a few solar radii of the Sun.

p_w can be negelected relative to p, which greatly simplifies the solution, (Axford and McKenzie, 1993).

From the observed quantities at large distances from the Sun, assuming that p tends to zero at infinity, we deduce that J/Am $\sim 2 \times 10^8/\text{cm}^2\text{sec}$ and $E \sim \text{nm}V_\infty^3/2 \sim 1.2$ ergs/cm^{2}sec at 1 a.u., where m is the mean ion mass (m $\sim 1.9 \times 10^{-24}$ gm, with 5 % Helium).

Finally there are the plasma momentum and wave energy exchange equations:

$$\rho dV/dr = -dp/dr - dp_w/dr - GM_o\rho/r^2 \tag{8}$$

$$(1/A)d(AF_w)/dr = Vdp_w/dr - Q \tag{9}$$

We neglect the terms involving pw and note that $AF_{wo}/J \sim V_\infty^2/2 + GM_or_o$, $AF_c = A_oF_{co}$ and

$$F_{wo} \sim Z_o \int_{ro}^{\infty} [\exp -(r - r_o)/L][A(r)/A(r_o)]dr \tag{10}$$

On choosing L a complete solution can be found which has a correct magnetic field geometry and correct particle flux and asymptotic flow speed. Each such solution will correspond to a certain value of no, the coronal base

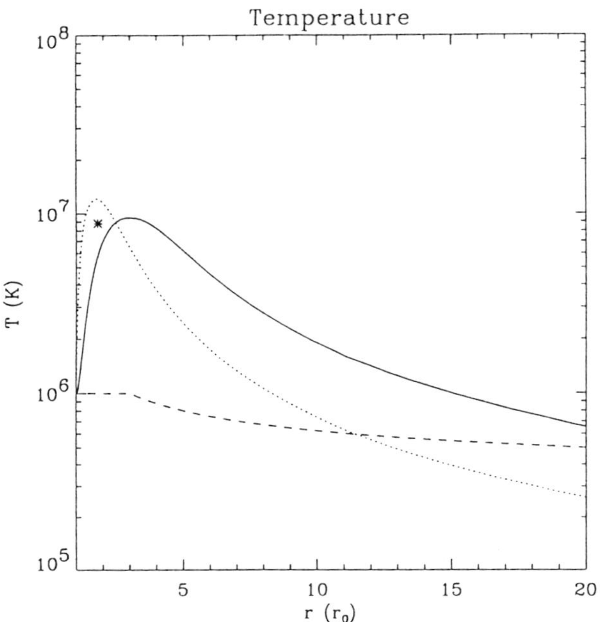

Figure 3. The variation of solar wind mean ion temperature in the range 1 - 20 R_s for L = 0.25 R_s (dotted line) and L = 0.5 R_s (full line). The electron temperature is given by the dashed line and is the same in each case. The mean ion temperature should be reduced by about 15 % to give the proton temperature as it allows for the presence of helium and heavier species with the same temperature per atomic mass as protons. As L decreases the mean ion temperature increases and the position of the maximum moves closer to the sun. The point shown corrresponds to the measurement by Kohl et.al. (1995), altered to correspond to mean ion temperature as described above.

density, and a particlular distribution of mean ion temperature, T(r). We find that for $2 \times 10^7 < n_o < 10^8/\mathrm{cm}^3$, we must have 0.25 R_s < L < 0.5 R_s and $T_{\mathrm{max}} \sim 9.5 - 12 \times 10^6$K. The sonic point occurs close to the Sun at about 2 R_s and the Alfvén point ($V = V_A$) occurs at r $\sim 8R_s$.

The significant features of these results are that with a single free parameter, namely the dissipation length L, we have been able to obtain excellent agreement with observations of the high speed solar wind. These include the correct speed and particle flux, a realistic magnetic field, rapid acceleration near the Sun associated with low coronal base densities and good agreement with densities measured in the outer corona (Figs. 2-4). The essential requirement is that the dissipation length is $\sim$ 0.25 - 0.5 R_s which implies that the wave frequencies must be higher than usually considered. The high Alfvén speed in the heated region places a constraint on the frequencies since one would expect that $L/< V_A >$ is an upper limit to the period of waves that can be dissipated within a distance L and this

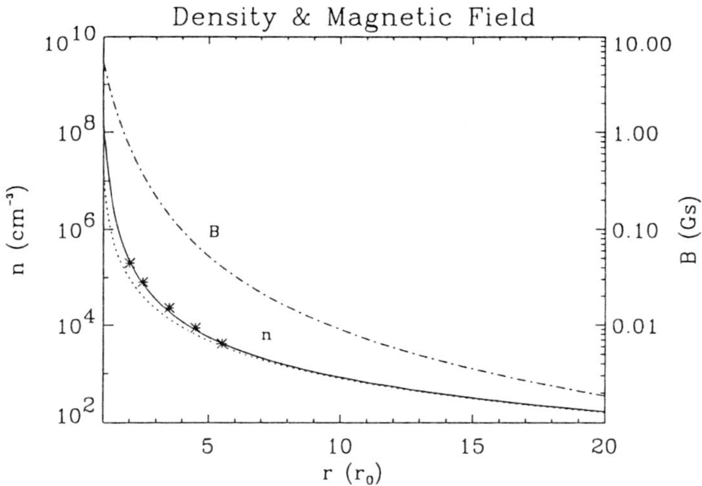

Figure 4. The variation of magnetic field (B) plasma number density (n) in the range 1 - 20 R_s for L = 0.25 R_s (dotted line) and 0.5 R_s (full line). The points shown are measurements of the coronal density obtained from the Spartan white light coronagraph (Fisher and Guhathakurta, 1994)

is $\sim$ 100 - 300 sec, which is somewhat greater than the Ferraro-Plumpton period. The waves responsible for heating the corona and ultimately accelerating the high speed solar wind are therefore not related to those with longer periods observed at larger distances from the Sun. The latter waves do not appear to have an energy flux at the Sun capable of producing the high speed wind and are neglected at this stage (e.g. Roberts, 1989).

Perhaps the most interesting aspects of the results is the prediction of the existence of high ion temperatures in the inner corona in high speed streams, namely maximum mean temperatures of 9.5 - 12 $\times$ 10^6 K corresponding to a proton temperature of 8 - 10 $\times$ 10^6 K. In the absence of strong wave pressure a gradients and with the electron temperature too small to produce a significant electron pressure gradient, such high temperatures are inevitable for the high speed solar wind to exist with rapid acceleration and low coronal base density. The temperatures quoted could be slightly over-estimated if for some reason the electron temperature exceeds 10^6 K, if the pressure of long period waves is not negligible and/or if the minor species take a greater share of the available wave energy than assumed here. However the range of temperatures quoted can probably be regarded as being a very good approximation to reality. Indeed we note that recent results from the Spartan 201 Coronal Spectrograph are consistent with our predictions, in that they indicate a proton temperature of 7.6 $\times$ 10^6 K at 1.8 R_s in a coronal hole on the basis of Lyman alpha line profile measurements (Kohl et.al., 1995). These observations also appear to indicate that the solar wind accelerates very quickly in coronal holes, consistent with our predictions.

In order to improve on this theory it will be necessary to consider the source spectrum and dissipation processes for high frequency waves. A beginning has been made (Axford and McKenzie, 1993) but much remains to be done. If, as expected on the basis of the obervations described above, the main dissipation process involves cyclotron heating, it will be necessary to assume that the heating term Q operates on the perpendicular temperature and hence a theory involving anisotropic ion pressures and temperatures is required (Leer and Axford, 1972). In this case the mirror effect may provide an enhancement of the acceleration rate, especially for minor ions if they extract more than their share of the wave energy.

The source of the plasma and the waves appears to lie in the supergranular network and should be considered together with the high speed wind. It appears reasonable to assume that small scale reconnection events associated with the strong and complex magnetic fields in the network are involved and that the wave periods are determined by the scale of the magnetic field structures and the Alfvén speed in this region. A wave source spectrum covering the range required (0.01 Hz - 10 kHz) does not seem to be unreasonable on this basis. The source of the plasma is ionization of neutral gas within the network which appears to take place without significant recombination losses: accordingly the mass flux in the wind and the density at the coronal base are controlled in this way rather than being independent of the source as is often argued.

References

Axford, W.I. (1977) in M.A. Shea et al. (eds.) *The Study of Interplanetary Travelling Disturbances*, D. Reidel, Dordrecht.

Axford, W.I., McKenzie (1993) In Cosmic Winds, University of Arizona Press

Bame, S.J.,Asbridge, J.R., Feldman, W.C., Gosling, J.T. (1977) J. *Geophys. Res.* **82**, pp. 1487-1492

Ferraro, V.C.A., Plumpton, C. (1966) *An Introduction to Magnetofluid Dynamics.* Clarendon Press, Oxford.

Fisher, R.R. and Gutathakurta, M. (1994) *Space Sci. Rev.* **70**, pp. 267-272

Gleeson, L.J., Axford, W.I. (1976) *J. Geophys. Res.* **81**, pp. 3403-3406

Habbal, S.R., Esser, R., Arndt, M.B. *Astrophys. J.* **413**, pp. 435-446

Kohl, J.L., Gardner, L.D., Strahan, L., Fisher, R.R. (1995) *Space Sci. Rev.* **72**, 29

Koutchmy, S. (1977) *Solar Phys.* **51**, pp. 399-407

Leer, E., Axford, W.I. (1972) *Solar Phys.* **23**, pp. 238-250

Mariani, F., Neubauer, F.M. (1990) In Schwenn, R, Marsch, E.K. (eds.) *Physics of the Inner Heliosphere*, Springer Verlag **1**, pp. 183-206

Roberts, D.A. (1989) *J. Geophys. Res.* **94**, pp. 6899-6905

Schwenn, R. (1990) In Schwenn, R., Marsch, E.K. (eds.) *Physics of the Inner Heliosphere*, Springer Verlag **1**, pp. 99-181

DECLINING PHASE CORONAL EVOLUTION: THE STATISTICS OF X-RAY ARCADES

A.H. MCALLISTER, A.J. HUNDHAUSEN AND J.T. BURKEPILE
High Altitude Observatory
PO Box 3000, Boulder, CO 80307, USA

P. MCINTOSH
HelioSynoptics
3885 Paseo del Prado, Boulder CO 80301, USA

AND

E. HIEI
Meisei University, Tokyo 191, Japan

Recent work has supported the link between the reforming, or brightening, coronal arcades seen by the *Yohkoh* soft X-ray telescope (SXT) [McAllister *et al.*, 1992; Hanaoka *et al.*, 1994] and CMEs, as observed by coronagraphs [Hiei, Hundhausen, and Sime, 1993; McAllister *et al.*, 1995].

In an informal survey comparing white light coronagraph observations from the High Altitude Observatory's Mauna Loa Observatory with the SXT images the authors noted that in the descending phase of this solar cycle many of the large scale SXT arcade events were associated with CME's, or were of a similar type to those that were. Furthermore there were several global field configurations that seemed to often produce such events. This led us to perform a statistical study of the locations of the coronal arcades seen in the SXT images during 1993 and the first half of 1994.

We made a catalog of all coronal arcade and other large scale coronal events that might be related to CMEs. The catalog was made more complete by including many events that took place on the backside of the Sun but, due to their scale and high latitude, were visible over the limb. From this catalog we selected coronal arcade events with a scale of more than 20°. Active region expansion events, both quiet [Uchida *et al.*, 1992] and explosive, and diffuse quiet sun events were excluded. This is not meant to imply that we believe they are unrelated to CMEs.

Y. Uchida et al. (eds.),
Magnetodynamic Phenomena in the Solar Atmosphere – Prototypes of Stellar Magnetic Activity, 123–124.

The remaining 240 arcades events were plotted on McIntosh synoptic charts (based on Hα, He10830, and magnetogram data) that show the location of coronal holes and magnetic neutral lines [McIntosh, 1979]. We compiled statistics for the heliographic location and type of neutral line (polar, mid-latitude(diagonal), or active region) for each arcade, keeping a separate record for the 73 large arcade events.

We found differing patterns of arcade scale and location in the two hemispheres. During the study period 52% of the events, but 81% of the large events, occurred in the southern hemisphere. In the north 83% of the events where over diagonal neutral lines while in the south 48% of the events (63% of large events) were over the polar crown. The locations of the arcades were not random, but were highly concentrated between large scale magnetic field concentrations (often marked by coronal holes) that showed relative apparent motion. This apparent motion was either of a colliding or shearing nature. There were three such 'active' locations during our period of study.

Our results suggest a picture of the global solar magnetic field evolution in the declining phase of the solar cycle in which the motion of the large scale photospheric magnetic field structures is a primary driver of evolution in the global corona. This evolution is mediated by coronal mass ejections and traced out by the coronal arcades seen by Yohkoh SXT.

A detailed treatment of this research is being written up as a full paper to be submitted to the Astrophysical Journal.

References

Hanaoka, Y., Kurokawa, H., Torii, C., Sekiguchi, H., Shibasaki, K., Takano, T., Shinohara, N., Irimajiri, Y., Choi, Y. -S., Koshiishi, H., Kosugi, K., Shiomi, Y., Sawa, M., Kai, K., Nakai, Y., Funakoshi, Y., Kitai, R., Ishiura, K., and Kimura, G. (1994) Simultaneous observations of a prominence eruption followed by a coronal arcade formation in radio, soft x-ray and Hα, *PASJ*, **46** pp. 205–216

Hiei, E., Hondhausen, A., and Sime, D. (1993) Reformation of a coronal helmet streamer by magnetic reconnection after a coronal mass ejection, *Geophys. Res. Lett.*, **20(24)** pp. 2785–2788

McAllister, A., Dryer, M., McIntosh, P., Singer, H., and Weiss, L. (1995) A large polar crown cme and a 'problem' geomagnetic storm: April 14-23, 1994. *Journ. Geophys. Res.*, submitted.

McAllister, A., Uchida, Y., Tsuneta, S., Strong, L., Acton, L., Hiei, E., Brunner, M., Watanabe, T., and Shibata, K. (1992) The structure of the coronal soft x-ray source associated with the dark filament disappearance of September 28, 1991 using the *Yohkoh* Soft X-ray Telescope. *Publ. Ast. Soc. Jap.*, **44** pp. L205–L210

McIntosh, P. (1979) Annotated atlas of Hα synoptic charts for solar cycle 20: 1964-1974. Technical report upper atmosphere geophysics report 70, NOAA World Data Center A for Solar-Terrestrial Physics, Boulder.

Uchida, Y., McAllister, A., Strong, K., Ogawara, Y., Shimizu, T., Matsumoto, R. and Hudson, H. S. (1992) Finding of continual expansion of active region corona with the *Yohkoh* Soft X-ray Telescope. *Proc. Ast. Soc. Jap.*, **44** pp. L115–L190

DEVELOPMENT OF A CORONAL HELMET STREAMER OF 24 JANUARY 1992

E. HIEI
Meisei University, Tokyo, Japan

AND

A.J. HUNDHAUSEN
High Altitude Observatory, Boulder, Colorado, U.S.A.

The coronal helmet streamer of 24 January 1992 was one of the most prominent events observed with the Soft X-ray Telescope (SXT) on board the Yohkoh satellite. The helmet was interpreted as the reformation of a closed magnetic structure through reconnection, following a prominence eruption and mass ejection (Hiei et al. 1993).

The coronal helmet streamer developed from a loop structure, as seen in Fig. 1. A faint X-ray coronal loop brightened at 09:20:18 UT, whichi was similar to limb flares observed in the Hα line. The maximum height of the

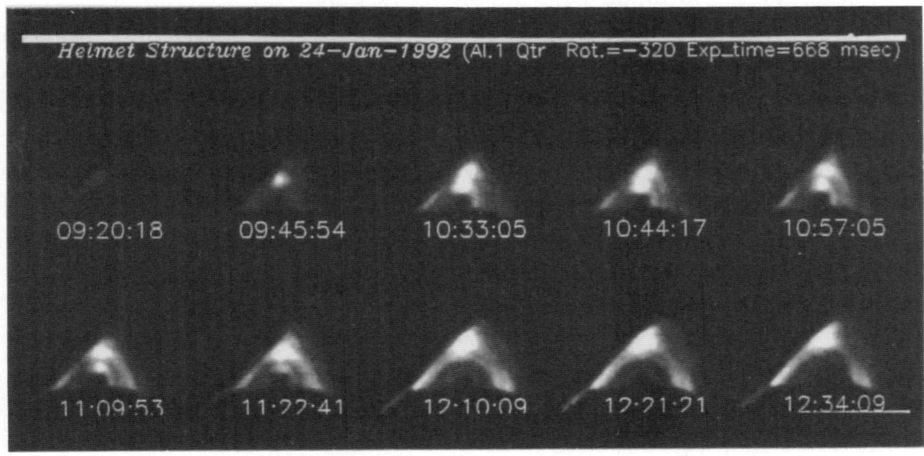

Figure 1. Development of loop structure

Y. Uchida et al. (eds.),
Magnetodynamic Phenomena in the Solar Atmosphere – Prototypes of Stellar Magnetic Activity, 125–126.

structure appeared higher and higher with time, but upon close examination of the SXT images, the structure was found to consist of at least three loops, which are seen in Fig. 1 in the image after 10:33:05.

The height of each loop top, projected in the plane perpendicular to the line of sight, was measured and plotted in Fig. 2. This shows that each loop appeared at a certain height and did not rise further, but rather shrank with time. About one hour later, another loop appeared at a greater height than that of the previous loop.

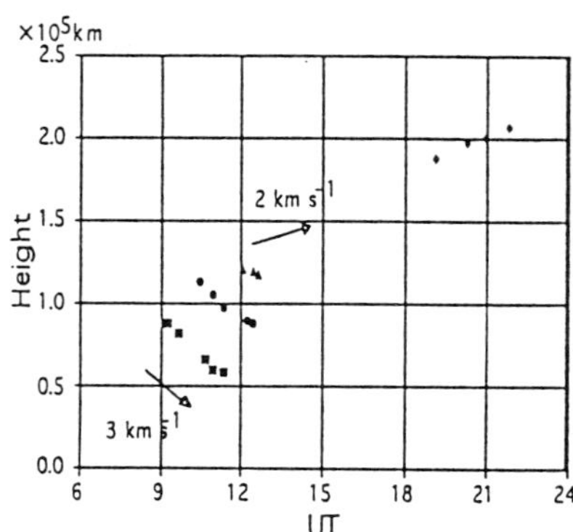

Figure 2. Height of loop top versus time

The loop structure after a two-ribbon flare had hitherto been thought to expand both in height and width with time, but in fact each loop in this event shrank; the downward velocity was first 3 km s^{-1} and then 0.5-1 km s^{-1} about 2 hours later. However, the velocity of rising of the loop structure as a whole was 2 km s^{-1} on the average.

The intensity at the top of the first loop became a maximum at 09:54, about 40 min after the first appearance of the loop, and then its brightness decreased with time. The 2nd loop reached maximum brightness at 11:22, about 50 min after its appearance.

The top of each loop brightened at first and then about one hour later the foot became bright. The first bright region of each loop was always at its top, and then the high temperature region moved down to its foot.

The observation showed the shrinkage of a loop with changing intensity, and the appearance of a discontinuous jump to the next loop. We infer from this observation that i) the energy release was not continuous, and the succeeding energy release occurred about one hour later for the next loop, ii) the duration of the energy release was about 40-50 min for each loop. The intensity of the loop tops became weak on a time scale of one hour due to conductive/radiative cooling. The time change of brightness and width of each loop will be discussed in a later paper. A shrinkage is also discussed by Svestka et al.(1987), and Forbes and Acton(1995).

References

Hiei, E., Hundhausen, A.J., and Sime, D.G. (1993) *Geophys. Res.*, L., **20**, p 2785

Svestka, Z., Fontenla, J.M., Machado, M.E., Martin, S.F., Neidig, D.F., and Poletto, G. (1987) *Sol. Phys.* **108**, 237

Forbes, T.G. and Acton, L.W. (1995) Submitted to *Astrophys. J.*

OBSERVATIONS OF SOLAR WIND BEYOND 5 R_S AND WITHIN 0.3 AU BY INTERPLANETARY SCINTILLATION

M. KOJIMA, H. WATANABE AND Y. YAMAUCHI
Solar-Terrestrial Environment Laoratory, Nogoya University, Toyokawa 442, Japan

AND

H. MISAWA
Upper Atmosphere & Space Research Laboratory, Tohoku University, Japan

Acceleration Phenomena beyond 20 Rs

Interplanetary scintillation (IPS) observations have found high-latitude streams which show large speed increases at a distance range of 0.1-0.3 AU and constant speeds beyond 0.3 AU. These streams tend to be observed more frequently in the minimum phase of solar activity than in the solar active phase, and their foot points are located in HeI polar coronal holes and their boundaries. We examined biasing effects in these observations using simulated IPS observations and found that IPS speed differences of 200 km/s within and beyond 0.3 AU cannot be caused by any combination of possible biasing effects. See Kojima et al. (1995) for more detail discussion.

Very Low Speed Winds From Active Regions

The relationship between low-speed solar winds observed within 0.4 AU with the IPS method and active regions on the sun was studied. It is already known that solar wind speed distributed along the neutral lines is generally low even in the solar active phase when the amplitude of the neutral line becomes large. However it should be noted that low-speed regions near the sun are distributed not only along the neutral line but also on active regions. In general, localized streamers from active regions were thought to disappear near the sun, but our results suggest the possibility that some streamers from active regions survive with a very low speed ($<$ 300 km/s) up to, at least, 0.2-0.4 AU. See Watanabe et al. (1995) for more

Y. Uchida et al. (eds.),
Magnetodynamic Phenomena in the Solar Atmosphere – Prototypes of Stellar Magnetic Activity, 127–128.

detail discussion.

Density Profile in Acceleration Region Beyond 2 Rs

An inner scale parameter has been used to measure the density profile at distances of 2-80 Rs. The inner scale is the scale size at which a power spectrum shows dissipation and is equal to a local proton Larmor radius. Therefore it is assumed that cyclotron damping plays an important role as an effective dissipation mechanism and that the inner scale can be expressed only with density N as $S_i = 684N^{-1/2}$. If the density expands as R^{-2}, the inner scale increases linearly with distance. This linear relation has been observed at R > 30 Rs; however, within 30 Rs it is lower than the values expected from the linear relation. We computed the radial distance dependence of the inner scale with the solar wind acceleration model. For the low-speed wind, the comparison of the radial distance profile found good agreement between observations and model calculation. Accordingly, the separation of the inner scale from the linear relation suggests a steeper density gradient than R^{-2} caused by the solar wind acceleration near the sun.

MHD Wave Related Velocity Fluctuation

Since velocity fluctuations in the solar wind are expected to be caused by MHD plasma fluctuations, measurements of the velocity fluctuations are key to revealing the acceleration process of solar winds. Skewness of a cross-correlation function is related to the velocity fluctuation parallel to bulk flow direction. We compared the scans of the observed cross correlation function with that of modeled CCFs in which velocity fluctuation was parameterized. Preliminary analyses show following results: high-speed winds (V_{SW} ≥500km/s beyond 0.3 AU) indicate increase of velocity fluctuations near the sun; low speed winds (V_{SW} ≤400Km/s beyond 0.3 AU) have small velocity fluctuations at all distances. See Misawa et al. (1995) for more detail discussion.

References

Kojima, M., H. Misawa, H. Watanabe, and Y. Yamauchi, Acceleration Phenomena of High-speed Wind Observed at 0.1–0.3 AU with Interplanetary Scintillation, *Proc. Solar Wind 8*, to be published, 1995.

Misawa, H., and M. Kojima, Dependence of velocity fluctuations on solar wind speeds: A simple analysis with an IPS method, *Proc. Solar Wind 8*, to be published, 1995.

Watanabe, H., M. Kojima, H. Misawa, Y. Kozuka, and Y. Yamauchi, Source Regions of Very Low Speed Solar Winds, *Proc. Solar Wind 8*, to be published, 1995.

II. Wind and Mass-Loss from the Sun and Stars

II.2. Wind and Mass-Loss from Stars

HST AND IUE OBSERVATIONS OF STELLAR MASS-LOSS FROM K AND EARLY M EVOLVED STARS

G.M. HARPER
JILA, University Of Colorado at Boulder
Campus Box 440, Boulder, CO 80309-0440, USA

The presence of out-flowing circumstellar material from evolved stars was originally detected in the optical as blue-shifted absorption features in the spectra of α Her (Deutsch, 1956) and as scattering of the K I λ7699 resonance line in the extended atmosphere of Betelgeuse (α Ori, M2 Iab), see Bernat et al. (1978). These observations of material far above the stellar surface provide direct evidence for mass-loss. The optical Ca II H & K and Infrared Triplet, Hα and He I 10830Å diagnostics are now complemented by studies in the ultraviolet where there is a wealth of information from numerous atomic transitions, e.g., the IUE study of α Her (Thiering & Reimers, 1993). UV observations from IUE and HST/GHRS have placed important constraints on the winds from K and early M evolved stars.

1. Introduction

Winds from late-type stars can be broadly divided into two categories; winds from coronal stars like the Sun which are thought to have terminal velocities of the order of the surface escape speed, i.e., $V(\infty) \sim V_{esc}(R_*)$ and significant support at the base of the wind from thermal pressure, and winds from cool evolved stars which have $V(\infty) < V_{esc}(R_*)$ and show insufficient hot plasma for thermal support at the base. This latter group is the subject of this review. The mechanisms which drive the winds from the K and early M stars remain uncertain, and are an outstanding problem is stellar astrophysics. In later spectral types pulsation may play a role in the mass-loss process. General reviews on mass-loss from cool stars can be found in Dupree (1986) and Dupree & Reimers (1987).

The region of the Hertzsprung-Russell (H-R) diagram where the K and early M evolved stars reside, represents an evolutionary mix, with stars on their first and second crossing of the H-R diagram (Iben, 1967). It is thought

Y. Uchida et al. (eds.),
Magnetodynamic Phenomena in the Solar Atmosphere – Prototypes of Stellar Magnetic Activity, 131–138.

that magnetic processes control and/or modify the wind structure in the K giants, e.g., although photospheric oscillations have been observed on K giants ($14 - 60\,\mathrm{m\,s^{-1}}$), model calculations suggest that the acoustic power spectrum alone is not responsible for the observed mass-loss (Sutmann & Cuntz, 1995 and references therein). Kuin & Ahmad (1989) found from a study of the wind energy budget in the ζ Aurigae eclipsing binaries that the energy flux should propagate at greatly supersonic speeds which are not observed, and they proposed that the energy propagates at an Alfvénic velocity. Unfortunately there are no direct measurements of surface magnetic fields which can be used to observationally constrain proposed magnetic wind models. Haisch, Linsky & Basri (1980) examined the effects of Lyα radiation pressure on the atmosphere of α Boo (K2 III) and found that it could not sustain the stellar wind. Also a study of the global energy balance for low and intermediate mass stars (K1 III - M5 III) by Judge & Stencel (1991) suggests that radiation pressure on dust does not initiate winds from most cool stars.

In the following we focus on some of the important results concerning stellar wind velocity fields obtained with the International Ultraviolet Explorer (IUE) and the Goddard High Resolution Spectrograph (GHRS) on-board the Hubble Space Telescope (HST).

2. UV Spectral Diagnostics (IUE & GHRS/HST)

For most stars, only disk integrated spectra are available and potential signatures for mass-flows are emission-line shifts from the wind and wind scattering features in broad optically thick lines. The IUE and GHRS spec-

TABLE 1. Approximate resolution of the IUE and HST spectrographs

Mode (Dispersion)	$R = \lambda/\Delta\lambda$	$R(\ \mathrm{km\,s^{-1}}\ )$
IUE Low (G)	~ 300	$\Delta V \sim 1000$
IUE High (E)	$\sim 15,000$	$\Delta V \sim 20$
GHRS Low (G)	$\sim 2,000$	$\Delta V \sim 150$
GHRS Medium (G)	$\sim 20,000$	$\Delta V \sim 15$
GHRS High (E)	$\sim 80,000$	$\Delta V \sim 4$

trographs both cover 1150-3300 Ångstroms. The approximate resolutions of the IUE high and low dispersion modes, the GHRS low and medium resolution gratings and the GHRS echelle mode are given in Table 1. A good discussion of IUE spectra from main-sequence and evolved late-type stars can be found in Jordan & Linsky (1987).

Chromospheric line widths in evolved stars are mostly supersonic [FWHM $20 \rightarrow 32$ km s^{-1}, FWHM$^{8000\,K}_{sound}$ = 19 km s^{-1}] but are much narrower than transition region lines, which can be as large as 100 km s^{-1}. Wind absorption features from single stars are typically $10 < V_{wind} < 180$ km s^{-1}. The IUE low dispersion mode is insufficient to resolve any velocity information but IUE has been very useful in measuring line fluxes. The IUE high resolution mode can resolve the broad transition region lines and wind absorption features, while the widths of chromospheric lines are close to the IUE high resolution limit. GHRS echelle observations of C II] λ2325 multiplet in α Tau (K5 III) (which are sensitive to electron density, n_e) (Carpenter et al., 1991) are fully resolved with FWHM$^{CII]}$ = 24 km s^{-1} and show net down-flows of $\sim$ 4 km s^{-1}. The very high S/N and resolution of these data allowed Judge (1994) to detect multiple n_e components from the line profiles.

Some important diagnostic lines observed with IUE and GHRS are given in Table 2. The last column indicates the optical depth (τ) in the line forming region, for transition region resonance lines these are uncertain.

TABLE 2. Selected diagnostic emission lines

Ion	Wavelength (Å)	Excitation Region	τ
Mg II h & k	2795.53+2802.71	Chromosphere	$\gg 1$
Fe II	Numerous	Chromosphere	$0 \rightarrow \gg 1$
Si III]	1893.03	Transition Region	$\ll 1$
Si III	1206.50	Transition Region	> 1
C II]	2325 mult.	Chromosphere	$\ll 1$
C III]	1908.73	Transition Region	$\ll 1$
Si IV	1393.76+1402.77	Transition Region	?
C IV	1548.20+1550.77	Transition Region	?
N V	1238.82+1242.80	Transition Region	?

3. Results from IUE and HST/GHRS

3.1. TRANSITION REGION DIVIDING LINE

One of the first important results from IUE was the identification of a "transition region dividing line" near K2 III (Linsky & Haisch, 1979). Giants earlier than K2 III show the presence of C IV flux in IUE spectra while late-type giants typically show little or no detectable C IV emission.

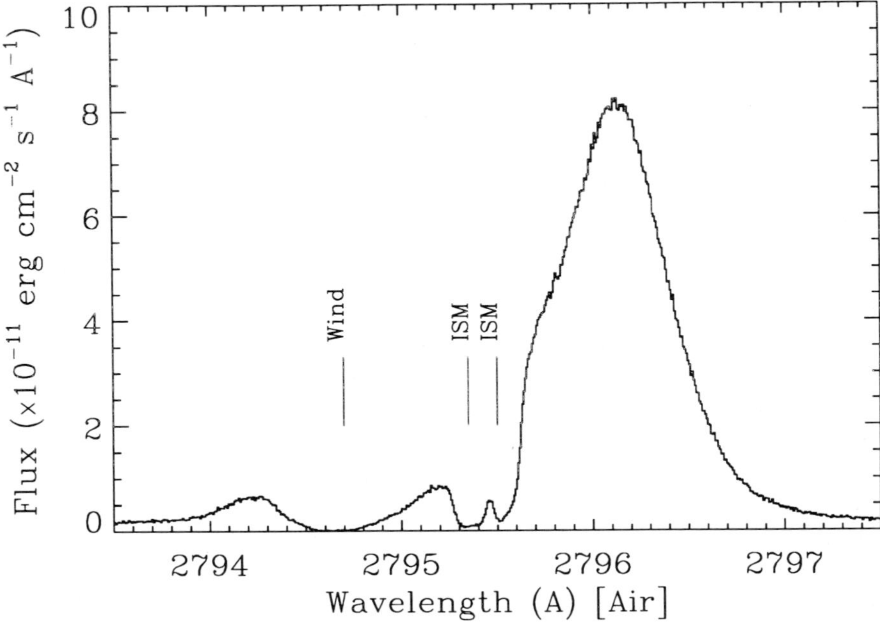

Figure 1. **Large aperture pre-COSTAR HST/GHRS echelle spectrum of Mg II k (λ2795.53) in α TrA. The figure is plotted in the photospheric rest frame. The broad wind feature is clearly visible, together with two ISM features not resolved in IUE high resolution spectra (Harper et al., 1995).**

X-ray observations have shown this division also extends to coronal emission (Ayres et al., 1981). Near this region of the H-R diagram there are changes in the morphology of the Mg II h & k profile (Stencel & Mullan, 1980). In self-reversed Mg II lines of earlier spectral types the blue-peak is similar or stronger than the red-peak like in the solar atmosphere. Later-type stars mostly show the opposite asymmetry (blue $<$ red), which model atmosphere calculations suggest indicates mass out-flow. A similar change in morphology occurs in the Ca II H & K profiles at slightly later spectral types (Stencel, 1978).

Shortly after the identification of the transition region dividing line, a class of stars known as the hybrid-chromosphere stars, e.g., α TrA (K4 II) were identified which are later than K2, showing wind absorption features **and** showing the presence of C IV emission (Hartmann et al. 1980, Reimers, 1982). These stars are mostly bright giants or supergiants which lie above the original dividing line for giants. Figure 1 shows a GHRS echelle spectrum of the Mg II k line in α TrA. The broad wind absorption is clearly visible as well as two ISM features near line center. GHRS and ROSAT PSPC observations of γ Dra (K5 III) show that Mg II wind features, C IV and X-ray emission do co-exist in a giant on the cool side of the dividing line (Brown et al., 1994). The presence of coronal and transition region plasma **and** Mg II wind absorption features are not mutually exclusive.

3.2. TRANSITION REGION LINE SHIFTS

If transition regions are associated with the base of a warm or hot wind then the possibility exists that emission lines might show line shifts indicative of the outflowing plasma. Analysis of IUE line shifts for coronal stars β Dra (G2 Ib-IIa), α Aur A (F9 III) and λ And (G8 III-IV) (Ayres et al., 1988) showed red-shifts (or down-flows) in Si III], C III], Si IV and C IV possibly weighted towards high densities. The Si III] and C III] are both optically thin so that the red-shifts are unambiguous indicators of down-flows (scattering of optically thick lines in out-flows can also appear red-shifted). IUE C IV observations for the hybrid-chromosphere star α TrA show no significant Doppler shifts (Ayres et al., 1984). HST observations have confirmed red-shifts of transition region lines for several stars (Linsky, Wood & Andrulis, 1994) and that C IV in α TrA shows no systematic redshift. Interestingly the HST has confirmed that active stars show C IV λ1548.2 profiles which are less red-shifted than the weaker λ1550.8 line. In α TrA λ1548.2 is blue-shifted while the other line is redshifted. The observed red-shifts of optically thin emission lines certainly do not provide evidence for mass outflows, however, the large line widths suggest that the atmospheres are very dynamic. The observed red-shifts may result from differences in the emissivity of up and down flowing material which form part of a net out-flow.

3.3. FE II LINE PROFILES

There are many Fe II emission lines which have optical depths which cover four orders of magnitude, thus making them ideal for sampling different heights in the stellar atmosphere. Observationally the profiles of Fe II lines can be summarized as follows: Lines of low optical depth ($\tau < 1$) are symmetric and centered on or close to the photospheric rest frame. They have line widths characteristic of the chromosphere. Moderately opaque lines ($\tau \sim 1$) are asymmetric with a depressed blue peak and the lines are opacity broadened with the emission wings centered on the photospheric rest frame. Very opaque lines ($\tau \gg 1$) show a deep wind absorption feature, the profile being highly asymmetric with very broad wings which are centered on the photospheric rest frame (Judge & Jordan, 1991). Figure 2 illustrates these trends with GHRS spectra of γ Cru (M3.4 III) (Carpenter,

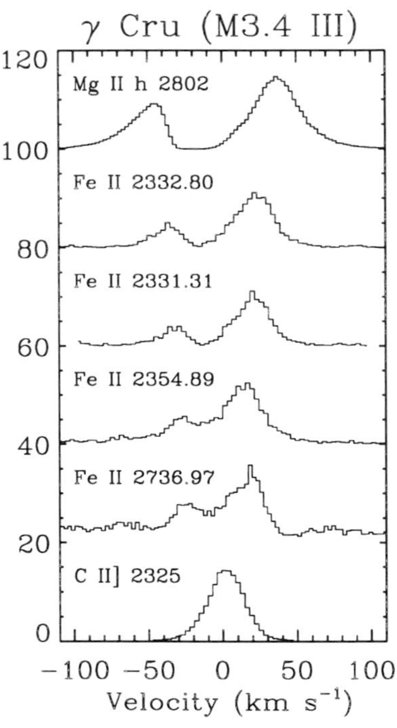

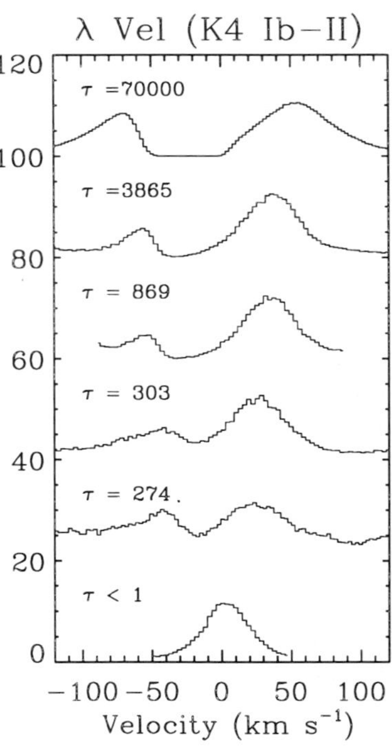

Figure 2. **Line profiles as a function of optical depth for γ Cru (M3.4 III) and λ Vel (K4 II), after Judge & Jordan (1991). The lowest panel shows an optically thin C II] line, the top panel shows the Mg II h line. The intermediate panels show a selection of Fe II lines.**

Robinson & Judge, 1995) and λ Vel (K4 II), the acceleration of the wind with increasing height (larger τ) can clearly be seen.

Qualitative analysis of IUE data for α Ori (M2 Iab) (Carpenter, 1984) shows that the velocity of the Fe II self-reversal first increases and then decreases with optical depth. The appearance of wind deceleration is an important result because in a steady-state wind, it marks the radius where the rate of energy input is less than that required to lift the gas through the gravitational potential. Note that more recent HST observations do not show the same flow patterns indicating time variability in α Ori (Carpenter et al., 1996). A quantitative study of the Fe II spectra will provide strong constraints on wind acceleration from single stars.

3.4. ECLIPSING BINARY SYSTEMS

The most detailed information on stellar winds has come from studies of cool evolved stars in eclipsing binaries such as ζ Aurigae, VV Cep and symbiotic systems. A review of IUE observations of these systems has been given by Hack & Stickland (1987). The spatial resolution obtained by observing the systems at different orbital phases has allowed two different approaches to estimate the wind velocity as a function of radius, $v(R)$. The first approach converts absorption line column densities into particle densities assuming spherical geometry, e.g., Eaton (1993), Vogel (1991) and Schröder (1985). The velocity can then be inferred by assuming a time-independent mass-loss rate. The second approach is to model line profiles which show P Cygni-type scattering. These models include the same assumptions but use both column density and velocity information directly, e.g., Kirsch & Baade (1994) and references therein. The velocity-law is frequently assumed to be of the form

$$v(R) = v(\infty)\left(1 - \frac{R_*}{R}\right)^{\beta}, \tag{1}$$

and results from the best studied systems are given in Table 3. The accelerations are found be relatively slow ($\beta \sim 2.5 \rightarrow 3.5$) compared to the semiempirical model of α Boo (K2 III) (Drake, 1985) and hot star winds where $\beta \sim 0.5$ (Caster, Abbott & Klein, 1975).

TABLE 3. Wind velocity parameters for ζ Aurigae and 32 Cyg.

System	$\dot{M}$ $M_\odot$ yr^{-1}	$v(\infty)$ km s^{-1}	v_{stoc} km s^{-1}	M_* $(M_\odot)$	R_* $(R_\odot)$	β
ζ Aur (K4 Ib)	6×10^{-9}	90	20	7	166	3.5
32 Cyg (K5 Iab)	1.5×10^{-8}	90	30	8	188	2.5

4. Conclusions

The absence of blue-shifted emission lines excited within stellar winds provide strong constraints on theoretical wind models. The absence of blue-shifted emission may be a result of line formation in a highly dynamic outflow or it may result from low particle densities or a gradual wind acceleration as suggested from the observations of ζ Aurigae systems. The Fe II wind diagnostics shown in Fig. 2 should reveal whether the apparent slow acceleration deduced from ζ Aurigae systems is also a property of single stars and a function of spectral-type.

I would like to thank Ken Carpenter for making his HST/GHRS γ Cru data available. Support for this work was provided by NASA under grants S-56460-D and GO-5307.02-93A.

References

Ayres, T. R., Linsky, J. L., Vaiana, G. S., Golub, L., and Rosner, R., 1981, ApJ, 250, 293
Ayres, T. R., et al., 1984, in Future of Ultraviolet Astronomy based on Six Years of IUE Research, Ed. J. M. Mead, R.D. Chapman and Y. Kondo, NASA Conf. Pub. No. 2349, p. 468
Ayres, T. R., Jensen, E., and Engvold, O., 1988, ApJS, 66, 51
Brown, A., Linsky, J. L., and Ayres, T. R., 1994, In The Soft X-ray Cosmos, Eds. E.M. Schlegel and R. Petre, A.I.P. Conf. Proc. 313, p.36
Bernat, A. P., Honeycutt, R. K., Kephart, J. E., Gow, C. E., Sandford, M. T. II, and Lambert, D.L., 1978, ApJ, 219, 532
Carpenter, K. G., 1984, ApJ, 285, 181
Carpenter, K. G., Robinson, R. D., Wahlgren, G. M., Ake, T. B, Ebbets, D. C., Linsky, J. L., Brown, A., and Walter, F. M., 1991, ApJL, 377, L45
Carpenter, K. G., Robinson, R. D., and Judge, P.G., 1995, ApJ, 444, 424
Carpenter, K. G., et al. 1996, In Preparation.
Castor, J. I., Abbott, D. C., and Klein, R.I., 1975, ApJ, 195, 157
Deutsch, A. J., 1956, ApJ, 123, 210
Drake, S. A., 1985, in Progress in Stellar Spectral Line Formation Theory, eds. J. E. Beckman and L. Crivellari (Dordrecht: Reidel), p.351
Dupree, A. K., 1986, Ann. Rev. Astr. Astrophys., 24, 377
Dupree, A. K., and Reimers, D., 1987, in Exploring the Universe with the IUE Satellite, Ed. Y. Kondo (Kluwer Academic Pub), 321
Eaton, J. A., 1993, ApJ, 404, 305
Hack, M., and Stickland, D., 1987, in Exploring the Universe with the IUE Satellite, Ed. Y. Kondo (D. Reidel Publishing Company) p. 445
Haisch, B. M., Linsky, J. L., and Basri, G. S., 1980, ApJ, 235, 519
Harper, G. M., Wood, B. E., Linsky, J. L, Bennett, P. D., Ayres, T. R., and Brown A., 1995, ApJ, 452, 407
Hartmann, L., Dupree, A. K., and Raymond, J. C., 1980, ApJ, 236, L143
Iben, I. Jr., 1967, Ann. Rev. Astr. Astrophys., 5, 425
Jordan, C., and Linsky, J. L., 1987, in Exploring the Universe with the IUE Satellite, Ed. Y. Kondo (Kluwer Academic Pub), 259
Judge, P. G., and Stencel, R. E., 1991, ApJ, 371, 357
Judge, P. G., 1994, ApJ, 430, 351
Judge, P. G., and Jordan, C., 1991, ApJS, 77, 75
Kirsch, T., and Baade, R., 1994, A&A,291, 535
Kuin, N. P. M., and Ahmad, I. A., 1989, ApJ, 344, 856
Linsky, J. L., and Haisch, B. M., 1979, ApJL, 229, L27
Linsky, J. L., Wood, B. E., and Andrulis, C., 1994, in Cool Stars, Stellar Systems, and the Sun, A.S.P., Conf. Ser. 64, p. 681
Reimers, D., 1982, A&A, 107, 292
Schröder, K.-P., 1985, A&A, 147, 103
Stencel, R. E., 1978, ApJL, 223, L37
Stencel, R. E., and Mullan, D.J., 1980, ApJ, 238, 221
Sutmann, G., and Cuntz, M., 1995, ApJL, 442, L61
Thiering, I., and Reimers, D., 1993, A&A, 274, 838
Vogel, M., 1991, A&A, 249, 173

PROMINENCES ON LATE TYPE ACTIVE STARS

P.B. BYRNE
Armagh Observatory
Armagh BT61 9DG, N. Ireland

Abstract. There is increasing evidence of prominence-like neutral hydrogen clouds suspended in the otherwise hot coronae of active late-type stars. We review this evidence here and discuss the nature of these clouds. In particular we discuss their possible role in rotational braking.

1. Introduction

The solar paradigm is widely used to explain a broad range of phenomena on late-type stars. These include stellar flares, sudden brightenings in broad-band radiation, from microwaves to hard X-rays (analogous to solar flares), modulation of global visible and near-IR light by the varying visibility of dark areas on the stellar surface as the underlying star rotates ("starspots", analogous to sunspots), and non-radiative heating of hot chromospheres and coronae. All of these phenomena arise from powerful magnetic fields, generated by the interaction of differential rotation and deep convection, and intensified locally into loop structures at the photosphere and above, in a manner familiar on the Sun.

Studying stellar counterparts of solar magnetic activity offers us many insights not available on the Sun. We observe stars which rotate at many times the solar rate ($P_{rot} \geq 0.2\,\mathrm{dy}$) and which possess convective regions extending almost to the core of the star. These result in more efficient magnetic field generation and, as a result, solar phenomena are scaled by orders of magnitude. Active late-type stars may also be observed in young open clusters and in the old disk field population giving an evolutionary perspective on their rotationally induced activity.

Such active stars include a "zoo" of objects, such as the FK Com, T Tau, W UMa stars and even the secondaries of cataclysmic variables, all of which display rotationally induced magnetic activity of the type described above. In this review we are going to discuss only the dKe/dMe stars, i.e. the

Y. Uchida et al. (eds.),
Magnetodynamic Phenomena in the Solar Atmosphere – Prototypes of Stellar Magnetic Activity, 139–146.

classical "Flare" stars, and the RS CVn sub-giant binaries. The former are rapidly rotating main sequence objects, while the latter are post-main sequence close binaries, in which tidal interaction has forced one, or both, components into rapid rotation. This selection is dictated largely by the range of available observational material rather than intrinsic interest.

On the Sun, prominences are intimately associated with solar magnetic active regions. They consist of neutral hydrogen gas trapped in a potential well near the apexes of loop arcades and embedded in a multi-million degree corona. They are seen as dark features via resonant scattering of the Hα-bright chromospheric radiation. When seen on the solar disk, large prominences are filamentary in structure and their total projected area is typically $\leq$1% of the solar disk (Allen, 1973). Their height above the photosphere is typically 0.05$R_{\odot}$ and they have lifetimes of several solar rotations.

2. AB Dor

AB Dor (= HD36705) is a bright (V$\sim$6.9) K0V star at a distance of $\sim$16pc. It rotates with a period, P_{rot}=0.514dy. Combining the directly measured $v_{rot}\sin i$ ($\sim$85km/s) and this period yields a radius, R=0.9$R_{\odot}$, close to normal for a K0 main sequence object. This large rotational velocity means that doppler broadening exceeds all other sources of spectral line broadening, even for the case of chromospheric Hα.

High-resolution spectra of AB Dor's Hα line show narrow absorption features crossing between $\pm v_{rot}\sin i$ in times short compared to P_{rot} (Cameron & Robinson, 1989a,b). These features cannot be at the stellar photosphere, since they would cross the rotationally broadened profile in a time $P_{rot}/2$ and so Cameron & Robinson interpreted them as clouds of neutral hydrogen supported high above the photosphere in a manner reminiscent of solar prominences.

Consideration of their disk crossing times, along with their projection geometry, led to the conclusion that they form at, or just beyond, the corotation radius, r_{co-rot} (where $\omega^2 r_{co-rot} = g$). Furthermore, Cameron & Robinson (1989b) showed that individual clouds could be identified on several rotations, suggesting that they have a lifetime of several days. They suggest that the neutral material forms at the apexes of magnetic loops which extend beyond r_{co-rot}, where centrifugal acceleration leads to condensation. Material is fed from the lower, hotter portions of the loops by thermal expansion and then centrifugal acceleration. The neutral component of the gas is tied to the magnetic field lines by collisions with the ionized component, persisting for a period of time comparable to the diffusion time. Clouds are seen to dissipate on a time scale of $\sim$2–4dy, while new clouds form at a rate $\sim$1–2/dy. Cloud masses have been estimated to be 10^{17}–10^{18}gr (Cameron et al. 1990), 2–3 orders of magnitude larger than typical solar prominences (Allen, 1973).

Cameron & Robinson suggest that the dissipation of the neutral clouds beyond r_{co-rot} is a potentially powerful means of removing rotational angular momentum from such stars on a time scale of $\sim 10^8$yr.

3. Open cluster rapid rotators

Stellar rotational velocities in the αPer (Stauffer et al. 1985, 1989) and the Pleiades (Stauffer et al. 1984, Stauffer and Hartmann, 1987) open clusters showed that, while many G stars rotate rapidly in the former, at an age $\tau \sim$50Myr, they were all slow rotators in the latter, whose age is $\tau \sim$70Myr. In contrast, many K stars in both clusters are rapid rotators, while they are slow rotators in the Hyades ($\tau \sim$600Myr)(Stauffer et al. 1987).

Assuming that these results reflect ageing and not the effect of stellar environments in the individual clusters, Stauffer et al. concluded that there was a rapid braking mechanism in operation in young, rapidly rotating stars which could brake G stars on a time scale $\leq$20Myr (the difference in age between αPer and the Pleiades), and somewhat longer in K stars.

The neutral hydrogen clouds discovered in AB Dor could provide such a mechanism. However, they would need to be ubiquitous in rapidly rotating stars and in cluster G/K stars in particular.

Cameron & Woods (1992) found evidence for Hα absorptions in four rapidly rotating G stars in αPer. In one of these it was possible to show that the clouds giving rise to the absorptions were a short distance outside the co-rotation radius. Therefore, while not providing conclusive evidence, these data are supportive of the braking hypothesis.

4. HK Aqr

HK Aqr (=BD–16°6218 =Gl890) is a 10th magnitude M2Ve star at a distance of ~20 pc, the 'e' suffix indicating Balmer Hα in emission, an indicator of extreme chromospheric activity. Young et al. (1984, 1990) discovered it to be an exceptionally rapidly rotating star with a period P~10.34 hr and $v_{rot}\sin i \sim$70km/s. Combining these two values suggests a stellar radius R~0.6R$_{\odot}$, normal for a main sequence star of HK Aqr's spectral type.

As with AB Dor, rotational broadening dominates over other sources of broadening, even in the case of the normally very broad chromospheric Hα emission (Byrne & Mathioudakis, 1993). Recent analysis of high resolution Hα profiles (Byrne et al. 1994) has shown that the emission line centroid varies in times short compared with a complete rotation.

In Fig. 1 (*upper panel*) we show a sequence of these spectra and (*lower panel*) the result of division of each spectrum by the mean Hα spectrum

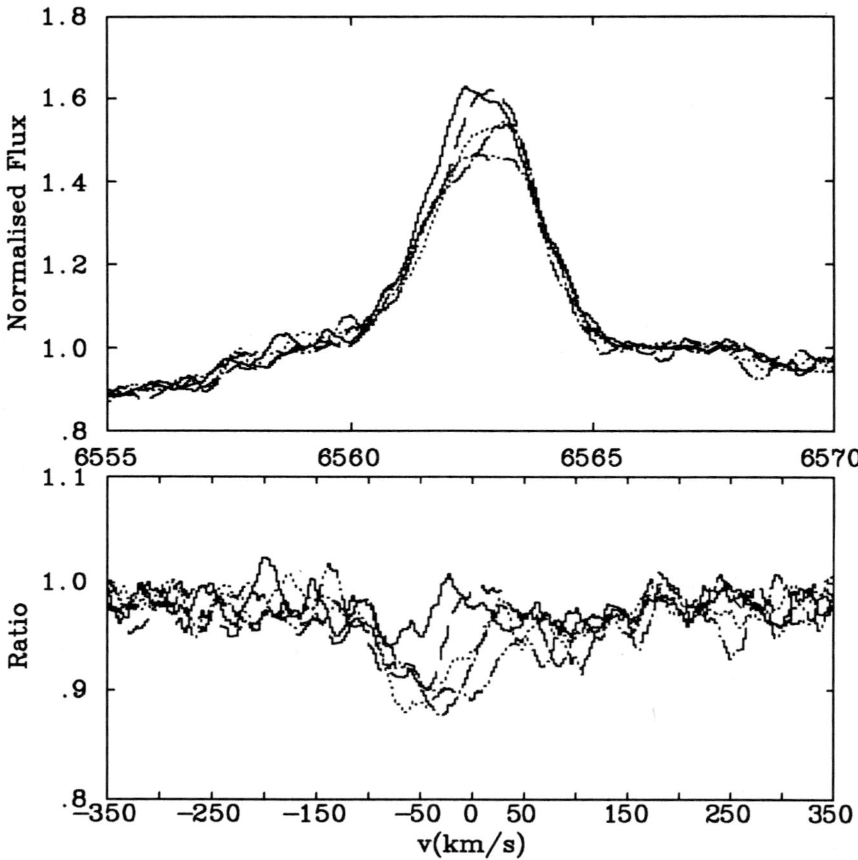

Figure 1. Example sequence of consecutive Hα emission profiles of HK Aqr on the night of 25 August 1991 (*upper panel*). The spectra are separated in time by ~5 min ($\Delta\varphi = 0.008$) and occur in the time sequence —-, — —, ..., -..- and -.-. Also shown (*lower panel*) is the result of dividing these same individual Hα profiles by the mean profile for the entire period of observation. This diagram is taken from Byrne et al. (1996a).

for the entire observing run. It is clear that the centroid variations arise from progressive absorption of the blue wing of the Hα emission. The full spectral sequence shows the absorption crossing the entire profile.

This interpretation is made even clearer in Fig. 2 where we have plotted the radial velocity with respect to line centre of these absorption features. They are seen to vary linearly in their velocity, while their range in RV is constrained to $\pm v_{rot}\sin i$. Furthermore, individual features can be identified over a period of at least 2 days (Byrne et al. 1996a). In these respects they are directly comparable to the absorption features seen in AB Dor and described above. Therefore we adopt the same interpretive framework of

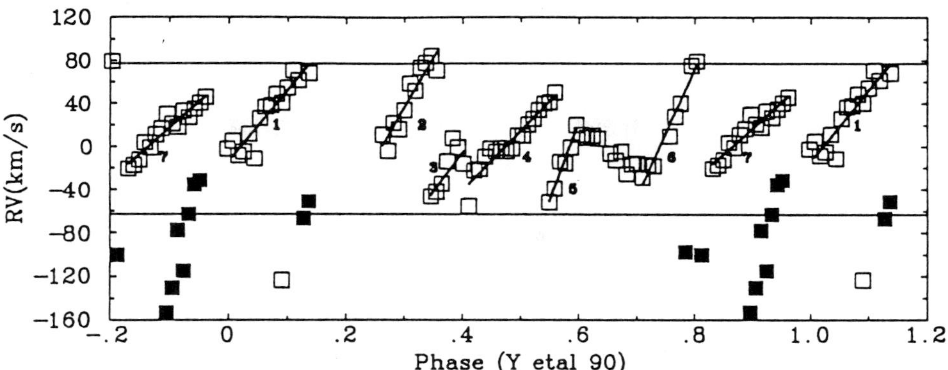

Figure 2. Radial velocities of gaussian fits to the excess/deficit in the ratios of individual Hα line spectra of HK Aqr on the night of 25 August 1991 to the overall mean. Ratios falling above or below a mean line defined by the ratio of nearby continuum regions, described in the text as in net absorption or net emission, are represented by *filled symbols* or *open symbols* respectively. The two horizontal lines represent $\pm v\sin i$ corrected for a mean velocity of $+7\,\mathrm{km\,s^{-1}}$.

neutral hydrogen clouds suspended in closed magnetic loops.

HK Aqr's Hα absorptions differ from AB Dor's in an important repect, however. Assuming the mass of HK Aqr can be derived from its spectral type, its corotation radius, r_{co-rot} can be determined and, from it, the crossing times for clouds held at that radius. This corresponds to $\Delta\varphi$=0.09. Perusal of Fig. 2 shows that all of the clouds detected on HK Aqr's Hα profile cross in times longer than this. Geometric arguments then force the conclusion that all of the clouds are at heights less that r_{co-rot}. Indeed, even assuming that they all cross the line-of-sight between the observer and the centre of the star, some may be at heights above the stellar photosphere comparable to large solar prominences.

The maximum scattered Hα flux is ~5–10% suggesting a minimal projected disk areal coverage of a similar magnitude. Obviously such structures are much larger than typical solar prominences.

So, observations of HK Aqr supports the contention that prominence-like neutral hydrogen features are present in rapidly rotating late-type stars, but suggest that their potential as a rotational brake is not universal.

5. II Peg

II Peg is a 6.72dy SB1 binary of spectral type K2IV with a measured $v_{rot}\sin i$ of 21km/s. Its Hα is strongly in emission and the width of its profile is determined by radiative transfer effects (see e.g. Byrne et al. 1995). Thus detection of cool clouds in its atmosphere cannot be achieved by looking

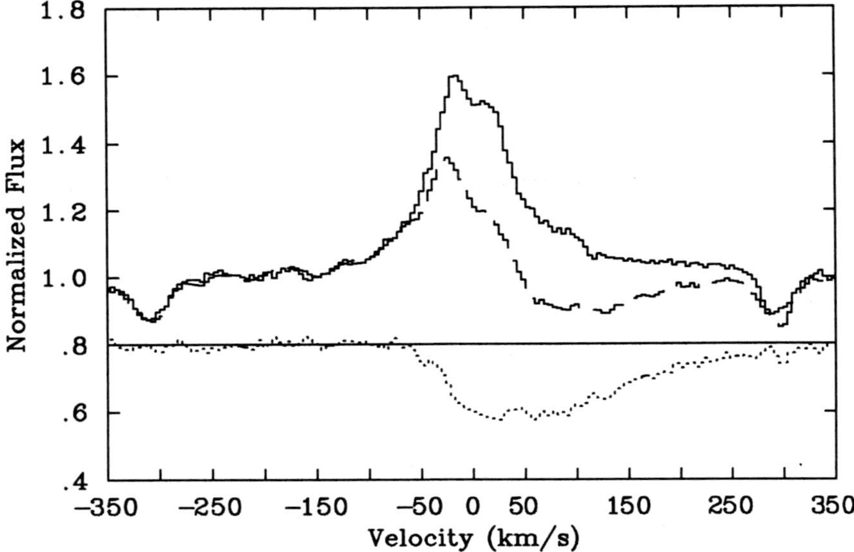

Figure 3. The mean time-averaged Hα profile of the RS CVn binary, II Peg, (—-) compared to its observed profile during an "absorption event" of the type described in the text (– –) in September 1993. The lower curve (.....) shows the result of dividing the "absorption event" profile by the mean time-averaged profile (Byrne et al. 1996b).

for their doppler signature crossing the line profile.

Nevertheless, evidence of Hα absorptions has been found (Byrne, 1987). Fig. 3 shows the mean Hα spectrum of II Peg compared to its profile during one of these Hα absorption "events". The profile is clearly is heavily absorbed, this time on the red wing of the line. The absorption profile (Fig. 3 *lower curve*) has a P-Cygni-like profile, sharply defined to the blue, but quasi-exponential to the red, suggesting a strong downflow associated with the material in the absorbing cloud. Although it is difficult to determine velocities definitively from profiles of an optically thick line, it appears that terminal velocities as high as 200–300km/s could be present. This is close to the escape velocity for II Peg (~260km/s) suggesting that perhaps the material originates in the otherwise invisible low-mass companion.

6. Flare-associated prominences?

Transient dips during broadband photometric flares (Flesch & Oliver, 1974, Giampapa et al. 1982, Doyle et al. 1988), during spectroscopic emission-line flares (Doyle et al. 1989, Houdebine et al. 1994) and during X-ray flares (Haisch et al. 1983) have been reported in the literature. By analogy with solar flares these have been ascribed to large-scale eruptive prominences or surges associated with the flare site.

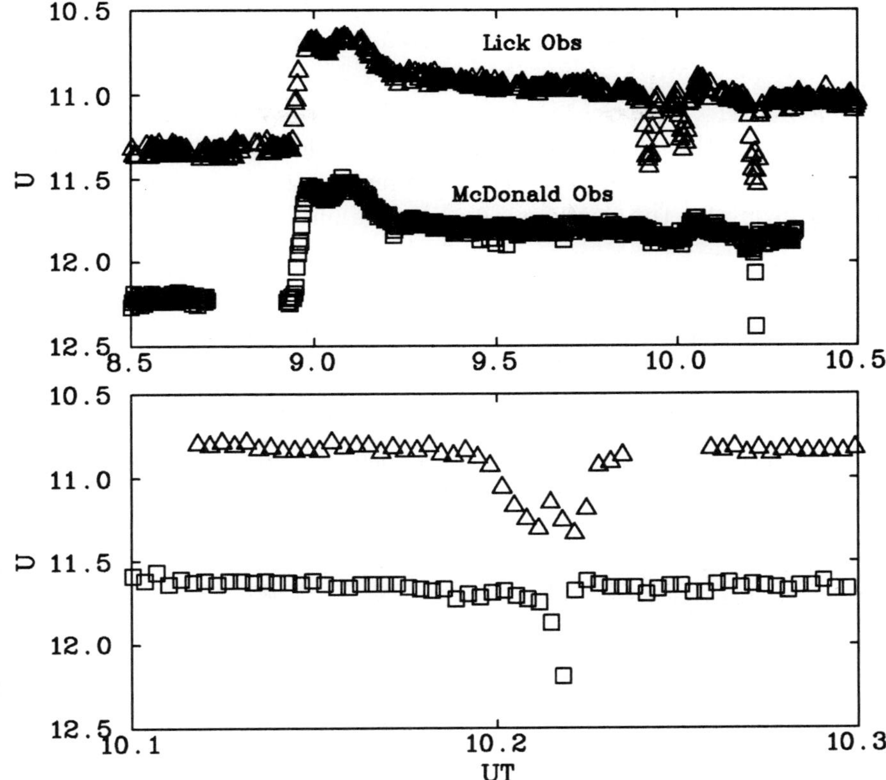

Figure 4. Broadband photometric "dips" during a flare observed simultaneously on the active flare star, AD Leo, at McDonald and Lick observatories. (adapted from Hawley et al. 1995).

Fig. 4 shows an example of a multi-site, simultaneous photometric observation of a broadband stellar flare on the dMe star, AD Leo (Hawley et al. 1995) which apparently shows a dip observed from both sites (at ~10:12UT). Closer examination of the data, however, shows that the profiles of the two dips are quite different (Fig. 4, *lower panel*), suggesting that both were caused by the passage of some light cloud or a temporary drift in telescope guiding. This illustrates the problems associated with accepting the reality of single-site observations of photometric evidence of flare-associated prominences.

7. Conclusion

There is an increasing body of evidence supporting the contention that clouds of neutral hydrogen exit in active late-type stars, held above their photospheres in magnetic loops. However, as with other stellar counterparts of solar magnetic activity, their scale appears to be larger than on the Sun, and the details of their physics subtly different.

Acknowledgements: Research at Armagh Observatory is supported by a Grant-in-Aid from the Department of Education for Northern Ireland. I acknowledge financial support from the conference organisors. I am grateful to Dr. S.L. Hawley for the data for Fig. 4 before publication.

References

Allen, C.W., 1973, Astrophysical Quantities, Athlone Press, London

Byrne, P.B.: 1987, in "Cool Stars, Stellar Systems and the Sun", eds. J.L. Linsky & R.E. Stencel, Springer, p. 491

Byrne, P.B., Mathioudakis, M., 1993, Physics of Solar and Stellar Coronae, eds. J.L. Linsky & S. Serio, Kluwer, p. 435

Byrne, P.B., Mathioudakis, M., Young, A., Skumanich, A., 1994, Cool Stars, Stellar Systems and the Sun, ed. J-P. Caillault, ASP Conf. Ser. No. 64, p. 375

Byrne, P.B., Panagi, P.M., Lanzafame, A.C., Avgoloupis, S., Huenemoerder, D.H., Kilkenny, D., Marang, F., Panov, K.P., Roberts, G., Seiradakis, J.H., van Wyk, F.: 1995, A&A, 299, 115

Byrne, P.B., Eibe, M.T., Griffin, I.P., Mathioudakis, M., Doyle, J.G., Nauta, M., 1996a, in preparation

Byrne, P.B., Amado, P., et al., 1996b, in preparation

Cameron, A.C., Robinson, R.D., 1989a, MNRAS, 236, 57

Cameron, A.C., Robinson, R.D., 1989b, MNRAS, 238, 657

Cameron, A.C., Duncan, D.K., Ehrenfreund, P., Foing, B.H., Kuntz, K.D., Penston, M.V., Robinson, R.D., Soderblom, D.R., 1990, MNRAS, 247, 415

Cameron, A.C., Woods, J.A., 1992, MNRAS, 258, 360

Doyle, J.G., Butler, C.J., Byrne, P.B., van den Oord, G.H.J.: 1988, A&A, 193, 229

Doyle, J.G., Byrne, P.B., van den Oord, G.H.J.: 1989, A&A, 224, 153

Flesch, T.R., Oliver, J.P., 1974, ApJ, 189, L127

Giampapa, M.S., et al. 1982, ApJ, 252, L39

Haisch, B.M., Linsky, J.L., Bornman, P.L., Antiochus, S.K., Golub, L., Vaiana, G.S., 1983, ApJ, 267, 280

Hawley, S.L., Fisher, G.H., Simon, T., Cully, S.L., Deustua, S.E., Jablonski, M., Johns-Krull, C.M., Pettersen, B.R., Smith, V., Spiesman, W.J., Valenti, J., 1995, ApJ, in press (Nov. 1 issue)

Houdebine, E.R., Foing, B.H., Doyle, J.G., Rodonó, M. 1994, A&A, 278, 109

Stauffer, J.R., Hartmann, L., 1987, ApJ, 318, 337

Stauffer, J.R., Hartmann, L., Soderblom, D.R., Burnham, N., 1984, ApJ, 280, 202

Stauffer, J.R., Hartmann, L., Burnham, N., Jones, B.F., 1985, ApJ, 289, 247

Stauffer, J.R., Hartmann, L., Latham, D.W., 1987, ApJ, 320, L51

Stauffer, J.R., Hartmann, L., Jones, B.F., 1989, ApJ, 346, 160

Young, A., Skumanich, A., Harlan, E., 1984, ApJ, 282, 683

Young, A., Skumanich, A., McGregor, K.B., Temple, S. 1990, ApJ, 349, 608

STELLAR CMES AND FLARE MASS MOTIONS

E.R. HOUDEBINE
ESA/ESTEC Space Science Department, The Netherlands

Abstract. We review evidence for mass motions during stellar flares and show that among them, some are possible signatures of Coronal Mass Ejections. So far, very little is known about stellar CMEs but we show that their potential implications are important. For instance, the CME driven mass loss rates, much larger on stars ($10^{-13} - 10^{-6} M_{\odot} yr^{-1}$) than on the sun ($\sim 10^{-14} M_{\odot} yr^{-1}$), may dominate the stellar wind as well as the mass balance of the interstellar medium, but also influence the angular momentum loss and stellar evolution.

1. Important Issues Relevant to Flare Mass Motions

The short time scale of flares and related phenomena has made the detection of flare mass motions and probable Coronal Mass Ejections (CMEs) quite difficult (here, we refer to CMEs as the discrete ejection of coronal plasma from the stellar gravity potential in its broader sense). Nevertheless, Doppler-Fizeau shifts in spectral lines have been reported on a number of occasions, mostly from ground-based spectroscopy with adequate time and spectral resolutions ($\delta t \sim$1min and $\lambda/\delta\lambda \geq 2000$).

We summarize all detections of flare mass motions in Table 1, with a total of 22 flares on dwarfs and RS CVn systems. The first detected signatures were red-shifted emissions during the impulsive phase, a well known and systematically observed motion in solar flares. This was detected in 10 flares out of 14 on dwarfs (71%). Including RS CVn systems, a redshift was reported in 80% (16 out of 20) of the cases for which there was a time coverage of the impulsive phase, and it is therefore the most frequently observed phenomenon. Blue shifts are less frequent on M dwarfs and are reported mostly during the gradual phase, with an occurrence of 10% (2 out of 20) during the impulsive phase and 19% (4 out of 21) during the gradual

Y. Uchida et al. (eds.),
Magnetodynamic Phenomena in the Solar Atmosphere – Prototypes of Stellar Magnetic Activity, 147–155.

phase. Blue-shifts were more frequently detected on RS CVn systems. More complex velocity fields were detected only in three cases (14%). Of course, those figures are *entirely dependent on the data quality and the resulting detectability threshold.* Nonetheless, they give us an idea of what are the dominant types of motions one can expect and their relative amplitudes.

Surprisingly, Doppler shifts are more frequently detected in *intermediate size flares* (typically 1 to 2 magnitudes on an M4 dwarf) *rather than very large flares.* The only motion seen among flares of the latter class is that of the 5.3 mag UV Ceti flare observed by Eason et al. (1992). The Balmer line FWHM also seems to drop for very large events (e.g. Houdebine 1992; Eason et al. 1992). This suggests that as the flare power increases, the contribution from the kernels and cool coronal structures is being overwhelmed by that from a stationnary back-heated chromosphere (Houdebine 1992).

2. Candidates to Coronal Mass Ejections

Among the observed mass motions, there is no definite evidence for coronal mass ejections. On the basis of the distance travelled by the plasmoid, we selected eight CME candidates detected as velocity shifts. The distances covered are between 60Mm and 10^4Mm; for comparison, the typical radius of a red dwarf is 200Mm. The strongest cases on dwarfs are for AD Leo (Houdebine et al. 1990) and AT Mic (Gunn et al. 1994) events where the distance is at least of the order of the stellar radius. There is a remarkable difference between mass flows on dwarfs and RS CVn systems; for the latter, even though the flows are often slow, they are sustained for a much longer period of time (typically hours) and much larger distances of several hundreds of Mm are covered.

Although impulsive phase downflows most often cover less than 10Mm, this is still about a factor of 5 larger than the chromosphere's height. This suggests that more than a single isolated dowflow takes place. In a couple of instances, Bopp and Moffet (1973) and Bookbinder et al. (1992), this distance and the flow velocity are abnormally high (respectively several tens of Mm and about $1000 km\ s^{-1}$) which shows that we may be observing ejected material with a negative projection along the line of sight for a flare occurring behind the limb. In particular, the latter authors observe profiles that show a surprisingly sharp edge on the blue wing of the C IV lines at 1545.5Å which is highly suggestive of an occulted flare. These two events may therefore be signatures of fast CMEs or chromospheric evaporations as reported in Houdebine et al. (1990).

More intriguing, red-shifted "downflows" on RS CVn systems apparently travel typical distances of a few hundreds of Mm, two orders of magnitude larger than what would be expected for a "chromospheric

Table 1: Compendium of all detected mass flows during flares on main-sequence dwarfs and RS CVn systems. We give the maximum observed velocities or the mean velocity, the estimated kinetic energy and the U-band magnitude (U)

Star	Authors	Lines	V (km/s)	Duration (min)	E_K (erg)	U	Mass (Kg)	δh (Mm)	Comments
			dMe Stars						
UV Ceti	Joy 1958		30	-	-	-	-	-	Suspected red-shift
Wolf 359	Greenstein & Arp	HI	-23	$\leq$5	-	-	-	$\leq$7	Blue shifts, imp.
	1969	CaII	-59	$\leq$5			-	$\leq$18	
AD Leo	Gershberg &	HI	-	-	-	-	-	-	Red asymmetry
	Shakovskaya 1971								Impulsive phase
AD Leo	Kulapova &	HI	-	-	-	-	-	-	Red asymmetry
	Shakovskaya 1973								Impulsive phase
UV Ceti	Bopp & Moffet	HI	$\leq$1100	2	-	2.3	-	$\leq$130	Red wing excess
	1973	CaII K	$\leq$600	2				$\leq$72	Around impulsive phase
Wolf 424	Robinson 1989	HI	-	-	-	-	-	-	Slight redshift
									Post-impulsive phase
EV Lac	Ambruster et al.	UV lines	-	-	-	-	-	-	Flux drop, cloud ?
	1986								CME ?
UV Ceti	Phillips et al.	HI	$\sim$100	$\leq$1	-	2.0	-	$\leq$6	Red wing excess
	1988								Impulsive phase
YZ CMi	Doyle et al.	HI	-	-	-	1.2	-	-	Slight red asymmetry
	1988								Impulsive phase
AD Leo	Houdebine et al.	HI	$\geq$-5800	3	$5\ 10^{34}$	2.1	$8\ 10^{14}$	900	Blue wing excess, imp
	1990								CME ?
AD Leo	Bookbinder et al.	CIV, SiIV	$\leq$1800	$\geq$2	-	-	10^6	$\sim$ 60	Redshift
	1992	HeII							CME ?
UV Ceti	Eason et al.	Hα	-70	$\geq$40	-	5	-	$\geq$160	Blue shift, CME?
	1992								
AU Mic	Woodgate et al.	Lyα	$\leq$3700	$\sim$3 s	-	-	-	$\leq$11	Weak flare
	1992								Proton beams ?
AD Leo	Houdebine et al.	CaII	-120	$\leq$1	$6\ 10^{32}$	2.1	10^{13}	$\leq$7	Blue shifted abs.
	1993ab	HI, HeI	$\leq$100	3 min			10^{13}	$\leq$18	Red-shift, imp. phase
		CaII	$\geq$-140	$\geq$14			$2\ 10^{12}$	$\geq$90	Blue-shift, grad. phase
		CaII	-60	$\geq$4					Blue-shift abs., CME?
YZ CMi	Gunn et al.	HI	$\geq$-250	$\sim$4	-	-	-	$\sim$60	Blue wing excess, CME?
	1994a								Post impulsive phase
AT Mic	Gunn et al.	HI	$\geq$-620	$\geq$13	$3\ 10^{29}$	-	10^{12}	$\geq$250	Blue wing excess, CME?
	1994b	CaII K	$\geq$-410	-	-	-			Post impulsive phase
AU Mic	Linsky & Wood	Si IV	$\sim$ 40	$\leq$3	-	-	-	$\leq$7	Weak flare, blue &
	1994		$\sim$ -30	$\leq$3	-	-	-	$\leq$5	red shifts
VB 10	Linsky 1995	UV	-	-	-	-	-	-	Blue wing excess
			RS CVn Stars						
UX Ari	Simon et al.	MgII, FeII	-45	$\leq$30	-	-	-	$\leq$80	Blue shift
G5/K0	1980		$\leq$860	$\leq$30			-	$\leq$1500	Red wing excess
λ And	Baliunas et al.	MgII	25	$\geq$5 h	-	-		$\geq$450	Redshift
G8 IV	1984								
V711 Tau	Linsky et al.	MgII k	90	$\sim$60	10^{32}	0.1	-	$\sim$300	Redshift
G5/K1	1989								
AR Lac	Neff et al.	HI	$\geq$30	$\sim$4 h	-	-	-	$\sim$400	Redshift
	1989								
V471 Tau	Mullan et al.	UV lines	$\geq$-800	$\geq$3.3 h	-	-	10^{12} Kg/s	10^4	Blue shifted abs.
K2/Wd	1989								CMEs, no flare
II Peg	Doyle et al.	MgII	-25	$\geq$4 h	-	-	-	$\geq$360	Red shift, imp. phase
K3IV	1989		-25						Blue-shifted abs.
HR 1099	Foing et al.	Hα	$\geq$-470	$\sim$60	$2\ 10^{38}$	2.3	$3\ 10^{18}$	1700	Blue wing excess, CME?
K1/G5	1994		$\leq$280	$\geq$40			-	670	Red wing excess
			50-90	$\geq$20			-	80	Red-shifted abs.
HR 1099	" "	Hα	$\geq$-70	$\sim$60	$2\ 10^{37}$	0.61	$1.5\ 10^{17}$	$\sim$220	Blue wing excess, CME?

condensation". This further underlines the *different natures of flares on RS CVn systems and on solar type stars.* We interpret those different properties (velocity, distance covered, mass, duration) as evidence for *inter-system flares* rather than traditional solar type "surface flares". Yet on another scale, the magnetodynamic CME-like events on cataclysmic variables reach velocities of a few thousand $km\ s^{-1}$ and are detectable for periods of hours to days (Cordova & Mason 1982). This highlights that magnetically triggered mass motions can be somewhat more energetic than solar phenomena and urges us to comprehend magnetodynamic phenomena on a wider scale.

A couple of other types of possible CME signatures were put forward respectively by Ambruster et al. (1986) and Mullan et al. (1989). The former observed a sudden flux drop in UV lines that they interpreted as an obscuring cloud of cool plasma ejected from lower atmospheric levels, whereas the latter took advantage of the primary eclipse on V471 Tau to detect fast moving ejected plasmoids in the corona of the K2 dwarf inducing a mass loss rate of $\sim 10^{12} kg\ s^{-1}$, i.e., a large solar prominence every second.

3. Flare Kernels and Chromospheric/Photospheric Evaporation

Because direct measurements of mass losses are rather difficult, we may devise other methods that could help us to obtain further constraints. Most notably, what are the mechanisms that carry the mass from the photosphere to the corona? One possibility is through flaring events, or in other words, explosive chromospheric evaporation.

Stellar flare kernels can leave singular spectral signatures that are evidence for very high plasma densities (Houdebine 1992). For a flare on AD Leo, Houdebine et al. (1993a, b) obtained a kernel area of $\sim 10^{14} m^2$ and a transition region column mass of $\sim 1 g\ cm^{-2}$ which shows that the atmosphere has been "burnt" down to the photosphere. This implies that a mass of $\sim 10^{15} kg$ has been evaporated from the chromosphere and the photosphere, which is comparable to the $\sim 8\ 10^{14} kg$ inferred for the fast mass ejection suspected to be the result of explosive evaporation (Houdebine et al. 1990). Coincidentally, it is indeed interesting to note that the masses as well as the momentum of the upward and downward flows are balanced (Houdebine et al. 1993b).

This evaporated mass is much larger than typical solar values. If it is indeed injected into the corona, it will have major implications for both the physics of the corona and the stellar wind. For this reason, we attempt to estimate the total mass evaporated during flares.

Using flare kernel calculations (Houdebine 1992), we applied the same method for other flares on read dwarfs (Table 2). Furthermore, with this data we found new empirical correlations between the U-band flux U ($erg\ A^{-1}\ cm^{-2}\ s^{-1}$) and peak flux U_p ($erg\ s^{-1} A^{-1}$), the Hδ line width $FWHM_{H\delta}$ (Å), the column mass at the transition region M ($g\ cm^{-2}$) and the evaporated mass m (kg) namely:

$$log(U) = 0.1363\ FWHM_{H\delta} + 24.74 \quad (1)$$

$$M = 0.1301\ FWHM_{H\delta} - 0.5297 \quad (2)$$

$$log(m) = 9.82[log(U_p)]^{0.537} - 42.57 \quad (3)$$

Although important assumptions have been made to calculate the parameters in Table 2, the gross estimates obtained when using these formulae will tell us if we are pointing to an interesting mechanism.

Table 2: Estimates of chromospheric and photospheric evaporations during stellar flares

Star	Authors	U-band	log(M) g cm^{-2}	Area cm^2	Mass (Kg)
YZ CMi	Zarro & Zirin 1988	(2.2)	-0.28	2 10^{18}	1.1 10^{15}
YZ CMi	Doyle et al. 1988	1.2	-0.08	3 10^{17}	2.1 10^{14}
UV Ceti	de Jager et al. 1989	8.0 (W-band)	-0.03	1 10^{18}	9.9 10^{14}
AD Leo	Hawley & Pettersen 1991	5	-1.19	4 10^{19}	2.7 10^{15}
AD Leo	Houdebine 1992	2.1	0.0	1 10^{18}	1.0 10^{15}
AD Leo	Houdebine 1992	0.9	-0.67	1 10^{18}	2.2 10^{14}
Prox Cen	Houdebine 1992	0.28	-1.70	1 10^{16}	2.2 10^{11}

When combining the evaporated mass-U peak flux relation to the flare occurrence formula (Kunkel 1968, 1975), one obtains a mean evaporation rate;

$$\dot{M} = 5.39\ 10^{-70}\ F_o^{2.5a} \int_{F_{min}}^{F_{max}} \frac{exp_{10}[9.821(log(F))^{0.537}]}{F^{1+2.5a}} dF \quad M_{\odot}\ yr^{-1} \quad (4)$$

where a is a constant close to 1 and F_o is the U-band flare flux ($ergs\ A^{-1}\ s^{-1}$) for flares with a frequency of occurrence of $1h^{-1}$ (Kunkel 1975). In the above integral, all the power is at low energies: for instance, for AD Leo, we have a=0.91 and F_o=1.9 10^{25} $ergs\ A^{-1}\ s^{-1}$, which yield $\dot{M} = 8\ 10^{-12}\ M_{\odot}\ yr^{-1}$ when integrating between 10^{25} and 10^{28}, but the figure becomes $\dot{M} = 1\ 10^{-10}\ M_{\odot}\ yr^{-1}$ when integrating down to a value of 10^{24}. Observations of Proxima Cen are evidence that flares exist at a much smaller energies, down to 10^{23} $ergs\ A^{-1}\ s^{-1}$ or less. For this larger energy range, equation (4) yields $\dot{M} = 1.2\ 10^{-9}\ M_{\odot}\ yr^{-1}$.

One can see the importance of the coronal mass flows generated during flares. The mass evaporation rates are in a range overlapping with the mass loss rates (c.f. §5). However, with our present knowledge of flare physics and statistics, the mass evaporation scenario seems unable to feed a 5 10^{-10} $M_{\odot}$ yr^{-1} stellar wind. Nonetheless the low energy distribution of flares which is poorly known can be a potential source of coronal plasma.

4. Spectral Line Asymmetries

Another possibility to carry the mass up to coronal heights is in a form of a more or less continuous flow. Applying the mass flux conservation for a mass loss rate of $10^{-10}M_{\odot}$ yr^{-1} (see next section) and a stellar radius of $0.5R_{\odot}$ actually implies a minimum flow velocity between $0.1km$ s^{-1} in the low chromosphere to several tens of km s^{-1} in the transition region (using Houdebine & Panagi (1990) and Houdebine & Doyle (1994) model chromospheres). Therefore, if such a mass loss is taking place and evaporation is continuous, then asymmetries or shifts should be detectable in emission lines.

Excess emission in the blue wing of H_{α} was noticed for AU Mic (Houdebine 1990; Robinson et al. 1990). More recently, we compared the spectra of stars with identical spectral types (Houdebine et al. 1995); the spectrum of a low activity star was subtracted to those of Gl 815AB, Gl 490A, Gl 867 and St 497. This last star shows a clear excess in the blue wing, whereas for the others a more detailed investigation is required.

Applying simple physics to the excess emission in the blue wing for St 497 gives a corresponding mass flux of 6 $10^{-10}M_{\odot}yr^{-1}$ (for a temperature of formation of 10,000K and an H_{α} emissivity from Houdebine & Doyle 1994 model atmosphere I-2. This value is comparable with that found by Mullan et al. (1989) from the IR/mm excess. It is important to note that a mass flux an order of magnitude less would hardly be detectable.

5. Impact of CMEs on the Stellar Wind and Mass Loss

Thermal pressure gradients are the main driving mechanism for the Solar wind, and its mass flux grows exponentially with the coronal temperature (Parker 1963). In active stars, the much higher coronal temperatures implies that the corona would "explode outwards into space" if it is isothermal, and the mass loss rate would be orders of magnitude larger (Mullan et al. 1992). Another contributor to the wind is the accumulation of CMEs which can provide up to 15% of the mass flux in the near-ecliptic plane of the Sun (Webb & Howard 1994). In stars, the proportions of these two leading mechanisms are unknown, but there are indications that CMEs may generate a mass flux orders of magnitude larger than on the Sun (Mullan et al. 1989, Houdebine et al. 1990).

We give below (Table 3) estimates of the mass loss per year using various methods. For the flare associated CME discussed in Houdebine et al. 1990, we took a plasma temperature of 50,000 K (Houdebine 1992) which yield a mass loss of 3 10^{-12} $M_{\odot} yr^{-1}$. Mullan et al. (1989) noted several highly blue-shifted absorption features in the UV spectrum of the V471 Tau binary system and inferred a mass loss rate of $\sim 10^{-11} M_{\odot} yr^{-1}$. In other active binaries, direct measurements of angular momentum loss suggest a mass loss in the range $10^{-13} - 10^{-6} M_{\odot}/yr$ (Hall et al. 1980, Hall & Kreiner 1980). Mullan et al. (1992) also derived a mass loss rate from the abnormal IR and mm excess in active dMes. They found a rather large figure of 5 $10^{-10} M_{\odot} yr^{-1}$ for YZ CMi (dM4e).

Table 3: Estimated mass-loss rates due to magnetic activity

Star	Authors	Method	Mass loss ($M_{\odot}$ yr^{-1})
AU Mic (dM2e)	This work	Flare statistics	6 10^{-10}
CR Dra (dM2e)	This work	Flare statistics	1.4 10^{-9}
UV Ceti A (dM6e)	This work	Flare statistics	9 10^{-12}
AD Leo (dM4e)	This work	Explosive evaporation	10^{-11}-10^{-9}
St 497 (dM1e)	This work	H_{α} blue wing excess	6 10^{-10}
AD Leo (dM3e)	Houdebine et al. 1990	Flare associated CMEs	3 10^{-12}
V471 Tau (K2/Wd)	Mullan et al. 1989	CMEs	1 10^{-11}
YZ CMi (dM4e)	Mullan et al. 1992	IR/mm excess	5 10^{-10}
dMe	Mullan et al. 1992	Corona thermal expansion	$\geq 3 10^{-11}$
RS CVn Systems	Hall et al. 1980 Hall & Kreiner 1980	Angular momentum loss	$10^{-13} - 10^{-6}$

So far, too little is known about solar CMEs as a function of other magnetic activity indicators so as to extrapolate to higher activity levels in stars. Nevertheless, with simple considerations based on the solar case we may work out a figure that we can compare to the above values. Assuming that; (i) 60% of CMEs are flare associated and 40% are not (Wagner 1984), (ii) the CME kinetic energy is comparable to the flare radiated energy, (iii) the maximum flare integrated energy is $\sim 1\% L_{bol}$ (Kunkel 1973), (iv) the CME average velocity is as on the Sun $< V > \sim 350 km\ s^{-1}$ (Hundhausen et al. 1994), then one can show that for active M dwarfs:

$$\dot{M} \sim 7\ 10^{-9}\ \frac{L_{bol}}{L_{\odot}}\ M_{\odot} yr^{-1} \qquad (5)$$

For high luminosity, early type M dwarfs such as AU Mic ($log(\frac{L_{bol}}{L_{\odot}}) = -1.1$) and CR Dra ($log(\frac{L_{bol}}{L_{\odot}}) = -0.70$), equation (5) yields a mass loss of respectively 6 10^{-10} $M_{\odot} yr^{-1}$ and 1.4 10^{-9} $M_{\odot} yr^{-1}$. According to the variation of the "absolute flare incidence" as a function of absolute magnitude (Kunkel 1975), this rate should diminish for later spectral types such that for one of the faint UV Ceti component ($log(\frac{L_{bol}}{L_\odot}) = -2.9$) we obtain $\dot{M} = 9\ 10^{-12}\ M_\odot yr^{-1}$. Those figures lie in the same range or even higher than those inferred from other methods and highlight the potential importance of coronal mass ejections.

6. Conclusion

The improved performance of spectroscopic facilities has made possible the detection of stellar flare mass motions and probable coronal mass ejections. The exceptionally high flaring rate of active dwarfs strongly supports the concept of permanently ongoing large mass flows in the corona. There are some arguments in favour of a CME driven wind (high flare frequency, closed magnetic topology in the corona?) as well as some influence of magnetic activity on momentum loss and stellar evolution. Indeed, if a mass loss of $5\ 10^{-10} M_\odot yr^{-1}$ applies, then a mass of $0.10 M_\odot$ will be lost in a 200 million years, which is a significant fraction of an M dwarf mass (typically 20%) over a period substantially shorter than the typical time scale for the decline in activity.

References

Ambruster, C.W., Pettersen, B.R., Hawley, S., Coleman, L.A. & Sandman, W.H. 1986, *8 years of UV astronomy with IUE*, ESA SP-263

Baliunas, S.L., Guinan, E.F. & Dupree, A.K. 1984, ApJ 282, 733

Bopp, B.W. & Moffet, T.J. 1973, ApJ 185, 239

Bookbinder, J.A., Walter, F.M. & Brown, A. 1992, Cool Stars, Stellar Systems and the Sun, eds. M.S. Giampapa & J.A. Bookbinder, PASPCS 26, 27

Cordova, F.A. & Mason, K. 1982, ApJ 260, 716

Doyle, G.J., Byrne, P.B. & van den Oord, G.H.J. 1989, A&A 224, 153

Doyle, G.J., Butler, C.J., Byrne, P.B. & van den Oord, G.H.J. 1988, A&A 193, 229

Eason, E.L.E., Giampapa, M.S., Radick, R.R., Worden, S.P. & Hege, E.K. 1992, AJ 104, 1161

Gershberg, R.E. & Skakhovskaya, N.I. 1971, Astr. Zh. 48, 934

Greenstein, J.L. & Arp, H. 1969, Astrophys. Letters 3, 149

Gunn, A.G., Doyle, J.G., Mathioudakis, M., & Avgoloupis, S. 1994a, A&A 285, 157

Gunn, A.G., Doyle, J.G., Mathioudakis, M., Houdebine, E.R. & Avgoloupis, S. 1994b, A&A 285, 489

Hall, D.S. & Kreiner, J.M. 1980, Acta Astron. 30, 387

Hall, D.S. Kreiner, J.M. & Shore, S.N. 1980, Close Binary Stars: Observations and Interpretations, eds. M.J. Plavec, D.M. Popper & R.K. Ulrich, 383

Hawley, S.L. & Pettersen, B.R. 1991, ApJ 378, 725

Houdebine, E.R. 1990, PhD dissertation of the Université d'Orsay-Paris XI

Houdebine, E.R. 1992, Irish Astron. J. 20, 213

Houdebine, E.R. & Panagi, P.M. 1990, A&A 231, 459

Houdebine, E.R., Foing, B.H. & Rodonó, M. 1990, A&A 238, 249

Houdebine, E.R., Foing, B.H., Doyle, J.G. & Rodonó, M. 1993a, 274, 245

Houdebine, E.R., Foing, B.H., Doyle, J.G. & Rodonó, M. 1993b, 278, 109
Houdebine, E.R., Foing, B.H. & Doyle, J.G. 1995, in preparation
Houdebine, E.R. & Doyle, J.G. 1994, A&A 289, 185
Hundhausen, A.J., Burkepile, JT & St. Cyr, O.C. 1994, J. Geophys. Res. 99, 6543
de Jager, C. et al. 1989, A&A 211, 157
Joy, A.H. 1958, PASP 70, 505
Kulapova, A.N., & Skakhovskaya, N.I. 1973, Izv. Krymsk. Ap. Obs. 48, 31
Kunkel, W.E. 1968, IBVS No. 315
Kunkel, W.E., 1973, ApJS 213, 25
Kunkel, W.E., 1975 in Variable Stars and Stellar Evolution, eds. V.E. Sherwood & L. Plaut, 15
Linsky, J.L. 1995, these proceedings
Linsky, J.L., Neff, J.E., Brown, A., Gross, B.D., Simon, T., Andrews, A.D., Rodonò, M. & Feldman, P.A. 1989, A&A 211, 173
Linsky, J.L. & Wood, B.E. 1994, ApJ 430, 342
Mullan, D.J., Sion, E.M., Bruhweiler, F.C. & Carpenter, K.G. 1989, ApJ 339, L33
Mullan, D.J., Doyle, G.J., Redman, R.O. & Mathioudakis, M. 1992, ApJ 397, 225
Neff, J.E., Brown, A. & Linsky, J.L. 1989, Solar and Stellar Flares, Poster Papers, Publ. Catania Obs., eds. B.M. Haisch & M. Rodonó, 111
Parker, E.N. 1963, Interplanetary Dynamical Processes, 74
Phillips, K.J.H., Bromage, G.E., Dufton, P.L., Keenan, F.P. & Kingston, A.E. 1988, MNRAS 235, 573
Robinson, R.D. 1989, Solar and Stellar Flares, Poster Papers,Catania Astrophysical Observatory Special Publications, Eds. B.M. Haisch & M. Rodono, 83
Robinson, R.D., Cram, L.E. & Giampapa, M.S. 1990, ApJ 74, 891
Simon, T., Linsky, J.L. & Schiffer, F.H. 1980, ApJ 239, 911
Wagner, W.J. 1984, Ann. Rev. A&A 22, 267
Webb, D.F. & Howard, R.A. 1994, J. Geophys. R. 99, No. A3, 4201
Woodgate, B.E., Robinson, R.D., Carpenter, K.G., Maran, S.P., & Shore, S.N. 1992, ApJ 397, L95
Zarro, D.M. & Zirin, H. 1985, A&A 148, 240

X-RAY EMISSION FROM O-STARS

S. KITAMOTO, T. SUZUKI, K. TORII AND Y. OHNO
Department of Earth and Space Science, Osaka University
1-1, Machikaneyama-cho, Toyonaka, Osaka, 560, Japan

1. Introduction

The X-ray emission mechanism of OB stars is still an unresolved problem. One possible emission mechanism involves hot gases produced by periodic shocks ploughing through the wind (Lucy & White 1980; Lucy 1982). Coronal models are viable alternatives: A photoionized hot corona can explain the observed soft X-ray spectra of OB stars. In this case, however, the origin of the hot corona is a problem (Stewart & Fabian 1981; Waldron 1984). A crucial key point to distinguish the above two models might be high quality measurements of soft X-ray spectra in order to confirm emission lines and absorption edges. So far, some detections of line emission have been reported (Cassinelli & Swank 1983; Corcoran et al. 1994; Kitamoto & Mukai 1996). These emission lines definitely indicate that the X-ray emission is thermal. But still the origin of the high temperature plasma is an open problem. In this work, we report *ASCA* SIS (Tanaka, Inoue & Holt 1994) observation results of four O stars: ζ Ori, λ Ori, δ Ori and ζ Pup.

2. Observation and Results

The *ASCA* SIS is a CCD camera installed at the focal plane of a high-throughput thin-foil X-ray telescope. It has a high sensitivity up to 10 keV from 0.4 keV and has excellent energy resolution. All observations are roughly one-day observations, and exposure times are ranging from $\sim$ 20 ksec to $\sim$ 40 ksec.

The X-ray energy spectra of the four stars are shown in Figure 1. Prominent Mg and Si K lines can be recognized in the spectra of δ Ori and ζ Ori. Significant Mg and Si K lines can be seen in the ζ Pup spectrum.

We fit energy spectra by an ionization-equilibrium-thin-thermal plasma model (e.g. Kaastra 1992). Absorption by a neutral gas is also taken into

Y. Uchida et al. (eds.),
Magnetodynamic Phenomena in the Solar Atmosphere – Prototypes of Stellar Magnetic Activity, 157–158.

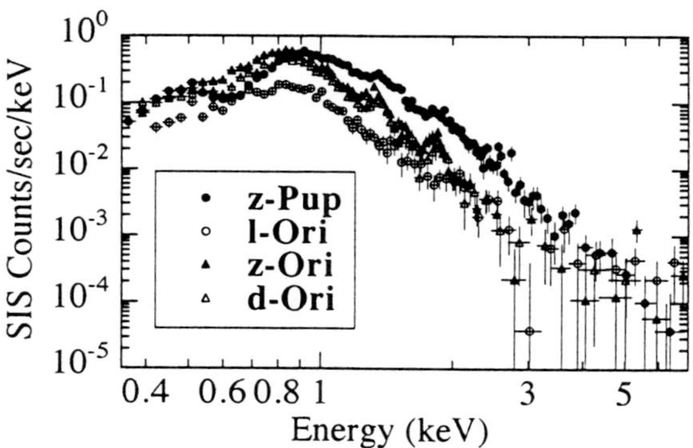

Figure 1. Energy Spectra of four O stars observed by *ASCA* SIS.

account. We assume He to be at cosmic abundance. Also, we assume that C, N and O; Ne and Na; Mg and Al; Si, S, Ar, and Ca; Fe and Ni abundance ratios are the same to the cosmic values, respectively. Since we assume that the He abundance is cosmic, the absolute values obtained are meaningless. Only the relative abundances ratio to the cosmic value are valuable.

All stars are roughly explained by a single component with a temperature from 0.3 keV to 0.7 keV, although to get an acceptable fit more complex models are required. All the interstellar absorptions are less than 10^{21} H cm^{-2} and also no prominent O-K absorption feature is observed.

Since the absolute values of the resultant abundances are meaningless, the ratios to the obtained Si abundance are studied. For the light elements, those are consistent with the cosmic value, with the one exception of Mg in λ Ori. However, all of the iron abundances relative to the Si abundance are significantly smaller than the cosmic value. This result seems to be inconsistent with the FIP effect observed in the solar corona (e.g. Drake et al. 1995)

Reference

Cassinelli, J.P. Swank, J.H. (1983), *ApJ*, **271**, 681
Corcoran M.F., et al. (1994), *ApJ*, **436**, L95
Drake J.J. et al. (1995),*Science*, **267**, 1470
Kaastra J.S., (1992), *An X-Ray Spectral Code for Optically Thin Plasma, (Internal SRON-Leiden Report, updated version 2.0)*
Kitamoto S., and Mukai K. (1996), *PASJ*, submitted
Lucy L.B. (1982), *ApJ*, **255**, 286
Lucy, L.B. White, R.L. (1980), *ApJ*, **241**, 300
Stewart G.C. Fabian A. (1981), *MNRAS*, **197**, 713
Tanaka Y., Inoue H., Holt S.S. (1994), *PASJ*, **46**, L37
Waldron W.L. (1984), *ApJ*, **282**, 256

III. Production of Superhot Plasma and High-Energy Particles in the Sun and Stars

III.1. Solar Flares

EVIDENCE OF MAGNETIC RECONNECTION IN SOLAR FLARES

S. TSUNETA
Institute of Astronomy, The University of Tokyo,
Mitaka, Tokyo 181, Japan

Abstract. The detailed analysis of the 1992 February 21 flare indicates that magnetic reconnection is responsible for heating of the thermal plasmas produced in the flare. The plasma density enhancement presumably associated with slow shocks is observed. The *Yohkoh* observations of this flare fit well with the Petschek model, and the structure of the slow shock is determined with the compressible Petschek theory.

1. Introduction

There are strong pieces of evidence from *Yohkoh* observations that magnetic reconnection is essentially responsible for the flare energy release (*eg.* Tsuneta *et al.* 1992). Evidence so far accumulated is summarized as follows: (1) The height and the foot-point separation of the soft X-ray loop increase as a function of time. This is due to the rise of the X-point location along the neutral sheet. (2) The loop top region has a cusp-like structure, implying the location of a reconnection site at the top of the flare loop. It would be hard to create a sharp cusp-like structure at the loop top without invoking the singular X-point structure. (3) The outer loops have higher temperatures in the decay phase. This is consistent with a flare energy supplied from the reconnection process near the top of the loop as shown in Figure 1.

These qualitative pieces of evidence encourage us to make more quantitative analysis of the data. We perform the detailed analysis of the well-observed 1992 February 21 flare with the compressible Petschek theory (Petschek 1964).

Y. Uchida et al. (eds.),
Magnetodynamic Phenomena in the Solar Atmosphere – Prototypes of Stellar Magnetic Activity, 161–168.

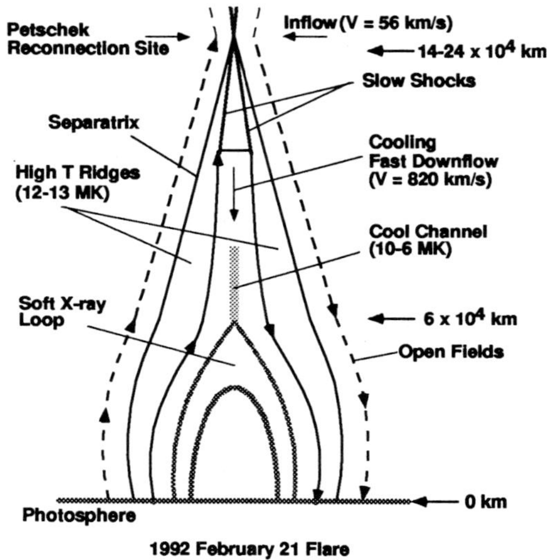

Figure 1. Inferred magnetic structure of the reconnection site. (Tsuneta 1996).

2. Overall Magnetic Structure

The observations of the 1992 February 21 flare show that the outer loops systematically have higher temperatures, reaching the peak (12 MK) far outside the apparent bright X-ray loop where the X-ray intensity is only 1–5 % of the peak. This is strong evidence that a reconnection point is located higher in the corona, and that the attached standing isothermal slow shocks heat the downstream plasma (Cargill and Priest 1983, Forbes and Malherbe 1991, Ugai 1992, Yokoyama 1995). The bright soft X-ray loops are the reconnected flux tubes subsequently filled with evaporated plasma. There is an inflow toward the X-point and the slow shock from the active region corona. The fast outflows come from the reconnection site upward and downward. The cool channel located in between the high temperature ridges is a cooling outflow from the reconnection site. The height of the X-point (diffusion region) is determined such that the hot outflow ($>$ 13 MK) is cooled to the observed temperature (6–7 MK) of the loop-top by conduction along the reconnected field lines. Figure 2 shows the observed plasma parameters across the slow shock (y direction in Figure 3) and along the vertical line (x direction).

3. Petschek Reconnection

Figure 3 (a) shows the schematic structure of the slow shock region. The anti-parallel field lines reconnect at the diffusion region, and the slow shock

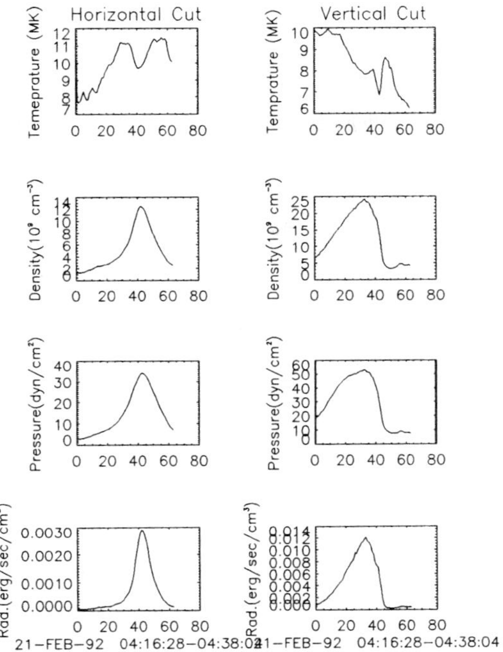

Figure 2. Temperature, plasma density, pressure, and radiative loss along the horizontal (y in Figure 3) and vertical (x in Figure 3) directions. The two humps in temperature horizontal cut corresponds to the high temperature ridges. The low temperature region in between the high temperature ridges is the cool channel with sharp increase of the plasma density, pressure and the radiative loss flux. The vertical cut shows decreasing temperature and increasing density and the pressure with lower height. See Tsuneta (1996) for details.

structure attached to the diffusion region is formed. We assume that the acute half angle of the slow shocks $\alpha \ll 1$, and that of the magnetic separatrix lines $\beta \ll 1$. [This assumption is supported by the observations (Tsuneta 1996).]

The inflow velocity toward the X-point and the slow shocks is $V_{\rm u}$, and the outflows both outward and downward have the Alfvén speed of the upstream either in the Sweet-Parker or in the Petschek theories. Mass conservation is given by

$$n_{\rm u} V_{\rm u} L \cos\alpha = n_{\rm d} V_{\rm d} L \sin\alpha, \tag{1}$$

where L is the cross-sectional length of the inflow as shown in Figure 3 (a), and the suffixes u and d indicate the upstream and the downstream. We thus obtain

$$V_{\rm u} = V_{\rm A} \cdot \tan\alpha \cdot \frac{n_{\rm d}}{n_{\rm u}}, \tag{2}$$

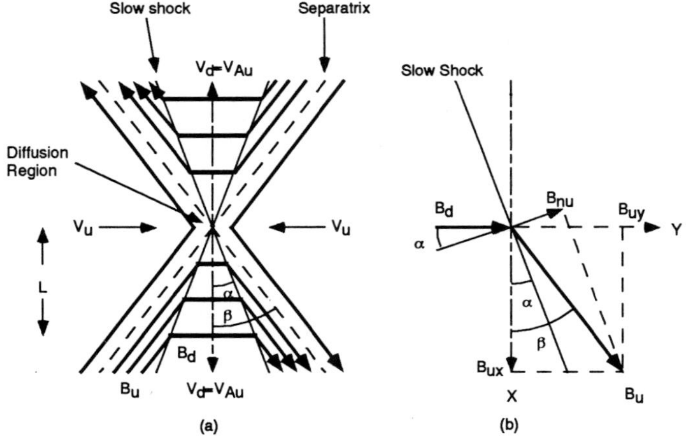

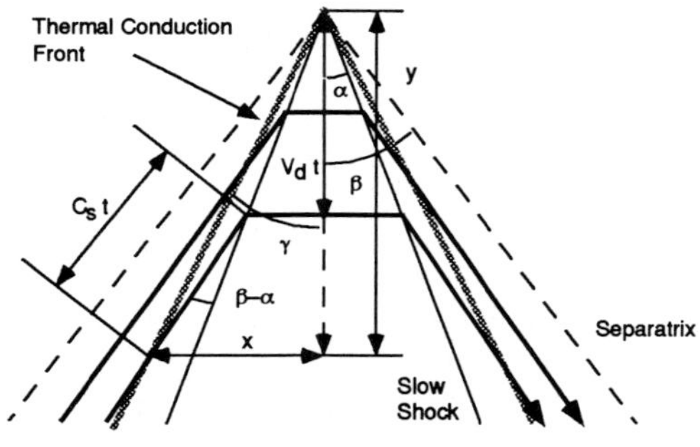

Figure 3. (a),(b) Magnetic structure of the reconnection site. (c) Thermal conduction front and the slow shock.

where V_{A} is given by

$$V_{\mathrm{A}} = \frac{B_{\mathrm{ux}}}{(4\pi m n_{\mathrm{u}})^{0.5}} \tag{3}$$

(m: proton rest mass). From the conservation of the magnetic field strength across the slow shock [Figure 3 (b)],

$$B_{\mathrm{nu}} = B_{\mathrm{uy}} \cos\alpha - B_{\mathrm{ux}} \sin\alpha = B_{\mathrm{d}} \cos\alpha. \tag{4}$$

From equation (4), we obtain

$$\tan\beta - \tan\alpha = \frac{B_{\mathrm{d}}}{B_{\mathrm{ux}}}, \tag{5}$$

where $\tan\beta = B_{uy}/B_{ux}$. If we assume that the slow shock is switch-off ($\alpha \sim 0$), the speed of the shock in a frame of reference at rest is given by

$$V_{shock} = \frac{B_d}{(4\pi m n_u)^{0.5}}. \tag{6}$$

From the steady state condition, the upflow velocity (in the shock frame of reference) V_u is

$$\frac{V_u}{V_A} = \frac{B_d}{B_{ux}}. \tag{7}$$

We thus obtain from equations (2), (5) and (7)

$$\tan\beta = (1 + \frac{n_d}{n_u})\tan\alpha. \tag{8}$$

Downstream magnetic field strength B_d is estimated from the conservation of the normal component of the magnetic field strength across the slow shock;

$$B_d = B_u \frac{\sin(\beta - \alpha)}{\cos\alpha}. \tag{9}$$

We assume that the slow shock is switch-off ($\alpha \sim 0$), and that the upstream speed $V_u \ll V_A$ and the downstream speed is smaller than the local sound speed. The momentum flux across the shock must be continuous, and we have

$$p_u + \frac{B_u^2}{8\pi} = p_d, \tag{10}$$

where p_u and p_d are upstream and downstream plasma pressures. Since the thermal conduction along the reconnected field lines is high, we expect the slow shock to be isothermal within the thermal conduction front as shown in Figure 3 (c) (*eg.* Yokoyama 1995). Since the conduction front propagates with speed of the local sound speed C_s along the reconnected field line, the acute half angle of the conduction front γ [Figure 3 (c)] is given by

$$\tan\gamma = \frac{x}{y} = \frac{\tan\alpha + (\frac{C_s}{V_A})\sin\beta}{1 + (\frac{C_s}{V_A})\cos\beta} \sim \frac{\alpha + (\frac{C_s}{V_A})\beta}{1 + (\frac{C_s}{V_A})}. \tag{11}$$

(The downstream side will be uniformly heated, unless $C_s \ll V_{Au}\tan\alpha$.) If the speed of the conduction front is faster than the Alfv'en speed, the conduction front is located closer to the separatrix. If otherwise, the conduction front is closer to the slow shock. Since $\frac{C_s}{V_A} \sim 1$ (Table 1), $\gamma \sim (\alpha + \beta)/2$. Since the slow shock is isothermal in any case, the jump condition for the plasma density is given as follows from equation (10):

$$\frac{n_u}{n_d} = 1 + \frac{1}{\beta_u}, \tag{12}$$

where plasma β_u is given by

$$\beta_u = \frac{p_u}{\frac{B_u^2}{8\pi}}. \tag{13}$$

These equations are a basic set of equations, that are useful in analyzing the soft X-data of the reconnection site. From the soft X-ray observations, we can obtain the temperatures, densities, and the pressures of both the upstream and the downstream. X-ray morphology may also provide the information on the the angles of the magnetic separatrix lines, the conduction fronts, and the slow shocks. For instance, equation (12) allows us to estimate the magnetic field strength of the upstream. The inflow velocity can be estimated from equations (2) and (8). Tables 1 and 2 list these observables and the derived parameters (see Tsuneta 1996 for details).

TABLE 1. Upstream-downstream parameters of magnetic separatrix lines and slow shocks (Tsuneta 1996)

Parameter	Upstream	Downstream
Temperature	< 7 MK	> 12–13 MK
Density	10^9 cm^{-3}	$< 5 \times 10^9$ cm^{-3}
Pressure	3 dyn cm^{-2}	< 20 dyn cm^{-2}
Magnetic field	20-30 G	5 G
plasma β	~ 0.25	~ 30
Flow velocity	56 km sec^{-1}	800 km sec^{-1}
Sound Mach number	~ 0.1	~ 1
Alfvén Mach number	~ 0.07	~ 5
Alfvén speed	800 km sec^{-1}	155 km sec^{-1}
Sound speed	550 km sec^{-1}	770 km sec^{-1}
Mass Flux	$\sim 1.7 \times 10^{35}$ sec^{-1}	$\sim 2.4 \times 10^{35}$ sec^{-1}

4. Discussion

Following three key questions need to be answered to propose that magnetic reconnection is responsible for the flare:

- (1) Is the plasma density jump, which is direct evidence for slow shocks, identified in the observed density and pressure maps?
- (2) Is the Petschek theory (structure) consistent with the data?
- (3) Can magnetic reconnection provide the energy that is required for the flare?

TABLE 2. Physical parameters of the reconnection region (Tsuneta 1996)

Parameter	Value
Cool Channel Temperature	$10 - 6$ MK
Cool Channel Density	$5 - 25 \times 10^9$ cm^{-3}
Cool Channel Pressure	$20 - 50$ dyn cm^{-2}
Density jump across the slow shock	< 5
Temperature jump across the slow shock	~ 1 $(\gamma \sim 1)$
Half angle of the slow shocks	0.8–1.8 degree
Half angle of the separatrices	5–11 degree
Kinetic energy of the outflow	5×10^{27} erg sec^{-1}
Shock heating rate	9×10^{27} erg sec^{-1}
Total energy	14×10^{27} erg sec^{-1}
Magnetic energy supply from the upstream	6×10^{27} erg sec^{-1}
Soft X-ray loop height	$\sim 6 \times 10^4$ km
X-point height	$\sim 8 - 18 \times 10^4$ km above the loop top
Slow shock length	$\sim$ a few 10^4 km

4.1. CAN DENSITY (PRESSURE) JUMP BE SEEN IN THE DATA?

Figure 2 horizontal cut shows the temperature and the density profiles along the horizontal (y in Figure 3) direction. The density enhancement is clearly seen, although the density profile is smooth and does not have a steep jump. We need to examine whether this smooth transition is the inherent structure of the slow shock, or is due to photon scattering to the dark reconnection site from the bright flare loop. In any case, the density enhancement of the downward region is seen, and we consider this to be evidence to support the slow shock structure. We plan to do the same analysis for other limb flares.

4.2. PETSCHEK THEORY (STRUCTURE) AND OBSERVATIONS

All the key parameters of the reconnection site are determined from the observations with Petschek model as shown in Tables 1 and 2. The Petschek picture (reconnection X-point with slow shock structure) appears to fit well with the data. The compressible Petschek theory is a useful guiding theory in analyzing the data. Although this does not rule out alternative interpretations or theories, it is not easy to explain the global features of these observations without the concept of magnetic reconnection and the associated slow shocks.

4.3. CAN RECONNECTION SUPPLY THE FLARE ENERGY?

From the derived parameters, we can show that total magnetic energy $E_{\rm B}$ fed to the reconnection site with the inflows;

$$E_{\rm B} = \frac{B_{\rm up}^2}{8\pi} S V_{\rm u} \tag{14}$$

is close to the estimated energy produced in the reconnection site (Tsuneta 1996). (S is the cross sectional area of the reconnection site.) One of the criticisms to the reconnection concept to explain solar flares is that magnetic reconnection is a local phenomena, and can not produce the huge amount of energy for flares. This concern is relieved now: the magnetic energy is brought with the inflows from vast area of the upstream regions. The total energy released in flares thus depends on (1) magnetic field strength of the upstream as well as (2) the size of the upstream volume that contains usable magnetic fields. If the upstream volume is small, these flares may be strong, but can be of short duration (and vice versa).

5. Conclusion

Yohkoh observations show that magnetic reconnection serves as highly efficient engine to convert magnetic energy into plasma kinetic and thermal energies with standing slow shocks. This implies an important role of magnetic reconnection for energetic phenomena in other astrophysical as well as magnetospheric contexts.

References

Cargill P. J. and Priest, E. R. (1983)*Solar Phys.*, **266**, 383.

Forbes, T. G. and Malherbe, J. M. (1991) *Solar Phys.* **135**, 361.

Petschek, H. E. (1964) *AAS-NASA Symposium on Solar Flares*, ed. W. N. Hess (NASA SP-50) p 425.

Tsuneta, S., Hara, H., Shimizu, T., Acton, L., Strong, K., Hudson, H., and Ogawara, Y. (1992) *Publ. Astron. Soc. Japan,* **44**, L63.

Tsuneta, S. (1996) *Ap. J.,* **456**, 840.

Ugai, M. (1992) *Phys. Fluids B* **4**, 2953.

Yokoyama, T. (1995) *Ph. D. Thesis*, National Astronomical Observatory, Tokyo.

NON-THERMAL PROCESSES AND SUPERHOT PLASMA CREATION IN SOLAR FLARES

T. SAKAO AND T. KOSUGI
National Astronomical Observatory, Mitaka, Tokyo 181, Japan

Abstract. We review new aspects of solar flare energy release and related non-thermal/thermal processes revealed by the Hard X-ray Telescope (HXT) aboard *Yohkoh*. Hard X-ray imagery of impulsive solar flares with HXT has shown that the bulk of hard X-rays above 30 keV is emitted near the footpoints of a flaring magnetic loop by electrons accelerated near the top of the loop, manifesting itself most frequently as "double sources" located on both sides of a magnetic neutral line. Besides these "double footpoint sources", a third hard X-ray source, located *above* the top of the flaring loop visible in soft X-rays, was discovered in some impulsive flares near the solar limb. This discovery provides us with clear evidence that solar flare energy release takes place outside the closed loop, most likely via magnetic reconnection in an open magnetic field configuration. Observations of superhot plasma creation with HXT are also briefly reviewed and their implication in the energy release and transfer in solar flares is discussed.

1. Introduction

During the impulsive phase of a solar flare, significant amount of the released energy (say, up to $\sim 1/2$) is converted into particle acceleration. At the same time, superhot plasmas with temperature exceeding 30 MK may be created. Since solar flare hard X-rays are emitted by collisions of highly energized electrons with ambient ions (Bremsstrahlung emission) and the emission is optically thin, the hard X-rays reflect these two processes (non-thermal and thermal) which operate following the energy release. Therefore, hard X-ray imagery of flares is one of the most powerful tools for investi-

Y. Uchida et al. (eds.),
Magnetodynamic Phenomena in the Solar Atmosphere – Prototypes of Stellar Magnetic Activity, 169–176.

gating the energy release, particle acceleration, superhot plasma creation, and subsequent hard X-ray emission which take place during the course of a flare.

The *Yohkoh* Hard X-ray Telescope (HXT; Kosugi *et al.*, 1991) has advanced characteristics relative to its predecessors (*SMM* HXIS and *Hinotori*). (1) It can image solar flare hard X-ray sources simultaneously in four energy bands covering 14 – 93 keV (with energy boundaries of 14-23-33-53-93 keV); it should be noted that the telescope takes images in the energy range above 30 keV for the first time, and that the simultaneous imaging enables us to observe both non-thermal and thermal phenomena at one time. (2) The basic temporal resolution is 0.5 s which enables us to study rapid fluctuations of hard X-ray emissions.

So far, HXT has observed more than 1,000 flares since the beginning of its routine observations in October 1991 (Kosugi *et al.*, 1995). In this manuscript, we summarize the new observational findings with HXT of solar flare energy release and related non-thermal/thermal processes.

2. Non-Thermal Processes — Energy release and particle acceleration

2.1. EVIDENCE FOR ELECTRON PRECIPITATION

One of the new findings with HXT is that the bulk of hard X-rays above 30 keV in impulsive flares is emitted from the footpoints of flaring magnetic loop(s) by electrons accelerated near the top of the loop. Sakao *et al.* (1996; see also Sakao, 1994) have revealed that hard X-ray sources most frequently show the double-source structure ($\sim$ 40 % of the events examined) in the energy range above 30 keV, while the remainer showing a single- or multiple-source structure. The double sources are located on both sides of a magnetic neutral line. In many cases, they are also located at both ends of a flaring loop seen with SXT (the Soft X-ray Telescope; Tsuneta *et al.*, 1991) and/or the loop-like structure (possibly tracing the flaring loop) seen in low-energy ($\leq$ 20 keV) hard X-rays (Kosugi *et al.*, 1992; Hudson *et al.*, 1994). Hence, it can be said that the observed double sources reflect fundamental characteristics of solar flare hard X-ray emission; at least in impulsive flares, the bulk of hard X-rays above 30 keV is emitted near the footpoints of a flaring loop. (Note that the double-source structure has been observed previously in a limited number of flares; see, *e.g.*, Hoyng *et al.*, 1981.)

Detailed temporal as well as spectral study of the double sources seen in seven impulsive flares (Sakao *et al.*, 1996) has shown that (1) the double sources emit hard X-rays simultaneously to each other within a fraction of a second, (2) the brighter source of a double-source pair tends to be located in the weaker photospheric magnetic field region regardless of its polarity, and that (3) the brighter source tends to have a harder spectrum than the darker one.

Observational fact (1) strongly suggests that hard X-rays are emitted near the footpoints by accelerated electrons which stream down along the loop towards both ends and that the acceleration site is located near the top of the loop. No other candidate for the origin of hard X-rays (such as thermal conduction fronts or proton beams; Batchelor *et al.*, 1985; Simnett, 1995) seems to be able to explain the observed simultaneity if we consider a loop length of a few $\times 10^4$ km inferred from the separation of the double footpoint sources (typically $\sim 20''$ or $\sim 1.5 \times 10^4$ km).

Observations (2) and (3) can naturally be explained by the precipitation of energetic electrons towards the two footpoints. Accelerated electrons precipitating downwards to the footpoint with the stronger magnetic field (hence with the stronger field convergence) may undergo strong magnetic mirroring. Only a limited fraction of the electrons can reach deep in the corona or chromosphere where the bulk of hard X-rays are emitted, yielding a smaller amount of hard X-rays than the other footpoint. At the same time, because electrons cannot precipitate so deeply into the (darker) footpoint, hard X-ray emission from the darker footpoint may originate from a blend of thick- and thin-target processes due to the lower ambient plasma density at the emission site. On the other hand, at the brighter footpoint, precipitating electrons may not suffer a strong mirroring effect, emitting hard X-rays via thick-target interaction with the dense ambient plasma deep in the corona or chromosphere. As the thick-target process yields a harder spectrum than the thin-target process for the same spectrum of injected electrons (*e.g.*, Hudson *et al.*, 1978), the brighter footpoint shows a harder spectrum than the darker one.

Thus, we conclude that, at least in impulsive solar flares, (a) the acceleration site is located near the apex of a flaring loop, (b) the accelerated electrons precipitate towards both footpoints of the loop with preference to the footpoint with the weaker magnetic field convergence, and (c) emit the bulk of hard X-rays (above 30 keV) at the footpoints.

2.2. ENERGY RELEASE SITE IN SOLAR FLARES

The most important finding with HXT relevant to solar flare energy release is the discovery of an *above-the-loop-top* hard X-ray source. Through careful coalignment of hard X-ray images with soft X-ray images taken with SXT, Masuda (1994; see also Masuda *et al.*, 1994, 1995) found that in impulsive flares occurring near the solar limb there may exist a blob-like hard X-ray source in the energy bands above 23 keV located *above* the corresponding

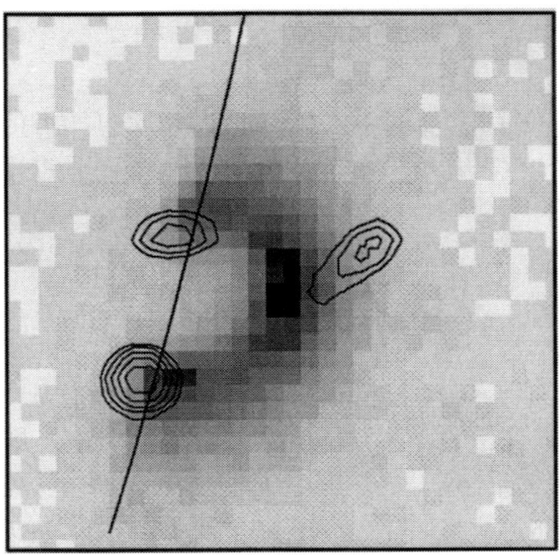

Figure 1. Above-the-loop-top hard X-ray source observed in the 13 January, 1992 flare (17:27 UT) near the west limb. The solar north is to the top, east to the left. The solar limb is shown in the map. The field of view of the map covers $78'' \times 78''$. *Contours:* Hard X-ray sources in the M2-band (33-53 keV) of HXT. The contour levels are 18, 25, 35, 50, and 71 % of the maximum brightness. *Gray scale:* The corresponding soft X-ray image taken with the Be filter of SXT.

soft X-ray loop (in at least three out of ten events). Hard X-ray emissions from this source are weaker than those from the two footpoints by a factor of ~ 5. A representative case is shown in figure 1 where we see the source located $\sim 8{,}000$ km above the soft X-ray flaring loop. The above-the-loop-top source emits hard X-rays impulsively, within the available temporal resolution (of several to ten seconds due to insufficient photon statistics). The source shows a relatively hard spectrum; if we assume a single-temperature thermal emission, temperature and emission measure of this source would be $T \sim 200$ MK and $EM \sim 10^{44} - 10^{45}$ cm^{-3}, respectively.

The discovery of the above-the-loop-top source is of crucial importance for understanding energy release in solar flares. Although it is widely believed that energy released during a flare originates from the non-potential magnetic field (Sweet, 1969), how and where the magnetic energy is released has long been controversial. It is clear in figure 1 that a certain energetic process is taking place *outside* the closed loop visible in soft X-rays. Also, if we take into account the conclusion that the particle acceleration site is located near the apex of a flaring loop (as mentioned in section 2.1), it is conceivable that this process is directly related to the particle acceleration. Finaly, the existence of the above-the-loop-top source reminds us of a magnetic field configuration with a current sheet above a closed loop (or a system of loops) sandwiched by anti-parallel magnetic fields, in which magnetic reconnection progresses.

Such a magnetic field configuration ("loop-with-a-cusp" structure), originally proposed for explaining Hα two-ribbon events with Hα filament eruption and solar wind breakthrough associated with flares (*e.g.*, Hirayama, 1974), has been observed in the soft X-ray corona with a large spatial scale (from $\sim 10^5$ km to over 10^6 km) involving long duration events (LDEs) and arcade formation (Tsuneta *et al.*, 1992a, 1992b). By showing that plasma temperature is the highest at the outer boundaries of the cusp and that the whole structure expands vertically as well as horizontally with time, Tsuneta *et al.* (1992a) claimed that the loop-with-a-cusp structure can be regarded as evidence for magnetic reconnection. The existence of the above-the-loop-top hard X-ray source implies that such a magnetic field configuration and subsequent (X-type) field reconnection may also be responsible at least in some impulsive flares with more violent energy release and on a smaller spatial scale than the soft X-ray events mentioned above. Following this discovery, Shibata *et al.* (1995) have found faint soft X-ray ejecta (which remind us of filaments and/or sprays seen in Hα events) during the impulsive phase in all the ten events studied by Masuda (1994). This observation may give more evidence for the "loop-with-a-cusp" structure with X-type magnetic reconnection being responsible for solar flare energy release.

One should notice, however, that the luminosity of the above-the-loop-top source is unexpectedly large. Hard X-ray emitting electrons need be trapped in this source (or the ambient plasma density would be arbitrarily large (Wheatland and Melrose, 1995)); otherwise the source would not show a blob-like structure nor so bright as observed. As of this writing, little theoretical interpretation has been presented for explaining this source (*e.g.*, Hudson and Ryan, 1995).

3. Superhot Plasma Creation in Solar Flares

So far we have concentrated on non-thermal aspects of solar flares. However, it should be recognized that thermal plasmas, together with non-thermal particles, represent in a non-negligible part of flare energetics (see, *e.g.*, Wu *et al.*, 1986). The *Yohkoh* HXT also reveals some fundamental aspects of thermal processes in solar flares.

Masuda (1994) noticed that later in the impulsive as well as in the gradual phases of a flare, there always appears a hard X-ray source (≤ 30 keV) with a soft spectrum located at the apex of the soft X-ray flaring loop in all of the ten limb flares he examined. Hard X-rays from this source show gradual temporal variations (hence the source is called the "loop-top gradual source") and are well characterized by thermal emission from a

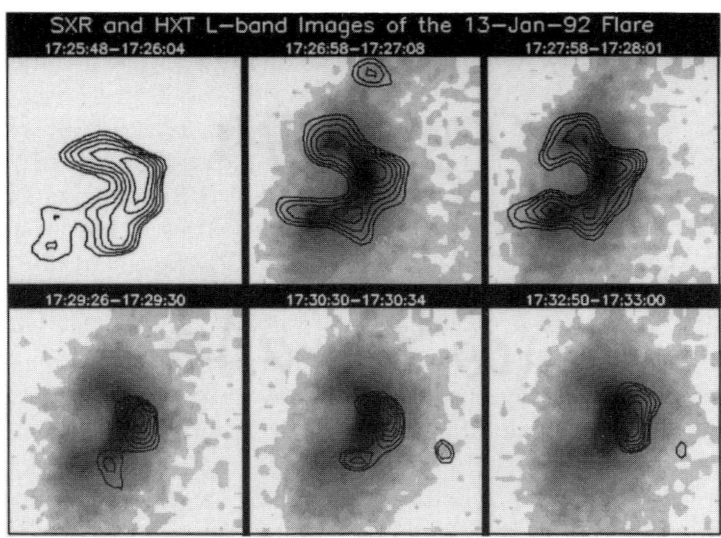

Figure 2. Superhot plasmas seen in the 13 January, 1992 flare. *Contours:* Hard X-ray sources in the L-band (14-23 keV) of HXT. The contour levels and the size of the field of view of each map are the same as those in figure 1. *Gray scale:* The corresponding soft X-ray images taken with the Be filter of SXT. At the time of the first HXT image (17:25:48-17:26:04 UT), the corresponding SXT image was not available. Solar north is to the top, east to the left.

high-temperature plasma with $T \sim 20 - 40$ MK and $EM \sim 10^{46} - 10^{49}$ cm^{-3} (electron density typically $\sim 10^{10}$ cm^{-3}).

Figure 2 shows hard X-ray sources in the L-band seen in the 13 January, 1992 flare. During the impulsive phase, the L-band source shows a loop-like structure tracing the flaring loop visible in soft X-rays. This may indicate emission from a thermal plasma filling the loop, possibly evaporated from the chromosphere due to the electron bombardment. On the other hand, after 17:29 UT (*i.e.*, after the peak of hard X-ray emission), the bulk of hard X-rays is concentrated near the loop top, above the brightest portion of the corresponding soft X-ray loop. We speculate that such a structural change indicates that some physical parameters in or above the flaring loop have changed, which made (direct) plasma heating near the loop top more effective than particle acceleration.

Another interesting example: in the 6 February, 1992 flare, M2-band HXT images indicate that a superhot plasma ($T \geq 40$ MK) is located at the apex of the flaring loop, without accompanying significant hard X-ray emission from the footpoints (see figure 2 of Kosugi *et al.*, 1994). The superhot plasma gradually expands towards the footpoints, most likely due to thermal conduction. Kosugi *et al.* (1994) argued that, unlike normal impulsive flares where the primary product of the energy release is accelerated particles, the flare energy released near the loop top is mainly consumed in direct heating of the plasma there. They also suggested that the plasma density inside the flare loop is responsible for the efficiencies of thermal/non-thermal processes. It is worthwhile mentioning that the superhot plasma of this flare, visible in hard X-rays, in fact shows the highest emission measure of $\sim 10^{49}$ cm^{-3}. This corresponds to a large plasma density of $\sim 10^{11}$ cm^{-3}, among the ten aforementioned limb flares.

From these examples, we suggest that at least a part of superhot plasmas at the loop top (manifesting itself as a "loop-top gradual source") is a product of direct heating caused by the energy release near the loop top and that such a heating process occurs *simultaneously* with non-thermal particle acceleration. Further studies on this issue is now under way.

4. Concluding Remarks

We have presented two new observations with *Yohkoh* HXT which directly address the energy release and acceleration/propagation of energetic electrons in impulsive solar flares (*i.e.*, non-thermal processes). HXT also provides observations of superhot plasma creation with clear spatial information with respect to the flare structure. Little attempt has been made in this manuscript to discuss possible relationships between the non-thermal and thermal processes which may take place simultaneously, near the loop top, during the course of a flare. It is of crucial importance to understand such non-thermal/thermal processes in a single context, which may provide us with a unified view on the roles of the two (apparently different) processes and on their energetics in solar flares.

ACKNOWLEDGEMENTS We are grateful to ISAS, NASA, SERC, and the *Yohkoh* Team for continuously supporting the mission. Hugh Hudson is acknowledged for careful reading of the manuscript. TS expresses his thanks to Satoshi Masuda for providing data as well as valuable and enjoyable discussions.

References

Batchelor, D.A., Crannell, C.J., Wiehl, H.J., and Magun, A. (1985) Evidence for collisionless conduction fronts in impulsive solar flares, *Astrophys. J.*, **295**, pp. 258–274.

Hirayama, T. (1974) Theoretical model of flares and prominences I: Evaporating flare model, *Solar Phys.*, **34**, pp. 323–338.

Hoyng, P., Machado, M.E., Duijveman, A., Boelee, A., de Jager, C., Fryer, R., Galama, M., Hoekstra, R., Imhof, J., Lafleur, H., Maseland, H.V.A.M., Mels, W.A., Schadee, A., Schrijver, J., Simnett, G.M., Svestka, Z., van Beek, H.F., van Tend, W., van der Laan, J.J.M., van Rens, P., Werkhoven, F., Willmore, A.P., Wilson, J.W.G., and

Zandee, W. (1981) Hard X-ray imaging of two flares in active region 2372, *Astrophys. J. Lett*, **244**, pp. L153–L156.

Hudson, H.S., Canfield, R.C., and Kane, S.R. (1978) Indirect estimation of energy disposition by non-thermal electrons in solar flares, *Solar Phys.*, **60**, pp. 137–142.

Hudson, H.S., Strong, K.T., Dennis, B.R., Zarro, D., Inda, M., Kosugi, T., and Sakao, T. (1994) Impulsive behavior in solar soft X-radiation, *Astrophys. J. Lett.*, **422**, pp. L25–L27.

Hudson, H., and Ryan, J.. (1995) High-energy particles in solar flares, *Ann. Rev. Astron. Astrophys.*, **33**, pp. 239–282.

Kosugi, T., Makishima, K., Murakami, T., Sakao, T., Dotani, T., Inda, M., Kai, K., Masuda, S., Nakajima, H., Ogawara, Y., Sawa, M., and Shibasaki, K. (1991) The hard X-ray telescope (HXT) for the Solar-A mission, *Solar Phys.*, **136**, pp. 17–36.

Kosugi, T., Sakao, T., Masuda, S., Makishima, K., Inda, M., Murakami, T., Ogawara, Y., Yaji, K., and Matsushita, K. (1992) The hard X-ray telescope (HXT) onboard Yohkoh: Its performance and some initial results, *Publ. Astron. Soc. Japan*, **44**, pp. L45–L49.

Kosugi, T., Sakao, T., Masuda, S., Hara, H., Shimizu, T., and Hudson, H.S. (1994) Hard and soft X-ray observations of a super-hot thermal flare of 6 February, 1992, in "Proc. of Kofu Symposium", eds. Enome, S. and Hirayama, T., NRO Report No.**360**, pp. 127–129.

Kosugi, T., Sawa, M., Sakao, T., Masuda, S., Inda-Koide, M., Yaji, K., and Sato, J. (1995) "The Yohkoh HXT Databook" (National Astronomical Observatory, Japan; available upon request from the authors).

Masuda, S. (1994) Hard X-ray sources and the primary energy release site in solar flares, Ph.D. thesis (University of Tokyo).

Masuda, S., Kosugi, T., Hara, H., Tsuneta, S., and Ogawara, Y. (1994) A loop-top hard X-ray source in a compact solar flare as evidence for magnetic reconnection, *Nature*, **371**, pp. 495–497.

Masuda, S., Kosugi, T., Hara, H., Sakao, T., Shibata, K., and Tsuneta, S. (1995) Hard X-ray sources and the primary energy-release site in solar flares, *Publ. Astron. Soc. Japan*, **47**, pp. 677–689.

Sakao, T. (1994) Characteristics of solar flare hard X-ray sources as revealed with the hard X-ray telescope aboard the Yohkoh satellite, Ph.D. thesis (University of Tokyo).

Sakao, T., Kosugi, T., and Makishima, K. (1996) Double footpoint hard X-ray sources in solar flares: II — Statistical approach —, *Publ. Astron. Soc. Japan* submitted.

Shibata, K., Masuda, S., Shimojo, M., Hara, H., Yokoyama, T., Tsuneta, S., Kosugi, T., and Ogawara, Y. (1995) Hot-plasma ejections associated with compact-loop solar flares, *Astrophys. J. Lett.*, **451**, pp. L83–L85.

Simnett, G.M. (1995) Protons in flares, *Sp. Sci. Rev.*, **73**, pp. 387–432.

Sweet, P. (1969) Mechanisms of solar flares, *Ann. Rev. Astron. Astrophys.*, **7**, pp. 149–176.

Tsuneta, S., Acton, L., Bruner, M., Lemen, J., Brown, W., Caravalho, R., Catura, R., Freeland, S., Jurcevich, B., Morrison, M., Ogawara, Y., Hirayama, T., and Owens, J. (1991) The soft X-ray telescope for the Solar-A mission, *Solar Phys.*, **136**, pp. 37–67.

Tsuneta, S., Hara, H., Shimizu, T., Acton, L.W., Strong, K.T., Hudson, H.S., and Ogawara, Y. (1992a) Observation of a solar flare at the limb with the Yohkoh soft X-ray telescope, *Publ. Astron. Soc. Japan*, **44**, pp. L63–L69.

Tsuneta, S., Takahashi, T., Acton, L.W., Bruner, M.E., Harvey, K.L., and Ogawara, Y. (1992b) Global restructuring of the coronal magnetic fields observed with the Yohkoh soft X-ray telescope, *Publ. Astron. Soc. Japan*, **44**, pp. L211–L214.

Wheatland, M.S., and Melrose, D.B. (1995) Interpreting Yohkoh hard and soft X-ray flare observations, *Solar Phys.*, **158**, pp. 283–299.

Wu, S.T., de Jager, C., Dennis, B.R., Hudson, H.S., Simnett, G.M., Strong, K.T., Bentley, R.D., and Bornmann, P.L. (1986) Flare energetics, in "Energetic Phenomena on the Sun", eds. Kundu, M.R., and Woodgate, B.E., SMM Flare Workshop Proceedings, NASA CP-2376, (Greenbelt, Maryland), Chapter 5.

MASS MOTIONS IN FLARES

R.D. BENTLEY
Mullard Space Science Laboratory
University College London
Holmbury St. Mary
Dorking, Surrey RH5 6NT
United Kingdom

1. Introduction

Mass motions of material are seen in many wavelength and at many different layers of the solar atmosphere during flares. In this paper we try to summarize observations of mass motions in the corona made in soft X-rays by crystal spectrometers on the SMM, P78-1, Hinotori and Yohkoh spacecraft. Such observations are key to our understanding of energy release and of what is happening to the hot plasma during a flare.

We will consider two categories of mass motions seen in the soft X-ray spectral lines: a) the blueshifted component, and b) line broadening.

2. Blueshifted Component

At the onset of some flares, an excess emission is seen on the blue (short-wavelength) side of soft X-ray emission lines such as those of helium-like ions of Ca XIX and Fe XXV. The emission is from Doppler shifted material that is moving towards the observer in what appears to be directed motion – it is seen at a time that is closely related to emission in hard X-rays. The material is thought to be flowing from the chromosphere up into the corona and the bluewing is considered to be the signature of *Chromospheric Evaporation* resulting from precipitating energetic electrons.

In this section, in an attempt to better characterize the blue component, we discuss the velocity, timing, location, temperature dependence of the component, and also look at the momentum balance of the component with material observed at other wavelengths.

Y. Uchida et al. (eds.),
Magnetodynamic Phenomena in the Solar Atmosphere – Prototypes of Stellar Magnetic Activity, 177–184.

2.1. VELOCITY OF THE BLUESHIFTED COMPONENT

The velocity range of the blueshifted component varies considerably from flare to flare. In some flares the range is only few tens of km/s, while in others it is several hundred. The maximum velocity does not appear to be sensitive to the magnitude of the peak hard X-ray flux (i.e. total energy of the precipitating energetic electrons). There is usually a rest component present at flare onset, but in a few flares only the blueshifted component is seen. This is seen to "decelerate" to rest position during impulsive phase. Very rarely, the blue component is completely separate from the rest component (Bentley *et al.*, 1986).

The material that produces the blueshifted component is thought to be flowing up the legs of flaring loops within the corona. Since the loop legs should normally be radial to the solar surface, loops that are located near the solar limb would not be expected to present a velocity component to the observer and only small blueshifts should be measured. Such a centre-to-limb (*cosine*) variation in the size of the blueshifted component is in fact observed and is demonstrated for Yohkoh BCS observations in the survey paper of Mariska *et al.* (1993). By assuming that the leg of the loop is radial to the surface, the absolute velocity of the component can be determined by multiplying the observed velocity by the cosine of the angle from Sun centre.

Velocities quoted for the blue component are determined by model dependent codes - a technique that fits two Gaussian components is most frequently used (e.g. Antonucci *et al.*, 1982; Fludra *et al.*, 1989). However, as the flare progresses there are inherent uncertainties in determining what proportion of the signal should be attributed to the moving and rest components, particularly when the blue shifted component is of very low amplitude. There have been various attempt to improve the fitting procedures, e.g. by fitting three components (Antonucci *et al.*, 1990), or by coupling the blue and rest components together in some way (Fludra *et al.*, 1989). However, the basic difficulty remains that such techniques poorly describe the problem since the blue component clearly consists of more than a simple Gaussian distribution of velocities. The different methods lead to different conclusions about the size and evolution of the blueshifted and rest components. When the second (blue) component should be "switched off" later in the flare presents a particular difficulty.

It is hoped that a new approach, called Velocity Differential Emission Measure (VDEM; Newton *et al.*, 1995), will go some way in helping to resolve the problem of characterizing the velocities of the blueshifted component.

2.2. TIMING OF THE BLUE COMPONENT

Whether the hard X-rays peak before the soft X-ray bluewing has been a matter of controversy. Not all investigators have agreed on the timing (e.g. Bentley *et al.*, 1994, Doschek *et al.*, 1993; Cheng *et al.*, 1994; Plunkett *et al.*, 1994), but views are converging. Bentley *et al.* (1994) find that the principal blue component peaks almost the same time or shortly after a hard X-ray burst. The burst is not necessarily the largest hard X-ray spike in an event, but it is usually the first significant one.

Differences in the conclusions reported by investigators may have arisen because of the analysis techniques used. In some cases, the intensity of the blue-shifted component is divided by that of the resonance line intensity in the same spectrum. While this helps "normalize" two intensities with respect to each other, the intensity of resultant quantity is very dependent on the rate of increase of the resonance line - for a blue component of fixed size, the time that the divided quantity appears to peak will occur earlier if the resonance peak rises more quickly (i.e. the peak appears time-shifted earlier for more rapidly rising flares).

In addition, there appears to an obsession that the largest hard X-ray peak will produce the largest blueshift. Since this is not always the case, and in fact the blue-wing appears to result from the first significant spike in an event, blueshifts have been erroneously perceived to peak before the hard X-rays. Occasionally, weak blue-wing emission is seen before the onset of the impulsive phase. However, this is of very low intensity compared to the main blueshifted component and appears to be associated with low-level hard X-ray activity.

Winglee *et al.* (1991) have studied the interrelationship between hard and soft X-rays. For all flares they show that the hard X-rays have a power-law spectrum that breaks down during the rise phase. They also find that the peak blueshift occurs is when the (downward) break in hard X-ray power-law spectrum vanishes.

At some point, the blue component ceases although the hard X-ray emission has not always finished. Why this happens is not clear, some investigators suggest that the loop has filled to capacity, while Winglee *et al.* think it is caused by the disappearance of driving electric field.

(Savy, 1995) reports that in some simple cases, the blue component follows a similar time profile to the hard X-ray flux.

2.3. LOCATION OF THE UPFLOWS

Because crystal spectrometers flown up to now have not been capable of imaging, the exact location of the upflow cannot be determined from their observations. For instance, from the Yohkoh BCS it is only possible to roughly tie the source to a flare site. However, from other observations it is possible to draw more definite conclusions about where the upflow is occurring.

The centre-to-limb variation in the size of the blueshifted component suggests material is moving within a loop and occasionally loops are observed to be "filling" in SXT images. However, loops in which this can have a large blueshifted component and it is the less obvious loops (viewed from above) that may be more easily identified with upflows.

A comparison of the pixel lightcurves from SXT with the time evolution of the blue-wing seen in BCS is a good way to identify the site of the blueshifted component - if a comparison is also made with the hard X-ray time profile, additional information may be derived (Savy, 1994??; Strong *et al.*, 1994). When making such comparisons, care must be taken in deciding exactly what area in the SXT image should be included. Depending on how areas are selected in the SXT image, if a source moves or grows so that it extends out of the selected area, the time evolution as plotted can be affected. Taking the whole flare will usually give a better match with HXT and BCS. In some simple flares, the hard X-ray flux time profile matches the lightcurve of the Ca XIX blue-wing and the time derivative of the SXT fluxes (Savy, 1995)

2.4. TEMPERATURE DEPENDENCE OF THE BLUE COMPONENT

The maximum velocity of the blueshifted component is different for spectral lines formed at different temperatures. It is generally observed that soft X-ray lines formed at high temperature show larger blueshifts than lines formed at lower temperatures (Antonucci *et al.*, 1990). Zarro (1993) studied S XV and a comparison of the velocities observed with those of Ca XIX and Fe XXV is given in Table 1.

TABLE 1. Temperature dependence of the blueshifted component

Ion	Peak Temperature	Maximum Velocity
Fe XXV	50 MK	$\leq$800 km s^{-1}
Ca XIX	30 MK	$\leq$500 km s^{-1}
S XV	15 MK	$\leq$200 km s^{-1}

The velocity-temperature relationship suggests a link between plasma upflow motions (implied by the blueshifts) and plasma heating that is occurring in flares.

Fludra *et al.* (1989) report that the mean temperatures of the moving and stationary plasmas agree although at the onset of the rise phase, the temperature of the upflowing plasma may be *higher*. They conclude that this implies that the flare energy deposited lower in the loop, rather than at the top.

2.5. MOMENTUM BALANCE OF COMPONENTS

In some flares, the momentum of the upflow (blueshift) seen in soft X-rays closely matches the downflow (redshift) seen in Hα (Canfield *et al.*, 1990). For example, for the flare at 22:35 UT on November 15, 1991, the balance derived by Wulser *et al.* (1994) are shown in Table 2.

TABLE 2. Momentum Balance in Flare of 15 November, 1991

Velocity Component	Derived Momentum
Total Hα downflow momentum	3.6 10^{21} gm cm s^{-1}
Upflow momentum in Ca XIX	2.9 10^{21} gm cm s^{-1}

For a simple, compact flare these figures provide interesting information, but for very extended, complex flares such calculations must be viewed with caution. The Ca XIX channel from the Yohkoh BCS is integrated over whole flare, while the Hα downflows are measured at discrete sites. In extended sources we cannot therefore be certain the the two momenta should be balanced against each other.

In some Hα downflows are observed in areas that do not correspond to where the intense hard X-ray signal is observed. If we believed that hard X-rays are a signature related in some way to the soft X-ray upflows, the momenta cannot be balanced in these circumstances.

However, in the example above Wulser *et al.* give compelling arguments why the two velocity components are in fact related.

3. Line Broadening

At the onset of many flares, the line-widths of soft X-ray emission lines are in excess of what is expected from by atomic theory. The line-broadening is thought to be non-thermal in origin and is probably caused by non-directed mass motions, or *turbulence*, in the flare region.

We will look at the timing, and location of the source of line broadening.

3.1. TIMING OF LINE BROADENING

Line broadening is greatest near flare onset and is typically observed in time correlation with hard X-ray emission (Tanaka *et al.*, 1982 Bentley *et al.*, 1994). The turbulence usually continues as long as the hard X-rays are present and it has normally disappeared by the peak of the flare, although there have been reports of turbulence detected in decay phase (e.g. Fludra *et al.*, 1989).

The coincidence in timing suggests that the non-thermal broadening are a direct consequence of the same process that produces the hard X-rays.

3.2. LOCATION OF SOURCE OF LINE BROADENING

Unlike the blueshifted component, the non-thermal line broadening (v_t = 100 $\sim$ 200 km s^{-1}) does not vary across the disk (Saba *et al.*, 1986; Mariska *et al.*, 1993). This suggests that it is not directed in nature, but rather consists of many randomly oriented components.

The broadening is thought to be closely related with energy release (Antonucci *et al.*, 1982), and energy release is normally indicated by impulsive hard X-rays (Winglee *et al.*, 1991). From flare observations in hard X-rays by the Yohkoh HXT, energy is normally released at the loop footpoints, but in some events Masuda *et al.* (1994) also found a hard X-ray source at the top of the loop. Whereas the footpoint release site is thought to be where precipitating energetic electrons are impacting the chromosphere, Masuda *et al.* suggest that the loop-top site is caused by shocks resulting from magnetic reconnection.

Two teams of Yohkoh investigators have recently studied line-broadening by looking at flares on the limb.

Khan *et al.* (1995) examined four flares over an 11 hour period as an active region rotated over the limb - in all flares the footpoints were obscured by the solar limb. They report that several spectral characteristics, including non-thermal broadening, are virtually indistinguishable no matter how much of the lower parts of the loop is obscured. They conclude that the broadening is either independent of height, or the source is placed near the loop top, and that the energy release driving flares is therefore sited in the corona.

Mariska *et al.* (1995) looked at eight flares from different active regions near or beyond the solar limb. In four of the flares, the footpoints (as observed by HXT) were occulted, while in the other four they were clearly visible. They report that the peak non-thermal velocities are less if the footpoints are obscured. Also, they find that observed hard X-ray emission is softer when footpoints are obscured.

There are many different models that try to explain non-thermal broadening and observations clearly important in deciding the validity of each model. Mariska *et al.* seem to be consistent with the model of Alexander *et al.* (1990) which predicts a variation with occulted height, but Khan are not inconsistent with other theories.

4. Conclusions

Observations from Yohkoh are providing a new insight as to the nature of the blueshifted component and line broadening seen in soft X-rays. For the first time, the enhanced spatial and temporal resolution of the Yohkoh instruments are allowing us to determine where and how the material that form the blue-wing originates, and where the turbulence is located.

Similarities between the time profiles of the hard X-ray burst, and the soft X-ray blue-wing and non-thermal broadening indicate that the processes that cause the phenomena are linked. More work needs to be done on non-thermal broadening, but interesting things are already beginning to emerge.

Much of what has been discussed here has concerned the signatures of mass motion in soft X-rays, and hence the corona. Although it is well known that such motions occur at other wavelengths and other layers of the atmosphere, this is beyond the scope of this paper. While we should not view coronal phenomena in isolation, in many cases we do not know how to relate them to phenomena seen in different layers of the atmosphere. It is hoped that the Solar-B mission may go some way to making this possible.

References

Alexander, A., (1990), Soft X-ray line broadening in solar flares, *A&A*, **236**, pp. L9–L12.

Antonnuci, E., Gabriel, A.H., Acton, L.W., Culhane, J.L., Doyle, J.G., Leibacher, J.W., Machado, M.E., Orwig, L.E., and Rapley, C.G., (1982), Impulsive Phase of flares in soft X-ray emission, *Solar Phys.*, **78**, pp. 107–123.

Antonucci, E., Dodero, M.A. and Martin, R., (1990), High-velocity evaporation during the impulsve phase of the 1984 April 24 flare, *ApJ (Suppl.)*, **73**, pp. 137–146.

Bentley, R.D., Lemen, J.R., Phillips, K.J.H., and Culhane, J.L., (1986), Soft X-ray observations of high-velocity features in the 29 June 1980 flares, *A & A*, **154**, pp. 255-262.

Bentley, R.D., Doschek, G.A., Simnett, G.M., Rilee, M.L., Mariska, J.T., Culhane, J.L., Kosugi, T., and Watanabe, T., (1994), The Correclation of solar flare hard X-ray bursts with Doppler blueshifted soft X-ray flare emission, *ApJ (Letters)*, **421**, pp. L55–L58.

Canfield, R.C., Zarro, D.M., Metcalf, T.R., and Lemen, J.R., (1990), Momentum Balance of four solar flares, *ApJ*, **348**, pp. 333–340.

Cheng, C.-C., Rilee, M. and Uchida, Y., (1994), Thermal and nonthermal energizations in solar flares: Soft X-ray spectroscopic and hard X-ray observations, *Proc. of Kofu Symposium*, eds. Enome, S., and Hirayama, T., **NRO Report No. 360**, pp. 213–216.

Doschek, G.A., *et al.*, (1993), The 1992 January 5 flare at 13.3 UT: Observations from Yohkoh, *ApJ*, **416**, pp. 845–856.

Fludra, A.F., Lemen, J.R., Jakmiec, J., Bentley, R.D., and Sylwester, J., (1989), Turbulent and directed plasma motions in solar flares, *ApJ*, **344**, pp. 991–1003.

Khan, J.I., Harra-Murnion, L.K., Hudson, H.S., Lemen, J.R., and Sterling, A.C., (1995), Yohkoh Soft X-ray Spectrocsopic Observtions of the Bright Loop-top Kernels of Solar Flares, *ApJ (Letters)*, in press.

Mariska, J.T., Doschek, G.A., and Bentley, R.D., (1993), Flare plasma dynamics observed with the Yohkoh Braagg Crystal Spectrometer, I. Properties of the Ca XIX resonance line, *ApJ*, **419**, pp. 418–425.

Mariska, J.T., Sakao, T., and Bentley, R.D., (1995), Hard and Soft X-ray Observations of Solar Limb Flares, *ApJ*, in press.

Masuda, S., Kosugi, T., Hara, H., Tsuneta, S., and Ogawara, Y., (1994), A loop-top hard X-ray source in a compact solar flare as evidence for magnetic reconnection, *Nature*, **Vol. 371**, pp. 495–497.

Newton, E.K., Emslie, A.G., and Mariska, J.T., (1995), The Velocity Differential Emission Measure diagnostic of bulk plasma motion in solar flares, *ApJ*, **447**, pp. 915–922.

Plunkett, S.P., and Simnett, G.M., (1994), Temporal correlation of solar hard X-ray bursts with chromospeheric evaporation, *Solar Phys.*, **155**, pp. 351–371.

Saba, J.L.R., and Strong, K.T., (1986), *Adv. Space Res.*, **6**, pp. 37.

Savy, S.K., (1994), ???

Savy, S.K., (1995), ???

Strong, K.T., Hudson, H.S., and Dennis, B., (1994), Evidence for Impulsive soft X-ray bursts during flares, *X-ray Solar Physics from Yohkoh*, eds. Uchida, Y., Hudson, H.S., Watanabe, T., and Shibata, K., University Academy Press, Inc., pp. 65–69.

Tanaka, K., Watanabe, T., Nishi, K., and Akita, K., (1982), High-resolution solar flare X-ray spectra obtained with rotating spectrometers on the Hinotori satellite, *ApJ (Letters)*, **254**, pp. L59–L63.

Winglee, R.M., Kiplinger, A.L., Zarro, D.M., Dulk, G.A., and Lemen, J.R., (1991), Interpretation of soft and hard X-ray emissions during solar flares. I. Observations, *ApJ*, **375**, pp. 366–381.

Wulser, J.-P., Canfield, R.C., Acton, L.W., Culhane, J.L., Phillips, A.T., Fludra, A., Sakao, T., Masuda, S., Kosugi, T. and Tsuneta, S., (1994), Multispectral Observations of Chromospheric Evaporation in the 1991 November 15 X-class solar flare, *ApJ*, **424**, pp. 459–465.

Zarro, D.M., (1993), Flare Dynamics Observed in S XV, *UV and X-ray of Laboratory and Astrophysical Plasmas*, eds. Silver, E. and Kahn, S., Cambridge University Press, pp. 603–606.

ENERGY BUILD-UP PROCESSES OF SOLAR FLARES STUDIED BY OPTICAL OBSERVATIONS

H. KUROKAWA
Kwasan and Hida Observatories, Kyoto University
Kamitakara, Gifu 506-13, Japan

Abstract.
The flare energy build-up processes, or the processes of the magnetic shear development are studied in detail by analyzing the high resolution Hα pictures of a rapidly-evoluving and flare-productive active region NOAA 6233. From the analysis, it is tentatively concluded that the flare energy is already stored in a twisted magnetic flux rope which is formed in the convection zone, and that the successive emergence of the twisted rope develops strong magnetic shear configuration and strong flare activity on the solar surface.

1. Introduction

The most important task assigned to the optical observation in the flare study is to study the flare energy build-up process, because the flare energy is stored in the solar magnetic field, which is supplied to the corona through the photosphere from the convection zone. For the study of the flare energy build-up process, it is essential to examine which type of magnetic field configuration in a sunspot region produces strong flare activity, and how it is formed. This paper, therefore, summarizes our recent studies of evolutional and morphological characteristics of flare-productive active regions by using high resolution Hα images.

Many previous optical observations showed that the strong flare activity occurs selectively along the strong magnetic shear region (Zirin and Tanaka 1973, Hagyard et al. 1984), and there is no longer any doubt that magnetic shear is the most important necessary condition for the occurrence of a

Y. Uchida et al. (eds.),
Magnetodynamic Phenomena in the Solar Atmosphere – Prototypes of Stellar Magnetic Activity, 185–194.

strong flare. An essential question for the study of the flare energy build-up process is, therefore, *how does the magnetic shear develop?*

Two types of processes for the development of magnetic shear were found in our previous paper (Kurokawa 1987): (A) collision of two sunspots of opposite magnetic polarities and (B) successive emergence of twisted magnetic flux ropes. These two processes are examined here in more detail in a typical example of flare-productive active region NOAA 6233, in which five magnetic shear neutral lines developed from 26 through 31 August, 1990. We show that process (B) is more effective for the energy build-up of strong flares.

2. Hα Fine Structures as Tracers of Chromospheric Magnetic Field

The observational data mainly used in this analysis are high resolution Hα images obtained with the 60 cm Domeless Solar Telescope at Hida Observatory of Kyoto University. Longitudinal magnetograms observed with the Okayama magnetograph of National Astronomical Observatory of Japan are also used to determine the magnetic neutral lines in the active region.

The frozen-in condition is considered to be satisfied in the solar atmosphere (Alfvén and Fälthammar 1963). Many observations show that Hα dark fibrils and threads stream in a direction connecting opposite magnetic polarities from the comparison between Hα fine structures and the photospheric magnetograms (Foukal 1971, Kawakami et al. 1989). The high resolution Hα filtergram observation, therefore, provides us with the most useful information on the evolutional changes of magnetic field configuration from the photosphere through the lower corona by means of the following basic rules.

(1) Hα dark and elongated features such as fibrils, threads, filaments, and superpenumbra fibrils give the direction of the transverse magnetic field (Foukal 1971, Kawakami et al. 1989). The term "thread "or "chromospheric thread"is used here for an Hα dark thin loop whose legs land both on the edges of Hα plagettes. When only one end of a dark loop is found to be associated with the plagette, we call it a "fibril"(Foukal 1970).

(2) Magnetic field lines are nearly vertical to the solar surface in the center of sunspot umbra and Hα plage.

(3) Hα threads and arch filament system (AFS) loops connect opposite magnetic polarities.

(4) An Hα arch filament system (AFS) is a clear sign of the emergence of a new bipolar magnetic region.

(5) An Hα filament (prominence) lies along the magnetic neutral line or the polarity boundary line, and consists of multiple thin fibrils whose

directions are nearly parallel to the filament. This means that the formation of a filament is an indication of the development of the magnetic shear.

(6) The spatial resolution of the best Hα images is about one order of magnitude better than that of the currently available best vector magnetograms, and they can show twisted fine structures and their changes along threads, AFS loops and filaments (Kurokawa 1991).

3. Development of Magnetic Shear in NOAA 6233

Figures 1 and 2 demonstrate the evolution of the active region NOAA 6233 and magnetic shear developments in it from 25 August through 2 September, 1990. The left row of the pictures shows the sunspot growth and decay in Hα-5.0 Å, the right row, the evolution of the transverse magnetic field in Hα line center. Sunspots p_A and p_B, which have the preceding magnetic polarity, already existed in the region when it rotated onto the north-east visible hemisphere. New bundles of magnetic flux tubes, whose cross sections at the photosphere are many sunspots, successively emerged from 24 through 30 August. The several bipolar sunspot pairs, which were found to be connected by Hα AFS loops at their emerging stages, are named p_1-f_1, p_2-f_2, etc. in figures 1 and 2. The magnetic polarities of these sunspots were determined by Okayama magnetograms. Almost all p-polarity sunspots except p_5 finally merged into p_1 to form a single large p-sunspot as seen in the picrure of 2 September. On the other hand, all f-polarity sunspots collided or interacted with p_A, p_B or p_5, and developed magnetic shear or were cancelled with them. Five neutral lines or polarity boundaries are examined in details to study the developing process of magnetic shear and flare activity in the following.

3.1. CANCELLATION BETWEEN P_A AND F_1-F_2-F_3 SUNSPOTS

When f_1 sunspot approached and collided with p_A, some magnetic shear developed between them from 25 through 27 August as seen in the sheared penumbra and Hα filaments running along the neutral line. Between 08 UT and 22 UT of 27 August, however, the magnetic shear disappeared and instead the short penumbral and Hα threads directly connected f_1-f_2 and p_A as seen in the pictures of 2247 UT, 27 August and 23 UT, 28 August. The reason of this relaxation of the magnetic shear is not clear, because we do not have any data between 08UT and 22 UT of 27 August. It may have some connection to the X3 flare of 2100 UT, 27 August, which occurred in the neighborhood of the neutral line.

After the diasppearance of the magnetic shear, the cancellation of different magnetic polarities in the sunspots p_A and f_1-f_2-f_3 rapidly progressed. Notice the rapid decrease of these sunspot areas from 28 through 30 August. On 2 September these sunspots have nearly been cancelled out. It is noteworthy that no flare activity was found during this cancellation.

3.2. MAGNETIC SHEAR BETWEEN F_2 AND P_B

New magnetic flux continued to emerge on 27 August in the same direction as the p_1-f_1. The new bipolar regions are marked p_2-f_2 and p_3-f_3 in figures 1 and 2. The sunspots f_2 and f_3 merged into f_1 and they successively collided with and were cancelled by p_A.

Most noticeable in the evolution of the NOAA 6233 region is a drastic change of magnetic field configuration or a rapid development of strong magnetic shear found between p_B and f_2 around the end of 27 August; while some Hα dark loops are seen to connect p_B and f_2 (figure 1) in the pictures of 0120 UT and 0755 UT of 27 August, new Hα AFS loops are found to be emerging along the magnetic neutral line between p_B and f_2 in the picture of 2247 UT 27 August. It means that the direction of the magnetic lines of force changed by nearly 90 degrees between p_B and f_2 during less than 14 hours. Such a drastic change of the direction of the strong magnetic field, or a strong magnetic shear, was caused by the emergence of the new bipolar magnetic flux p_5-f_5 along the neutral line between p_B and f_2. This type of magnetic shear development is classified as type B (Kurokawa 1987).

Even after the X3 flare of 2100 UT 27 August, the active emergence of the new magnetic flux p_5-f_5 continued till late on 28 August and further developed the magnetic shear. Notice the sunspot growth along p_5-f_5 in the pictures of 224727 UT 27 August and 234247 UT 28 August as well as conspicuous Hα AFS loops emerging between p_5 and f_5 in the picture of 234534 UT 28 August. This additional development of magnetic shear caused five M-class and two C-class flares during 28 August. Wang (1992) studied the evolution of vector magnetic fields in this region, and concluded that the magnetic shear of this area was built-up by the photospheric flow motion. Our study, however, clearly shows that the motive power for the magnetic shear development in this area is not the photosperic flow but the emergence of a new bipolar magnetic flux. The emerging magnetic flux produced the photospheric flow motion found by Wang (1992).

Our result demonstrated above can easily explain the magnetic shear increase associated with the X-class flares found by Wang et al (1994). The successive emergence of a twisted flux rope can develop the magnetic shear even after the occurrence of an X-class flare.

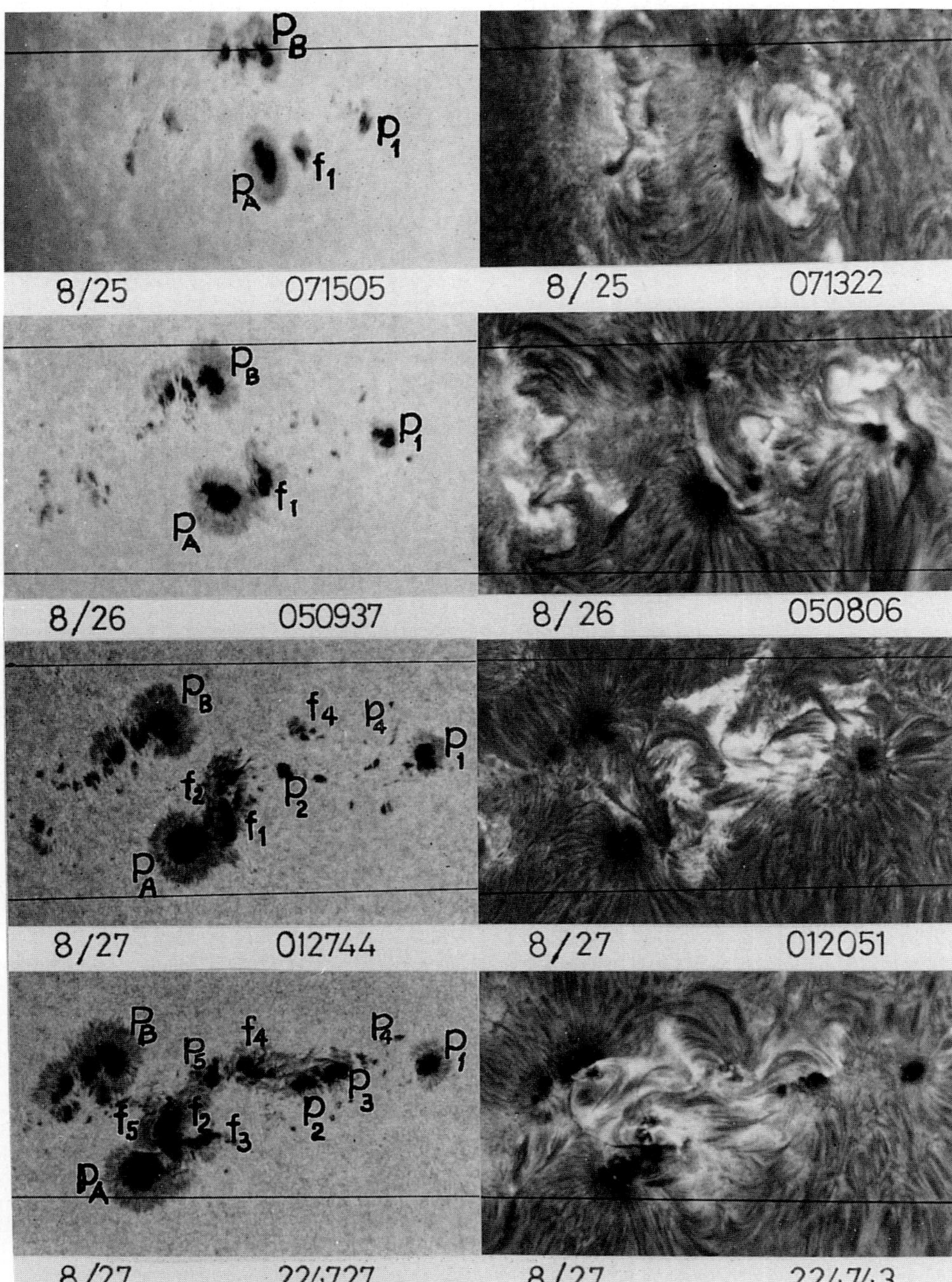

Figure 1. Development of magnetic shear in NOAA 6233 from 25 through 27 August, 1990. The left row shows the sunspot growth in Hα-5.0Å, and the right row, the Hα transverse magnetic fields in Hα line center. The p_A and p_B are preexisting sunspots, and p_1-f_1, p_2-f_2, etc. are newly-emerging bipolar regions.

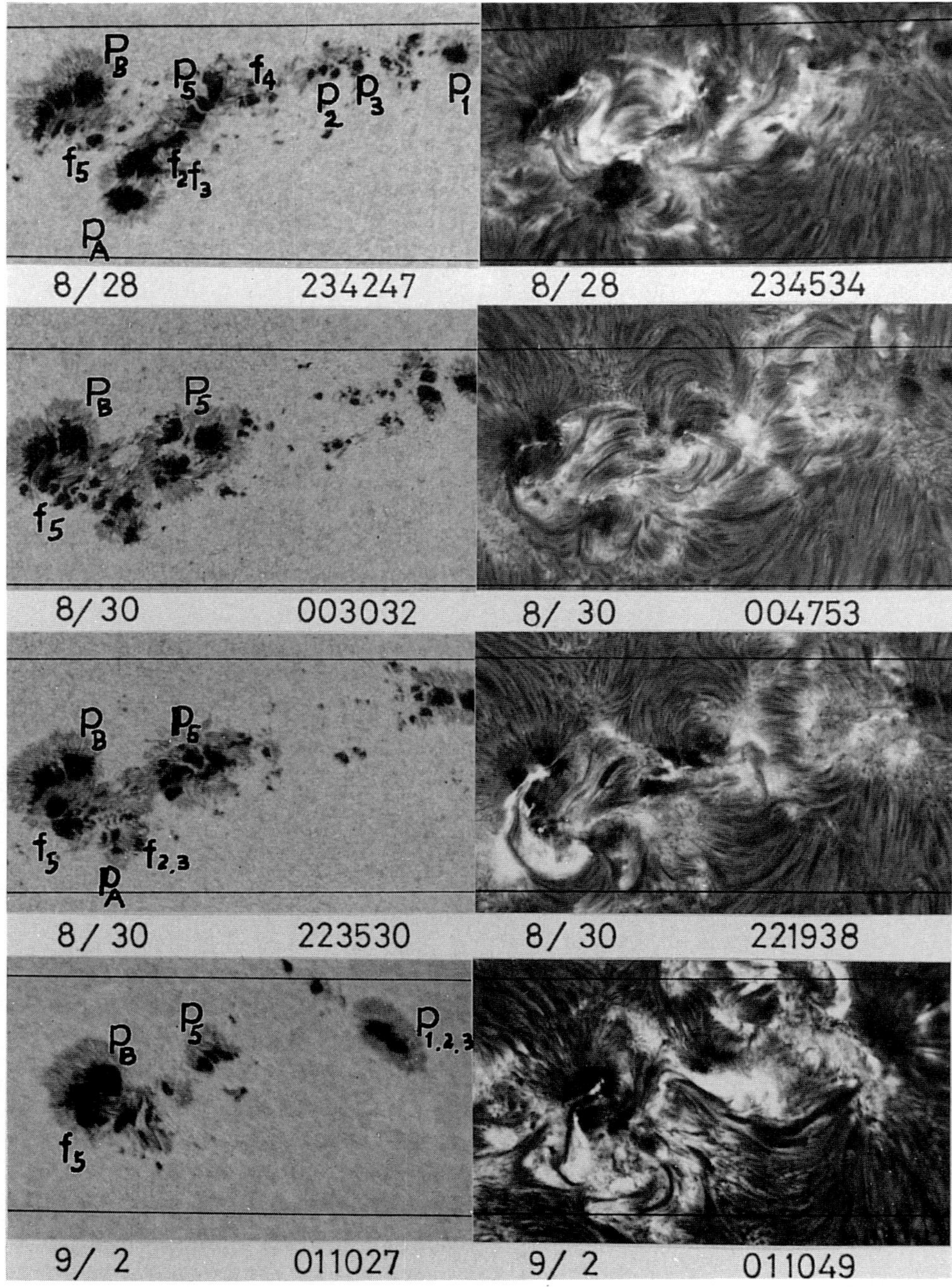

Figure 2. Continued from figure 1. Development of magnetic shear and cancellation of sunspots in NOAA 6233 from 28 August through 2 September, 1990. The left row shows the sunspot growth and decay in Hα-5.0Å, and the right row, the Hα transverse magnetic fields in Hα line center. Notice the cancellation between p_A and f_2-f_3 and the magnetic shear developments between p_B and f_5.

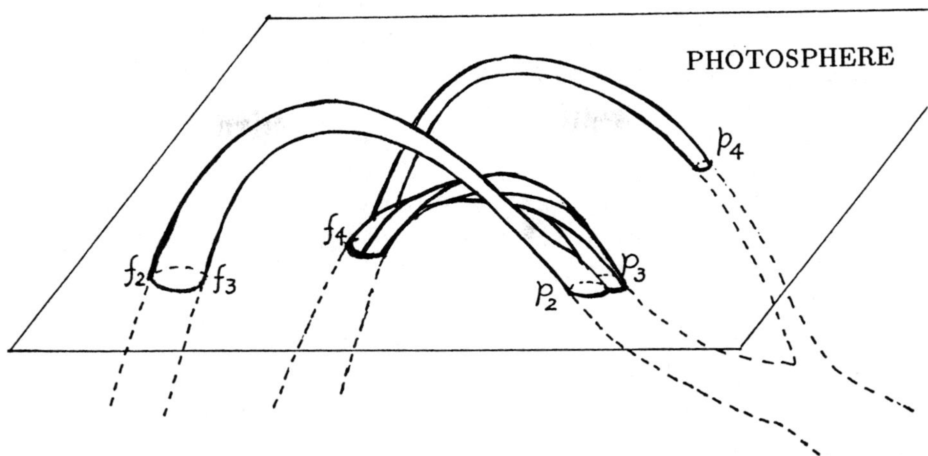

Figure 3. Schematic drawing of an emerging magnetic flux rope including p_2p3-f_2f_3, f_4-p_3 and p_4-f_4 branches. The emergence of the twisted flux branch f_4-p_3 developed the magnetic shear between f_4 and p_3.

3.3. MAGNETIC SHEAR BETWEEN F_4 AND P_2-P_3

Another rapid development of magnetic shear between f_4 and p_2-p_3 is also remarkable during 27 August. Notice the rapid growth of sunspots p_2, p_3 and f_4 and the associated development of sheared Hα filament between f_4 and p_2-p_3, comparing the pictures of 01 UT and 22 UT of 27 August. There may be two possible explanations of the magnetic shear development of this area: One is the collision of the growing sunspots of opposite magnetic polarities. The other is the emergence of a twisted magnetic flux rope. In the former scenario, the sunspots p_2-p_3 and f_4 are not originally connected below the photosphere, but they meet and reconnect with each other above the photosphere, and the opposite proper motion or shear motion between them develops the magnetic shear. In the latter scenario, f_4 and p_3 are originally connected by a branch of twisted magnetic flux rope below the photosphere. In the course of the emergence of a main trunk of flux rope including p_2-f_2, p_3-f_3 and p_4-f_4 branches, the twisted branch f_4-p_3 also emerges and the magnetic shear develops between them. A schematic drawing of the emerging twisted f_4-p_3 branch is shown with p_2p_3-f_2f_3 and p_4-f_4 branches in figure 3. We tentatively conclude that the latter scenario is real for this case, because the magnetic shear developed simultaneously with the rapid growth of f_4 and p_3 sunspots. Notice, in the picture of 2247 UT 27 August (figure 1), the conspicuous development of sheared penumbral filaments connecting f_4 and p_3, which corresponds to the emerging twisted flux loops.

3.4. MAGNETIC SHEAR BETWEEN P_B AND F_5

In the picture of 224727 UT 27 August (figure 1), when the following spot f_5 of the newly-emerging pair p_5-f_5 just began to contact the p_B sunspot, the penumbral filaments of p_B are still radially oriented. At 234247 UT 28 August (figure 2), however, the southern part of the penumbra of p_B has already reconnected to the intruding f_5 sunspots with sheared penumbral and Hα thin structures. They formed a sheared δ-type sunspot, but it was fairly stable and only produced several small flares.

3.5. MAGNETIC SHEAR AT THE EAST BOUNDARY OF F_5

The p_5-f_5 sunspots continuously grew till the end of 30 August, and Hα boundary filaments were formed at the east of the f_5 sunspots as seen in the picture of 234534 UT 28 August (figure 2) by the reconnection between the growing f_5 and the preexisting old p-polarity magnetic field. The several boundary filaments merged into a strongly-sheared filament from 29 through 30 August, when the f_5 further grew and intruded. Several small flares occurred along the filament.

4. Summary and Conclusion

A rapidly-evolving and flare-productive active region NOAA 6233 was studied in detail by using high resolution Hα pictures obtained with the 60 cm Domeless Solar Telescope at Hida Observatory, Kyoto University.

Magnetic shear developments were found along five magnetic neutral lines or polarity boundary lines in the NOAA 6233 region from 26 through 31 August, and their developing processes were carefully examined. The two types of processes found in our previous work (Kurokawa 1987) were again confirmed. The magnetic shears along the two regions between p_B and f_2 and between f_4 and p_3 were found to develop by the successive emergence of twisted branches of magnetic flux ropes. The other three magnetic shears between p_A and f_1, between p_B and f_5, and at the east end of f_5 were found to develop by the collision and reconnection between the opposite magnetic polarities. The former process of the magnetic shear development matches type B, and the latter type A of Kurokawa (1987) respectively.

Main flare activity including one X-class and six M-class occurred from late on 27 August through 29 August at the p_B-f_2 and f_4-p_3 regions where the magnetic shear rapidly developed by the process of type B. This conclusion agrees with our previous works (Kurokawa 1987, 1989, 1992, 1994). Wang (1992) also studied the magnetic shear in the same region as our

p_B-f_2, but he concluded that the magnetic shear was built-up by the photospheric flow motion. He failed to correctly notice the active emergence of the new magnetic flux p_5-f_5 between p_B and f_2. Our study clearly shows that the new flux p_5-f_5 produced the photospheric flow motion as well as the magnetic shear. In addition, the magnetic shear increase after the X3 flare of 2100 UT found by Wang et al. (1994) is easily explained by the continuous emergence of p_5-f_5 flux. Our result indicates that the successive emergence of a twisted flux rope can develop magnetic shear even after the occurrence of an X-class flare.

We tentatively conclude that the flare energy is already stored in a twisted magnetic flux rope which is formed in the convection zone, and that the successive emergence of the twisted rope develops strong magnetic shear and produces strong flare activity. This conclusion is compatible with the results of Tanaka(1991) and Leka et al.(1996).

It is still completely unknown, however, what triggers the explosive release of energy stored in the twisted magnetic loops. We need more observations of much higher spatial resolution simultaneously with optical and soft X-ray instruments to answer this exciting question.

Acknowledgement

The author would like to express his hearty thanks to Prof. T. Sakurai for his supplying us with magnetograms of NOAA 6233 obtained at Okayama Observatory of NAOJ.

References

Alfvén, H. and Fälthammar, C.-G. (1963): *Cosmical Electrodynamics, Fundamental Principles,*2nd ed. (Oxford University Press, London, p102.

Foukal, P. (1970): *Solar Phys.* **19**, 59.

Foukal, P. (1971): *Solar Phys.* **20**, 298.

Hagyard, M. J., Smith, J. B., Teuber, D., and West, E. A. (1984): *Solar Phys.* **91**, 298.

Kawakami, S., Makita, M., and Kurokawa, H. (1989): *Publ. Astron. Soc. Japan* **41**, 175.

Kurokawa, H. (1987): *Solar Phys.* **113**, 259.

Kurokawa, H. (1989): *Space Science Rev.* **51**, 49.

Kurokawa, H. (1991): *Lecture Notes in Physics* **87**, 39.

Kurokawa, H., Nakai, Y., Funakoshi, Y., Kitai, R. (1991): *Adv. Space Res. vol.***11** No.5, (5)233.

Kurokawa, H., Kitai, R., Kawai, G., Shibata, K., Yaji, K., Ichimoto, K.,

Nitta, N., Zhang, H. (1994): in S. Enome and T. Hirayama (eds.) *Proc. of Kofu Symp., NRO Report* **No.360**, p.283.

Leka, K.D., Canfield, R.C., McClymont, A.N. and van Driel-Gesztelyi, L. (1996): *Astrophy. J.* in press.

Tanaka, K. (1991): *Solar Phys.* **136**, 133.

Wang, H. (1992): *Solar Phys.* **140**, 85.

Wang, H., Ewell, M.W., Zirin, H., and Ai, G. (1994): *Astrophys. J.* **424**, 436.

Zirin, H. and Tanaka, K. (1973): *Solar Phys.* **32**, 173.

HIGH-QUALITY IMAGES BASED ON THE STEER ALGORITHM DECOVOLUTION FROM THE CORRELATION DATA OBSERVED WITH NOBEYAMA RADIOHELIOGRAPH

SHINZO ENOME
Nobeyama Radio Observatory, NAOJ
Minamimaki, Minamisaku, Nagano 384-13, Japan

Abstract. After the Makuhari meeting of IAU Colloquium 153 remarkable developments have been made at Nobeyama on high-quality imaging, which applies the Steer algorithm for deconvolution. Considering the importance of this achievement, it is briefly introduced in this report with farther application to a high-resolution narrow field mapping. Some preliminary results are presented for both cases as well as dual- frequency images at 17 and 34 GHz.

1. Introduction

Everyone knows that if there is a lens, it will produce an image behind it for the scenery in front of it at the focal plane . But it is not a simple common sense, even of physicists or astronomers, why a lens produces an image. If we look into a standard text book on optics, for example one by Born and Wolf 1970), it is written, after many pages of preparation and introduction, that a lens is physically equivalent to a real-time and analog correlator and a Fourier transformer. A simple application of this theorem to obtain the point spread function of a circular aperture is a well-known Bessel function of 0-th order.

Radioastronomers have long been using, in principle, this theorem to synthesize radio images of heavenly bodies (e.g. Ryle, 1954?). It is now very clear that digital correlators and fast computer are fundamental tools for radio telescopes or radio interferometers to obtain images efficiently. As easily understood, apertures of multi-element interferometers are often sparse,

Y. Uchida et al. (eds.),
***Magnetodynamic Phenomena in the Solar Atmosphere – Prototypes of Stellar Magnetic Activity*, 195–201.**

and, hence, their point spread functions or beam patterns are associated with high side-lobe levels. calibration, therefore, is a very important procedure to estimate the beam pattern as accurate as possible. This procedure will correspond to grinding the lens surfaces as accurate as 1/16 lambda in optics. To observe heavenly bodies with high side-lobe-level beams is inevitable in radio astronomy, and this characterizes the peculiarity of radio astronomy observations. Probably this situation will never occur in optical telescopes. This peculiarity produced a unique field in image processing in radio astronomy of CLEAN (Hoegbom, 1974). The original CLEAN algorithm of deconvolution of assumes the observed image as an assemble of point sources. This algorithm has been very effective, when it was proposed, since the spatial resolution of then-available interferometers were low and the assumption of point source assembly was well satisfied. As spatial resolutions and sensitivities increased, this simple assumption was found not effective for diffuse and halo-type extended sources. Deconvolution of these extended source with the standard CLEAN algorithm will be deteriorated by "stripes" or "corrugation" from side-lobe effect (Steer et al., 1984). A number of "modified" CLEAN algorithms (Cornwell, 1983, Schwab, 1984, Steer et al., 1984) has been proposed to avoid or reduce instability of these "strips" or "corrugations", among which we have found the deconvolution with the Steer algorithm CLEAN is most promising for radio images of the Sun with the Nobeyama Radioheliograph (Koshiishi, 1996).

2. Processing of Correlation Data

The fundamental idea in the Steer algorithm to estimate components of images in the relevant dirty map or the residual map is to introduce a new concept of "trim contour", which corresponds to a level surrounding the peak at some fraction of the peak level and slightly higher than the largest side-lobe level. If we specify a trim contour for a dirty map or for a residual map, we can draw a contour or a feature surrounding the peak, which will cut out a component feature of the relevant image. This algorithm has been tested in numerical simulations (Steer et al., 1984).

Koshiishi (1996) has applied the Steer algorithm to CLEAN dirty maps synthesized from the correlations observed with the Nobeyama Radioheliograph. It has been found that the CLEAN criterion can be reduced as low as three times sigma of the noise level on the images, of which the major component is due to the receiver noise. This criterion is about ten times lower than the current value, which means we can obtain images with 1000 K noise level for the quiet Sun brightness of about 10,000 K, and faint and extended structures are well reproduced in the new images. This new method will be applied to all the correlation data obtained so far in three years, which will enhance development of studies of quiet features around polar regions, in and around active regions, and faint discrete features such as polar holes, bright points, dark filaments etc.

A number of preliminary steps of data processing has been taken before he attain this performance. Details have been described in Koshiishi's doctoral thesis (Koshiishi, 1996). These include accurate alignment of the solar disk, evaluation of rms noise level, adoption of CLEAN BOX and unCLEAN BOX etc.

With this new Deep-CLEAN method Koshiishi has confirmed polar-cap brightenings on both poles outside of 0.9 solar radii with extended structure of 1000 K and patchy structures of 2000 K. He has also found a plateau structure elongates toward middle or equatorial latitude with 500 K above the average disk temperature. This plateaus in some case associate with a soft X-ray low-brightness channel or coronal hole, and in other case does not. Evolution of the polar-cap brightenings is examined with respect to solar cycle (Koshiishi, 1996).

Fujiki (1996) has been working on 16d high-resolution narrow field imaging, which uses spatial Fourier components of higher harmonics of 16d components, where d denotes the fundamental spacing of 86.6459 lambda, which corresponds to 40 arcminute of the interferometer field of view and, therefore, the 16d narrow field to 2.5 arcminute. This imaging naturally involves contamination of the image due to aliasings from outside the relevant field of view, but on the other hand it attains a higher resolution up to 10 arcsec, whereas in the case of wide field imaging the nominal resolution is 15 to 20 arcsec. It is, therefore, suitable and recommended for the mapping of flaring sources, where very high brightness will be expected. The concept of CLEAN BOX is applied to this narrow field imaging to reduce the effect of aliasing. The Steer algorithm is also included in this case with improved quality.

3. Examples of Images

One of spectacular events observed with Nobeyama Radioheliograph is the prominence eruption on July 30-31, 1992. The newly processed images (Fig. 1) show fine structure of twisting or untwisting motions of threads within the erupting prominence.

The flare on September 6, 1992 is processed with the 16d high-resolution imaging method (Fig. 2). The small frames show a sequence of radio bursts with 2.5 arcminute frame size for two minute interval during twenty minutes to cover the start, impulsive, and decay phases of the event. The flare started over the pre-existing S-component or above a sunspot. With the increase of the brightness a remote compact source of about one arcminute

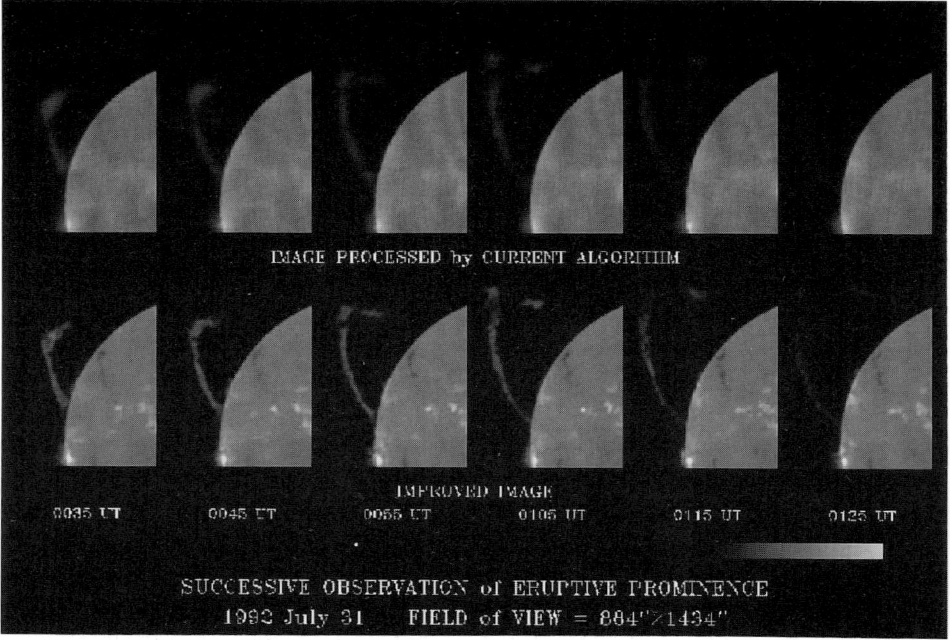

Figure 1. An eruptive prominece on 1992 July 31 with the current algorithm and with the Steer algorithm by H. Koshiishi(1996).

north-west started to brighten up. A further increase of brightness at the orginal point was accompanied by an expansion of the source to the north-east, which probably interepted as an evaporation or brightening of a loop originating from the spot. After the maximum phase a diffuse extended source appeared to the north-east of second source or the expanded source. The shape of this diffuse source is somewhat elongated along a line from the south-east to the north-west. One might imagine another loop over the first loop. This is a very naiive discription of behavior of the flare from the newly obtained images. The scientific analyses based on these or further improved images will be done with close and critical examination or evaluation of methods and sample images. The current version of the 16d imaging program assures the detection of 80000 K compact sources (Fujiki, 1996)

The flare on November 10, 1995 was observed with the newly-equipped dual-frequency operation mode at 17 and 34 GHz (Takano, 1996). This upgrading was completed in the1995 autumn. The images were synthesized with the 16d and 8d high-resolution imagings for 17 and 34 GHz maps (Fig. 3). It is remarkable that two sources are seen at 17 GHz, whereas one northern source is observed at 34 GHz and in soft X-ray (SXT/*Yohkoh*).

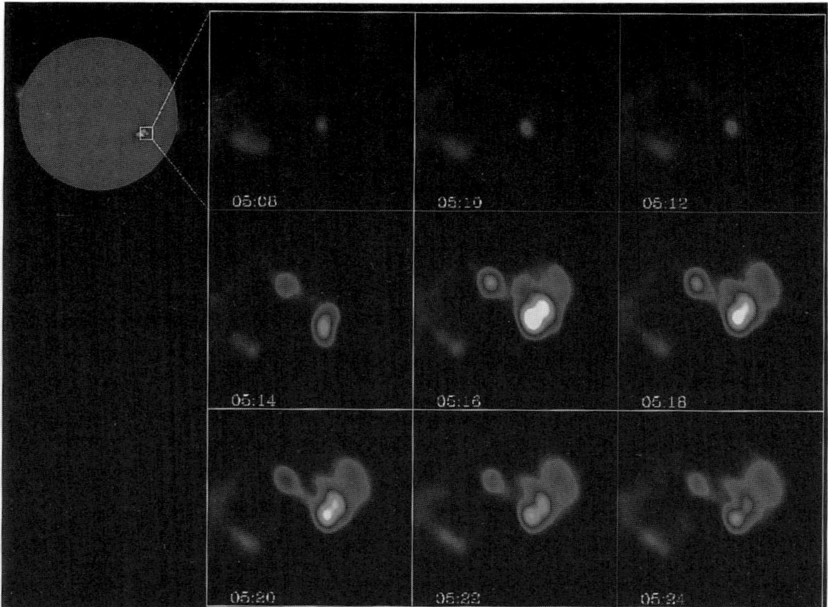

Figure 2. Narrow field high resolution imaging of the 1992 September 6 Flare with the Steer algorithm by K. Fujiki(1996). The field of view for the small panels is 2.5 × 2.5 arcminute.

The southern source at 17 GHz is emitted by the gyro-resonance mechanism at the third harmonics of the gyro frequency, which suggest the existence of 2000 Gauss fields at the coronal temperature. It is also to be noticed that the northern source has a loop structure at all frequencies and energy, and the 34-GHz image is almost same as the SXT image. This proves that the imaging performance of 5 arcsec spatial resolution at 34 GHz is probably satisfied.

4. Conclusions

New imaging methods and examples of maps processed with these methods are introduced and presented. A brief description of their performances is also given, but it should be reminded that there is a room to improve these performances. These programs will be opened shortly. The author would like to express thanks to Hideki Koshiishi for his laborious efforts and patience to complete his doctoral thesis.

Dual–Frequency Images of a Flare on November 10, 1995

T.Takano and Nobeyama Radioheliograph Group

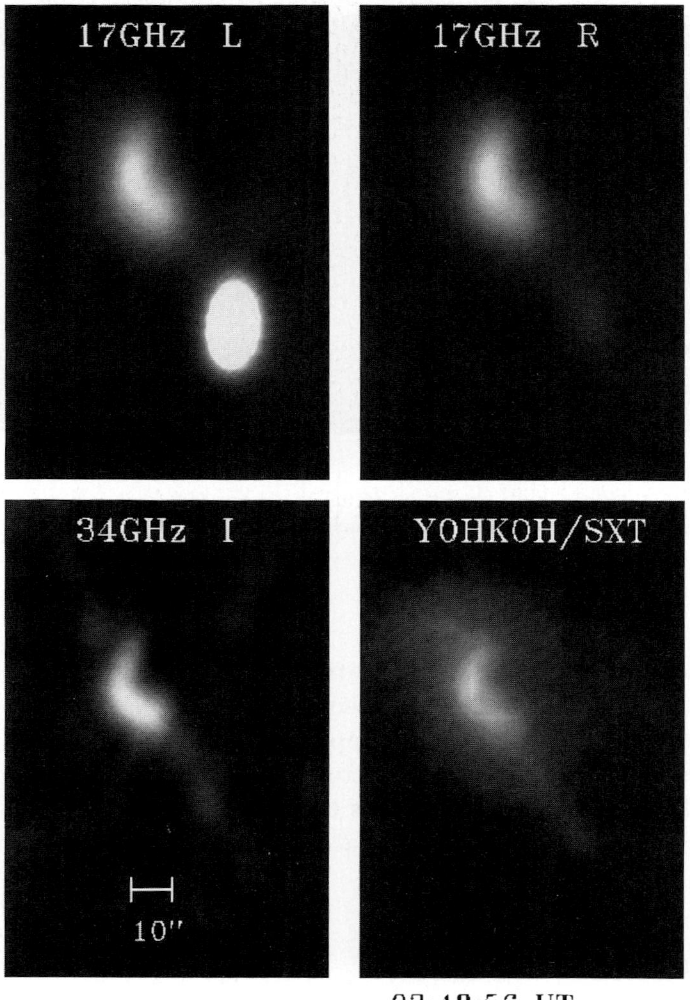

Figure 3. Dual-frequency images of the 1995 November 10 flare together with the image by *Yohkoh*/SXT (Soft X-ray Telescope)

References

Born, M. and Wolf, E. (1970) *Principle of Optics*, Pergamon Press, New York.

Cornwell, T. J. (1983) *Astron. Astrophys.* **121**, pp.281-285

Fujiki, K. (1996) Private communication

Hoegbom, J. (1974) *Astron. J. Suppl.* **15**, pp.417-426

Koshiishi, H. (1996) Doctoral Thesis "A Deep-CLEAN Imaging Method Applied to the Nobeyama Radioheliograph - Data Processing and Imaging Performance for Diffuse and Faint Features -", University of Tokyo, Department of Sciences, School of Astronomy.

Ryle, M. (1952) *Proc. Roy. Soc. A.*, **211**, pp. 351-375

Schwab, F. R. (1984) *Astron. J.*, **89**, pp.1076-1081

Steer, D. G., Dewdney, P. E., and Ito, M. R. (1994) *Astron. Astrophys.*, **137**, pp.159-165

Takano, T. (1996) Private communication

LOOP-TOP HARD X-RAY SOURCE IN SOLAR FLARES

S. MASUDA
Solar-Terrestrial Environment Laboratory, Nagoya University
3–13, Honohara, Toyokawa, Aich 442, Japan

AND

T. KOSUGI, K. SHIBATA, H. HARA AND T. SAKAO
National Astronomical Observatory Japan
2–21–1, Osawa, Mitaka, Tokyo 181, Japan

Extended Abstract

Where does the primary energy release take place in solar flares? Although this is a very fundamental question, no clear observational result had been achieved yet. Hard X-ray imaging observations of solar flares give us hints to answer this question. Actually a new type of hard X-ray source, which may be a very important hint, was observed with the Hard X-ray Telescope (HXT) on board *Yohkoh*. We summarize the characteristics of this hard X-ray source and other related observational results below.

We analyzed three impulsive solar flares on 13 January, 1992 (17:29 UT), 4 October, 1992 (22:21 UT), and 17 February, 1993 (10:35 UT) occurring near the limb. Hard and soft X-ray images of these flares, taken simultaneously with HXT and the Soft X-ray Telescope (SXT) aboard *Yohkoh*, accurately coaligned. It is clearly revealed that, in addition to double-footpoint sources, a hard X-ray source exists well above the corresponding soft X-ray loop structure around the peak time of the impulsive phase (Masuda *et al.* 1994, 1995). This type of hard X-ray source shows an intensity variation similar to the double-footpoint sources and its spectrum is as hard as that of the footpoint sources. These observational results suggest that the magnetic reconnection that is responsible for the primary flare energy release occurs above the soft X-ray flaring loop. This "loop-top" hard X-ray source may represent the reconnection site itself or the site where the downward plasma stream, ejected from a reconnection point far above the hard X-ray source, collides with the underlying closed magnetic loop.

Y. Uchida et al. (eds.),
***Magnetodynamic Phenomena in the Solar Atmosphere – Prototypes of Stellar Magnetic Activity*, 203–404.**

Magnetic reconnection originates not only downward flow but also upward flow. Recently hot-plasma ejections, accompanied with impulsive solar flares, are found (Shibata *et al.* 1995). In each of the three flares discussed above, such a phenomenon is observed. The hot-plasma ejections seem to start around peaks of spikes in hard X-ray light curve. In the case of the 4 October 1992 flare, ejections took place twice, corresponding two intense hard X-ray spikes. So it is plausible that the ejection is related to impulsive energy release, *i.e.*, magnetic reconnection.

We also found a hard X-ray source located at site remote from the intense double footpoint sources in the 13 January 1992 flare. The position correspond to a footpoint of a large diffuse loop which exists above a compact flaring loop. If this hard X-ray footpoint source resulted from precipitation of high energy electrons along the loop, the electrons are energized near the top portion of the large diffuse loop. This energization could be identified with a result of a collision between the upward plasma stream from the reconnection site and the large diffuse loop.

References

Masuda, S., Kosugi, T., Hara, H., Tsuneta, S., and Ogawara, Y., (1994) A loop-top hard X-ray source in a compact solar flare as evidence for magnetic reconnection, *Nature*, **371**, pp. 495–497

Masuda, S., Kosugi, T., Hara, H., Sakao, T., Shibata, K., and Tsuneta, S., Hard X-ray Sources and the Primary Energy Release Site in Solar Flares, (1995), *Publ. Astron. Soc. Japan*, **47**, pp. 677–689

Shibata, K., Masuda, S., Shimojo, M., Hara, H., Yokoyama, T., Nitta, N., Tsuneta, S., and Kosugi, T., X-ray Filament/Plasmoid Ejections in Impulsive Limb Flares, (1995), *Astrophys. J. Lett.*, **451**, pp. L83–L85

RADIO IMAGING OBSERVATION OF A SOLAR FLARE CUSP

K. SHIBASAKI
Nobeyama Radio Observatory,
Nobeyama, Minamimaki, Minamisaku,
Nagano, 384-13, Japan

A Gradual Rise and Fall type solar radio event on the east limb was observed by the Nobeyama Radio Heliograph. The event started 23 UT on May 1, 1993 and lasted more than four hours. Radio images synthesized every three minutes showed a cusp shape in the later phase of the event. The peak radio flux was 6 sfu and the GOES X-ray classwas C6. In this event, we could clearly observe the flare cusp formation process due to its favorable location (on the limb) and its large size (2 arc min.).

In the beginning of the event, a bright source appeared in the corona, then a bright bar was formed which connected the source and the lower atmosphere. The lower edge of the bar became brighter and the brighter part gradually elongated upward along the bar. Another bar was gradually formed starting from the initial source position toward the lower atmosphere away from the original bar. These two bars formed a cusp shape. After the flux peak, the top part of the cusp was bright and the whole shape were symmetric.

Radio emission mechanism of this event are mainly optically thin free-free emission. This is supported by the flat frequency spectrum of the total flux at 2 - 17 GHz and no detectable circular polarization (less than 0.5 percent). In this case, radio brightness is proportional to EM/sqrt(T). Initial radio brightening need EM enhancement in the corona. Plasma temperature increase, which is expected in magnetic field reconnection above the cusp top, causes radio brightness decrease. Plasma compression in the corona will be the cause of initial radio brightening. A bar with a bright foot is interpreted as plasma evaporation from the chromosphere into the cusp. High density plasma from the chromosphere can explain the bright foot. The elongation toward the cusp top is the high density plasma flow or diffusion. This flow continued to the other leg through the cusp top. We could not detect the evaporation from the other foot. After the evapora-

Y. Uchida et al. (eds.),
Magnetodynamic Phenomena in the Solar Atmosphere – Prototypes of Stellar Magnetic Activity, 205–206.

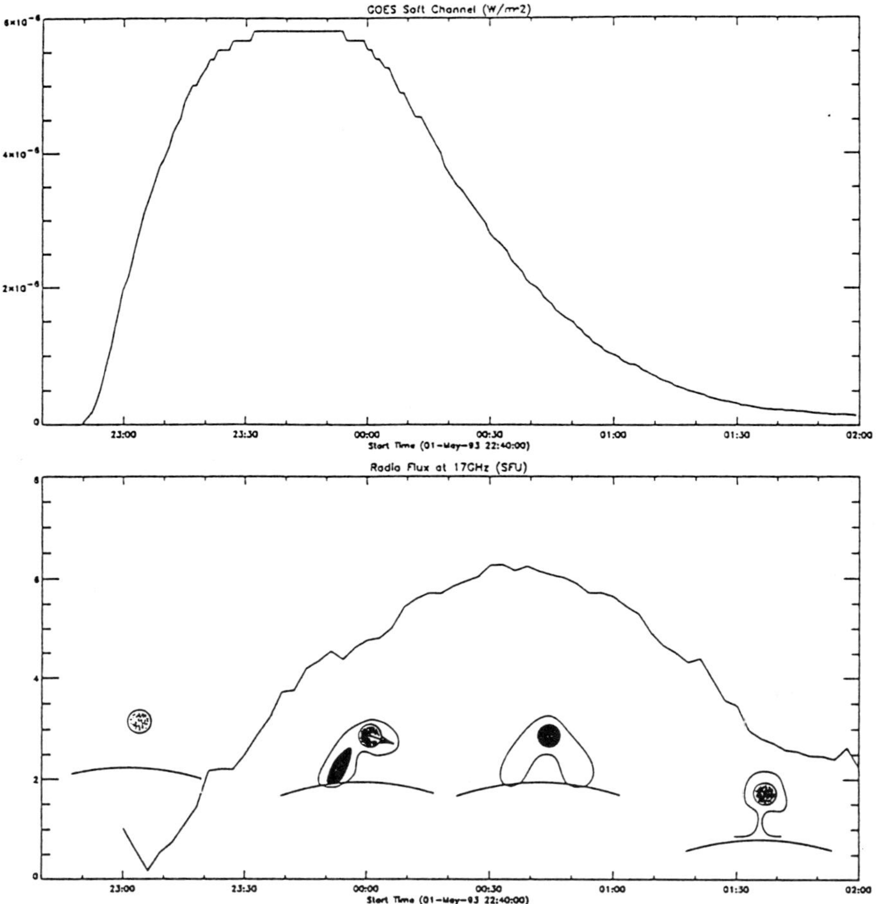

Figure 1. Top) GOES soft X-ray total flux in soft channel. Bottom) 17 GHz total flux and the sketch of source shapes.

tion stopped, the bright foot disappeared and the whole shape became a symmetric cusp.

Initial radio brightness increase is more than 4 times that of the background. To explain this enhancement by adiabatic compression of coronal plasma, the volume compression ratio should be larger than 8. The chromospheric evaporation from one leg suggests large difference of heat conductivity between two legs, assuming that the heat dump occurred at the cusp top. Another possibility is that heat dump occurred not at the top of the cusp, but at or close to the foot where evaporation continued. In this case, we need another flare scenario which explains heat dump near the foot.

FLARES AND PLASMA FLOW CAUSED BY INTERACTING LOOPS

Y. HANAOKA
Nobeyama Radio Observatory, NAOJ
Minamimaki, Minamisaku, Nagano 384-13, Japan

Remote brightenings, defined as small brightenings distant from the main flare sites during flares, have been observed by many authors. The coronal configuration underlying the phenomenon is understood to be multiple loops (Nakajima *et al.*, 1985; Machado *et al.*, 1988; and Benz, 1993). However, there remain important questions. Is the remote brightening only a by-product of a single loop flare? Does an interaction between loops in the multiple loop structure causes a flare?

We present here new observations of flares and plasma flows caused by interacting loops. The active region NOAA 7360 was observed in 1992 December with various instruments including the *Yohkoh* satellite. In this region, a small loop emerged near one of the footpoints of a pre-existing large coronal loop, and the subsequent interaction caused flares, microflares, and plasma flows. The magnetic field structure of the region is shown in Figure 1.

Four flares were observed by *Yohkoh* during the flux emerging activity. They show brightenings occurring first in a small loop, and then the large loop. The brightenings in the large loop show typical phenomena in flares; impulsive brightenings of both footpoints of the loop in hard X-rays and microwave occur first, and then gradually the loop brightens in soft X-rays. Therefore, two flare-like brightenings occur sequentially. However, the brightenings in the large loop do not occur spontaneously, but must be triggered by the brightenings in the small loop. There must be interactions between the loops to cause these flares.

As well as the flares, many microflares occurred in the small loop. More than half of them are accompanied by plasma ejection phenomena from the small loop into the large loop. The large loop is filled with the ejected plasma with the velocities of about 1000 km s^{-1}. These ejection phenomena are considered as X-ray jets observed by *Yohkoh* (e.g. Shibata *et al.*, 1992).

Y. Uchida et al. (eds.),
Magnetodynamic Phenomena in the Solar Atmosphere – Prototypes of Stellar Magnetic Activity, 207–208.

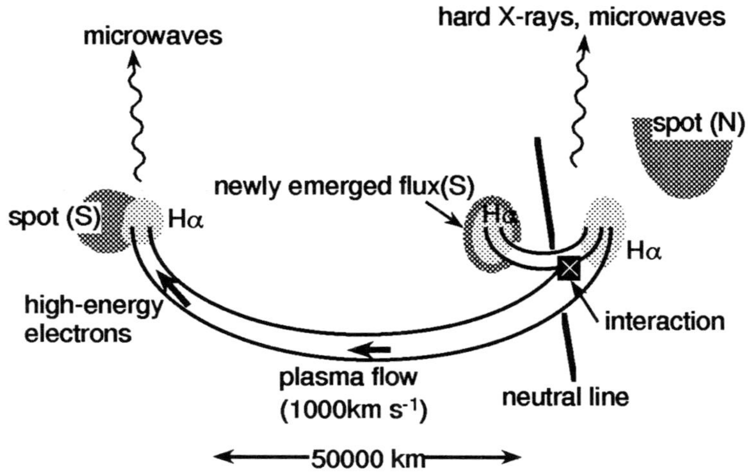

Figure 1. Schematic drawing of the magnetic field and the activities in NOAA 7360. Interactions between the small loop and the large loop cause plasma flows into the large loop, accelerations of electrons which produce hard X-ray and microwave emission, and Hα brightenings in flares.

The associated occurrences of the microflares and the jets suggest that they are also caused by interactions between the loops.

The recurrent occurrences of the homologous flares and microflares mean that the magnetic field structure in this region inevitably causes the activities due to loop-loop interactions; the flares and jets occur under the common magnetic field structure.

Our observations give a new insight into the previous observations of the remote brightenings. The events with the typical velocity of about 1000 km s^{-1} are probably explained as jets. The events caused by high-energy electrons are probably not a by-product of the flare at the main flare site, but a part of a newly occurring flare. Therefore they need not be categorized as special phenomena, 'remote brightenings'.

The full description of the results of our analyses will appear in a journal (Hanaoka, 1996).

References

Benz, A. O. (1993) *Plasma Astrophysics,* Kluwer, Dordrecht, Holland.
Hanaoka, Y. (1996) *Solar Phys.* in press.
Machado, M. E. *et al.* (1988) *Astrophys. J.* **326,** 451.
Nakajima, H. *et al.* (1985) *Astrophys. J.* **288,** 806.
Shibata, K. *et al.* (1992) *Publ. Astron. Soc. Japan,* **44,** L173.

ELECTRON TIME-OF-FLIGHT MEASUREMENTS

MARKUS J. ASCHWANDEN
University of Maryland, Astronomy Department, College Park, MD 20742, USA (e-mail: markus@astro.umd.edu)

Summary

High-precision time delay measurements of hard X-ray emission in solar flares can be used to infer the energy-dependent time scales of particle acceleration, propagation, and energy loss. A systematic time delay of $\tau = t_{25keV} - t_{50keV} \approx 20$ ms between the 25 keV and 50 keV HXR-producing electrons was discovered in a statistical study of 640 flares observed by *BATSE/CGRO* (Aschwanden, Schwartz, & Alt 1995). The magnitude and sign of the observed delay provides a diagnostic of the predominant timing mechanism (Aschwanden & Schwartz 1995), i.e. propagation time-of-flight differences in the thick-target model ($0 < \tau \lesssim 100$ ms), energy loss time differences in the thin-target model (-1 s $\lesssim \tau < 0$), trapping time differences ($-10 \lesssim \tau \lesssim -1$ s), or convolution delays with thermal effects ($+1$ s $\lesssim \tau \lesssim +10$ s). In the flare of 1992 Jan 13 (Fig.1), which became famous for the discovery of a hard X-ray looptop source (Masuda 1994), the delays of the spiky HXR pulses (Fig.2) are charachteristic for time-of-flight differences. The inferred electron time-of-flight distance was found to exceed the path difference between the HXR looptop and footpoint sources (Aschwanden et al. 1995).

References

Aschwanden,M.J., Schwartz,R.A., & Alt,D.M. 1995, Electron Time-of-Flight Differences in Solar Flares, *Ap.J.*, **447**, 923-935.

Aschwanden,M.J. and Schwartz,R.A. 1995, Accuracy, Uncertainties, and Delay Distribution of Electron Time-of-Flight Measurements in Solar Flares, *Ap.J.*, **455**, 799-814.

Aschwanden,M.J., Hudson,H.S., Kosugi,T. & Schwartz,R.A. 1995, Electron Time-of-Flight Measurements During the Masuda Flare 1992 Jan 13, *Ap.J.*, in press.

Masuda,S. 1994, Ph.D. Thesis, National Astronomical Observatory, University of Tokyo, Mitaka/Tokyo

Y. Uchida et al. (eds.),
Magnetodynamic Phenomena in the Solar Atmosphere – Prototypes of Stellar Magnetic Activity, 209–210.

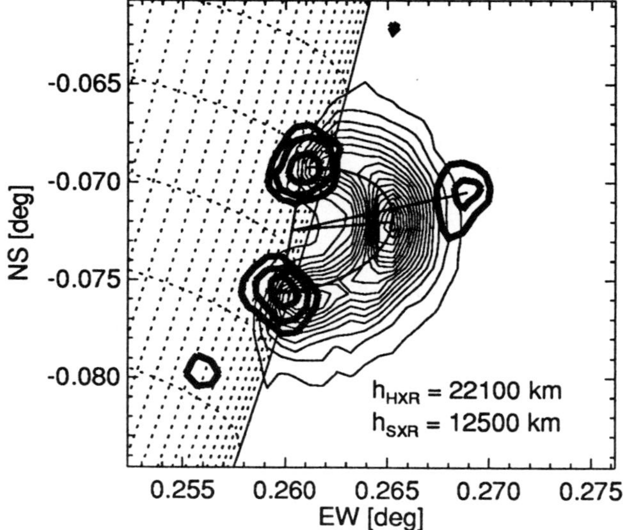

Figure 1. Loop geometry of Masuda's limb flare on 1992 January 13, 1728 UT, observed by *Yohkoh/SXT* (1728:07 UT, Be 119 filter; thin contours) and HXT (1728:04-1728:40 UT, M1 channel, 23-33 keV; thick contours). The apex of the soft X-ray loop is measured at a height of $h_{SXT} = 12,500 \pm 1750$ km, while Masuda's "above-looptop" HXR source is measured at a height of $h_{HXR} = 22,100 \pm 6000$ km.

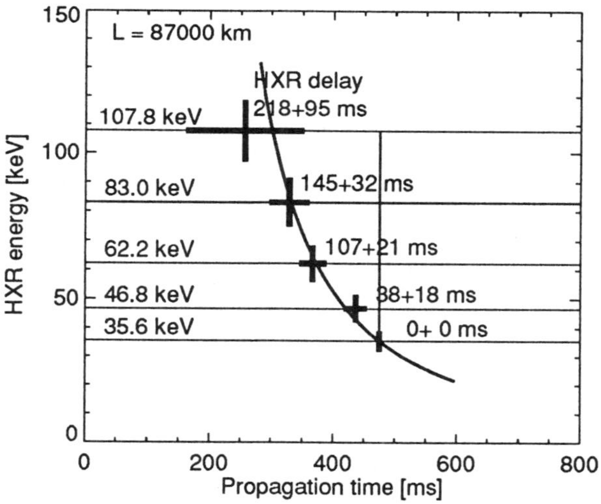

Figure 2. HXR time delay measurements from *BATSE/CGRO* between 5 energy channels in the 31-122 keV range (crosses). A theoretical model (thick line) of propagation time differences is fitted to the energy-dependent HXR delays, yielding a propagation path length of $L = 87,000 \pm 11,000$ km, which corresponds to an altitude of $h_{Acc} = 44,000 \pm 6000$ km for the acceleration site, after correction for loop curvature, helical twist of field line, and electron pitch angle.

HOT AND COOL POST-FLARE LOOPS: FORMATION AND DYNAMICS

B. SCHMIEDER, P. HEINZEL, L. VAN DRIEL-GESZTELYI AND J. E. WIIK
Observatoire de Paris, Section Meudon, 92195 Meudon, Frnace

AND

J. LEMEN
Lockheed Palo Alto Research Loboratory,
Palo Alto, CA 94304-1911, USA

Using coordinated observations of a large system of post-flare loops on 25-26 June 1992, we studied the gradual formation and dynamics of the loops and the relationship between hot X-ray and cool Hα loops. *Yohkoh/SXT* observed these loops for more than 14 hours (after the larger X-class flare of June 25 at 20:11 UT), and between 7:00 - 9:40 UT we have obtained also Hα spectra with the MSDP on Pic-du-Midi. Form a coalignment analysis between X-ray and Hα images we see that the cool loops are situated just below the hot ones and that the whole system grows up with the velocity of about 1 km s^{-1} (Schmieder *et al.* 1995 a). This is explained by gradual reconnection process as suggested by Forbes and Malherbe (1986).

For the whole gradual phase, we have determined the temperature and the emission measure *EM* from the SXT data (using two Al-filters). *EM* was also computed for the cool Hα loops. Estimating the thickness of cool loops to be around 2000 km (MSDP images) and taking into account the fact that SXT loops are more extended, we computed the electron densities n_e in hot and cool plasmas. We found $n_e^{hot} \leq n_e^{cool} \simeq 2 \times 10^{10}$ cm^{-3}, at the time of the Hα observations. However, from the temporal evolution of the *EM* we see that $n_e^{hot} \simeq 5 \times 10^{10} - 10^{11}$ cm^{-3} at around 0:00 UT and $n_e^{hot} \simeq 5 \times 10^{9} - 10^{9}$ cm^{-3} at the end of the gradual phase (around 13:00 UT). The *EM* itself decreased by almost two orders of magnitude during the same period. Using these densities, we computed lower limits of the cooling time which increases from several minutes at the beginning to more

Y. Uchida et al. (eds.),
Magnetodynamic Phenomena in the Solar Atmosphere – Prototypes of Stellar Magnetic Activity, 211–212.

than 2 hours at the end of the gradual phase of the flare (Schmieder *et al.* 1995 b).

As a next step, we analysed in detail the dynamics of cool Hα loops, using the MSDP data. After a geometrical reconstruction of the true shape ot the loops we have derived the flow velocities of the cool plasma blobs. At the upper part of the loops we observed a free-fall motion, while in the lower parts of the legs the blobs seem to be decelerated and the flow velocity appeared to be almost constant. Several possible mechanisms for such a deceleration are discussed in a recent paper by Wiik *et al.* (1995) devoted to the dynamics of cool post-flare loops.

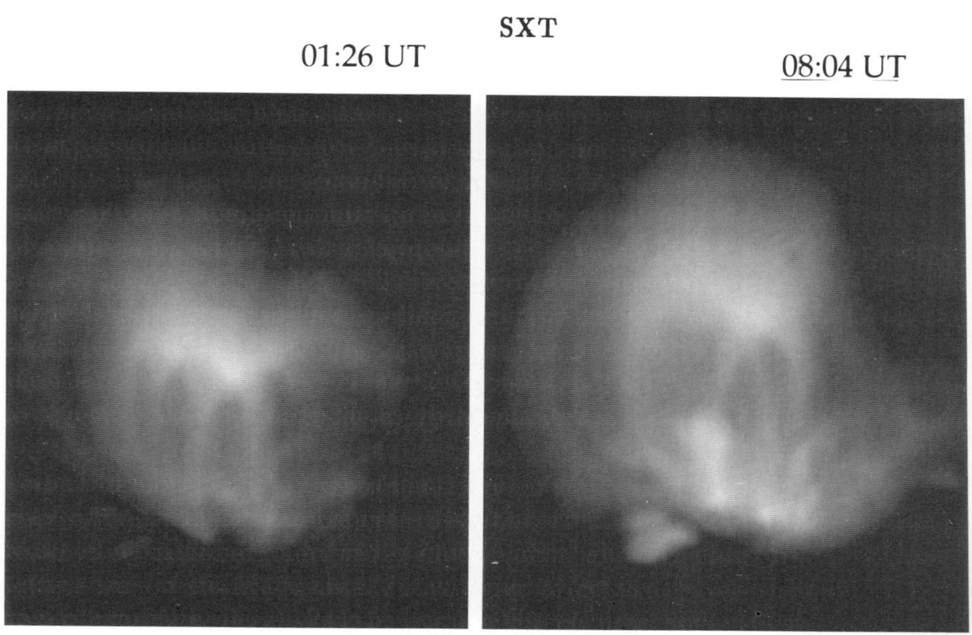

Evolution of post-flare loops on June 26 1992

References

Forbes, T.G. and Malherbe, J.M. (1986) Reconnection Process, *Astrophys. J.*, **302**, L 67.
Schmieder, B., Heinzel, P., Wiik, J.E., Lemen, J.R., Anwer, B., Kotrč, P., Hiei,E. (1995a) Relation between Cool and Hot Post-flare loops of June 1992, *Solar Phys.*, **156**, 337.
Schmieder, B., Heinzel, P., van Driel-Gesztelyi, L., Lemen, J. R. (1995b), Gradual Evolution of Cool and Hot Loops, *Solar Phys.*, submitted.
Wiik, J. E., Schmieder, B., Heinzel, P., Roudier, T. (1994), Dynamics and Inhomogeneities in Cool Flare Loops, *Solar Phys.*, submitted.

ULTRAVIOLET (1200 TO 1800 Å) EMISSION DURING THE IMPULSIVE PHASE OF A CLASS 3B-X3 SOLAR FLARE OBSERVED WITH SOLSTICE

P. BREKKE
Institute of Theoretical Astrophysics, University of Oslo, P.O. Box 1029 Blindern, N-0315 Oslo 3, Norway

AND

G. J. ROTTMAN, J. FONTENLA AND P. G. JUDGE
High Altitude Observatory, National Center for Atmospheric Research P.O. Box 3000, Boulder CO 80307-3000, USA

Abstract.

An observation of the ultraviolet spectrum (1200 to 1800 Å) during the impulsive phase of a very extended 3B-X3 class solar flare on 27 February 1992 has been obtained with the Solar-Stellar Irradiance Comparison Experiment – SOLSTICE (Rottman *et al.* 1993). This flare shows a dramatic enhancement of lines formed in the solar transition region. The full disk irradiance (i.e. the *Sun as a star*) of the resonance lines of C IV and Si IV increased by a factor of 15-20 during the impulsive phase of the flare which is comparable to flux enhancement during stellar flares.

The observation is compared with ground based Hα, magnetogram, and microwave observations as well as hard X-ray measurements from the Ulysses spacecraft. By taking into account the emitting area the radiance of the flare increased by a factor of 20,000 or more relative to the non-flaring background. Such enhancement far exceeds previous published values (e.g. OSO 8, Skylab, and SMM) where transition region lines increased about a factor of 100. These results were probably affected by limited dynamical range and temporal resolution. Thus, our SOLSTICE observation may be the first measurement of the UV enhancement during large flares (Brekke *et al.* 1996).

In addition to C IV and Si IV, the the Si III multiplet near 1295 Å also shows remarkable enhancement. Most other allowed lines such as C II, Si

Y. Uchida et al. (eds.),
Magnetodynamic Phenomena in the Solar Atmosphere – Prototypes of Stellar Magnetic Activity, 213–214.

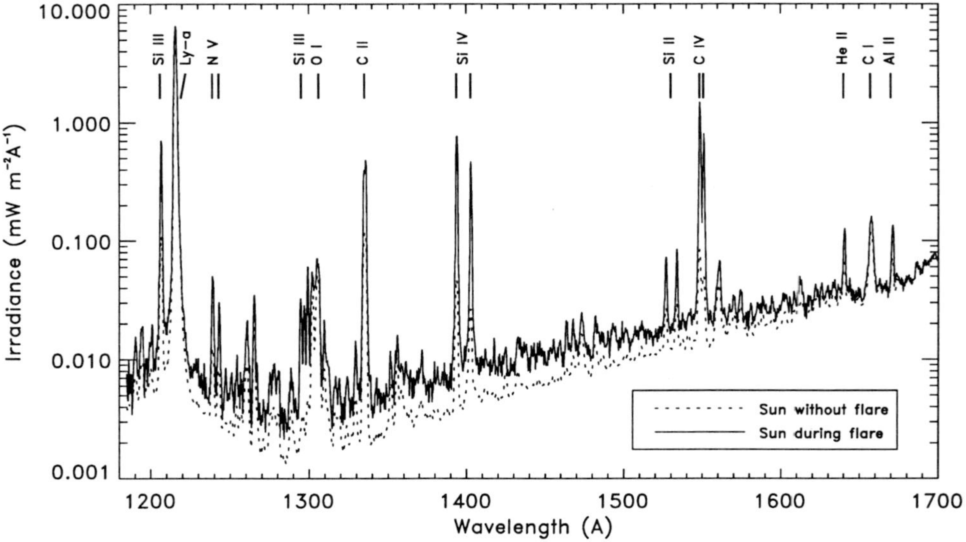

Figure 1. The SOLSTICE spectrum recorded during the impulsive phase of the February 27 1995 flare compared to a "normal" spectrum recorded the previous day.

III(1206 Å) , N V, and He II show moderate enhancements. The weakest enhancement of all is the H I Lyman-α, for which the irradiance increases only 10%. Some of the differences between the various enhancements are probably due to the observations timing since the scanning spectrometer observes different spectral features with up to 4 minutes time difference. During the flare the strong transition region lines are systematically red-shifted by 50 km s^{-1}.

At present SOLSTICE is the only UV instrument that could observe such large flux enhancements as presented in this paper. Also, SOLSTICE gives us an unique possibility to observe "the Sun as a star" and to relate the observations with results from stellar UV instruments. Thus, we plan to operate SOLSTICE in a "flare-mode" to obtain both spectra and light curves at discrete wavelengths. TRACE, to be launched in 1997, is another instrument with sufficient dynamic range to study flares.

References

Brekke, P., Rottman, G. J., Fontenla, J., and Judge, P. G. 1996, *Astrophysical Journal*, in press

Rottman, G. J., Woods, T. N., and Sparn, T. P. 1993, *JGR*, **90**, 10,667

III. Production of Superhot Plasma and High-Energy Particles in the Sun and Stars

III.2. Stellar Flares

OPTICAL EMISSION FROM STELLAR FLARES

C.J. BUTLER
Armagh Observatory
College Hill, Armagh, BT61 9DG, N. Ireland

1. Introduction

Optical observations of stellar flares on late-type stars are common, particularly in the U and B bands of the Johnson system. The reason for this preference is the large proportionate increase in the flux in the ultraviolet during flares on late-type stars, as compared to the quiescent stellar continuum which peaks in the near infra-red. Thus the contrast with the photosphere of the quiet star makes detection of flares much easier in the near ultraviolet than at longer wavelengths. Whilst single band, U or B, observations have provided very useful information on the statistics of flares, they are inadequate in their lack of spectral information. Only from spectral information can we hope to gain insight into the processes that give rise to the optical emission and thereby ultimately, to realistic flare models.

Stellar flare light curves are characterised by two distinct phases: the initial *impulsive* phase during which the optical continuum is strong, and the *gradual* phase which follows as the flare plasma cools in a roughly exponential decay. In general, it is during this second phase, that the optical emission lines of HI and CaII reach their peak. In overall energy terms, the continuum emission dominates the optical flux, though the proportion of line emission to continuum rises as the flare progresses through the gradual phase.

2. The Optical and Ultraviolet Continuum Emission

Several recent sets of observations of flares in the UBVRI bands have appeared which give limited spectral information over the whole optical and near infra-red range. We shall discuss these, individually, below. Initially, we look at attempts to explain the observed continuum shape as emission from a uniform mass of high temperature plasma. In this context the mech-

Y. Uchida et al. (eds.),
Magnetodynamic Phenomena in the Solar Atmosphere – Prototypes of Stellar Magnetic Activity, 217–225.

anisms which have been suggested for the origin of the optical continuum are the following:

- Black-body radiation
- Free-free and bound-free emission from hydrogen in the optically thick and optically thin cases
- Optical synchrotron from energetic electrons

Firstly in Figure 1, we show a rare coodinated observation of a flare on AD Leo during which simultaneous IUE and UBVR observations were obtained by Pettersen et al. (1986). The broad-band fluxes suggest a pseudo black-body curve with a peak in the near ultraviolet somewhere between 2,000 and 3,000 Å. Indeed, as we see in Figure 1, a black-body curve with a temperature of 13,000 K, gives a reasonable fit to the data. This is an unexpected result, as we know we are dealing with matter in a highly perturbed state, i.e. one we expect to be far removed from thermodynamic equilibrium. The apparently rather reasonable fit by a black-body curve to the AD Leo data, may be telling us something about the nature of the plasma which is responsible for the ultraviolet-optical continuum radiation.

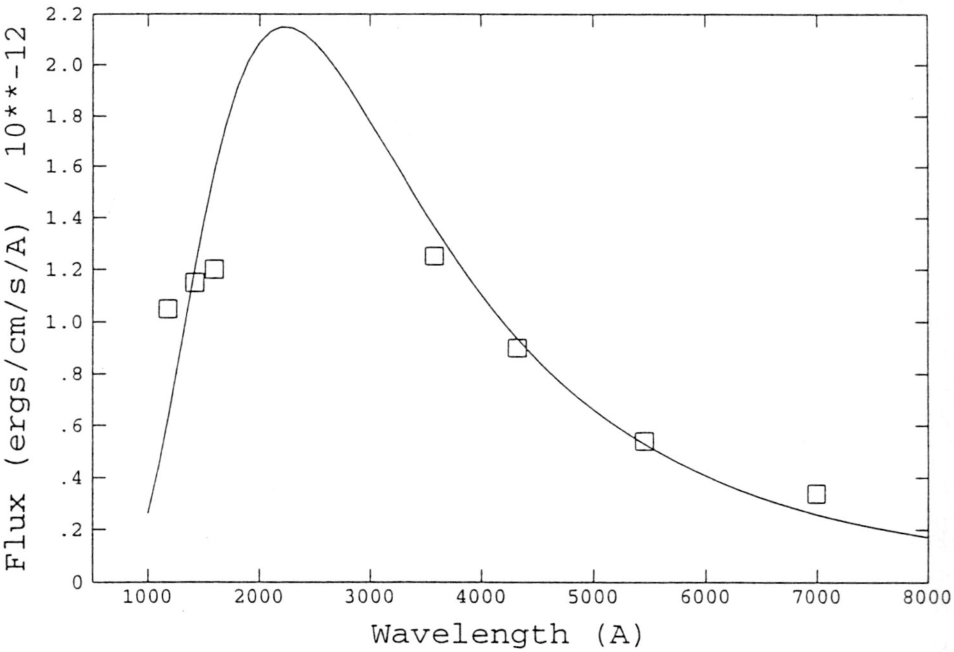

Figure 1. Broad-band fluxes during a 15 minute interval near the maximum of a flare on AD Leo observed by Pettersen et al. (1986). The curve is the black-body distribution for T_e = 13,000 K.

One might suspect, for instance, that it is pointing to an extremely compact and dense source in which thermodynamic equilibrium could be established in a very short time. Indeed, from the results of Houdebine (1992), discussed later, this seems a real possibility.

We note, however, with reference to Figure 1, that the slope of the continuum observed in the IUE SWP range is not consistent with the black-body curve and, both the IUE and optical data suggest a rather flatter spectral distribution. Due to the inability of IUE to observe in both the LWP (1950-3250 Å) and SWP (1160-2000 Å) ranges simultaneously, it was not possible to define the spectrum of the AD Leo flare at the flux peak. Even after a decade and a half of almost continual IUE operation, it seems, we have not been able to obtain coverage of flares in the full IUE and optical ranges, simultaneously. It remains an important goal in flare studies, to fill this vital gap in our knowledge.

A recent study by Alekseev et al. (1994) gives details of a large flare ($\Delta U \sim 3.7^m$) on EV Lac. Detailed (U-B), (B-V), (V-R) and (V-I) colour curves were derived throughout the flare. They show a dramatic blue enhancement at the very earliest phase of the flare, followed by a prolonged

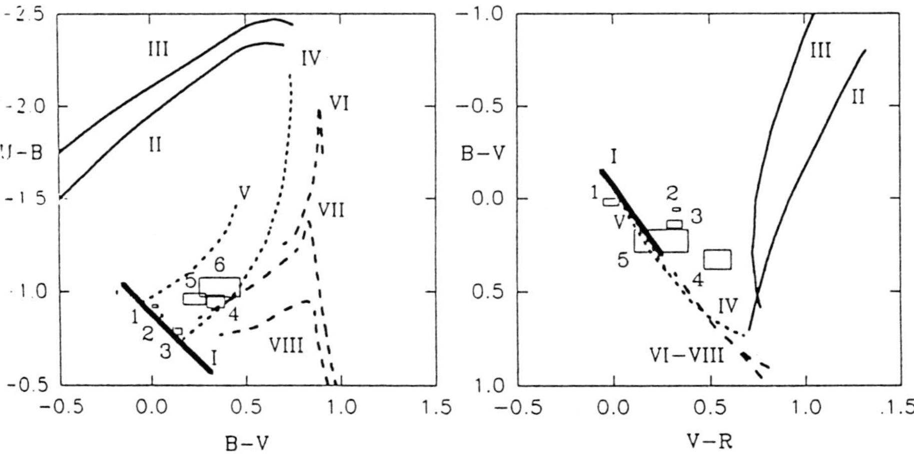

Figure 2. Two-colour diagrams for a large flare on EV Lac reported by Alekseev et al. (1994). The observations are represented by rectangles, the size of which corresponds to 2σ in the value of the respective colour indices. The position in the two-colour diagrams of the following emission models are shown: I - Black-body, T_e = 6,000 - 20,000 K; II & III - Optically thin Balmer continuum emission, T_e = 10,000 K, $n_e = 10^{12}$ & 10^{14} cm^{-3}; IV & V - Optically thick Balmer continuum emission, T_e = 10,000 & 15,000 K; VI, VII & VIII - Emission arising from proton beams of 1, 3 & 5 MeV.

period when the flare colours were almost constant in (U-B), (B-V) and (V-R). In Figure 2, (their Figure 5), we show the position of the error boxes for the EV Lac flare in the two colour diagrams. In addition Alekseev et al. plot the loci of various model predictions for continuum radiation. The models they chose were: (1) black body radiation, (2) Balmer continuum from a hydrogen plasma in the optically thin and optically thick cases, and (3) heating by high energy protons. The various stages of the flare are numbered, with (1) corresponding to the early impulsive phase, (2) to the peak of the U-band emission, and (3), (4), (5) and (6) to successive stages of the decline. In both two-colour diagrams, the first phase sits firmly on the black-body line. Similarly, in the (U-B) - (B-V) diagrams, the next two phases lie close to the black-body line. However, the diagram involving (V-R) shows some differences from this picture. This was interpreted as due to a greater sensitivity in the R-band to plasma emission at the early stages of the flare. Overall, Alekseev et al. (1994) concluded that a black-body distribution gave a good fit to the broad-band colours of flare light in the early stages of this flare but that during the later stages the colours were more consistent with optically thick Balmer continuum radiation.

A large ($\Delta U \sim 3.8^m$) flare on the dMe star, Gliese 234, was observed in UBVRI by Doyle et al. (1989). The observations were made using an automatic filter change photometer operating sequentially in the various colours and are therefore not truly simultaneous. However, by interpolation of the light curves, approximate colours can be derived during the progress of the flare. On the assumption that the flare emission is the result of free-free or bound-free radiation from a hydrogen plasma, broad-band colours were derived theoretically for the range of optical and near infra-red wavelengths observed. It was found that the broad-band BVRI colours during the early stages of the flare were reasonably well fitted by both emission mechanisms for a plasma with a temperature of 1.4 10^4K. However, the U-band flux was significantly higher than that predicted by either mechanism. Optical synchrotron radiation with $\gamma = 4.5$, was also able to fit the observed colours, though again it underestimated the U-band flux. However, the crucial factor in their interpretation turned out to be, not the shape of the continuum spectral distribution, but the volume of plasma that would be required to produce the observed fluxes. In the case of free-free emission the volume of plasma required was of the order of the volume of the star itself. This makes it unlikley that free-free emission can be the prime mechanism. By similar arguments it was found that optical synchrotron emission requires an unrealistic number of high energy electrons to produce the observed flux. Bound-free emission, on the other hand, is more efficient and appeared to be a likely candidate in this case.

3. Non-LTE Computations of Flares in Stellar Atmospheres

The studies discussed so far have considered the optical radiation from a stellar flare on a rather simplistic basis, as arising from a plasma at a single temperature and density. Though useful in pointing the general direction we may expect to see progress, it is necessary to move to more complex atmospheric models for an improved understanding of the true situation. Two studies have recently appeared that attempt to compute the effect on a NLTE model atmosphere of a sudden release of energy such as occurs in a flare, one by Houdebine (1992) and the other by Hawley and Pettersen (1991) and Hawley and Fisher (1992).

In these studies, several possible models have been considered to explain the optical continuum flux. They are:

- Conduction from the hot overlying corona.
- Irradiation of the chromosphere/photosphere by soft X-rays from the corona.
- Emission from a downward moving condensation.
- Reprocessing of the EUV/UV radiation from a condensation to produce optical emission.

One of the flares that has been most carefully studied in recent years is the large ($\Delta U \sim 4.5^m$) flare observed by Hawley and Pettersen on AD Leo on 12 April, 1985, discussed previously. For this flare we show, in Figure 3(a), the observed continuum fluxes at several phases during the flare.

Hawley (1991) and Hawley and Fisher (1992) computed the continuum emission arising from conduction from the hot corona and the excitation of the quiescent chromosphere by downwards directed soft X-rays. They computed the optical emission arising from these processes in five different models with a *coronal apex temperature* which varied from 3 10^6K to 20 10^6 K. The spectrum of the continuum emission predicted is shown in Figure 3(b). In all cases the emission is much redder that that normally observed in stellar flares.

From these studies, they were able to establish: (1) that conduction from the corona was unlikely to be a major contributor to the optical continuum in the AD Leo flare, (2) that downward directed soft X-rays from the corona can explain the total optical continuum and line flux but not the observed spectrum of the emission. Therefore coronal soft X-rays could not be the dominant source of excitation of hydrogen in this flare.

Initially, Hawley and Pettersen (1991) proposed that the observed excess blue emission might arise from a multitude of metal lines in the near ultraviolet, however, in a follow-up paper by Hawley and Fisher (1992), it was proposed that the observed, approximately black-body, continuum emission could arise from reprocessing in the photosphere of strong EUV/UV emission from the upper chromosphere. A basically similar conclusion was reached, independently, by Houdebine (1992) following an earlier suggestion by Houdebine (1990) and Houdebine and Butler (1990).

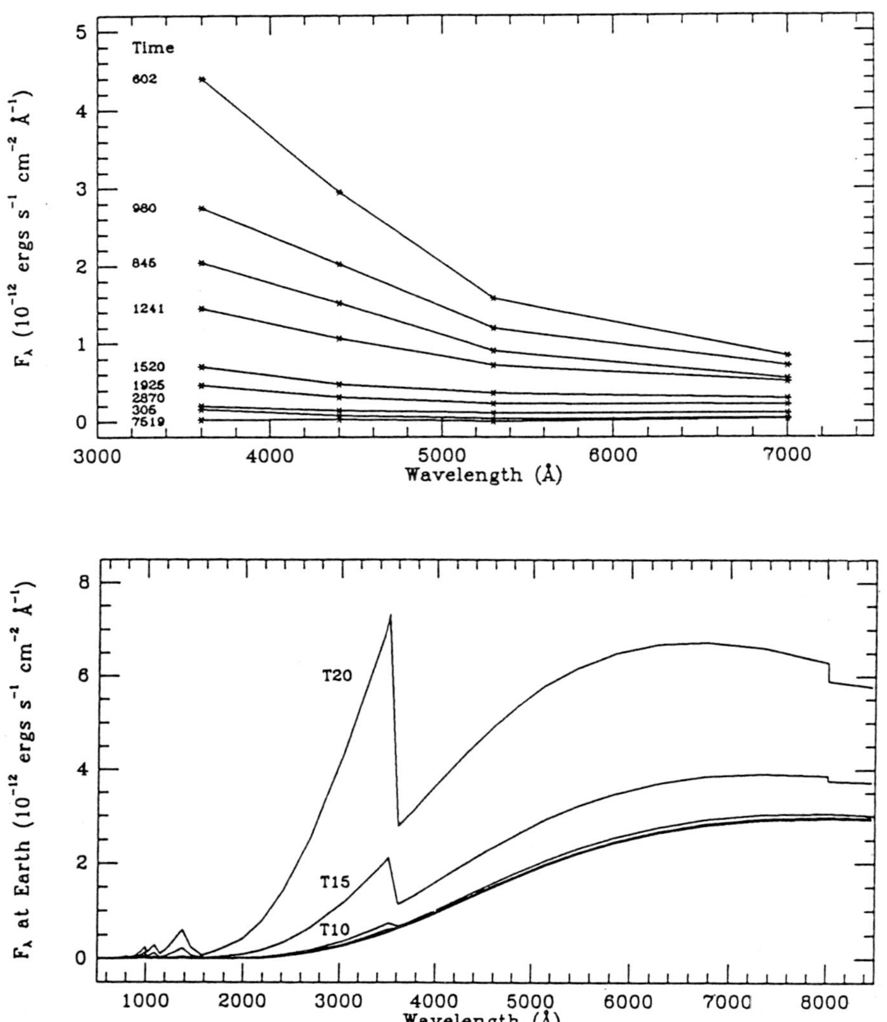

Figure 3 (**a**) Continuum fluxes in UBVR for the AD Leo flare, (from Hawley, 1991). Note the large ultraviolet flux during the early stages of the flare. (**b**) The predicted continuum spectra for the AD Leo flare from models by Hawley and Fisher (1992).

One of the most promising models of flare optical continuum emission is that proposed by Katsova et al. (1981) and Katsova and Livshits (1988), whereby a *flare kernel*, formed near the footpoints of a magnetic loop, is driven downwards into the upper chromosphere with an equal and opposite momentum to material evaporated into the corona. Alternatively, in proton beam models, the momentum could be imparted by these more massive particles. Canfield et al. (1990) has presented evidence for such downward moving *kernels* or *condensations* in solar flares.

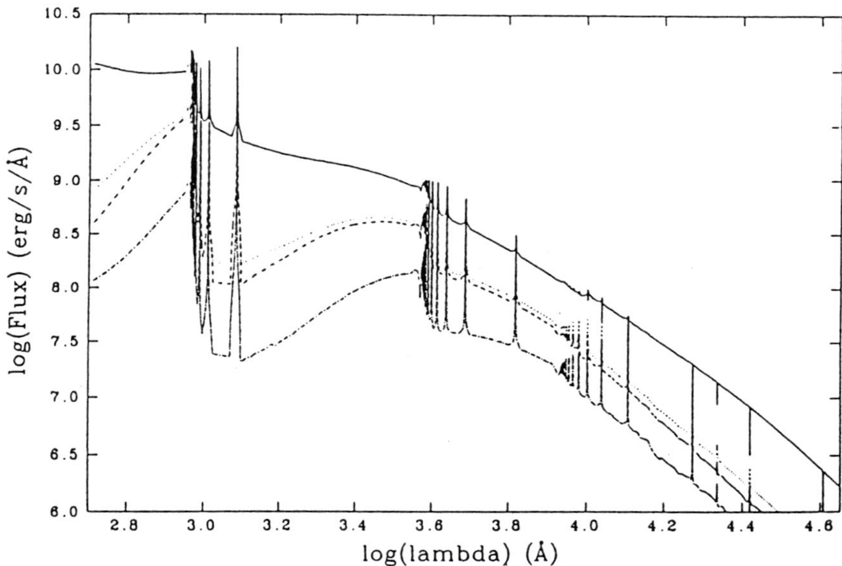

Figure 4 The merged hydrogen spectrum of a flare kernel from models by Houdebine (1992). The higher density models are at the top. Note that for high density models the continuum emission becomes increasingly similar to a black-body spectrum with a temperature in the range 10,000 to 20,000 K.

Houdebine (1992) has considered, in some detail, the optical emission that would arise from such flare kernels and concludes that many of the observed features are explained by his model. In Figure 4, (Figure 18 of Houdebine, 1992), we show the computed spectrum of hydrogen emission from a flare kernel. The spectrum becomes increasingly like a black-body curve as the density of the kernel rises, The model is also able to explain the strong ultraviolet continuum flux and gives Balmer decrements similar to those observed. With increasing kernel density, the continuum dominates the line emission and the Balmer jump virtually disappears. Such behaviour has been seen in some of the more energetic flares (e.g. AD Leo, Hawley and Pettersen, 1991).

4. Conclusions

In this short review, we have discussed the observations of continuum emission from stellar flares and the various models that have been proposed for its origin. Whilst we still have relatively few broad-band observations of flares over the entire optical to near infra-red range, those we have are beginning to lead to a consistent picture. The continuum emission, though similar to black-body in the early phases of a flare, becomes increasingly dominated by optically thick hydrogen emission as the flare progresses. The observations support the contention that more than one emission mechanism is invloved during the progress of a flare.

With the recent application of NLTE codes, it has proved possible to probe further; both the processes involved in the heating of the plasma, and the emission mechanisms which subsequently dominate the production of optical radiation. From the limited evidence we have, it seems that the *hot kernel* model shows the greatest promise of explaining the continuum emission, particularly if we include the reprocessing of ultraviolet and extreme ultraviolet emission by the adjacent quiet atmosphere. The studies that have been carried out by Katsova and Livshits (1988), Houdebine (1992) and Hawley and Fisher (1992), have given a new insght into the origin of stellar flare optical emission. However, we shall need to apply these models to a much greater panoply of flares before we can convincingly accept a generally applicable flare model. For this, we need a much larger data bank of simultaneous UBVRI observations, obtained by multichannel techniques. The normally available, sequential UBVRI observations, are not sufficiently accurate, particularly in the impulsive phase when the flux is changing rapidly on a time-scale comparable to the interval between successive colours. More observations of flares, using a truly multichannel system, such as that described by Barwig et al. (1987), are badly needed.

5. Acknowledgements

Research at Armagh Observatory is grant-aided by the Department of Education for Northern Ireland.

6. References

Alekseev, I.Yu., Gershberg, R.E., Ilyin, I.V., Shakovskaya, N.I., Avgoloupis, S., Mavridis, L.N., Seiradakis, J.H., Kidger, M.R., Panferova, I.P. and Pustil'nik, L.A. (1994) *A&A* **288**, 502.

Barwig, H. Schoems, R. and Buckenmayer, C. (1987), A&A *175*, 327.

Canfield, R.C., Zarro, D.M., Metcalf, T.R. and Lemen, J.R. (1990), Ap. J. *348*, 333.

Doyle, J.G., Van den Oord, G.H.J. and Butler, C.J. (1989), *A&A* **208**, 208.

Hawley, S.L. (1991) *Mem. Soc. Astron. Ital.* **62**, 271.

Hawley, S.L. and Pettersen, B.R. (1991) *Ap. J.* **378**, 725.

Hawley, S.L. and Fisher, G.H. (1992) *Ap. J. Suppl.* **78**, 565.

Houdebine, E.R. (1990), Ph.D. Thesis, University of Orsay, Paris XI, September 1990.

Houdebine, E.R. (1992), *Irish Astron. J.* **20**, 248.

Houdebine, E.R. and Butler, C.J. (1990) *Flare Stars in Star Clusters, Associations and the Solar Vicinity*, proceedings IAU Coll. 137, Mirzoyan et al. (eds), Kluwer.

Katsova, M.M. and Livshits, M.A. (1988) *Activity in Cool Star Envelopes*, proceedings of the Midnight Sun Conference, Tromso, 1987, O. Havnes et al. (eds), Kluwer, 143.

Katsova, M.M. Kosovichev, A.G. and Livshits, M.A. (1981), *Astrofizika* **17**, 285.

Pettersen, B.R., Hawley, S.L. and Anderson, B.N. (1986) *New Insights in Astrophysics*, ESA SP-263, 157.

FLARES ON ACTIVE BINARY SYSTEMS

S. CATALANO

Catania Astrophysical Observatory
Viale A. Doria 6, I-95125 Catania, Italy

Abstract

Flares in the Sun and in binary stars seem to have the same physical nature but very different energetic power. The total energy released by flares on RS CVn binaries is in the range 10^{35} - 10^{38} erg while typical energies of solar flares are $\approx 10^{28}$ - 10^{32} erg. Flares in binaries are usually longer–lived than those observed in the Sun and in dMe stars. The relevance of the binary configuration in the flare energy balance and storage is analysed in the framework of loop models. Evidences for modifications of active region configurations concurrently with flare eruptions, in some very active RS CVn binary, are presented. It is argued that such manifestations may be the result of a dynamic global coupling of the magnetic fields between the flaring site and the rest of the activity complex.

1. Introduction

Following the first detection of a soft X-ray flare on the RS CVn binary HR 1099 by White et al. (1978), numerous flares have been observed on RS CVn and Algol binaries during various satellite missions, as well as at radio and optical wavelengths. Although the flare phenomenon for decades has been known to occurr in stars other than the Sun, but at quite similar energy levels, the flares observed in close binaries appear to be much more powerful events which may not share too many similarities with solar flares.

Since other speakers at this colloquium will discuss observations of stellar flares in different types of stars and in the wide range of physical conditions available on any given star, I will mainly concentrate on some specific problem related to the unusually large amount of energy released in close binary star flares, and on the condition for the flare occurrence.

Y. Uchida et al. (eds.),
Magnetodynamic Phenomena in the Solar Atmosphere – Prototypes of Stellar Magnetic Activity, 227–234.

2. Which Binaries do Flare?

The obvious immediate answer is: all close binaries. By close binaries I refer to short period ($P_{orb} \sim 0.5 - 50$ days) binary systems which contain non-degenerate components: the RS CVn Binaries, Algols, and W UMa systems.

RS CVn binaries

RS CVn binaries are very active systems with prominent X-ray and radio luminosity $L_X = 10^{29.5}$-$10^{31.5}$ erg s^{-1}, $L_{6cm} = 10^{15.5}$-10^{17} erg s^{-1}Hz^{-1} respectively. Chromospheric and transition region lines manifest as strong emission features in their spectra.

To date tenths of X-ray flares have been observed in several RS CVn systems and many of them have been found flaring more than one time. Enhancements of a factor of ten is frequently observed at the flare peak, though intensity increases by factors of 35-40 have also been observed. To these cases should belong the flares detected on σ Gem $E_X = 3.5 \times 10^{38}$ erg, (Pye and Mc Hardy 1980), BD +61 1211, $E_X = 4\times10^{37}$ erg, (Charles et al. 1979) and UX Ari $E_X = 1\times10^{37}$ erg, (Tsuru et al 1989). These very energetic events have time scale of about ten hours, with cases of longer durations like the 3 day X-ray flare on HR 5110 (Graffagnino et al. 1995).

Spectral observations yield constraints on the temperture and the volume emission measure of the emitting plasma, which are in the range 10^7 - 10^8 K and 10^{52} - 10^{54} cm^{-3} respectively. Although not very energetic, the February 17, 1980 flare on HR 1099 has the highest maximum temperature ($T_{max} = 1.3\times10^8$ K) of any stellar flare ever observed (Agrawal and Vaidya 1988).

The flare activity at radio wavelengths is one of the most outstanding characteristics of RS CVn binaries. Microwave flares exhibit a large variety of time scales, as short as 4-5 minutes (Brown and Crane 1978), or as the very long-lasting single flare on UX Ari, $\approx$ 50 hr (Gibson, Hjellming and Owen 1975). States of continuous flaring lasting for weeks or months (Feldman et al. 1978, Umana et al. 1995a) appear to be a peculiar characteristic of RS CV n systems. The extreme case is represented by UX Ari which has been found in one of such high activity states for about 4 months, from October 7 1992 to February 6, 1993 (Elias et al. 1995, Neidőfer et al. 1993, Umana et al. 1995b).

Chromospheric and transition region flares are frequently detected in the ultraviolet emission lines, and in Hα. Although typical time-scales for the rising and decay time are 1-2 hr and 3 hr, longer durations, e. g. 20 hr for the 3 October 1981 flare on HR 1099 (Linsky et al. 1989) and 6 days for the 24-29 September 1989 Hα flare on HK Lac (Catalano & Frasca 1994) have been ascertained.

Optical, or white light, flares are very rare events in RS CVn binaries. The best documented are the 14 and 15 December 1989 flares observed on HR 1099. The integrated losses in the 3100-5900 Å continuum were estimated to reach 2.85×10^{38} erg in the 15 December flare only. A similar amount of energy has been estimated to be associated with the mass flow deduced from the Hα emission profile behaviour (Foing et al. 1994).

Algol systems

Although Algol systems are different from RS CVn systems, mainly from the evolutionary point of view, they exhibit some of the RS CVn activity phenomena. Algols also contain an early K subgiant/giant secondary, but the primary is typically of spectral type A or B. Furthermore, Algols are semi-detached systems, wherein the evolved secondary is tranferring mass to the primary.

Of the two main X-ray flares on Algol, the first one, observed with the EXOSAT satellite (White et al. 1986) yielded a total energy of 2.5×10^{34} erg in the 0.1 -10 KeV band. The second one, observed with the GINGA satellite (Stern et al. 1992), resulted much more powerful, the total energy being about 2.3×10^{35} erg with a peak luminosity of 10^{31} erg s^{-1}. The total duration was about 15 hr while the EXOSAT flare duration was about 2.4 hr. Before the large GINGA flare a smaller *precursor* lasting about 5 hr reached a peak intensity of about 1.5 times the quiescent flux. It seems that Algol X-ray flares are characterized by an energy output of $2\times10^{34}-2\times10^{35}$ erg i.e. about one-two order of magnitudes lower than in typical RS CVn flares.

The quiescent radio properties of Algols and RS CVn's are very similar, as regard the average luminosity and spectra (Umana et al. 1990). Several flares have been detected on Algol at radio wavelengths, all of them exhibiting about the same structure with a decay time twice as long as the rise time. They appear to preferably occur outside the eclipses, suggesting that the source of flares is located between the two stars (Catalano 1990). Observation of flares in the optical and UV region are strongly biased by the high continuum of primary component which dominates the spectrum at these wavelengths.

W UMa systems

W UMa systems are very short period ($P_{orb} \leq 1$ day) contact binaries with both components of similar spectral type, G main sequence, filling their Roche lobe. The best observation is the detection of an X-ray and radio flare during an EXOSAT and VLA coordinated monitoring on VW Cep (Vilhu et al. 1988). The duration is shorter in the x-ray band ($\sim$ 1 hr) and longer in the radio band ($\sim$ 3 hr). The flare appeared as a gradual rise in the soft X-ray band, followed by an impulsive nonthermal radio burst and impulsive X-ray emission in the harder band a few minute later. This behaviour is similar to that of many solar flares.

Observation of optical flares in the W UMa systems were reported long ago in 44 i Boo B (Eggen 1948) and U Peg (Huruhata 1952). However the best observed is a flare in VW Cep (Egge and Pettersen 1983). The energy released in the U band during the 35 min duration was estimated to be 4×10^{33} erg.

From this short summary of binary star flare properties, it appears that RS CVn systems display the most powerful flares with energies up to 2-3×10^{38} erg, followed by Algol's with energies of $2\times10^{34-35}$ erg and by W UMa systems with 10^{32-33} erg, which are rather close to the solar flare energies, 10^{28-32} erg.

3. The Role of Binarity in the Flare Energetics

Although for many flares in binaries, comparison with solar two ribbon flares has been made (van den Oord 1988, Doyle et al. 1988), the time scale and energy involved require larger size scales and may be different magnetic configuration. Therefore the question is: i) to which extent is the binary nature important for the flare enrgetics, ii) what is the magnetic flux tube dimension, and iii) where is the pre-flare energy stored ?

Simon et al. (1980) proposed a speculative scenario in which large magnetic flux tubes, attached to both stars, occasionally interact and interconnect. Such kind of model that has been worked out more quantitatively by Uchida (1986) may be applied to RS CVn systems where both components are active stars, but not to Algols where the primary early type component is not active. Radio maps of RS CVn systems typically extends over the whole system with brighter compact sources associated with the two stars, while in the case of Algol the source is clearly associated with the late type secondary componet. A "core-halo" structure, the "core", of size smaller or comparable to the size of the star, being associated with the active component, was seen on UX Ari during an outburst (Mutel et al. 1985). VLBI scans during a flare on HR 1099 show that the source size expands during the cooling phase, reaching the size of the entire system (Trigilio et al. 1993).

This would suggest that flares erupt in closed loops anchored to the star but may expand along field lines open to the circumstellar space or interconnected between the two stars, like in the Uchida model. As a matter of fact, results from X-ray flare analysis, typically in the assumption that no further energy deposition occurs during the flare decay phase, and that the cooling is due to radiation and conduction loss, lead to more compact sources. Estimates, in a two-ribbon flare model, of loop height for Algol flares led to values of 0.2 R_* (van den Oord and Mewe 1989) and 0.8 R_* (Stern et al 1992). Similar values are found in the case of RS CVn flares, e.g. 0.2 R_* for a flare on σ CrB (van den Oord 1988) and on V711 Tau (Agrawal and Vaidya 1988), and $\sim$ 1 R_* on UX Ari (Tsuru et al 1989).

Flares are thought to be produced by the annihilation of a *portion* of the magnetic field energy stored in the field configuration. The shear of an arcade of loops leads to the magnetic energy increase of the system until the occurrence of a reconnection or an interconnection with another loop system. The energy storing structure should therefore be larger than the energy releasing volume.

Adopting a two ribbon flare model van den Oord(1988) has shown that the maximal storage of energy is obtained when the filament is located between both components of the binary system. The application of the model to very energetic flares has led to the requirement that the pre-flare energy has to be stored in a structure longer than the stellar size therefore requiring an intersystem filament, as in the case of the 2.4×10^{36} erg flare on II Peg (Doyle et al 1989). Similar requirements were needed to explain the optical and Hα flare in HR 1099 (Foing et al 1994) and the X-ray flare on HR5110 (Graffagnino et al 1995). However for the 1.3×10^{37} erg Hα flare on HK Lac Catalano and Frasca (1994) showed that the pre-flare energy could be stored in a filament, about 0.5 $R_\star$ in length, anchored on the active component. Catalano and Frasca noticed that the relative length of this structure, 0.5 R_*, is not much larger than that of the largest two-ribbon solar flares when scaled with Sun's radius. So the larger energy released in the HK Lac flare, compared with that in solar flares, can be accounted for by the larger intrinsic dimension (i.e. larger amount of heated plasma) of the structure involved.

HK Lac is a relatively wide system, so the interaction between the two components is rather weak, and there is enough space for large loops to develop without interaction with the field of the other companion. At present there are not enough observed cases to disclose whether the binary separation is a crucial parameter for the energy storage.

4. Flare Occurrence and Light-Curve Changes

The large amount of energy and the large magnetic structures involved in the binary flares raise also the question of possible visible evidence of the magnetic field disruption or of the pre-flare configuration that leads to the flare eruption. Catalano and Frasca (1994) showed that the big 24-29 September 1989 Hα flare on HK Lac occured concurrently with a significant change in the light curve that had been interpreted by Olàh et al. (1991)

as the development of a new spot group. Catalano and Frasca also showed that a previous light curve change had occurred in connection with the enhancement of Hα emission observed by Bopp and Talcott (1980) on September 29 1977. In both cases the flare developed on the same hemisphere of the new formed spot group. Investigating the time lag between the spot appearance and the flare eruption Catalano and Frasca concluded that the lag could be shorter than one stellar rotation. Light curve changes have also been observed in association with flare events in HR 1099. Zhai et al (1994) found an average brightening of all the system after the flare events of 14-15 December 1989. They interpreted this systematic brightening as the emergence of a bright spot (white light plage), which was needed to model the light curve. Furthermore a steady light curve evolution took place between November and December 1992 (Cutispoto et al. 1994) concurrently with the eruption of a series of violent flares (Neff et al. 1994).
One can argue that the flare eruption produces the light curve changes. However, in the solar case, spatially resolved observations show that numerous flares erupt early in the development of emerging magnetic flux region. The largest flares are found to erupt where old and new magnetic structures interact with one another. This appears to be the case for both HK Lac and HR 1099 flares, even with different manifestations. In the case of HK Lac, the emergence of new magnetic flux was marked by the birth of a new dark spot group (Olàh et al. 1991), while in the case of HR 1099 it was marked by the formation of a bright photospheric facula.

5. Condition for flare occurrence and flare rates

It is observed in the Sun that during its lifetime, an activity complex is refreshed by injection of new magnetic flux in the form of bipolar active regions, creating the condition for repeated flare events. The solar flare rate has been found to be correlated with the complexity index of active regions (Bornmann & Shaw 1994). For stellar active regions it is not possible to establish a complexity classification as for the solar ones (McIntosh 1990), but it is of interest to look for indicators that can give an idea of active region complexity. A quick search for useful activity parameters shows that light curves may provide the best index of complexity. In this report I would like to call the attention on a few cases of high flaring activity and on the concurrent behaviour of the light curve.

Within the discrete spot model approach, light curves are normally fitted by one or two large spots or spot groups, and very rarely more than two spots are needed. At least in two cases recurrent intense flares occurred when three spots were needed to fit the light curve complex shape. The first case refers to the strong white light and Hα flares of 14-15 December 1989 on HR 1099 (Foing et al. 1994). The light curve obtained at that time was well fitted by two dark spots plus a bright spot (Zhai et al 1994). The second outstanding case refers to the October, 7 , 1992 – February 6 1993 continuous radio flaring activity of UX Ari. Also in this case the light curve displayed a complex shape that required three dark spots to perform a good fit (Elias et al 1994). Moreover, Umana et al. (1995a) found a correlation between the mean V magnitude of HR 1099 and the radio emission flux, indicating that the strongest radio flares occur with higher probability when the average V brightness is smaller, i.e. at the times of maximum spot coverage.

Although these represent only few cases and a quantitative parameter scale cannot be established, they give representative indirect indications that active region complexity favour the onset of strong flare activity in RS CVn binaries, in a fashion very similar to the solar case.

6. Conclusions

The large energies and long time scales of binary star flares raise the problem whether the mechanisms for the energy release and transport are the same in binaries and in single stars including the Sun. Differences in the morphology of binary outburst light curves seem to be indicative of a different magnetic field configuration, and this combined with the derived loop heights (which in many cases have the same dimension as the binary separation), suggests that the outburst may be of inter-binary nature. However the binary configuration appears also relevant in the energy balance and storage. Solar flare eruptions take place shortly after (1-2 days) the emergence of new flux tubes within an activity *complex* or *nests*, and the level of flare activity appears to be still high during the subsequent rotation (Gaizauskas 1989). If the process in RS CVn stars is similar to that of the Sun it is of interest to investigate how long the interaction time between magnetic structure is and which stellar properties depends on?

Acknowledgements: Financial support by the Italian *Ministero dell' Univeristà, e della Ricerca Scientifica e Tecnologica*, the *GruppoNazionale di Astronomia of the CNR* and the *Regione Sicilia* is gratefully acknowledged.

References

Agrawal P. C., & Vaidya J., 1988, *MNRAS*, 235, 239
Bopp B. W. & Talcott J.C., 1980, *AJ*, 85, 55
Bornmann P. L.& Shaw D., 1994, *Solar Phys.*, 150, 127
Brown, F. N., Crane P. C., 1978, *AJ*, 83, 1504
Catalano S., 1990, in *Active Close Binaries*, C. Ibanoglu ed., Kluwer Academic Publisher, Dordrecht p. 411

Catalano S., Frasca A., 1994, *A &A*, 287, 575
Charles P. A., Walter F., & Bowyer S., 1979 *Nature* 282, 691
Cutispoto G., Rodonò M., Zhai D., Pagano I., 1994, in *Proceedings of the 4th MUSICOS Workshop*, L. Huang et al. eds., Beijing Observatory, p. 165
Doyle J. G., Byrne P. B., Van den Oord G. H. J., 1989, *A&A* 224, 153
Egge K. E., & Pettersen B.R., 1983, in *Activity in Red-Dwarf Stars*, IAU Colloquium No. 71, P. B. Byrne & M. Rodonò eds, Reidel Dordrecht , p. 481
Eggen O. J., 1948, *ApJ*, 108, 15
Elias N. M., et al., 1995, *ApJ*, 439, 983
Feldman P.A., et al., 1978, *AJ*, 83, 1471
Foing B. H., et al., 1994, *A&A*, 292, 543
Gaizauskas V., 1989, in *Solar and Stellar Flares*, B. Haish & M. Rodonò eds., Kluwer Academic Publisher, Dordrecht, p. 71
Gibson D. M., Hjellming, R. M., Owen F. N., 1975, *ApJ* 200, L99
Graffagnino V. G., Wonnacott D., Schaeidt S. , 1995, *MNRAS*, 275, 129
Huruhata M. , 1952, *PASP*, 64, 200
Linsky J. L., 1989, *A&A*, 211, 17
McIntosh P. S., 1990, *Solar Phys.*, 125,251
Mutel R. L., et al., 1985, *ApJ*, 289, 262
Neff J. E., Pagano I., Rodonò M.: 1994, in *Cool Stars, Stellar Systems and the Sun*, Eighth Cambridge Workshop J-P. Caillault ed., ASP Conf. Series, Vol. 64, 1994
Neidőfer J., Massi M., Ciuderi-Drago F., 1993 *A&A*, 278,L51
Olàh K., Hall D. S., Henry G. W., 1991, *A&A*, 251, 531
Pye J. P., Mc Hardy I. M., 1983, *MNRAS*,, 205, 875
Simon T., Linsky J. L., Schiffer F. H. III, 1980, *ApJ*, 239, 911
Stern R. A., et al., 1992, *ApJ*, 391, 760
Trigilio C., Umana G., Migenes V.: 1993, *MNRAS*, 260, 903
Tsuru T., et al., 1989, *PASJ*, 41, 679
Uchida Y., 1986, *Astrophys Space Sci.*, 118, 127
Umana G., Catalano S., Rodonò M., 1991, *A&A*, 249, 217
Umana G., Trigilio C., Tumino M., Catalano S., Rodonò M., 1995a, *A&A*, 298, 143
Umana et al. 1995b, private communication
Van den Oord G. H. J., 1988, *A&A*, 205, 167
van den Oord G. H. J., & Mewe R., 1989, *ApJ* 213, 245
Vilhu O., Caillault J-P., Heise H., 1988, *ApJ*, 330, 922
White N. E., et al., 1986, *ApJ*, 301, 262
White N. E., Sanford P. W. & Weiler E. J., 1978, *Nature*, 274, 569
Zhai D. S., Foing B. H., Cutispoto G., 1994, *A&A*, 282,168

STELLAR X-RAY FLARES

B. HAISCH
Lockheed Solar and Astrophysics Laboratory
Dept. 91-30, Bldg. 252
3251 Hanover St.
Palo Alto, CA 94304 USA

1. Introduction and Historical Overview

What is the importance of stellar X-ray flares to astrophysics, or even more, to the world at large? In the case of the Sun, changes in solar activity at the two temporal extremes can have quite significant consequences. Long-term changes in solar activity, such as the Maunder Minimum, can apparently lead to non-negligible alterations of the earth's climate. The extreme short term changes are solar flares, the most energetic of which can cause communications disruptions, power outages and ionizing radiation levels amounting to medical X-ray dosages on long commercial flights and even potentially lethal exposures for unshielded astronauts (Haisch, Strong and Rodonò 1991). Why does the Sun exhibit such behaviour? Even if we had a detailed knowledge of the relevant physical processes on the Sun — which we may be on the way to having in hand as evidenced by these *Proceedings* — our understanding would remain incomplete in regard to fundamental causation so long as we could not say whether the Sun is, in this respect, unique among the stars.

Since the 1970's we know that the Sun is not unique: Stars emit X-ray flares and this opens the opportunity to test for relationships between flaring and such stellar conditions as mass, age, rotation and the like.

The first stellar X-ray flare detections were the 1974 October 19 event on YZ CMi and the 1975 January 8 one on UV Ceti, both observed by the Dutch ANS satellite (Heise et al. 1975). This was followed by the 1975 July 22 EUV flare on Proxima Centauri detected by an astronaut-deployed telescope during the Apollo-Soyuz mission (Haisch et al. 1977). Two attempts were made to observe flares (on YZ CMi and Prox Cen) with the SAS-C satellite in 1975 and 1977, but without success (see Haisch 1983). The

Y. Uchida et al. (eds.),
Magnetodynamic Phenomena in the Solar Atmosphere – Prototypes of Stellar Magnetic Activity, 235–242.

HEAO-1 satellite in its 1977 all sky survey detected X-ray enhancements on AT Mic and AD Leo (Kahn et al. 1979). The *Ariel V* X-ray survey (2–18 keV) from 1974–1980 detected a number of transients which were later associated with flares (Pye and McHardy 1983). Substantial progress began to be made with the the capabilities of the *Einstein Observatory* (1978-1981); much more has followed via *Exosat*, *Ginga*, ROSAT, EUVE, ASCA and now XTE. The *Exosat* flare results have been nicely summarize by Pallavicini, Tagliaferri and Stella (1990). A review of ROSAT flare observations has been published by Schmitt (1994). The *Annual Reviews* article by Haisch, Strong and Rodonò (1991) is still reasonably current.

2. Observations of Hot Thermal Emission

The original "flare stars" were the dKe/dMe UV Ceti-type known since 1948 to undergo visual flaring episodes best observed in the Johnson $U-$ and $B-$bands where a 10000 to 20000 K optical flare spectrum has the advantage of enormous contrast over cool photospheric radiation. Flares in various wavelength regimes have now been observed on many different types of stars (see Pettersen 1989) and in particular it now appears that X-ray flares can be found "on all types of late-type stars" (Schmitt 1994). This conclusion is drawn from the ROSAT all-sky survey, carried out in 1990–91. During the survey, any individual star transited the X-ray telescope (XRT), Position Sensitive Proportional Counter (PSPC), 2-deg field-of-view once per satellite orbit for several days, depending upon the star's ecliptic latitude. The effective exposure time per orbit reached a maximum of 32 seconds. Some stars showed large enhancements during a given scan, interpretable as flares which could have started anytime after the previous scan and could have lasted up to the next scan. Examples of such X-ray flare snapshots are the two isolated single-scan outbursts on the F5 V star 36 Dra shown in Figure 1 of Schmitt (1994). Other flares are so long-lived that they are rather well-resolved even at this coarse resolution: for example the event on the G5 III star HR 3922 shown in Figure 2 of Haisch and Schmitt (1994). The all-sky survey provided a coarse but unbiased sampling of flare activity.

The benchmark for stellar X-ray flares is the Sun. Solar X-ray monitoring has been carried out since 1969 using the Geostationary Operational Environmental Satellites (GOES) series of the U. S. National Oceanic and Atmospheric Agency (NOAA). The flux at the earth in a ~ 1 to 8 Å band is used to classify flares as type C, M or X. A flare of class M1 would have a corresponding peak luminosity of $L_x = 2.8 \times 10^{25}$ ergs s^{-1} in that GOES band (the luminosity in a ROSAT-PSPC-like band would be about 16 times larger); a flare of class M2 would be twice as energetic; a flare of class X1 would have ten times the power, etc. (Sawyer, Warwick and Dennett 1986). At solar maximum the Sun undergoes $\sim$ 600 M-flares and $\sim$ 40 to 60 X-flares per year; at solar minimum there are typically one to two dozen M-flares per year with X-flares being quite unlikely to occur at all (see Table 1 in Haisch, Antunes and Schmitt 1995 for a summary over the past 25 years). How do stars compare to this?

Most flares have naturally been considerably more energetic: Owing to stellar distances and instrumental detection limitations, the most intrinsically powerful events are the ones observed. Moreover, while we can with considerable confidence extrapolate from the generally softer energy regimes of *Einstein*, *Exosat* and ROSAT to the GOES band, it is desireable to actually have overlapping measurements. The ASCA satellite, whose wavelength regime is $\sim$ 1 to 24 Å, has done this, monitoring Prox Cen over 23 satellite orbits in March 1994 and succeeding in detecting a number of low-level flares that correspond to solar M-flares (Haisch, Antunes and Schmitt 1995) at a rate that appears to be somewhat in excess of the Sun at maximum ($\sim$ factor of two). A two-temperature fit to the summed flare data using the Mewe-Kaastra (MEKA) plasma code yielded 0.63 ± 0.01 and 3.83 ± 0.67 keV, i.e. 7.3 and 44 MK. This average over several events would appear to be significantly higher than even the peak of the 1980 *Einstein* flare: 27 MK (Haisch et al. 1983). While a few rare solar flares have been observed with temperatures as high as 50 MK (Garcia and McIntosh 1992), what we would expect to measure if the Sun were observed by ASCA (or *Einstein*) would be a maximum temperature during the event of 25 MK or less (Doschek and Feldman 1987, Doschek and Tanaka 1987). These ASCA observations constitute the lowest level of credible stellar X-ray flare detection and, overall, provide a reassuring indication that solar and stellar flares are fundamentally the same phenomenon.

At the other extreme are such superflares as those on the RS CVn stars UX Ari observed by *Ginga* (Tsuru et al. 1989) and AR Lac detected by ROSAT (Ottmann and Schmitt 1994), both with $L_x \sim 2 \times 10^{32}$ ergs s^{-1}. A flare on the RS CVn star CF Tuc lasted 9 days and released $E_x \sim 5 \times 10^{36}$ ergs (Kürster 1994). Although less energetic in an absolute sense, X-ray flares on the dM6e star AZ Cnc and on the K0 V star 197890 ("Speedy Mic") attained X-ray to bolometric luminosity ratios, $L_x/L_{bol} \sim 0.08$ (Schmitt 1994), constituting significant perturbations on the total power output of these stars.

The most energetic flares in terms of spectral distribution are those observed by the Japanese *Ginga* satellite in the 1.5–37 keV range. These include the observations of the RS CVn binary systems II Peg (6.7-d period) by Doyle et al. (1991) and UX Ari (6.4-d) by Tsuru et el. (1989), and of the eclipsing binary Algol (2.87-d) by Stern et al. (1992). The large flares that were observed have peak luminosities in the range $L_x \sim 10^{31}$ to 2×10^{32} ergs s^{-1} and peak temperatures of $T \sim 65$ to 80 MK. The highest temperature measured during any flare is $T \sim 100$ MK, both on the dMe star EQ 1839.6+8002 (Pan et al. 1995) and possibly during one of the UX Ari flares (Tsuru et al. 1989).

At such high temperatures lines of highly ionized iron, primarily Fe XXV at 6.7 keV, are a detectable feature in *Ginga* spectra, and the behaviour of these lines is interesting. In the II Peg, UX Ari and Algol flares, the equivalent width of this feature, actually a composite of several lines, is significantly less than calculated from models assuming solar abundance and an optically thin emitting plasma. A depression due to opacity is certainly possible but cannot explain reductions of up to a factor of 5; the most likely explanation is that the flaring plasma has a different and sometimes measurably variable abundance (cf. Stern, these proceedings). Moreover such an abundance deficiency cannot be explained by the first ionization potential (FIP) effect, since at 7.9 eV Fe is an easily ionized element and would thus, if anything, be overabundant (see Drake, Laming and Widing 1995 for entry into the FIP literature).

3. Searching for Impulsive Emission

There is no indication of any impulsive, non-thermal emission in the *Ginga* flares. When higher spectral resolution with greater sensitivity becomes available, with the AXAF transmission grating spectrometers for example, it will be possible to separate such prominent high-temperature lines as Fe XXV (6.7 keV) from low ionization Kα lines, such as Fe II at 6.4 keV. Photoionization by X-rays above 7.11 keV or collisional ionization by particle beams of the ground-state electron result in an inner-shell transition of this sort. It is well known from solar flares that inner-shell ionization of neutral or near-neutral Fe occurs (see Bai 1979) and that such a Kα line originating in cool gas subject to photoionizing flux or a particle beam often tracks the hard X-ray flux (see Figure 4 of Emslie, Phillips and Dennis 1986). The model of Bai (1979) is one assuming photoionization as the mechanism and the process is termed Kα-fluorescence. His calculations suggested a $\sim$ 1 to 3 percent efficiency for conversion of continuum hard X-rays into Kα-fluorescence during the impulsive phase of a solar flare.

We applied this model in a very rough way to a stellar situation by equating the peak L_x for an observed flare on Prox Cen with a presumed impulsive HXR burst, a simple (if not simplistic) application of the "Neupert effect" (Neupert 1968, Dennis and Zarro 1993). This then allowed us to predict how many ASCA counts there would be during an impulsive event on Prox Cen at the onset of a flare equivalent to the 4 ct s^{-1} ROSAT

PSPC event measured during the all sky survey (see Figure 4 of Schmitt 1991). This admittedly optimistic model predicted about 100 to 300 counts for the ASCA detectors during an impulse Kα-fluorescence burst. None of the flares observed by ASCA during its monitoring of Prox Cen were as energetic in soft X-ray as assumed in the model (Haisch, Antunes and Schmitt 1995). Nonetheless, there was no indication of any impulsive Kα emission.

Theoretical calculations of the relationship between hard and soft X-rays do indicate that there should be a Neupert effect-like correlation between the hard X-ray time profile and the derivative of the soft X-ray light curve in the beam heating model but not in the superthermal model (Li, Emslie and Mariska 1993). Forthcoming observations of flares using XTE will search for rapid variations in the soft X-ray flux on Prox Cen, AR Lac and other stars. Another way to express the Neupert relationship is to equate the time integral of hard X-ray emission up to some point with the current value of the thermal soft X-ray emission. Hawley et al. (1995) have recently interpreted comparison of the time integral of $U-$band with EUV luminosity during a flare on AD Leo in this way.

With a single possible exception all X-ray emissions detected to date appear to originate in the thermal phase; X-ray impulsive emission has not yet been observed in stellar flares. The single exception is intriguing and owes its discovery to the availability of simultaneous optical photometry. UV Ceti was monitored by a high-speed multisource, multicolor photometer, specifically designed for stellar flare observations, operating at the Wendelstein Observatory in Bavaria in January 1992 during a ROSAT PSPC pointing. Two optical spikes were detected in the $U-$ and $B-$bands. The normally processed X-ray light curve showed a typical gradual event following the second optical burst. However, alerted to the presence of two optical bursts, a search was done for clustering of X-ray photons by examining the behaviour of a constant-count binned light curve. This is possible to do since every photon is time-tagged to $\sim$millisecond accuracy. The result was the presence of X-ray bursts with durations of $\sim$12 s. With this resolution the gradual phase event appears to consist of a dozen or so bursts, but more mysterious is the fact that both optical spikes are followed by X-ray bursts with a time delay of $\sim$30 s. The statistics of this, including the sticky issue of how to allow for the degrees of freedom permitted by varying the search windows, are presented in Schmitt, Haisch and Barwig (1993). While it is possible that these soft X-ray bursts could be the low energy tails of hard X-ray emission from particle beams, the time delay does not yet have a plausible explanation.

4. Discussion

Turning briefly to analytical tools, there are basically two models that have been applied to the thermal or decay phase of the X-ray light curves of stellar flares. The quasi-static cooling loop model has been developed most by van den Oord, Mewe and Brinkman (1988); the two-ribbon flaring model has been developed by Kopp and Poletto (1984). They make fundamentally different assumptions: the first model assumes that changing plasma conditions within a single loop give rise to the observed flare characteristics; the second assumes that over the course of the event *different* loops become activated. Both analyses have been applied to the same flare: viz. the Poletto et al. (1988) and the Reale et al. (1988) analyses of Prox Cen; also the Schmitt (1994) analyses of EV Lac. The bottom line, unfortunately, is that it does not yet appear to be possible to differentiate between low density, large structures and high density, compact ones on the basis of these models and the existing flare light curves. (Flare light curves are of course complicated by the situation that, without spatial resolution, one cannot separate out the combined signature of multiple, overlapping flares.)

Attempts to understand flares by studying ratios of emission in disparate wavelength regimes has a long history. Twenty years ago Mullan (1976) proposed a theory predicting ratios of X-ray to optical luminosity during flares. This was a laudable attempt to develop a quantitative relation between high temperature coronal plasma properties and the response of a heated chromosphere, but it was quickly found to be at variance with new flare observations (Haisch et al. 1977). Indeed, the ratio L_x/L_{opt} varies between zero and infinity over the course of a flare. Nevertheless, could such a type of ratio still have significance as an average over a flare? Butler et al. (1988) found what appeared to be a striking correlation between a chromospheric line (Hγ) and L_x applicable over four orders of magnitude to both solar and stellar flares, and this relationship was found to extend another two orders of magnitude recently (Butler 1993). This would be quite a significant result, however on looking at 370 solar flares in detail, classified simultaneously in Hα and as GOES-events, it became clear that while the proposed relationship is true on average over many flares, for any given flare the deviation from the relationship can be very large: a spread of up to three orders of magnitude in L_x for a given Hα emission (Haisch 1989). The conclusion is that ratios of coronal to chromospheric emission vary substantially both during a flare and from flare to flare; only the time- and event-averaged ratios show a credible correlation.

Flares and coronae are thought to go hand in hand, and indeed the concept that coronal heating is really a flare-like process is still a debated but viable concept. (The issue of coronal heating is so vast a topic that the place to enter the literature might be Zirker's 1993 "review of reviews"; also the Ulmschneider, Priest and Rosner 1991 volume; and recently Cargill 1994). The most current discovery on the stellar side relevant to this is the finding of a uniform soft X-ray to microwave relation ranging from solar microflares to stellar coronae, suggesting that "the heating mechanism of active stellar coronae is a flare-like process" (Benz and Güdel 1994).

This is relevant to the proposal of Kashyap et al. (1994) that the hybrid star α Trianguli Australis is an X-ray source on the basis of flaring, not steady emission. This late-type star (K4 II), and the other dozen or so luminosity class II or III K giants defining the hybrid star category appear to manifest both massive cool wind signatures and X-ray emission. It is thought that in most stars these two characteristics are mutually exclusive, owing to a phase transition across a dividing line in which coronae disappear (see Haisch, Schmitt and Fabian 1992 and references therein for entry into the stellar dividing line issues). The proposal that Kashyap et al. and Rosner et al. (1994) make is that X-rays from hybrid stars are solely due to flares, and that this is an indicator that large-scale magnetic dynamo activity is being replaced by much smaller scale magnetic structures. This would be an important discovery about stellar evolution, but it is not clear how to interpret the significance of their observation if all coronal heating is flare-like.

Although the majority of flare observations have come from such prolific sources as dKe/dMe stars and RS CVn binaries, it would not be surprising to detected a flare on any type of star that shows coronal emission. But what is one to make of a flaring Be star? A 30 ks ROSAT PSPC observation of λ Eri (B2e) shows a normal level of X-ray emission for a hot star presumably with a radiatively-driven wind. A substantial flare then takes place with flux levels enhanced by a factor of 6 and a duration of $\sim$50000 s. The following day the X-ray luminosity has returned to the preflare level and remains constant for the remainder of the observation, approximately half a day (Smith et al. 1993). The analysis of the event yields a peak luminosity, $L_x \sim 4 \times 10^{31}$ ergs s^{-1} and a temperature of 14 MK. IUE and optical spectra do not support any binary explanation, and Smith et al. argue for "violent magnetic activity on some B-type stars."

The possibility of having flares on Be stars, or in accretion disks, or in magnetic structures connecting two stars, all of which have been proposed, certainly make X-ray flare studies a challenging field.

References

Bai, T. 1979, *Solar Phys.*, **62**, 113–121.
Benz, A.O. and Güdel, M. 1994, *A&A*, **285**, 621–630.
Butler, C.J., Rodonò, M. and Foing, B.H. 1988, *A&A*, **206**, L1–L4.

Butler, C.J. 1993. *A&A*, **272**, 507–513.
Cargill, P.J. 1994, *ApJ*, **422**, 381–393.
Dennis, B.R. and Zarro, D.M. 1993, *Solar Phys.*, **146**, 177–190.
Doschek, G.A. and Feldman, U. 1987, *ApJ*, **313**, 883–892.
Doschek, G.A. and Tanaka, K. 1987, *ApJ*, **323**, 799–809.
Doyle, J.G. et al. 1991, *MNRAS*, **248**, 503–507.
Drake, J.J., Laming, J.M. and Widing, K.G. 1995, *ApJ*, **443**, 393–415.
Emslie, A.G., Phillips, K.J.H. and Dennis, B.R. 1986, *Solar Phys.*, **103**, 89–102.
Garcia, H.A. and McIntosh, P. 1992, *Solar Phys.*, **141**, 109–126.
Haisch, B. 1983, in *Activity in Red Dwarf Stars: Proc. IAU Coll. 71*, P.B. Byrne and M. Rodonò (eds.), p. 255–272.
Haisch, B. 1989, *A&A*, **219**, 317–319.
Haisch, B. and Schmitt, J.H.M.M. 1994, *ApJ*, **426**, 716–724.
Haisch, B., Linsky, J.L., Lampton, M., Paresce, F., Margon, R. and Stern, R. 1977, *ApJ*, **213**, L119–L124.
Haisch, B. et al. 1983, *ApJ*, **267**, 280–290.
Haisch, B., Strong, K.T., and Rodonò, M. 1991, *Ann.Rev.Astr.Ap.*, **29**, 275–324.
Haisch, B., Schmitt, J.H.M.M. and Fabian, A.C. 1992, *Nature*, **360**, 239–241.
Haisch, B., Antunes, A., and Schmitt, J.H.M.M. 1995, *Science*, **268**, 1327–1329.
Hawley, S.L. et al. 1995, *ApJ*, in press.
Heise, J., Brinkman, A.C., Schrijver, J., Mewe, R., Gronenschild, E., den Boggende, A. and Grindlay, J. 1975, *ApJ*, **202**, L73–L76.
Kahn, S.M., Linsky, J.L., Mason, K.O., Haisch, B.M., Bowyer, C.S., White, N.E., and Pravdo, S.H. 1979, *ApJ*, **234**, L107–111.
Kashyap. V., Rosner, R., Harnden, F.R. Jr., Maggio, M., Micela, A. and Sciortino, S. 1994, *ApJ*, **431**, 402–415.
Kopp, R.A. and Poletto, G. 1984, *Solar Phys.*, **93**, 351–361.
Kürster, M. 1994, *ASP Conf. Series*, **64**, 104–106.
Li, P., Emslie, A.G. and Mariska, J.T. 1993, *ApJ*, **417**, 313–319.
Mullan, D.J. 1976, *ApJ*, **207**, 289–295.
Neupert, W.M. 1968, *ApJ*, **153**, L59.
Ottmann, R. and Schmitt, J.H.M.M. 1994, *A&A*, **283**, 871–883.
Pallavicini, R., Tagliaferri, G. and Stella, L. 1990. *A&A*, **228**, 403–425.
Pan, H. C., Jordan, C., Makishima, K., Stern, R. A., Hayashida, K. and Inda-Koide, M. 1995, in Proceedings of the IAU Colloquium No 151 'Flares and Flashes' (Lecture Notes in Physics, Springer-Verlag).
Pettersen, B.R. 1989, *Solar Phys.*, **121**, 299–312.
Poletto, G., Pallavicini, R. and Kopp, R.A. 1988, *A&A*, **201**, 93–99.
Pye, J.P. and McHardy I.M. 1983, *MNRAS*, **205**, 875.
Reale, F., Peres, G., Serio, S. Rosner, R. and Schmitt, J.H.M.M. 1988, *ApJ*, **328**, 256–264.
Rosner, R., Musielak, Z.E., Cattaneo, F., Moore, R.L. and Suess, S.T. 1994, *ApJ*, **442**, L25–L28.
Sawyer, C., Warwick. J.W. and Dennett, J.T. 1986, *Solar Flare Prediction*, (Colorado Assoc. Univ. Press, Boulder).
Schmitt, J.H.M.M. 1991, in *Cool Stars, Stellar Systems and the Sun*, ASP Conf Series vol. 26, 83–92.
Schmitt, J.H.M.M. 1994, *ApJSuppl*, **90**, 735–742.
Schmitt, J.H.M.M., Haisch, B., and Barwig, H. 1993, *ApJ*, **419**, L81–L84.
Smith, M.A., Grady, C.A., Peters, G.J. and Feigelson, E.D. 1993, *ApJ*, **409**, L49–L52.
Stern, R., Uchida, Y., Tsuenta, S. and Nagase, S. 1992, *ApJ*, **400**, 321–329.
Tsuru, T. et al., 1989, *PASJ*, **41**, 679–695.
Ulmschneider, P., Priest, E.R., and Rosner, R. (eds.) 1991, *Mechanisms of Chromospheric and Coronal Heating*, (Berlin: Springer).
Van den Oord, G.H.J., Mewe, R., and Brinkman, A.C. 1988, *A&A*, **205**, 181–196.
Zirker, J.B. 1993, *Solar Phys.*, **148**, 43–60.

HARD X-RAY EMISSIONS FROM PROTOSTAR CANDIDATES

K. KOYAMA, Y. TSUBOI AND S. UENO
Department of Physics, Faculty of Science, Kyoto University, Sakyo-ku, Kyoto 606-01, Japan

Abstract. We have carried out a systematic survey of hard X-rays from the core of molecular clouds with the ASCA satellite and discovered hard X-rays from protostar candidates (class I objects) in the cores of the dark clouds R CrA and ρ Oph. The X-ray spectra of protostars have higher temperatures and larger absorptions than those of T Tauri Stars (TTSs). The protostars exhibited flare-like activity with the fast rise and slow decay, which is similar to TTSs. The temperature increases rapidly and decreased gradually. We found striking feature in the X-ray spectrum of R1: unusually flat spectrum and two emission lines at 6.2 and 6.9 keV. Possibly the unusual hard X-ray spectrum have close connection to the jet-like radio structure.

1. Pre-Main Sequence Stars

Infrared spectrum may be useful to understand the evolution of pre main sequence stars (PMSs). As a star evolves, we can classify the stars as class 1 to 3. Classes 2 and 3 may be called as TTSs which is divided to CTTS (class2) and WTTS (class3). Class 1 may be younger than TTS and can be called as protostars.

The Einstein and ROSAT satellites found TTSs are strong X-ray sources with a temperature of about 1 keV. As for protostars, whether it is also X-ray source or not has been unclear. Because protostar may be deep inside a molecular cloud, hence most X-rays in the *Einstein* and ROSAT band are invisible behind the large N_H.

Now we have the ASCA satellite, which has an imaging capability in high energies up to 10 keV, and can see deeply into the molecular cloud in which a protostar may reside.

Y. Uchida et al. (eds.),
Magnetodynamic Phenomena in the Solar Atmosphere – Prototypes of Stellar Magnetic Activity, 243–246.

2. The ASCA Satellite

ASCA carries 4 telescopes with grazing-incidence X-ray optics. The telescope has multi-nested (119 layers) thin-foil conical mirror. This technology provides us a large effective area in the high energy band. The effective area and energy band of XRTs are both larger than those of *Einstein* and ROSAT. The focal plane detectors are two CCD cameras (SIS) and two gas counters (GIS). The SIS has good energy resolution as is demonstrated by many SNR spectra. The GIS has a higher detection efficiency above $\sim$ 3 keV than SIS.

3. The R CrA Dark Cloud

The R CrA dark could is one of the nearest star-forming regions, at a distance of 130 pc. The central core of the molecular cloud is located near the variable star R CrA, and contains many infrared and radio continuum sources. It is referred to as the Cornet cluster (Taylor and Storey 1984). We made X-ray images of the cloud center with SIS in the 0.5-2 keV (the soft band) and 4-10 keV (the hard band) separately. The soft X-ray band image is similar to that of the *Einstein* and ROSAT images, showing several point sources around this molecular could. Most of the point sources are already identified as the Herbig Ae/Be or T-Tauri stars. However the cluster center is dark in the soft X-rays.

The image of the hard X-ray band is completely different form the soft X-ray band. The soft band sources disappear while new and bright hard sources appear at the cloud center associated with the positions of the class 1 IR sources. We find agreement of the hard X-ray peak positions to the class 1 source within an error of 20".

The detected sources are given in Table 1. With the hardness ratio, we can have a clear idea of the difference between TTS or class III and protostar class I sources. Class I sources definitely have harder spectra and larger absorption than those of class IIIs. Most of the hard X-ray photons came from the most dense cloud core, including R1, where double radio peaks were found on both sides.

The spectra of WTTS shows a temperature of about 1 keV while the hard sources shows very flat continuum and strong emission line at about 6.2 and 6.9 keV. The temperature of this source was higher than 10 keV. I also note that form this source, the radio jet source R1, we observed a beautiful X-ray flare.

TABLE 1. X-ray Source List

No	$(\alpha_{1950}, \delta_{1950})$	Candidate(<20")	Class
1	18 58 19.4, -37 02 47	IRS2=TS13.1	I
2	18 58 24.7, -37 01 36	IRS5=TS2.4	I
3	18 58 28.3, -37 02 19	IRS1=TS2.6=HH100IR=VSS15	I
4	18 58 28.6, -37 00 54	IRS6=TS2.3	I
5	18 58 31.9, -37 01 29	IRS7=R1	I
6	16 24 08.4, -24 30 15	EL29=YLW	I
7	16 24 20.0, -24 22 58	WL6=YLW14	I
8	16 24 26.0, -24 32 51	YLW16B=IRS46	I

Ref. Taylor and Storey(1984), Wilking et al.(1992), Wilking, Taylor Storey (1986), Zinnecker and Preibisch(1984), Walter and Kuhi (1981), Andre and Montmerle(1994), Elias(1978), Lada and Wilking(1984), Wilking, Lada, and Young(1989).

4. The ρ-Oph Cloud

The ρ-Oph dark cloud is also one of the wellknown nearest star forming region, located 165 pc form our solar system. We also made separate X-ray images using SIS in the 0.5-2 keV (the soft band) and 2-10 keV (the hard band) near the center of the dark cloud. The soft X-ray band sources are found around the molecular could which are classified as T-Tauri stars or class II or III sources. Again the central part of the cloud is dark in the soft X-rays.

In the hard X-ray band, 5 new sources appear near the cloud center. Two of the sources coincide with the class I objects, which are EL29 and WL6. From EL 29, we found a hard X-ray flare. The plasma temperatures of the flare and the quiescent states were about 7 keV 2.5 keV, respectively. Even in the hard X-rays, we found no point source at the center of the cloud. However we found excess emission. We made an x-ray spectrum from this central region using the GIS detector. The ASCA GIS spectrum shows optically thin thermal plasma emission with kT=3.6 keV and 0.3 of the cosmic abundance. Therefore, many unresolved X-ray sources with a mean temperature of 4 keV could be lying near the could core, significant fraction of which could also be protostars.

5. Summary

(1) We detected absorbed hard X-rays from class I sources. The non detection of class I sources with the previous X-ray instruments is undoubtedly due to the heavy absorption.

(2) X-ray spectra of protostars have higher temperatures than TTSs.
(3) We found flare-like activity from protostars with a fast rise and slow decay, which is similar to that of TTSs. The temperature also rapidly and decreases gradually.
(4) The radio-jet source R1 showed a striking X-ray spectrum with very flat shape and emission lines at 6.2 and 6.9 keV. A "crazy" idea is that these lines are blue and red-shifted iron K-shell emission in the analogy of SS433.

References

Andre, Ph. and Montmerle, T. (1994) *Astrophys. J.*, **420**, 837

Brown, A. (1987) *Astrophys. J. (Letter)*, **322**, L 31

Dame, T. M., Ungerechts, H., Cohen, R. S., Geus, E. J., Grenier, I. A., May, J., Murphy, D. C., Nyman, and Thaddeus, P. (1987) *Astrophys. J.*, **322**, 706

Elias, J. H. (1978) *Astrophys. J.*, **224**, 453

Gottlieb, C. A., Gottlieb, E. W., Litvak, M. M., Ball, J. A. and Penfield, H., (1978) *Astrophys. J.*, **219**, 77

Koyama, K., Maeda, Y., Ozaki, M., Ueno, S., Kamata, Y., Tawara, Y., Skinner, S., and Yamauchi, S. (1994) *Publ. Astron. Soc. Japan*, **46**, L 125

Lada, C. L. and Wilking, B. A. (1984) *Astrophys. J.*, **287**, 610

Loren, R. B., Sandqvist, Aa. and Wootten, A. (1983) *ApJ*, **270**, 620

Montmerle, T., Koch-Miramond, K., Falgarone, E., and Grindlay, J. E. (1983) *Astrophys. J.*, **269**, 182

Taylor, K. N. R. and Storey, J. W. V. (1984) *Mon. Not. R. Aster. Soc.*, **209**, 5p

Walter, F. M. and Kuhi, L. V. (1981) *Astrophys. J.*, **250**, 254

Wilking, B. A., Taylor, K. N. R., and Storey, J. W. V. (1986) *Astronomical J.*, **92**, 103

Wilking, B. A., Lada, C. J., and Young E. T., (1989) *Astrohys. J.*, **340**, 823

Wilking, B. A., Greene, T. P., Lada, C. J., Meyer, M. R. and Young, E. T. (1992) *Astrophys. J.*, **397**, 520

Zinnecker, H. and Preibisch, T. H. (1994) *Astronomy and Astrophsics*

X-RAY FLARES AND VARIABILITY OF YOUNG STELLAR OBJECTS

THIERRY MONTMERLE AND SOPHIE CASANOVA
Service d'Astrophysique
CEA/DAPNIA/SAp
Centre d'études de Saclay
91191 Gif-sur-Yvette Cedex
France

Abstract. The most well-known "Young Stellar Objects" are the T Tauri stars, i.e., young, solar-type stars, first detected in X-rays by *Einstein*. With *ROSAT* and *ASCA*, the family of X-ray emitting young stars has been extended to protostars. X-ray emission is only related to surface magnetic phenomena, independently of the amount of circumstellar material. Although other forms of variability have been observed, solar-type flares are the main form of X-ray activity (at levels typically $10^2 - 10^3$ higher than on the Sun). However, general X-ray properties bring no clearcut proof that the classical dynamo mechanism operates in YSOs.

1. What are "Young Stellar Objects"?

The surface activity on the Sun is particularly visible in soft X-rays, as shown by *Yohkoh* (this Colloquium). Since the Sun arrived on the main sequence, $\sim 4.5 \times 10^9$ yrs ago, this activity has been relatively constant, with only a slow decline, as inferred from recent *ROSAT* results on a selected sample of main sequence solar analogs of various ages (Güdel & Guinan 1996). But what happened before, in the pre-main sequence phase ?

To answer this question, we should first consider young stars known as "T Tauri stars" (for reviews, see, e.g., Bertout 1989). Their ages are $\tau \sim 10^5 - 10^7$ yrs, with masses $M_\star \sim 0.5 - 2M_\odot$, radii $R_\star \sim 2 - 3R_\odot$, and bolometric luminosities $L_\star \sim 0.1 - 10L_\odot$; they are "cool stars" (spectral types G to M, i.e., $T_{eff,\star} \sim$ 3,000 - 5,000 K). They have deep convective outer layers, a very important factor to explain their surface activity.

Y. Uchida et al. (eds.),
Magnetodynamic Phenomena in the Solar Atmosphere – Prototypes of Stellar Magnetic Activity, 247–258.

Observationally, the T Tauri stars are characterized by emission-line spectra (in particular Hα). They also display correlated UV and IR excesses with respect to a cool blackbody stellar photosphere. The IR excess is currently associated with a circumstellar disk (e.g., André 1996, Sargent 1996), and the UV excess and emission lines are thought to be produced by some form of interaction between the (magnetized) star and the disk.

The access to new wavelengths has revealed objects closely related to T Tauri stars, forming a family usually referred to as "Young Stellar Objects", or "YSOs". (*i*) In the IR, many objects display an excess much larger than for the visible T Tauri stars, interpreted in terms of an extensive circumstellar envelope (e.g., Wilking, Lada, & Young 1989; Adams, Lada, & Shu 1987; André & Montmerle 1994). Their age is $\sim 10^4 - 10^6$ yrs. (*ii*) In the X-ray domain, a new category of young stars has been discovered by *Einstein*: the so-called "weak-line T Tauri stars" (WTTS for short). These stars basically share all the characteristics of the previously known T Tauri stars (dubbed "classical" T Tauri stars, or CTTS), except for the emission lines, IR excess, etc. The WTTS are now understood as being diskless, or at most with faint disks with no interaction with the star. The transition CTTS $\rightarrow$ WTTS corresponds to the dissipation of the circumstellar disk, which is not understood.

2. X-ray emission from T Tauri stars: generalities

Whether or not there is a difference in the X-ray properties of CTTS and WTTS is somewhat debated. Present results show that, close to molecular clouds, CTTS and WTTS are similar (Feigelson et al. 1993, hereafter FCMG; Casanova et al. 1995, hereafter CMFA), while WTTS may be brighter than CTTS farther away (Neuhaüser et al. 1995). At any rate, there is no indication of X-ray emission associated with the circumstellar disks of CTTS, which means that the X-ray emission of T Tauri stars is a *purely stellar* phenomenon.

The X-ray emission mechanism is thermal bremsstrahlung, from a hot plasma ($n_e \sim 10^{10} - 10^{12}$ cm^{-3}, $T_X \approx 1$ keV). However, combined *ROSAT* - *ASCA* observations have shown that spectral fits with a single temperature are generally poor, and that two temperatures and/or non-solar abundances give much better fits (e.g., Carkner et al. 1996).

Another important feature of the X-ray emission from TTS (CTTS and WTTS alike) is the strong time variability. Already well established with *Einstein* (e.g., Montmerle et al. 1983), flaring activity is now seen to be commonplace in TTS of both kinds, and light curves basically support the idea of enhanced, solar-like flares. This early X-ray result is strongly supported by optical and near-UV results, which show, for instance, the existence of large starspots, covering up to 30 % of the stellar surface (e.g., Montmerle et al. 1993).

3. The new observational picture

In recent years, a number of programs have been undertaken to study various regions of star formation, and in particular low-mass stars, with the *ROSAT* and *ASCA* satellites. Combining them has obvious advantages : the best features of *ROSAT* are the large field of view of the PSPC instrument (2^0), which allows to map molecular clouds with one, or at most two, exposures; its improved sensitivity, and its good angular resolution. On the other hand, *ASCA* has a smaller field of view, a poorer angular resolution, but excellent spectral properties (high resolution, wide energy range).

Observing nearby clouds ($d \sim 150$ - 300 pc) allows a sensitive study of YSOs in X-rays. Works published so far include the Cha I cloud ($d \simeq 200$ pc: FCMG), the ρ Oph cloud ($d \simeq 160$ pc: CMFA), the L1551 cloud in Taurus ($d \simeq 150$ pc: Carkner et al. 1996), and the IC348 cluster ($d \simeq 300$ pc: Preibisch, Zinnecker, & Herbig 1996). Other regions are under study, either by us or by other authors (see, e.g., Caillault 1995).

In each of these regions, several tens of sources are found. Based on cross-identifications with existing catalogs, optical plates, IR surveys, etc., or on follow-up observations (for instance, search for Hα emission, and Li absorption which is an indicator of youth), it can be shown that almost all the X-ray sources are indeed YSOs.

4. Studies of X-ray time variability

Because both *ROSAT* and *ASCA* have a low-altitude orbit, they cannot in general observe a given region in the sky continuously. Their detectors must be shut off every time they pass within the so-called South Atlantic anomaly. This produces gaps of at least half an hour in the data. Since flaring events, for instance, usually have durations of that order, a complete light curve is in general impossible to obtain. The exposures performed between the gaps are usually referred to as "orbits".

4.1. EXAMPLES AND CHARACTERISTICS OF T TAURI FLARES

4.1.1. *LkHα 332-20*

This star, located in the Cha I cloud, is a K2 CTTS, of age $\sim 5 \times 10^6$ yrs. It has been observed with the *ROSAT* PSPC (Feigelson et al. 1993), in three orbits of duration $\sim$ 2,000 sec, separated by gaps of duration $\sim$ 3,000 sec (See Fig. 1). The maximum observed luminosity occurred at the end of the second orbit, at a level $\sim$ 20 times higher than the "quiescent" (?) state

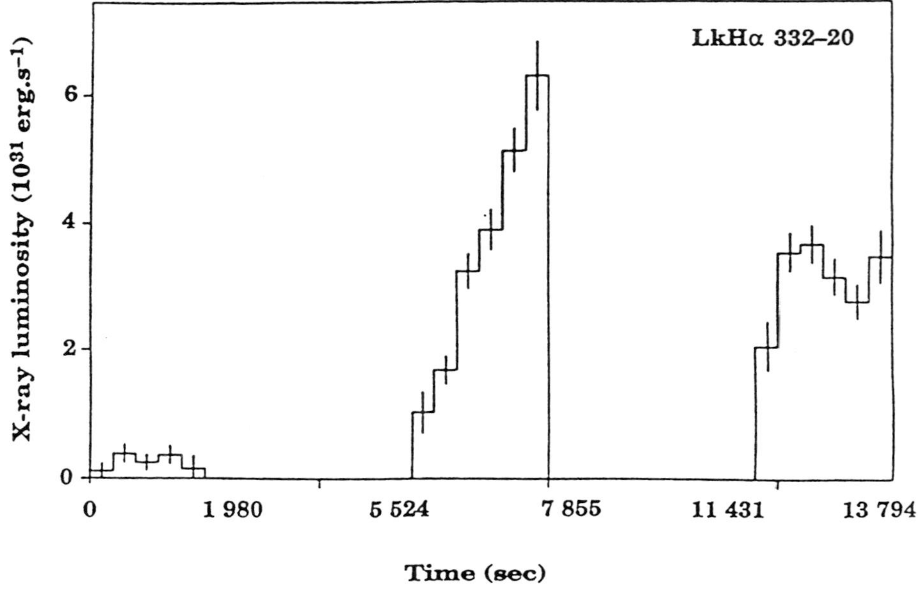

Figure 1. Light curve of the CTTS LkHα 332-20, obtained with *ROSAT* (Casanova 1994).

of the first orbit, $L_{X,max} \sim 3.5 \times 10^{31}$ erg s^{-1}. This luminosity, which is a lower limit to the true value since the count rate was still rising when the exposure was interrupted, is among the largest detected (ROX-20, a WTTS discovered by *Einstein*, Montmerle et al. 1983; see below). Then the luminosity decreased by a factor $\sim$ 2 during the second gap. Looking at the data, the event can be approximately described as having occurred shortly before the second orbit, with a rise time of $\approx$ 2,300 sec, and an e-folding decrease timescale of $\approx$ 9,000 sec.

Each exposure can be analysed separately. Although the light curve is far from complete, the time behaviour and the overall spectral shape suggest a flare event with thermal emission from a hot plasma. We take a Raymond-Smith model, and look for χ^2 fits to the spectrum as a function of the column density N_H and the X-ray temperature T_X. The best-fit values are summarized in Table 1.

The flaring hypothesis is supported by the temperature behaviour: heating from 0.8 keV to $\gtrsim$ 2.2 keV, followed by cooling, down to $\sim$ 1.5 keV. Assuming cooling to be radiative, one obtains a plasma density $n_e \sim 10^{11}$ cm^{-3}, and from the luminosity a loop size $l \sim 5 \times 10^{10}$ cm ($\sim 0.3R_\star$). The column density N_H is well constrained at the higher temperatures, which match better the 0.4 - 2.4 keV energy range of the PSPC; there is

TABLE 1. *Spectral fits:* LkHα 332-20

Orbit	T_X(keV)	logN_H(cm^{-2})	L_X (10^{30} erg s^{-1})
1	0.8	20 - 22	1.7
2	> 2.2	21.3 ± 0.1	35
3	~ 1.5	21.2 ± 0.1	29

no constraint for the lowest temperature (Casanova 1994).

4.1.2. *LkHα 92*

This star is located in the IC 348 young star cluster. It is a $\sim$ K-type CTTS. A strong flare has been detected during a *ROSAT* PSPC observation. The light curve (Preibisch, Zinnecker, & Schmitt 1993; their Fig. 1) comprises 5 different orbits. This observation is more conclusive than in the case of LkHα 332-20, because the maximum luminosity $L_{X,max}$ was reached during the second orbit, and the decrease observed in the course of the second orbit smoothly matches an exponential behaviour between $L_{X,max}$ and the quiescent state observed in the first and last orbits. The parameters derived from spectral fits with a thermal spectrum are comparable to those of LkHα 332-20, with a somewhat higher $L_{X,max} \sim 5 \times 10^{31}$ erg s^{-1}, but with a significantly higher –and unusually high– maximum temperature (4.3 ± 1.6 keV).

4.1.3. *Flaring sources in the ρ Ophiuchi cloud*

This region was observed with the *ROSAT* PSPC in three observing sessions, lasting from $\sim$ 1 h to $\sim$ 20 h, separated by 4 days and 6 months, respectively, and for a total of 21 orbits (Casanova 1994, Casanova et al. 1995). Fig. 2 shows the time behaviour of a sample of eight of the most variable sources, which comprises a variety of T Tauri stars : CTTS, WTTS, and also an embedded source without optical conterpart. In spite of their morphological differences, the light curves are very similar : there is evidence for a more or less quiescent level, and for large variations over timescales of a few hours, with a rapid rise (duration $\tau_r \lesssim 1$ hr) followed by a slow decrease (duration $\tau_d \sim$ several hrs); the variation amplitude is typically a factor $\lesssim 10$. This is very reminiscent of the two flares reported above, albeit with smaller maximum observed luminosities: $L_{X,max} \lesssim 10^{31}$ erg s^{-1}.

These sources however have a good S/N ratio. What about weaker sources ? Because the statistics are too low, it is meaningless to draw light curves. It is nevertheless possible to apply a Kolmogorov-Smirnov (K-S) test of variability (using for instance the ASURV statistical package, La

Valley, Isobe & Feigelson 1992) to the whole sample of 93 *ROSAT* sources. Figure 3 shows the K-S variability indicator k, vs. the S/N ratio. Sources with 99% probability of variation have $k > 1$. On the other hand, sources with $S/N \lesssim 4$ may be statistical background fluctuations. There is a good correlation between k and S/N, even when the statistical confidence for existence and/or variability is poor. This suggests that the ρ Oph X-ray sources are all variable, WTTS and CTTS alike (in a statistical sense), the S/N ratio being the only limiting factor to detect the variability of individual sources.

4.2. OTHER TYPES OF VARIABILITY

The analysis of the very large database resulting from the ρ Oph PSPC observations reveals other types of variability which cannot be explained by flares, i.e., for which τ_r is not $<< \tau_d$.

4.2.1. *SR 13*

This star is an M2.5 CTTS with an estimated age $\sim 10^6$ yrs. Its period P_{rot} is poorly determined, but is on the order of several days (see Casanova 1994). Fig. 4 shows the light curve of SR 13 during the last 12 orbits of the PSPC observation, i.e. for an overall duration of $\sim$ 19 hrs. The luminosity rises slowly during the first 5 orbits ($\sim$ 7 hrs), then declines roughly back to the initial level after 7 orbits ($\sim$ 12 hrs). The spectral analysis (including an average of the first, "pre-event", 9 orbits) yields Table 2.

TABLE 2. *Spectral fits:* SR 13

Orbits	T_X(keV)	$\log N_H$(cm^{-2})	L_X (10^{30} erg s^{-1})
1 - 9	~ 1.1	$20.8^{+1.0}_{-0.3}$	1.2
10 - 14	≥ 1.1	20.8 ± 0.2	3.2
15 - 21	~ 1	20.8 ± 0.2	2.4

The maximum luminosity is thus only $\sim$ 2.5 times higher than the "pre-event" luminosity. More remarkably, there is essentially no change in the plasma temperature, which remains $T_X \sim 1$ keV. This last result clearly confirms that we are not dealing with a flare. It is possible to model the light curve in terms of a plasma loop initially "rising" over the star's horizon, then cooling with an e-folding time $\sim 10^4$ sec as it moves towards the observer following the star's rotation. A good fit to the X-ray light curve is then obtained for $P_{rot} \sim 3$ days. The plasma parameters are $n_e \sim 5 \times 10^{10}$ cm^{-3}, and $l \sim 3 \times 10^{10}$ cm, or $\sim 0.2 R_\star$ (see Casanova 1994 for details).

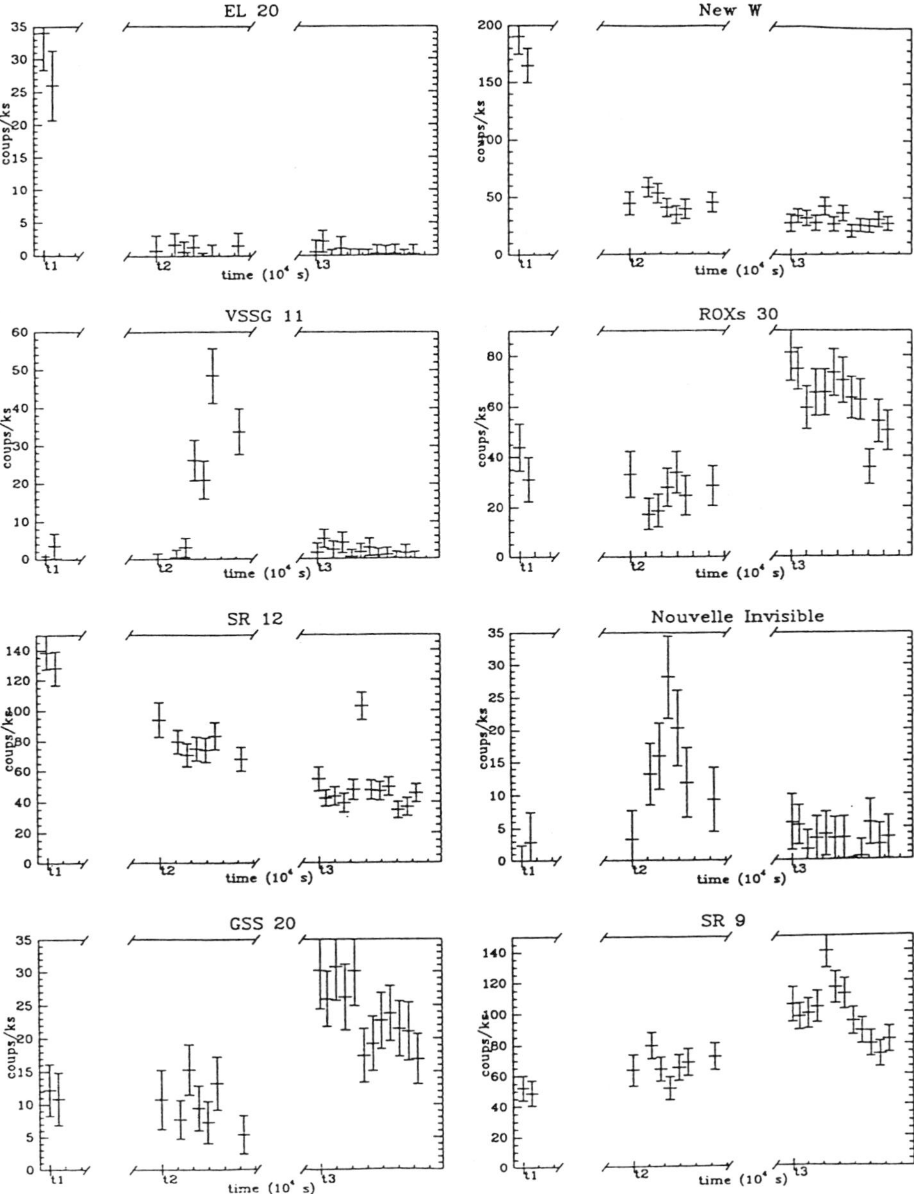

Figure 2. Light curves of eight TTS in the ρ Oph cloud, as observed with *ROSAT* (Casanova 1994). They are all WTTS, except EL 20 and SR 9; one has no optical counterpart.

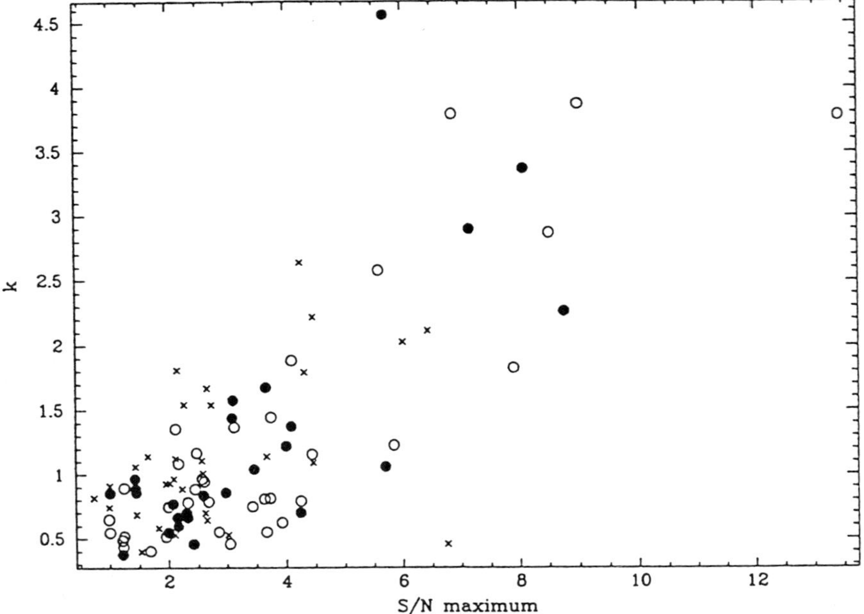

Figure 3. Kolomogorov-Smirnov variability indicator k, as a function of the S/N ratio, for the ρ Oph *ROSAT* sources (Casanova 1994). All sources show some variability; dots: CTTS; circles: WTTS; crosses, new WTTS.

4.2.2. *DoAr 21*

This star is a K0 WTTS, with an estimated age $< 10^6$ yrs. Eruptions have been observed in the past with *Einstein* and in the radio with the VLA (Feigelson & Montmerle 1985). In the course of the *ROSAT* observations reported here, however, no significant variations were seen within the three observing sessions (see Fig. 5). No variability was detected by *ASCA* either (Koyama et al. 1994), with a significantly higher plasma temperature than average for a T Tauri star ($T_X \sim 2.4$ keV). However, over the $\sim$ 8-month interval between the second and the third observing session, the luminosity decreased by a factor $\sim$ 2, which is quite typical of TTS.

Such a behaviour is difficult to explain in simple terms like the preceding examples. Over several years of monitoring with the VLA, the radio flux of DoAr 21 (as that of other T Tauri stars) shows changes of a factor of $\sim$ 2, with an occasional flaring event (Stine et al. 1988). Also, the star was discovered in 1949 via its relatively strong Hα emission, which has disappeared since (e.g., Montmerle et al. 1983, Feigelson & Montmerle 1985). It seems that DoAr 21 undergoes more or less erratic changes in activity, but

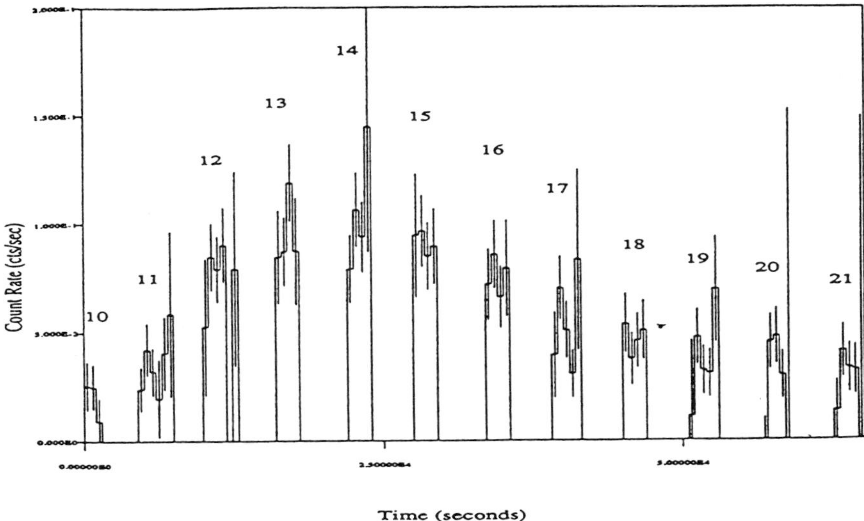

Figure 4. Light curve of the CTTS SR 13, over the last 11 *ROSAT* orbits (Casanova 1994). This variability is not due to flaring, but rather to an eclipsing plasma.

the existing time and wavelength coverage of this star is not adequate to understand its behaviour (binarity has been searched, but seems excluded).

5. The case of protostars

Going one step backward in time, what is the status of the X-ray emission of stars *younger* than T Tauri stars? Some advances have been made recently with *ROSAT* and *ASCA*. In the ρ Oph core, CMFA found as possible counterparts to *ROSAT* sources a few so-called "Class I" IR sources (interpreted as evolved protostars surrounded by a tenuous circumstellar envelope); Koyama et al. (1996) found *ASCA* sources in the R CrA cloud also associated with Class I protostars. In both cases, however, the angular resolution is not good enough to exclude confusion with other IR sources.

Whatever the exact nature of the *ASCA* sources, a strong variability has been observed in the R CrA cloud core. This event is described in detail by Koyama (this conference). The timescales involved ($\tau_r << 10^5$ s, $\tau_d \sim 10^5$ s) suggest a "normal" flare ; however, the very high temperature reached ($\sim$ 10 keV), and the fact that the 6.7 keV line appears double, are most unusual. A tantalizing conclusion is that such high temperatures are some signature of a very young object. In this respect, it is interesting to note that the very young CTTS XZ Tau shows, in addition to its "normal"

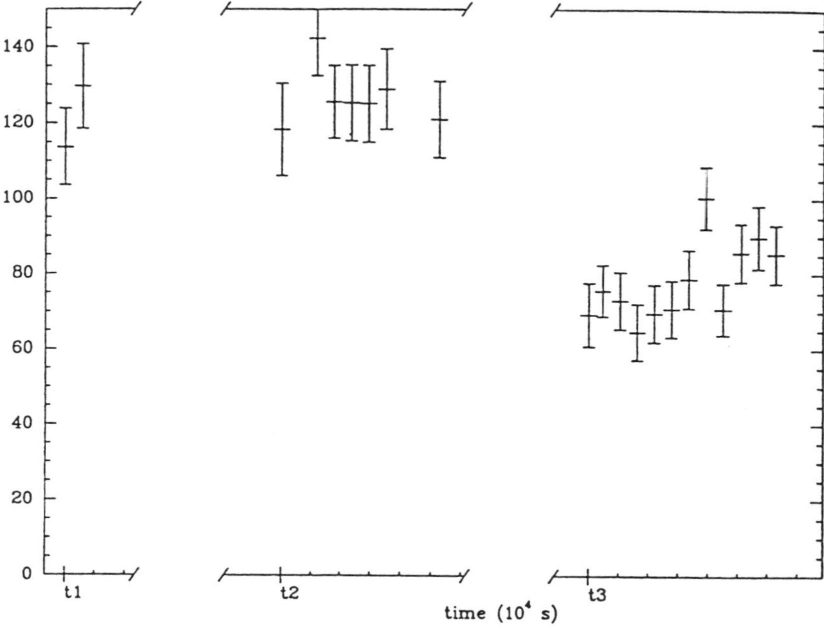

Figure 5. Light curve of the WTTS DoAr 21, over three *ROSAT* orbits (Casanova 1994). Non-flaring random variability of a factor ~ 2 is common among TTS.

≈ 1 keV spectrum, a high-energy tail extending up to several keV (Carkner et al. 1996).

6. Concluding remarks

6.1. X-RAYS ARE GOOD PROXIES FOR MAGNETIC FIELDS

The most likely interpretation of the existence of X-ray emission from YSOs, and in particular T Tauri stars, is that their surface is covered with large volumes of hot plasma, independently of the presence of circumstellar matter. Best determined in the case of flares, the plasma properties (densities, temperatures, etc.) are in general quite comparable to the solar ones, but with an activity level enhanced several orders of magnitude with respect to the Sun. This solar analogy, in view of the *Yohkoh* results, strongly support the idea that the plasma is magnetically confined. Then it can be shown that the high luminosities reached (up to $\sim 10^5 L_{X,\odot}$) may be explained by the large size of the magnetic loops ($l \lesssim 1 - 3R_\star$) (Montmerle et al. 1993).

6.2. IS THERE EVIDENCE FOR THE DYNAMO MECHANISM ?

Since YSOs are strongly convective, one is then particularly anxious to use X-rays to test the paradigm of magnetic field generation, the dynamo mechanism. Unfortunately, the picture is so far fairly confusing. On the one hand, the strong correlations between the X-ray luminosity and global stellar parameters (such as $L_X \simeq 10^{-4} L_{bol}$) are not predicted by the classical dynamo mechanism (see FCMG and CMFA for details, and Montmerle et al. 1994). Also, there is at best a weak evidence of a correlation between X-ray luminosities (or surface fluxes) and rotation velocities ; while such a correlation seems to exist for a sufficiently large sample of TTS (e.g., Bouvier 1990), it is not the case for TTS belonging to the same star-forming region. On the other hand, if the generally higher level of X-ray activity of WTTS with respect to CTTS over large areas in the sky, advocated by Neuhaüser et al. (1995), is confirmed, it may be consistent with the current scenario of CTTS → WTTS rotation evolution.

In this scenario, the circumstellar disk of CTTS is magnetically anchored to the star (see, e.g., Hartmann, Hewett, & Calvet 1994), so that momentum exchange between the star and the disk ensures a constant rotation velocity ; when the disk disappears, angular momentum conservation of the newborn WTTS implies an increase of the rotation velocity as the star contracts towards the main sequence, which, according to the classical dynamo, implies in turn an increase in the X-ray activity.

However, the real picture must be more complicated, because the same kind of X-ray correlations between global stellar parameters still hold at this stage. So, even if in principle T Tauri stars are a particularly interesting site to study the generation of stellar magnetic fields, its true mechanism so far remain largely unclear.

References

Adams, F.C., Lada, C.J., & Shu, F.H. 1987, ApJ, 321, 788.
André 1996, *Disk-like Structures around Young Stars,* Ed. Frontières, in press.
André, P. & Montmerle, T. 1994, ApJ, 420, 837.
Bertout, C. 1989, Ann. Rev. Astr. Ap., 27, 351.
Bouvier, J. 1990, AJ, 99, 946.
Caillault, J.-P. 1995, *ROSAT Science Symposium,* AIP Conf. Ser., in press.
Carkner, L., et al. 1996, ApJ, in press.
Casanova, S. 1994, Ph.D. thesis, Université Paris VII
Casanova, S., Montmerle, T., Feigelson, E.D., & André, P. 1995. ApJ, 439, 752.
Feigelson, E.D., Casanova, S., Montmerle, T., & Guibert, J. 1993, ApJ, 416, 623.
Feigelson, E.D., & Montmerle, T. 1985, ApJ, 289, L19.
Güdel, M., & Guinan, E.F. 1996, IAU Coll. 152, in press.
Hartmann, L., Hewett, R., & Calvet, N. 1994, ApJ, 426, 669.
Koyama, K., et al. 1994, PASJ, 46, L125.
Koyama, K., Ueno, S., Kobayashi, N., & Feigelson, E.D. 1996, Nature, in press.

La Valley, M., Isobe, T., & Feigelson, E.D. 1992, BAAS, 24, 839.
Montmerle, T., et al. 1983, ApJ, 269, 182.
Montmerle, T., et al. 1993, *Protostars & Planets III*, U. of Arizona Press, p. 689.
Montmerle, T., et al. 1994, *Cosmic Magnetism*, Cambridge, p. 33.
Neuhaüser, R. et al. 1995, A&A, 297, 391.
Preibisch, T., Zinnecker, H., & Herbig, G.H. 1996, A&A, in press.
Preibisch, T., Zinnecker, H., & Schmitt, J.H.M.M. 1993, A&A, 279, L33.
Sargent, A.I. 1996, *Disks and Outflows around Young Stars*, Springer, in press.
Stine, P.C., Feigelson, E.D., André, P., & Montmerle, T. 1988, AJ, 96, 1394.
Wilking, B.A., Lada, C.J., & Young, E.T. 1989, ApJ, 340, 823.

RADIO EMISSION FROM SOLAR AND STELLAR FLARES

T. S. BASTIAN
National Radio Astronomy Observatory
P. O. Box 'O', Socorro, NM 87801, USA

1. Introduction

Flare activity on stars is an important aspect of the general question of stellar activity. The proximity of the Sun allows us to study the flare phenomenon in considerable detail. To what extent does our knowledge of solar flares carry over to those on other stars? In this brief review I consider radio emission from the Sun, classical flare stars, close binaries, and pre-main sequence objects. Simple scaling arguments are presented which show radio flares on stars are unlike those on the Sun in important respects. The reasons for this are discussed.

2. Solar Flares

2.1. FLARE CLASSIFICATION

Although more complex flare classification schemes have been proposed (e.g., Tanaka 1987; Bai and Sturrock 1989) it has been convenient to divide solar flares into two broad classes, referred to as i) *compact*, *impulsive*, or *confined* flares, and ii) *long-duration*, *extended*, or *two-ribbon* flares (Pallavicini *et al.* 1979).

Compact or impulsive flares are characterized by size scales of $\sim 10^9$ cm, energy densities of $10^2 - 10^3$ erg-cm^{-3}, and time scales of 10s of sec. Long-duration or extended flares are large events characterized by a size scale of $\sim 10^{10}$ cm, greater heights ($\gtrsim 5 \times 10^4$ cm), energy densities of $10 - 10^2$ erg-cm^{-3}, and time scales of 10s of min to many hrs. Furthermore, long-duration events are associated with coronal mass ejections (CMEs; see Kahler 1992) – which are, in turn, associated with metric radio bursts of type II and type IV – and with fast particles in the interplanetary medium (solar energetic particle events: SEPs).

Y. Uchida et al. (eds.),
Magnetodynamic Phenomena in the Solar Atmosphere – Prototypes of Stellar Magnetic Activity, 259–266.

It is widely held that compact flares involve the acceleration of electrons and ions in closed magnetic loops, with a relatively small fraction of the accelerated particles escaping. In the case of long-duration events (LDEs), a second phase of particle acceleration is involved. Shock waves propagating through the corona and the interplanetary medium (IPM) produce large numbers of fast protons; associated processes produce relativistic electrons. Energetic particles events in the IPM are referred to as "solar energetic particle" events, or SEPs. Because shocks can subtend a large solid angle, a large fraction of the corona can become involved in second-phase acceleration.

2.2. RADIO EMISSION FROM SOLAR FLARES

The radio emission from solar flares is due to a variety of mechanisms. It was first studied extensively at meter-wavelengths and a number of radio-burst types were identified. These include the type II (shock associated) bursts, classical type III bursts (electron beams), and type IV flare continuum. Plasma radiation mechanisms predominate although gyrosynchrotron emission may also play an occasional role. Meter-wavelength emissions are described in detail in McLean & Labrum (1985).

At decimetric wavelengths (300 MHz – 3 GHz) and meter wavelengths, a variety of plasma emission mechanisms come into play (Benz 1996). Of particular interest are type IIIdm bursts, due to discrete beams of electrons, which accompany flares in large numbers and are correlated with hard X-ray (HXR) emission, at times on a one-to-one basis (Aschwanden *et al.* 1995). The cyclotron maser instability (CMI) may be relevant to the so-called "millisecond spike" bursts (Melrose & Dulk 1982).

Of more relevance to observations of stellar flares are microwaves. In microwaves (3 – 30 GHz) incoherent gyrosynchrotron radiation from mildly relativistic electrons predominates. Peak brightness temperatures rarely exceed $10^8 - 10^9$ K on the Sun; i.e., the characteristic energies of the emitting electrons is 10s to 100s of keV. Microwave emission is well-correlated with HXR emission (Kosugi, Dennis, and Kai 1988); both emissions result from essentially the same electron distribution, although the HXR and microwave sources are not necessarily cospatial. Microwave and HXR emissions are of considerable interest because a significant fraction of the flare energy goes into the electrons emitting in these wavelength regimes during the impulsive phase.

2.3. COMPARISON WITH OPTICAL AND X-RAY EMISSION

At optical wavelengths, continuum emission from solar flares is generally meager and most emission is in spectroscopic lines. Chromospheric Hα emission originates in so-called "ribbons" which straddle the magnetic neutral line and correspond to conjugate magnetic footpoints of the flaring coronal loops. Soft X-ray (SXR) emission shows impulsive emission from magnetic footpoints (Hudson *et al.* 1994), and more gradual emission along the flaring magnetic loops; that is, the SXR source spans the Hα ribbons, and delineates the coronal loops. Similarly, flare-associated microwave emission shows footpoint emission which evolves to one or more structures which span the Hα ribbons (e.g., Bastian and Kiplinger 1991). Hence, the optical, soft X-ray, and incoherent radio emission are characterized by a comparable spatial scale, the spatial scale of the flaring magnetic coronal loop system.

2.4. EXTENSION TO STELLAR RADIO FLARES

We assume that, as on the Sun, the spatial scales characterizing optical, soft X-ray, and radio flares on stars are similar: $l_O \approx l_X \approx l_R$. Take a characteristic size scale of $l_R \sim 5 \times 10^9$ cm on classical flare stars, and $l_R \sim 5 \times 10^{10}$ on RS CVn binaries (see, e.g., Bastian 1994). Assuming the observed radio emission is incoherent gyrosynchrotron radiation, with a maximum brightness temperature of $\lesssim 10^{10}$ K, the flux density expected from a classical flare star at a distance of 5 pc, and a RS CVn binary at a distance of 50 pc, is $S_\nu \approx 0.1\nu_9^2$ mJy, where ν_9 is the observing frequency in GHz. Hence, over the microwave frequency range, one might expect to observe flares with peak flux densities of a few to a few 10s of mJy from both classes of object. In general, this is not the case. There are now many examples of bursts on classical flare stars with peak flux densities exceeding 100 mJy at frequencies $\nu_9 \lesssim 5$ (Bastian & Bookbinder 1987; Abada-Simon *et al.* 1994; Stepanov *et al.* 1995) and, in one case, approaching 1 Jy (Bastian *et al.* 1990). In the case of active binaries and pre-main-sequence (PMS) objects, radio flares are also observed with peak flux densities of several 100 mJy to well in excess of 1 Jy (e.g., Feldman *et al.* 1978). Since the flux density observed from a source at a frequency ν in a given polarization is $S_\nu = k_B T_B (\nu/c)^2 \Omega$, where k_B is Boltzmann's constant, T_B is the brightness temperature, and Ω is the solid angle subtended by the source, the following must be true: stellar radio flares are much *brighter* than solar radio flares ($T_{B*} \gg 10^{10}$ K), or the radio source size on stars is much *larger* than the size of the SXR source, or both.

3. Stellar Flares

3.1. CLASSICAL FLARE STARS

The fact that radio bursts on classical flare stars – also known as dMe (dwarf M emission line) stars or UV Ceti-type stars – are large on frequencies $\nu_9 \lesssim 5$ is due to the fact the *brightness* of the emission is extremely high, with $T_B \gtrsim 10^{16}$ K in some cases. Furthermore, radio bursts on flare stars are usually highly circularly polarized, often 100% polarized. Dynamic spectral observations have shown that some bursts display the presence of narrowband and/or drifting structures, the presence of quasiperiodic oscillations, and of the "sudden reduction" phenomenon (Bastian *et al.* 1990). The correlation between radio bursts on frequencies $\nu_9 \lesssim 5$ and flares in the optical, UV, and X-ray wavelengths is poor or entirely absent (Kundu *et al.* 1988). These properties indicate that coherent emission mechanisms are responsible for the observed radio bursts.

A variety of emission mechanisms has been proposed to account for the observed features of radio emission from flare stars on frequencies $\nu_9 \lesssim 5$ (see Lang 1996). Chief among them are the cyclotron maser instability (CMI) or various plasma radiation mechanisms. Melrose (1993) has argued in favor of the CMI as the most likely mechanism because it more easily accounts for the high brightness temperature, the high degree of circular polarization, and the short time scales sometimes observed. On the other hand, Kuijpers (1989), Bastian *et al.* (1990), and Stepanov *et al.* (1995) have argued that plasma radiation mechanisms may also play a significant role. Mechanisms which have no solar analog may also be relevant; for example multiple Raman up-conversion of photons excited at the electron plasma frequency or its harmonic may produce radiation with a maximum well above the plasma frequency if the energy density of Langmuir waves is sufficiently high (Russell, Newman, and Goldman 1985).

A key goal, therefore, is to identify the radio emission mechanisms responsible for the high brightness bursts observed at frequencies $\nu_9 \lesssim 5$ on classical flare stars. Bastian (1996) has pointed out that, given the extremely high brightness temperatures observed on classical flare stars, induced scattering processes such as induced Compton scattering (e.g., Coppi *et al.* 1993) or stimulated Raman scattering (e.g., Levinson and Blanford 1995) may play important roles in modifying the spectral and angular properties of the emitted radiation. A detailed understanding of the propagation of the emitted radiation through the coronal medium is therefore required.

Observations at $\nu_9 > 5$ are few in number, but peak flux densities are a few 10s of mJy, consistent with the simple scalings in §2.4. The degree of circular polarization is low to moderate and indications are that the degree of correlation between microwave bursts and the optical counterpart is excellent (e.g., Rodonò *et al.* 1990). Benz, Güdel, & Schmitt (1996) find there is a general correlation between radio variability at 8.4 GHz and that in SXRs on a time scale of 1 hr although the issue of whether a more detailed correlation exists was difficult to establish because the ROSAT SXR observations were interrupted by periodic Earth occultations. In contrast to observations with $\nu_9 \lesssim 5$, existing observations are consistent with the idea that high-frequency radio bursts on flare stars are due to incoherent gyrosynchrotron emission. In this sense they may be analogous to those on the Sun, albeit in a scaled up fashion.

3.2. ACTIVE BINARIES AND PMS OBJECTS

Unlike the classical flare stars, radio sources on active binaries and PMS objects have been spatially resolved through direct imaging using VLBI techniques (see review by Mutel 1996). These observations demonstrate that the source is very large: comparable in scale to the binary separation in the case of RS CVn binaries. The inferred source brightness is $10^8 - 10^{10}$ K and the degree of circular polarization is generally small during flares. These properties are consistent with gyrosynchrotron and/or synchrotron radiation being the relevant emission mechanism during flares. Hence, unlike the Sun and classical flare stars, radio flares on active binaries and PMS objects are intense because the source is *large*, much larger than the optical and SXR sources.

Why should the radio source be so large? One possibility may be the natural scale associated with active binaries: the binary separation. Uchida and Sakurai (1983) proposed an "interacting magnetospheres" model for RS CVns. Van den Oord (1988) has explored the question of whether a filament could be supported between the two stars and whether sufficient energy would be available to power the observed flares. He finds that there is, although the details of how such a configuration might destabilize and flare are left unspecified. Neither scenario accounts for the large radio sources of similar or greater size and intensity seen on PMS objects.

The most recent models of radio flares on RS CVns, which may also have some bearing on PMS objects, involve an impulsive injection of fast electrons into a magnetic trap, presumably a large magnetic loop or loop system (Chiuderi-Drago and Franciosini 1993; Franciosini and Chiuderi-Drago 1995). The population of fast electrons radiates via the gyrosynchrotron process and evolves through radiative losses and escapes from the trap. Good agreement has been obtained for the flare amplitude, the radio spectrum, and its temporal evolution (Mutel *et al.* 1987). Whether closed magnetic configurations exist on the required spatial scales has not been demonstrated. Furthermore, the details of the initial particle injection have not been addressed. If the source of the fast electrons is associated with the optical/SXR flare, the electrons suffer severe adiabatic losses as they expand into a large loop. An alternative idea to models involving an impulsive injection of fast electrons into a magnetic trap is the stellar analog to a solar long-duration event, a possibility I briefly explore here.

In the case of the Sun, there appear to be fundamental differences between impulsive and long-duration or gradual flares, which lead to two classes of SEP events (e.g., Cane, McGuire, and von Rosenvinge 1986; Reames 1992): those associated with impulsive flares and those associated with gradual events (see Gosling 1993 for a review). Impulsive SEPs are most often observed in association with optical and impulsive X-ray flares. They are characterized by a rapid rise and decay over a period of hours. The electron to proton ratio is high and the ionization states of ions are characteristic of flare temperatures ($\sim 10^7$ K). Impulsive SEPs are usually detected only when the observer is magnetically well-connected to the site of the impulsive flare. Most events produce modest fluxes of energetic particles in the heliosphere. These properties have led workers to conclude that impulsive SEPs are associated with flare events and that the energetic particles are accelerated in or near the flare.

Gradual SEPs, as their name implies, have slower rise times ($\sim$ 1 day) and decay times of days. Energetic particles associated with gradual events are proton-rich and have elemental abundances similar to those found in the corona and solar wind. Almost all gradual SEPs are associated with CMEs that drive shocks in the solar wind. Gradual SEPs can arise from events which occur anywhere on the solar disk. Gradual SEPs are interpreted as the product of shock acceleration of coronal and solar wind particles. Ellison and Ramaty (1985) have shown that diffusive shock acceleration by moderate to strong shocks in the corona can account for many of the observed properties of electrons, protons, and alphas in SEPs. Kahler *et al.* (1986) have demonstrated that a CME and the associated shock *alone* can produce relativistic electrons and ions. It is worth noting that aside from the (shock-associated) type II radio bursts, and type IV flare continuum at meter wavelengths, no other radio signatures are associated with second-phase electron acceleration in LDEs on the Sun.

I suggest that gradual SEPs may dominate radio flares on RS CVns and, possibly, PMS objects as well. The scenario is as follows: 1) a fast coronal mass ejection occurs on an active star, driving a shock through the stellar corona, out into the asterosphere[1]; 2) shock acceleration of protons and ions occurs; associated processes accelerate electrons to relativistic energies (e.g. Melrose 1990); 3) fast electrons stream behind and in front of the shock and radiate via the gyrosynchrotron and/or synchrotron process; 4) unlike the solar case, the radiation is visible at microwavelengths because the asterosphere around an RS CVn binary, an Algol system, or a PMS object is more strongly magnetized ($\sim$ 10 G) than is the heliosphere; 5) the duration of the radio flare is determined by the shock duration and

[1]I use the term *asterosphere* in analogy to the Sun's *heliosphere*

associated electron acceleration.

This scenario is appealing because it eliminates the need for extremely large magnetic traps, yet accounts for the large source size observed. Furthermore, it allows for continuous, as opposed to impulsive, electron acceleration. The proposed scenario has at least one important consequences: the source size should evolve (expand) in time. There are two examples where source expansion may have been observed. Trigilio, Umana, and Migenes (1993) report VLBI observations of a large flare on HR 1099. The 5 GHz observations were consistent with source expansion during the rise of a flare to maximum. Similarly, Phillips *et al.* (1991) present VLBI observations of the weak-lined T Tauri star HD 283447 which shows a morphology change from compact to extended structure on a time scale of order 1 day. Interestingly, the radio emission from HD 283447 has shown a linearly polarized component (R. Phillips, private comm.), indicative of synchrotron radiation by fully relativistic electrons in weak fields.

4. Concluding Remarks

While flares on dMe stars resemble those on the Sun at radio frequencies $\nu_9 \gtrsim 5$ GHz to the extent that they appear to correlate well with optical events and are due to incoherent gyrosynchrotron emission, the opposite is true for $\nu_9 \lesssim 5$. As yet unidentified coherent emission mechanisms predominate at low frequencies. Unlike the solar case, their extreme brightness may drive stimulated scattering processes such as induced Compton or stimulated Raman scattering, further complicating identification of the underlying emission mechanism or mechanisms. With the upgrade of the Arecibo 305 m telescope, broadband spectroscopy should enable us to determine whether induced scattering processes are relevant.

Radio flares on RS CVn and pre-main-sequence objects are observationally similar. Unlike the Sun, they produce extremely large sources. The flare emission is likely due to incoherent gyrosynchrotron or synchrotron emission. Whether the particles are confined to extremely large magnetic traps or are acclerated over an extended period of time in eruptive events remains to be seen. Direct imaging is now possible on a routine basis using phase referenced VLBI techniques. It is hoped improved time-resolved imaging will resolve these issues in the coming years.

The National Radio Astronomy Observatory is a facility of the National Science Foundation operated under cooperative agreement by Associated Universities, Inc.

References

Abada-Simon, M., Lecacheux, A., Louarn, P., Dulk, G. A., Belkora, L., Bookbinder, J.

A., & Rosolen, C. (1994) A&A, **288**, 219
Aschwanden, M. J., Montello, M. L., Dennis, B. R., & Benz, A. O. (1995) *Astrophys. J.*, **440**, 394
Bai, T., and Sturrock, P. A. (1989) *Ann. Rev. Astron. Astrophys.*, **27**, 421.
Bastian, T. S. (1994) *Sp. Sci. Rev.*, **68**, 261.
Bastian, T. S. (1996) in *Radio Emission from the Stars and the Sun*, Barcelona, July 1995, A. R. Taylor and J. M. Paredes (eds.), in press.
Bastian, T. S., & Bookbinder, J. A. (1987) Nature, **326**, 678
Bastian, T. S., Bookbinder, J. A., Dulk, G. A., and Davis, M. (1990) *Astrophys. J.*, **353**, 265.
Bastian, T. S., and Kiplinger, A. L. (1991) in *Proc. 3rd Max '91 Workshop*, R. M. Winglee and A. L. Kiplinger (eds.)
Benz, A. O. (1996) in *Radio Emission from the Stars and the Sun*, Barcelona, July 1995, A. R. Taylor and J. M. Paredes (eds.), in press.
Benz, A. O., Güdel, M., & Schmitt, J. H. H. M. (1996) in *Radio Emission from the Stars and the Sun*, Barcelona, July 1995, A. R. Taylor and J. M. Paredes (eds.), in press.
Cane, H. V., McGuire, R. E., and von Rosenvinge, T. T. (1986) *Astrophys. J.*, **301**, 448.
Chiuderi-Drago, F., and Franciosini, E. (1993) *Astrophys. J.*, **410**, 301.
Coppi, P., Blandford, R. D., & Rees, M. J. (1994) *MNRAS*, **262**, 603
Ellison, D. C., and Ramaty, R. (1985) *Astrophys, J.*, **298**, 400.
Feldman, P. A., Taylor, A. R., Gregory, P. C., Seaquist, E. R., Balonek, T. J., & Cohen, N. L. (1978) A&A, **83**, 1471
Franciosini, E., and Chiuderi-Drago, F. (1995) *Astron. Astrophys.*, **297**, 535.
Gosling, J. T. (1993) *J. Geophys. Res.*, **98**, 18937.
Hudson, H. S. *et al.* (1994) in X-ray Solar Physics from Yohkoh, Y. Uchida, K. Shibata, T. Watanabe, & H. Hudson (eds).
Kahler, S. W., Cliver, E. W., Cane, H. V., McGuire, R. E., Stone, R. G., and Sheeley, N. R., Jr., (1986) *Astrophys. J.*, **302**, 504.
Kahler, S. W. (1992) *Ann. Rev. Astron. Astrophys.*, **30**, 113.
Kosugi, T., Dennis, B., and Kai, K. (1988) *Astrophys. J.*, **324**, 1118.
Kuijpers, J. (1989) Solar Phys, **121**, 163.
Kundu, M. R., Pallavicini, R., White, S. M., & Jackson, P. D. (1988) A&A, **195**, 159
Lang, K. R. (1996) this proceedings.
Levinson, A., and Blandford, R. (1995) *MNRAS*, **274**, 717.
McLean, D. J., and Labrum, N. R. (1985) *Solar Radiophysics*, Cambridge University Press, Cambridge, 516 pp.
Melrose, D. B. (1990) *Aust. J. Phys.*, **43**, 703.
Melrose, D. B. (1993) *Proc. ASA*, **10**, 254
Melrose, D. B., & Dulk, G. A. (1982) *Astrophys. J.*, **259**, 844
Mutel, R. L. (1996) this proceedings.
Mutel, R. L., Morris,.D. H., Doiron, D. J., and Lestrade, J.-F. (1987) *Astronom. J.*, **93**, 1220.
Pallavicini, R., Serio, s., Vaiana, G. S. (1977) *Astrophys. J.*, **216**, 108.
Phillips, R. B., Lonsdale, C. J., and Feigelson, E. D. (1991) *Astrophys. J.*, **382**, 261.
Reames, D. V. (1992) in *Eruptive Solar Flares*, Z. Svestka, B. V. Jackson, and M. E. Machado (eds.), Springer-Verlag:New York.
Rodonò, M., *et al.* (1989) Proc. IAU Colloq. 104 – Poster Papers, B. M. Haisch and M. Rodonò (eds.), p. 53.
Russell, D., Goldman, M., and Newman, D. (1985) *Phys. Fluids*, **28**, 2162.
Stepanov, A. V., Fürst, E., Krüger, A., Hildebrandt, J., Barwig, H., and Schmitt, J (1995) *Astron. Astrophys.*, in press.
Tanaka, K. (1987) *Pub. Astron. Soc. Jpn.*, **39**, 1.
Uchida, Y., and Sakurai, T. (1983) in *Activity in Red Dwarf Stars*, P.B. Byrne and M Rodonò (eds.), Reidel: Dordrecht
Trigilio, C., Umana, G., and Migenes, V. (1993) *MNRAS*, **260**, 903.
van den Oord, G. H. J. (1988) *Astron. Astrophys.*, **205**, 167.

RADIO EVIDENCE FOR NONTHERMAL MAGNETIC ACTIVITY ON MAIN-SEQUENCE STARS OF LATE SPECTRAL TYPE

KENNETH R. LANG
Dept. of Physics & Astronomy, Tufts University, Medford, MA 02155, U.S.A

Abstract. The Very Large Array (VLA) and other radio interferometric arrays provide information about the magnetic fields, electrons, temperatures and both thermal and nonthermal particle acceleration on the Sun and other nearby magnetically active stars of late spectral type. The long-lived, nonthermal radio emission of these stars is correlated with their thermal X-ray luminosity. The nonthermal particle acceleration process is probably related to coronal heating in these stars, and to magnetic activity resulting from internal rotation and convection. Thermal radiation dominates the quiescent radio output of the Sun, which is radio underluminous when compared to other radio-underluminous active stars. However, different thermal and nonthermal radio structures are detected at different wavelengths and in different places on the Sun. Peculiar nonthermal radio sources and nonthermal noise storms, are respectively found above solar active regions and in more extended large-scale coronal loops; they may both be fed by global reservoirs of high-energy particles. High-velocity electrons can be nearly continuously accelerated in the magnetically confined coronae of nearby active stars of late spectral type, providing nonthermal incoherent gyrosynchrotron or synchrotron radiation and/or nonthermal coherent radiation.

1. Introduction

This paper is a shortened version of a longer article of the same title (Lang, 1995). The longer version extends the discussion of quiescent radio radi-

Y. Uchida et al. (eds.),
Magnetodynamic Phenomena in the Solar Atmosphere – Prototypes of Stellar Magnetic Activity, 267–276.

ation, and includes flaring radio emission from both the Sun and other magnetically active stars of late spectral type. These topics are briefly included in the Summary of this shorter paper. Radio emission from pre-main-sequence T Tauri stars, dwarf M flare stars, and RS CVn stars has been recently reviewed by Lang (1994).

2. Overview of Stellar X-ray and Radio Radiation

2.1. X-RAY RADIATION FROM NEARBY ACTIVE STARS

We now know that virtually all single, nearby main-sequence (dwarf) stars of late spectral type (F, G, K, M) emit detectable, quiescent X-ray radiation. They are usually more active and thousands of times more energetic than the Sun (spectral type G2V), with absolute X-ray luminosities of up to $L_x = 10^{30} ergs^{-1}$, compared with the solar value of $L_{xo} = 2 \times 10^{27} ergs^{-1}$ (at the maximum in the 11-year activity cycle, and perhaps twenty times less at the minimum).

The X-ray coronae of late-type stars have temperatures of 1 to 10 million degrees that are attributed to the thermal radiation of a hot, tenuous gas. Moreover, their absolute X-ray luminosity increases with the square of the equatorial rotation velocity of the star (Pallavicini et al., 1981). A close correlation between the quiescent thermal X-ray emission and rotation speed for cool, single main-sequence stars of spectral type F through M has been demonstrated by Hempelmann et al. (1995). In contrast, the quiescent X-ray luminosity of these main-sequence stars is not closely related to either stellar age or stellar mass.

The greater speed of rotation presumably results in enhanced magnetism by internal dynamo action, resulting in greater coronal heating and more luminous, thermal X-ray radiation. The late-type RS CVn binary stars are, for example, tidally locked into rapid synchronous rotation, with exceptionally intense X-ray emission attributed to unusually strong magnetic activity. The amplified dynamo action is related to both rapid rotation and deep convection zones.

2.2. CORRELATIONS BETWEEN THE THERMAL X-RAY EMISSION AND NONTHERMAL RADIO EMISSION OF NEARBY ACTIVE STARS

Güdel and Benz (1993) have shown that there is a strong correlation between the quiescent X-ray luminosity, L_x, and the quiescent radio luminosity, L_r, of a variety of late-type, main-sequence stars, given by:

$$\log L_x = \log L_r + 15.5.$$

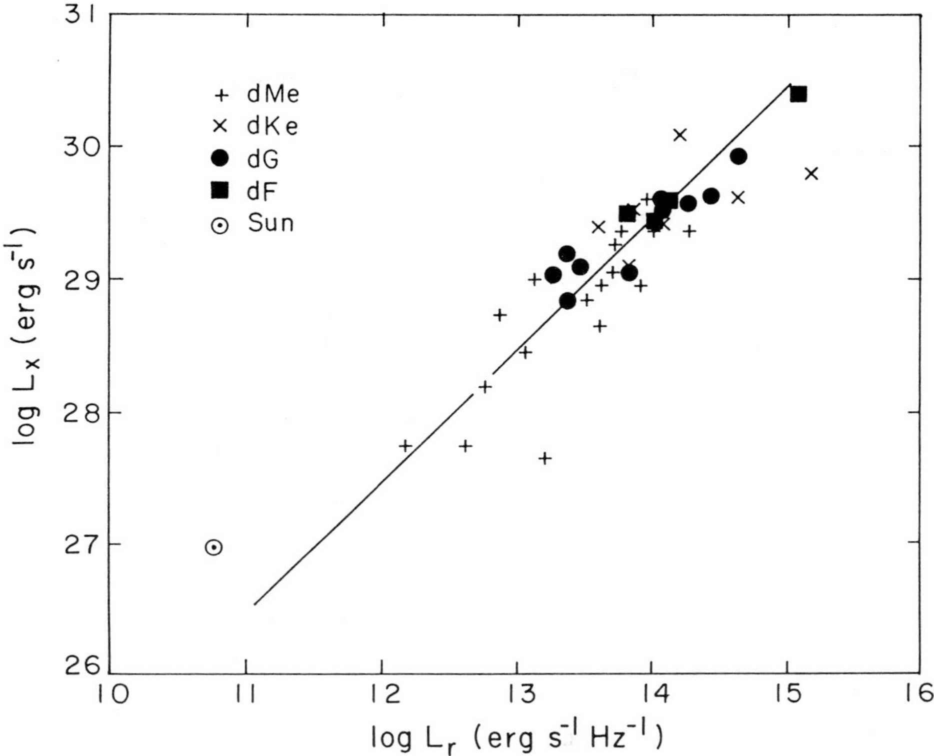

Figure 1. Comparison of the soft X-ray absolute luminosity, L_x, and the absolute radio luminosity, L_r, of main-sequence (dwarf) stars of late spectral type, F, G, K and M. The solid line denotes the relation $\log L\_x = \log L\_r + 15.5$. The Sun is radio underluminous when compared with the other magnetically active stars of late spectral type. (Adapted from Gdel, 1992; Güdel et al., 1993; Güdel, Schmitt and Benz, 1994, 1995).

The intriguing correlation is independent of stellar age, spectral class, binarity, rotation or photospheric/chromospheric activity. This relationship is shown in Figure 1, where more recent data for main-sequence stars of spectral type F, G, K, and M are included.

In order to be detected at radio wavelengths, most main-sequence stars of late spectral type have to be emitting quiescent radio radiation by nonthermal processes. The nonthermal radio radiation is attributed to electrons accelerated to nearly the speed of light in the presence of strong magnetic fields. Thus, there appears to be an intimate connection between particle acceleration, required for the quiescent nonthermal radio radiation from stars, and the coronal heating mechanisms needed to create the quiescent thermal X-ray radiation.

The Sun is probably radio underluminous when compared with other magnetically active, late-type stars because the solar radio emission is dominated by thermal processes (bremsstrahlung and gyroresonance), but the Sun nevertheless does emit quiescent, nonthermal radio radiation. We next discuss this relatively new area of solar radio astronomy.

3. Long-Lasting, Nonthermal Radio Radiation from the Sun

Some of the more intense solar radio sources at 2 to 6 cm wavelength occur above the magnetic neutral line that separates regions of opposite magnetic polarity in the underlying photosphere. They can be found within solar active regions with a multipolar (delta) configuration of the photospheric sunspots. These so-called peculiar sources have low circular polarization, steep radio spectra and high brightness temperatures of up to $Tb = 10^7$ degrees (Akhmedov et al., 1986), and they cannot be explained by conventional thermal processes. The steep radiation spectra are incompatible with the thermal bremsstrahlung that normally occurs at the apex of coronal loops (above the magnetic neutral line), and the magnetic fields at the loop tops are an order of magnitude weaker than those required to give intense thermal gyroresonance emission at these wavelengths. The low circular polarization is also incompatible with thermal gyroresonance similar to that found above the strong magnetic fields of sunspots.

The "peculiar" coronal radio sources that appear above magnetic neutral lines suggest long-lasting, nonthermal coronal heating and appear to require nearly continuous acceleration of energetic nonthermal electrons by a yet unknown process. In some cases, they may be due to the gyrosynchrotron radiation of mildly relativistic electrons (Alissandrakis et al., 1993); in other situations nonpotential magnetic fields and localized heating by neutral current sheets are implied (Lang et al., 1993).

Noise storms are another form of long-lasting (hours to days), nonthermal radio radiation; they are the most common type of activity observed on the Sun at metric wavelengths. Noise storms are highly circularly polarized (up to 100 percent), have brightness temperatures of $Tb = 10^7$ to 10^9 degrees, and require nonthermal electrons accelerated to energies of a few keV to tens of keV.

High-resolution VLA observations at 90cm wavelength indicate that noise storms do not lie directly above either the 20-cm or soft X-ray loops that connect sunspots within individual active regions. They are instead located within large-scale loops that link widely separated active regions, even in opposite hemispheres of the Sun (Willson et al., 1995), or connect active regions to more distant areas on the Sun (Lang and Willson, 1987; Krucker et al., 1995).

Noise storms are not always associated with the brightest active regions, and instead seem to be connected with active regions that have

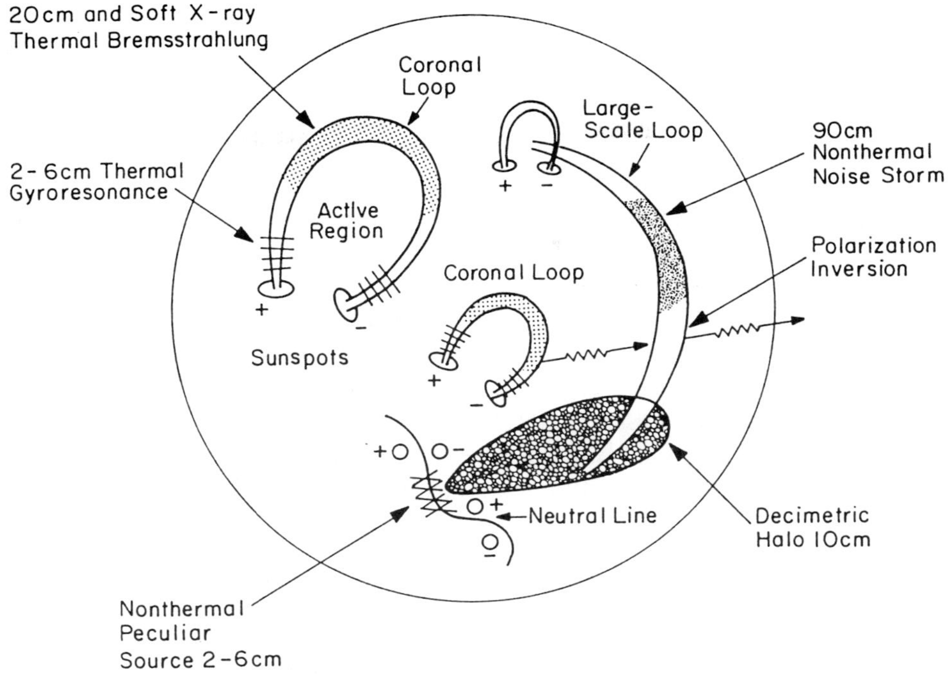

Figure 2. Long-lived (hours to days), quiescent radio radiation from the Sun. Both thermal and nonthermal sources can be detected, depending upon the wavelength and on the location.

weaker thermal microwave emission and fainter thermal soft X-ray radiation than those associated with other visible active regions (Bogod et al., 1995). When circularly polarized radiation from sunspot-associated radio sources passes through the large-scale magnetic loops, the polarization inversion may be used to infer the magnetic field strength, suggesting that anomalously-high, nonpotential magnetic fields are located at the source of the noise storms (Willson et al., 1995). Although the storm onset might be triggered by evolving magnetic fields, including the emergence of new magnetic flux (Raulin et al., 1991), the exact acceleration mechanism for the electrons emitting nonthermal radio noise storms remains a mystery (Krucker et al., 1995). The noise-storm particles seem to be accelerated to nonthermal energies within the corona near the site of storm emission, and the relationship between the thermal and nonthermal coronal structures is unclear.

The accelerated particles may be trapped in high-lying radiation belts forming an extended decimetric halo. These halo sources have nonthermal spectra, brightness temperatures of $Tb = 10^6$ degrees, and a relatively large radio flux because of their large size and high temperature. They are extended structures that can link more than one active region and lie above the coronal loops detected at 20cm or X-ray wavelengths within the individual active regions. The decimetric halo sources may outline large-scale coronal loops that act as a reservoir from which energetic particles drain down to lower-lying, nonthermal magnetic structures, feeding both noise storms and peculiar sources (Bogod et al., 1995).

To sum up, what you see on the quiescent radio Sun depends on how you look at it, and you can tune in different coronal structures at different wavelengths and in different places. Most of the intense, long-lived radio sources detected at short centimeter wavelengths emit thermal radiation within solar active regions, but by different mechanisms with sizes that generally increase with wavelength (Figure 2). Long-lasting, "peculiar" nonthermal radio sources are also found in active regions at 2 to 6 cm wavelength (also see Figure 2), but they do not usually dominate solar radio emission in this spectral region. At longer 90 cm wavelength, long-lived nonthermal noise storms within large-scale coronal loops may dominate the solar radio output; this emission is not localized within individual active regions and may be related to more extensive, global reservoirs of nonthermal particles.

4. Long-Lasting, Nonthermal, Coherent Radio Radiation from Late-Type Active Stars

Coherent, nonthermal radiation is suggested by observations of relatively intense, narrow-band radiation lasting from several hours from several dwarf M flare stars (Lang and Willson, 1986; White, Kundu and Jackson, 1986; Kundu et al., 1987; Lang and Willson, 1988). YZ Canis Minoris has, for example, emitted 100 percent circularly polarized radio radiation for hours with a fractional bandwidth of about 0.02 (Lang and Willson, 1988; Lang, 1994). Nonthermal gyrosynchrotron emission is not expected to have either a narrow bandwidth or high circular polarization, but the high brightness temperatures still require a nonthermal process. The narrow-bandwidth and high circular polarization may be due to coherent mechanisms like electron-cyclotron masers or coherent plasma radiation. (Maser is an acronym for microwave amplification by stimulated emission of radiation.)

The coherent radiation processes provide constraints on the physical conditions in the stellar coronae. An upper limit to the electron density, Ne, is given by the requirement that the observing frequency must be greater than the plasma frequency for the radiation to propagate out and reach the observer; at 20cm wavelength this requires an electron density less than, or equal to, $Ne = 2.5 \times 10^{10}\mathrm{cm}_{-3}$. If an electron-cyclotron maser emits at the second harmonic of the gyrofrequency, then a coronal magnetic field strength of $H = 250$ Gauss is required to explain the 20cm radiation; lower magnetic field strengths are needed if coherent plasma radiation dominates.

The RS CVn stars can emit long-lived (hours) radio emission that is not resolved with VLBI; these core sources are smaller than either visible component of these binary systems and have brightness temperatures greater than $Tb = 10^{10}$degrees. Long-lasting radio radiation with high circular polarization has been observed from RS CVn stars (Brown and Crane, 1979; Mutel et al., 1987), and Willson and Lang(1987) have shown that variable radio emission lasting for hours can have fractional bandwidths less than 0.5. So, the long-lasting, coherent radio radiation detected from dwarf M stars may be operating in the core RS CVn radio sources, with similar coherent maser or coherent plasma radiation processes (Lestrade et al., 1988).

Different radiation mechanisms may dominate at different wavelengths or for various stellar components, as they do on the Sun. At short centimeter wavelengths, the long-lived radio emission from late-type active stars may be due to nonthermal, incoherent gyrosynchrotron radiation, while nonthermal, coherent processes may be required to explain the radio radiation at longer radio wavelengths or for shorter wavelengths and core components.

5. Summary

This is the Summary for the longer version of this paper (Lang, 1995).

5.1. INTRODUCTION

High-resolution X-ray and radio observations, respectively using telescopes in space and ground-based arrays, have transformed our understanding of the million-degree solar corona, showing that it is spatially inhomogeneous and time variable on every detectable scale. The radio data provide diagnostic information on magnetic fields, electrons, temperatures and both thermal and nonthermal particle acceleration on the Sun. Magnetic activity seems to play the dominant role in structuring and modulating the million-degree gas, including powerful, transient solar flares.

5.2. STELLAR X-RAY AND RADIO RADIATION

Virtually all single, nearby main-sequence stars of late spectral type (F, G, K, M) emit detectable, quiescent X-ray radiation that is attributed to the thermal emission of a million-degree stellar corona; the X-ray luminosity increases with a star's rotation velocity, suggesting that enhanced magnetism plays a role in coronal heating. In order to be detected at radio wavelengths, nearby cool active stars must emit the radio radiation of nonthermal electrons accelerated to high velocity in strong magnetic fields. Particularly intense, X-ray emitting, main-sequence stars have a nonthermal radio luminosity that increases with the thermal X-ray luminosity, suggesting a relationship between nonthermal particle acceleration and thermal heating of late-type stars. In contrast, the Sun is radio underluminous when compared with other nearby magnetically active stars of late spectral type.

5.3. QUIESCENT RADIO EMISSION FROM THE SUN

Thermal radiation dominates the long-lasting, quiescent radio emission of the Sun at short centimeter wavelengths, but different thermal structures are detected at different wavelengths and in different places on the Sun; they include thermal gyroresonance in the intense magnetic fields above sunspots (at 2cm and 6cm), thermal bremsstrahlung of the X-ray emitting coronal plasma at the apex of coronal loops within individual active regions (at 20cm), and large-scale thermal coronal structures that extend across a substantial fraction of the solar disk (at 90cm). Nonthermal, quiescent solar radio sources apparently require the nearly continuous acceleration of energetic nonthermal electrons; they include the "peculiar" sources (2cm to 6cm) above the magnetic neutral line in the underlying photosphere, long-lasting (hours to days) noise storms located within large-scale loops (90cm) that link widely separated active regions or connect active regions to more distant areas on the Sun, and extended, large-scale radiation belts (decimeter halo - 10cm) that may act as global reservoirs from which energetic particles drain down to lower-lying, nonthermal magnetic structures. Nonthermal processes may dominate solar radio radiation at the longer meter wavelengths.

5.4. QUIESCENT RADIO EMISSION FROM MAGNETICALLY ACTIVE STARS OF LATE SPECTRAL TYPE

Nonthermal gyrosynchrotron or synchrotron radiation from high-speed electrons in intense magnetic fields may explain the quiescent, long-lasting radio emission from some late-type active stars. The high brightness temperatures, high degrees of circular polarization, and/or narrow-bandwidths of the long-lived (several hours) radio radiation from some dwarf M stars and some RS CVn stars may be explained by nonthermal coherent mechanisms like electron-cyclotron masers or coherent plasma radiation, providing constraints on the electron density, Ne less than or equal to $10^{10}\mathrm{cm}^{-3}$, and the magnetic field strength, $H = 250$ Gauss, in the radio source.

5.5. FLARING RADIO EMISSION FROM THE SUN

Radio, soft X-ray and hard X-ray observations have been used in support of a canonical model of solar flares in which nonthermal particles are accelerated at the apex of coronal loops within individual active regions, giving rise to nonthermal gyrosynchrotron radio emission; the high-speed electrons then move down along the loop legs to give rise to hard X-ray radiation when striking the underlying chromosphere. Other observations indicate that extended coronal loops, much larger than an active region in size, play an important role in particle acceleration and flare triggering on the Sun. Nonthermal, coherent emission processes are required to account for millisecond spike bursts observed during solar flares at short centimeter wavelengths.

5.6. FLARING RADIO EMISSION FROM MAGNETICALLY ACTIVE STARS OF LATE SPECTRAL TYPE

Rapid radio flares from dwarf M stars require sources much smaller than the visible stars in size, and suggest brightness temperatures exceeding 10^{15} degrees. Such high temperatures require a coherent radiation mechanism, and the high circular polarization (100 percent) indicates an intimate connection with stellar magnetic fields. The available radio and X-ray evidence for flares on other late type stars, including RS CVn binaries, suggests that they also have highly structured magnetic coronae with sizes much smaller than the visible stars and coherent nonthermal radio bursts.

Acknowledgments Radio astronomical studies of the Sun and other active stars at Tufts University are supported by NASA grant NAGW-2383 and by a recent NASA PSP Coronas award NRA-94-SSS-PSP-004. The VLA-RATAN 600 solar observations are supported by NSF grant ATM-9317795 and NATO collaborative research grant CRG 921394.

References

Akhmedov, Sh. B., Borovik, V. N., Gelfreikh, G. B., Bogod, V. M., Korzhavin, A. N., Petrov, Z. E., Dikij, V. N., Lang, Kenneth R., and Willson, R. F. (1986) *Astrophys. J.*, **301**, 460.

Alissandrakis, C. E., Gelfreikh, G. B., Borovik, V. N., Korzhavin, A. N., Bogod, V. M., Nindos, A., and Kundu, M. R. (1993) *Astron. Astrophys.*, **270**, 509.

Alissandrakis, C. E., and Kundu, M. R. (1982) *Astrophys. Journal (Letters)*, **253**, L49.

Benz, A. O., and Güdel, M. (1994) *Astron. Astrophys.*, **285**, 621.

Bogod, V. M., Garimov, V., Gelfreikh, G. B., Lang, K. R., Willson, R. F., and Kile, J. N. (1995) "Noise Storms and the Structure of Microwave Emission of Solar Active Regions", Solar Physics, in press.

Brown, R. L., and Crane, P. C. (1979) *Astron. J.*, **83**, 1504.

Güdel, M. (1992) *Astron. Astrophys.*, **264**, L31.

Güdel, M. (1992) *Astron. Astrophys.*, **264**, L31.

Güdel, M., and Benz, A. O. (1993) *Astrophys. J. (Letters)*, **405**, L63.

Güdel, M., Benz, A. O., Bastian, T. S., Furst, E., Simnett, G. M., and Davis, R. J. (1989) *Astron. Astrophys.*, **220**, L5.

Güdel, M., Schmitt, J. H. M. M., and Benz, A. O. (1994) *Science*, **265**, 933.

Güdel, M., Schmitt, J. H. M. M., and Benz, A. O. (1995) "Microwave Emission from X-ray Bright Solar-Like Stars: The F-G Main Sequence and Beyond", *Astron. Astrophys.*, in press.

Hempelmann, A., Schmitt, J. H. M. M., Schultz, M., Rüdiger, G., and Stepien, K. (1995) Astron. Astrophys., *294*, 515.

Krucker, S., Benz, A. O., Aschwanden, M. J., and Bastian, T. S. (1995) "Location of Type I Radio Continuum and Bursts on Yohkoh Soft X-ray Maps", in press.

Kundu, M. R., Jackson, P. D., White, S. M., and Melozzi, M. (1987) *Astrophys. J.*, **312**, 822.

Lang, K. R. (1994) *Astrophys. J. Supp.*, **90**, 753.

Lang, K.R. (1995) "Radio Evidence for Nonthermal Magnetic Activity on Main-Sequence Stars of Late Spectral Type", submitted to it Solar Physics.

Lang, K. R., and Willson, R. F. (1986) *Astrophys. J. (Letters)*, **302**, L17.

Lang, K. R., and Willson, R. F. (1986b) *Adv. Space Res.*, **6**, No. 6, 97.

Lang, K. R., and Willson, R. F. (1988) *Astrophys. J.*, **326**, 300.

Lang, K. R., Willson, R. F., Kile, J. N., Lemen, J., Strong, K. T., Bogod, V. M., Gelfreikh, G. B., Ryabov, B. I., Hafizov, S. R., Abramov, V. E., and Svetkov, S. V. (1993) *Astrophys. J.*, **419**, 398.

Lestrade, J.-F., Mutel, R. L., Preston, R. A., and Phillips, R. B. (1988) *Astrophys. J.*, **328**, 232.

Mutel, R. L., Morris, D. H., Doiron, D. J., and Lestrade, J.-F. (1987) *Astron. J.*, **93**, 1220.

Pallavicini, R., Golub, L., Rosner, R., Vaiana, G. S., Ayres, T., and Linsky, J. L. (1981) *Astrophys. J.*, **248**, 279.

Raulin, J. P., Willson, R. F., Kerdraon, A., Klein, K.-L., Lang, K. R., and Trottet, G. (1991) *Astron. Astrophys.*, **251**, 298.

White, S. M., Kundu, M. R., and Jackson, P. D. (1986) *Astrophys. J.*, **311**, 814.

Willson, R. F., Kile, J. N., Lang, K. R., Donaldson, S., Bogod, V. M., Gelfreikh, G. B., Ryabov, B. I., and Hafizov, S. R. (1995) "Large-Scale Coronal Magnetic Fields - Noise Storms, Soft X-rays and Inversion of Radio Polarization", *Proceedings of the Committee on Space Research (COSPAR), Adv. Space Res.*, **17**, 265.

Willson, R. F., and Lang, K. R. (1987) *Astrophys. J.*, **312**, 278.

AN ANALYSIS OF X-RAY FLARES IN PLEIADES STARS

S. SCIORTINO, G. MICELA AND F. REALE
Istituto e Osservatorio Astronomico di Palermo, 90134, Italy
V. KASHYAP AND R. ROSNER
Dep. of Astronomy, Univ. of Chicago, Chicago, IL, USA
AND
F. R. HARNDEN, JR.
Harvard-Smithsonian CFA, Cambridge, MA, USA

In the assumption that stellar flares, in analogy to many solar flares, occur in coronal magnetic loops (see Serio 1995), Reale et al. (1995) have devised a method, calibrated with extensive hydrodynamics modeling, to derive quantitative constraints on loop length (L) and the characteristic time scale of heating (τ_{ht}) from quantities observed during the decay phase. We have applied this method to flares observed in the Pleiades. This work is based on those on solar flares by Serio et al. (1991) and Jackimiec et al. (1992) which showed that the *Slope* of the curve traced by flares in the density-temperature (n-T) plane during the decay depends only on the presence of significant heating during the decay (i.e. with e-folding time, τ_{ht}, at least, comparable to the loop thermodynamic decay time, τ_{tdh}). The resulting diagnostics has been succesfully applied to flares observed with SMM (Sylwester et al. 1993). Reale et al. (1993) have extended this modeling to stellar cases analyzing the dependence from gravity, g, and shown that the *Slope* in the (n-T) diagram depends also on the ratio of loop length, L, and pressure scale height $H = 2kT/g\mu$. The time evolution of soft X-ray flare spectra is non-parametrically described in terms of Spectral Shape Indexes (SSI) computed projecting the time resolved spectra observed during flare decay along a set of principal component axes derived from the analysis of a sample of spectra of hydrostatic loop models. For the ROSAT PSPC the value of the 1st SSI explains most of the variance among template loop spectra and is directly related to the *effective* (i.e. as measured by the detector) temperature of coronal plasma. The same data allow also to compute the intensity decay time, τ_{lc}. This is the other

Y. Uchida et al. (eds.),
Magnetodynamic Phenomena in the Solar Atmosphere – Prototypes of Stellar Magnetic Activity, 277–278.

Name	Sp	R_* $[10^{10}]$ (cm)	τ_{lc} $[10^3]$ (sec)	T_{max} $[10^7]$ (K)	EM_{peak} (cm^{-3})	E_{Flare} (erg)	L	H	Heating during Decay Phase & $\tau_{heating}(sec)$
							$(10^{10}$	cm)	
Hz 892	(K0)	5.92	2.8	9.6	1.9 (54)	5.5 (34)	5	42	May Be [~ 2000]
Hz 1100	K3	5.43	10	3.7	2.5 (53)	2.9 (34)	17	16	No [<<10000]
Hz 1516	(K7)	4.87	5	8.2	1.9 (54)	9.6 (34)	2	32	Yes [~ 5000]
Hz 2411	M3/M4e	2.78	13	6.3	3.2 (52)	4.1 (33)	< 20	17	?
AD Leo	dM3.5e	2.78	0.6	2.5	8.9 (51)	7.7 (31)	0.6	7	May Be [~ 600]

quantity required to characterize flare properties, given that loop emission intensity scales approximately with the square of plasma density. Hence, in the stellar context the (n-T) plane is replaced by an equivalent plane of **observable** quantities, namely 1st SSI, and (Cnt Rate)$^{0.5}$. The same model flare data points allow also to define two analytical curves in the $(\tau_{lc}/\tau_{tdh}\text{-}Slope)$ plane, one for $L/H << 1$ and the other when $L/H \sim 1$, from which one can estimate L with an uncertainty of about 30%.

We have inspected the X-ray light curves of all detected stars in the 4 PSPC images toward the Pleiades clusters (cf. Stauffer et al. 1994; Micela et al. 1996), identifying 3 Pleiades stars and one Hyades star (Hz 2411) showing intense flares with coverage of the flare decay phase appropriate for the outlined analysis method. To this sample we have added for comparison a flare seen on AD Leo, a nearby dMe star. The light curves of sample stars have been subdivided in few segments and background-subtracted spectra have been accumulated in each of them. Given the intensity of Pleiades stars we have been able to collect only hundreds of counts per segment (typical for any PSPC observation of young nearby open clusters). The temperature at the top of the flaring loop has been computed scaling the fitted single-temperature Raymond-Smith model at the observed flare maximum and is adopted to estimate the pressure scale height, H. Relevant data for sample stars are summarized in the table together with the main result of our analysis that reveals the occurrence of substantial heating during the decay phase of the flare on Hz 1516. A more extensive report of these results will be presented by Sciortino et al. (1995).

References

Serio, S., Reale, F., Jakimiec, J., Sylwester, B., Sylwester, J. 1991, A&A, 241, 197.
Jackmiec et al., 1992, A&A, 253, 269.
Sylwester et al., 1993, A&A, 267, 586.
Serio, S., 1995, this volume
Reale, F., Serio, S., Peres, G. 1993 , A&A, 272, 486.
Reale, F. et al. 1995 , A&A, to be submitted
Micela, G., et al., 1996, ApJS, January, in press.
Stauffer, J., et al., 1994, ApJS, 91, 625.
Sciortino, S. et al., 1995, A&A, to be submitted.

SIMULTANEOUS X-RAY, EUV, UV, AND RADIO OBSERVATION OF THE CORONA OF THE RS CVN BINARY HR1099

A. BROWN
CASA, U. of Colorado, Boulder, CO 80309-0389, USA
S. L. SKINNER
JILA, U. of Colorado, Boulder, CO 80309-0389, USA
R. T. STEWART
ATNF, PO Box 76, Epping, NSW 2121, Australia
AND
K. L. JONES
Dept. of Physics, U. of Queensland, Brisbane, Australia

HR1099 (V711 Tau; K1 IV + G5 IV) is one of the most coronally-active RS CVn binaries known. We have obtained multi-spectral-region observations of this star during 1994 August 24-28; a time interval equivalent to roughly 1.5 orbital periods (= 2.84 days) of the system. Included in these observations were 130 ksec of EUVE spectroscopy (70 - 400 Å, $\frac{\Delta\lambda}{\lambda} \sim 200$; emission lines of Fe XV-XXIV), 40 ksec of ASCA X-ray spectroscopy (0.6 - 8 keV, $\frac{\Delta\lambda}{\lambda} \sim 50$), 64 hours of IUE UV spectroscopy (1150 - 2000 Å, $\frac{\Delta\lambda}{\lambda} \sim 300$; 2000 - 3200 Å [optimised for Mg II] , $\frac{\Delta\lambda}{\lambda} \sim 10{,}000$), and 32 hr of VLA and 44 hr of AT radio continuum (3,6,13,20 cm) data.

In Fig. 1 we show the light curves from four spectral regions covering the time period of the ASCA observation. During this time a small (peak $L_X \sim 2\ 10^{31}$ erg s^{-1}) flare was seen. The EUVE DS and ASCA SIS light curves are almost identical. The ASCA spectra were fitted by a 2-temperature model; the cooler 0.85 keV component was constant throughout the observation, while the flare was confined to the hotter component whose temperature increased from 2.1 keV in quiescence to a flare peak of 2.9 keV. The EUVE spectra show the presence of continuous, increasing emission measure distribution over $6.3 \leq \log T \leq 7.2$. Two large EUV flares were detected; one on Aug. 24 lasted 12 hr, while the other on Aug. 26-27 lasted 24 hours. 3 cm radio flares are consistently seen at the very start of the EUV flares.

Y. Uchida et al. (eds.),
Magnetodynamic Phenomena in the Solar Atmosphere – Prototypes of Stellar Magnetic Activity, 279–280.

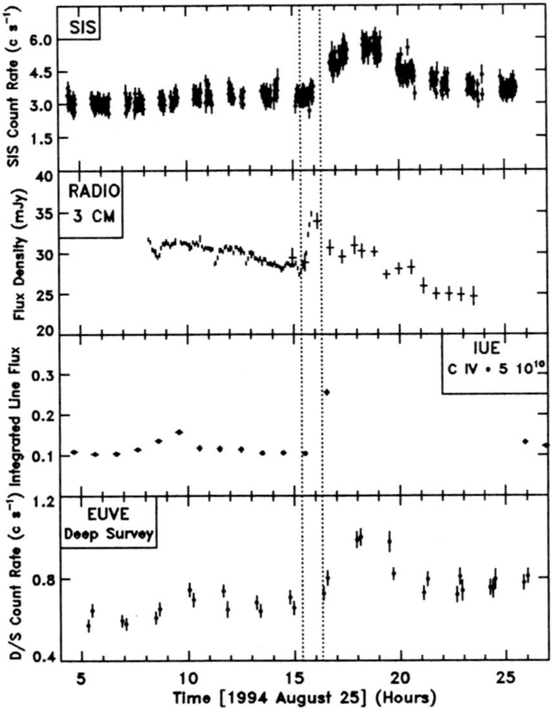

Figure 1. Variations of HR1099 in the X-ray, radio, UV, and EUVE spectral regions on 1994 August 25. Note the earlier flare response in the radio.

Our multiday observations of HR1099 show :
(i) that large flares are common on RS CVn binaries,
(ii) that soft X-ray (ASCA) and EUV (EUVE) observations are sensitive to the gradual phase of flares but insensitive to the impulsive phase,
(iii) that the ASCA flare response was confined to the hotter (20-30 MK) coronal component,
(iv) that the radio gyrosynchrotron emission (3 & 6 cm) does respond to the flare impulsive phase, while 20 cm emission is dominated by coherent processes that can show activity prior to or independent of thermal flaring,
(v) further evidence of the presence of high velocity (500 - 1000 km s^{-1}) chromospheric motions associated with flares,
and (vi) evidence that the onset of flares occurs at particular orbital phases, even though the photospheric starspots and active regions are moving relative to these phases.

ON THE ORIGIN OF INTENSE RADIO EMISSION FROM SOLAR AND STELLAR FLARES

A.V. STEPANOV
Pulkovo Observatory, St.Petersburg 196140, Russia

There is much in common between solar and stellar flaring radio emission. Indeed, very rich fine structures in radio radiation from both the Sun and red dwarfs are observed. There are quasiperiodical pulsations, sudden reductions, narrowband highly polarized spikes, etc. The brightness temperatures of these fine structures are high enough, $10^{10} - 10^{15}$ K, to suggest nonthermal emission mechanism, and moreover coherent ones. To interpret the intense microwave emission from the solar and stellar flares, the Electron Cyclotron Maser (ECM) emission is widely accepted. ECM requires, however, strong magnetic field condition which is not satisfied in many cases, especially in the solar corona, and it produces mainly x-mode radiation. There also is problem with the escape of ECM radiation from the source. ECM emission should be absorbed by thermal electrons at gyrolevels $s = 2,\ 3,\ 4$.

As in the solar case there are indications in favour of plasma radiation in the flaring radio emission from dMe stars. Fast-drift bursts, sudden reductions (Bastian et al. 1990), and a high degree of polarization in the ordinary sense (Lim 1993) are among them. The present work argues that intense microwave emission from solar and stellar flares can be understood in terms of plasma radiation.

The plasma radiation mechanism suggests the excitation of Langmuir waves with subsequent conversion of plasma waves into electromagnetic ones, resulting in radio emission at the fundamental and the second harmonic of the plasma frequency. Scattering of the plasma waves on the background ions gives an emission at the fundamental with brightness temperature $T_b \approx 3\,T(m_i/m_e)\exp(\tau)$ where $\tau > 1$ is the optical depth. This value can be as high as $T_b \approx 10^{14} - 10^{16}$ K (Zaitsev & Stepanov 1983). Second harmonic radiation is driven by a wave-wave coalescence process and prevails over the fundamental if the plasma turbulence level $W/nkT < 10^{-5} - 10^{-4}$.

Y. Uchida et al. (eds.),
Magnetodynamic Phenomena in the Solar Atmosphere – Prototypes of Stellar Magnetic Activity, 281–282.

Stepanov et al. (1995) have shown that radio spikes observed at 4.75 GHz during the UV Ceti flare on December 31, 1991 can be explained also in terms of plasma radiation. A number of plasma waves with similar phases can interfere and some irregular plasma density inhomogeneities occur due to the thermo diffusion (Genkin et al. 1990). Plasma waves modify the density fluctuations and hence change the radio emission characteristics at the fundamental. The result brightness temperature of the spikes with time scale of about 100 ms is of order of the effective temperature of the plasma turbulence and can be about 10^{12} K for UV Ceti flares.

The polarization of plasma radiation depends on the ratio of electron gyrofrequency – to plasma frequency $p = \omega_e/\omega_p$, and on the plasma wave spectrum. It is generally believed that the o-mode polarization is possible for the fundamental only due to the x-mode cut-off. Nevertheless the second harmonic can be polarized in the o-mode also. For the one-dimensional spectrum of plasma waves (along the magnetic field) the maximum of circular polarization degree is $\pi = 0.31\ p$, $p < 1$ (Zlotnik 1980). For the case $p > 1$ the emission can be highly polarized in the o-mode due to strong gyroabsorption of the x-mode at the levels $s = 2, 3, 4$ (Stepanov et al. 1995). This may be a reason for 80 % o-mode polarization in the remarkably intense radio emission from Rst 137B observed by Lim (1993).

It is not excluded that the plasma radiation plays an important role in the intense radio emission from the "radiation belts" of chemically peculiar stars and RS CVn binary systems.

References

Bastian, T.S., Bookbinder, J.A., Dulk, G.A., and Davis, M. (1990) *Ap.J.*, **353**, 265

Genkin, L.G., Erukhimov, L.M., and Levin, B.N. (1990) *Sol.Phys.*, **128**, 423

Lim J. (1993) *Ap.J.Lett.*, **405**, L33

Stepanov, A.V., Fürst, E., Krüger, A., Hildebrandt, J., Barwig, H., and Schmitt J. (1995) *A&A*, **299**, 739

Zaitsev, V.V. and Stepanov, A.V. (1983) *Sol.Phys.*, **88**, 297

Zlotnik, E.Ya. (1980) *A&A*, **101**, 250

MAGNETIC STRUCTURES AND GIANT FLARES IN HR1099

Summary Results from the MUSICOS 1989 Campaign

B.H. FOING, S. CHAR, T. AYRES, C. CATALA, D.S. ZHAI, S. JIANG, L. HUANG, J.X. HAO, E. HOUDEBINE, S. JANKOV, J. BAUDRAND, J. CZARNY, J.F. DONATI, P. FELENBOK, S. CATALANO, G. CUTISPOTO, A. FRASCA, M. RODONO, J.E.NEFF, T. SIMON, A. COLLIER-CAMERON, C.J. BUTLER

AND

THE MUSICOS 1989 CAMPAIGN COLLABORATION
ESTEC/SO, PB 299, 2200 AG Noordwijk, NL (solar::bfoing)

Abstract. We describe results from the December 1989 multi-site continuous observing campaign (MUSICOS 89 with 17 telescopes around the globe) dedicated to the study of active structures and flares on the RS CVn system HR1099=V711 Tau. Two exceptional white-light flares were detected on 14 and 15 December with remarkable photometric and spectral signatures during rise and decay. Equivalent colours, temperature excesses, projected areas (0.55 and 0.89 solar disc areas), radiative and kinetic energy budgets were derived for the two white-light flares. At IAU Colloq. 153, we presented the relation between Doppler imaged spots, extended emission and these giant flares, and discussed them in the context of models of magnetic interconnection and filament energy storage between the stars.

1. HR1099 MUSICOS 1989 Objectives and Operations

The main objectives of this campaign were to obtain information on magnetic structures on this RS CVn system, through photospheric spots Doppler imaging and chromospheric spectroscopic rotational modulation, and to monitor the spectral signature and dynamics of flares. The operations, participants and first results of the MUSICOS 89 campaign have been described in Catala and Foing (1990), Catala et al (1993) and Foing et al (1994). A total of 17 telescopes around the globe participated to the

Y. Uchida et al. (eds.),
***Magnetodynamic Phenomena in the Solar Atmosphere – Prototypes of Stellar Magnetic Activity,* 283–284.**

HR1099 campaign, with an almost continuous coverage from 14 to 17 December 1989, giving 22 spectra for Doppler imaging , 41 (+28) Hα spectra at high (+ medium) spectral resolution. We also measured the Ca II K chromospheric plage modulation from 5 to 20 December (Char et al 1990).

2. Results on HR1099 Magnetic Structures and Flares

A complete phase coverage over the 2.84 day period of HR1099 was achieved, allowing a Doppler imaging of photospheric spots (Jankov and Donati, 1995). Two exceptional white-light flares were detected photometrically, and spectroscopically in Hα. The 14 Dec. 13:00 UT flare was observed from China and Catania, and the 15 Dec. 1:00 UT flare from ESO, Hawaii, China and Catania (Foing et al 1990, 1994). Part of the flare decay was also measured with IUE in lines of low and high excitation, more than one day after the last white-light flare. The projected flare areas were estimated to 0.55 and 0.89 solar disc areas. We derived the radiative energy budget for these two events to $8x10^{36}$ and 10^{38} ergs integrated over the white-light events and in the range 310-590 nm, with peak intensities of radiative losses of 0.8 and 7 solar constants. These are the largest white-light flares detected on a solar-like star. The ratio of optical continuum emission over Balmer emission was 3-4 times larger than reported for other flares in the literature. Both flares were shown to occur near the limb, at positions consistent with active regions derived from photometric reconstruction and Doppler imaging. The kinetic energy budget was estimated from highly broadened and shifted Doppler components in the Hα profiles, and found approximately equal to the radiative losses (Foing et al 1994). We could relate the sites of white-light flares and coronal mass ejections to the Doppler imaged cool spots and hot faculae, and to the Hα emission extending well above the primary surface and near the inner Lagrangian points. We discussed at IAU 153 these results in the context of magnetic interaction and filament energy storage between the binary components.

References

Catala, C. and Foing, B.H. Eds., 1990, Proceedings 2nd MUSICOS Workshop on MUlti SIte COntinuous Spectroscopy, Paris-Meudon Obs. Publ.

Catala, C., Foing, B.H., Baudrand, J. et al, 1993 A&A, 275, 245

Char, S., Foing, B.H, Lemaire, P. et al, 1990 in Proc.2nd MUSICOS Workshop, p.69

Foing, B.H. et al, 1990 in Proceedings 2nd MUSICOS Workshop, p.117

Foing, B.H., Char S., Ayres, T. et al, 1994 A&A, 292, 543

Huang, L., Zhai, D.S., Catala, C. and Foing, B.H. Eds., 1995, Proc. 4th MUSICOS Workshop on MUlti SIte COntinuous Spectroscopy, Beijing/Toulouse/ESTEC Publ.

Jankov, S., Donati, J.F. in Proceedings 4th MUSICOS Workshop, p.143

van den Oord G.H., 1988 A&A, 205, 167

Zhai D.S., Foing, B.H., Cutispoto, G. et al, 1994 A&A, 282, 168

III. Production of Superhot Plasma and High-Energy Particles in the Sun and Stars

III.3. Models of Flares

ARCADE FLARE MODELS

T.G. FORBES
Institute for the Study of Earth, Oceans, and Space
University of New Hampshire, Durham, NH 03824, USA

Much of the recent work on magnetic arcades is motivated by eruptive phenomena such as coronal mass ejections, prominence eruptions, and large two-ribbon flares. Arcade models of these phenomena are based on a rapid release of magnetic energy stored in coronal currents. However, the mechanisms by which the energy release is triggered vary from model to model. Some models trigger the release by a loss of ideal-MHD equilibrium, while others use a non-ideal processes such as magnetic reconnection. Models which invoke reconnection require the presence of a current sheet prior to the eruption of the field, but models which are based on a loss of equilibrium do not. By examining the observational evidence for the existence of a current sheet prior to the eruption, it may be possible to determine the type of mechanism which triggers an eruption.

1. Introduction

Most models for eruptive flares and coronal mass ejections are based on the principle that the energy which drives them comes from magnetic energy stored in coronal currents (Svestka and Cliver 1992). The currents may form when a flux-tube emerges from the convection zone as shown in Figure 1a or when the footpoints of a pre-existing arcade are sheared, as shown in Figure 1b. The principle theoretical question that the storage models try to address is what causes the configuration to erupt (Figure 1c). However, there is a secondary question which is also important, namely, how does the reconnection of the erupted fields occur (Figure 1d).

During an eruption, magnetic field lines mapping from the ejected plasma to the photosphere are stretched outwards to form an extended, open field structure. This opening of the field creates an apparent paradox since the stretching of the field lines implies that the magnetic energy of the system is increasing whereas storage models require it to decrease. Aly (1991) and Sturrock (1991) have shown that for simply connected fields, the fully-opened state is in fact the highest energy state of the system. Therefore, to create an eruption with a storage

Y. Uchida et al. (eds.),
Magnetodynamic Phenomena in the Solar Atmosphere – Prototypes of Stellar Magnetic Activity, 287–294.

model, one must either start with a field which is not simply connected or avoid creating a field which is everywhere open.

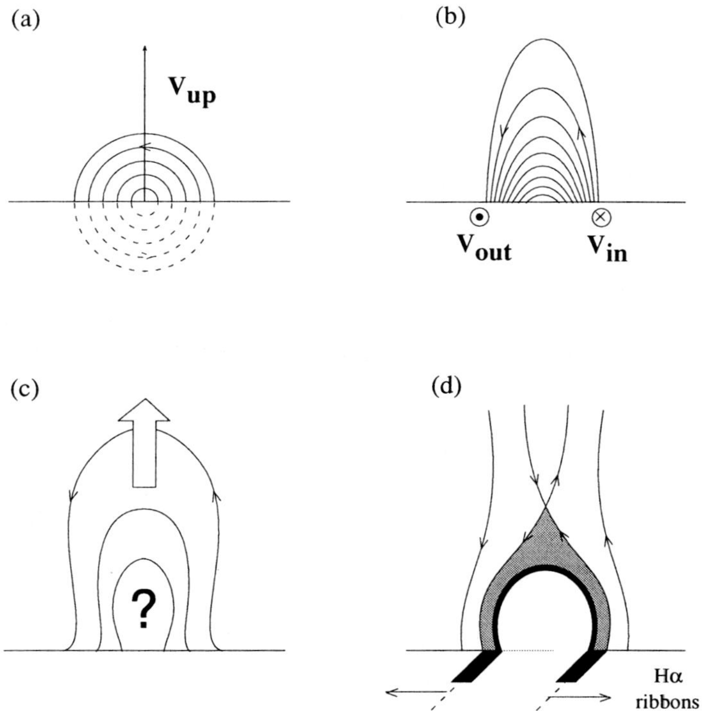

Figure 1. Stages in arcade models of eruptive flares.

2. Sheared Arcades

One class of arcade models that has received much attention is based on trying to create an eruption by shearing the footpoints of an arcade of loops (Mikic *et al.* 1988, Martinell 1990, Steinolfson 1991, Inhester *et al.* 1992). In two-dimensional force-free configurations with translation symmetry (*i.e.* sometimes referred to as 2 1/2 D configurations) shearing causes the arcade to smoothly expand outwards towards a fully opened state without ever producing an eruption. However, in two-dimensional configurations with an axial symmetry, it is not yet known for certain whether an arcade will maintain a stable equilibrium as its footpoints are sheared. Wolfson and Low (1990) have analyzed a spherically symmetric arcade of loops located along the equator of a sphere which is sheared by turning the northern and southern hemispheres in opposite directions. Their analysis, although not conclusive, suggests that an initially closed arcade of loops may erupt to form a partly opened configuration containing a current sheet.

Even if no loss of equilibrium or stability occurs, it is still possible to create an eruption of the arcade by invoking a resistive instability. Shearing an arcade leads to the formation of a current sheet which can undergo reconnection. If the reconnection occurs rapidly, say at a rate which is on the order of a few Alfvén time-scales, τ_A, then an eruption occurs (Aly 1994). Figure 2 shows a simulation of this process by Mikic and Linker (1994). From $t = 0$ to 540 τ_A, the arcade is sheared with the resistivity as near to zero as possible. After 540 τ_A the shearing is stopped and the resistivity is instantaneously increased to a value which gives an effective magnetic Reynolds number of about 10^4. This increase leads to reconnection and the formation of a flux-rope which is expelled outwards, away from the sun.

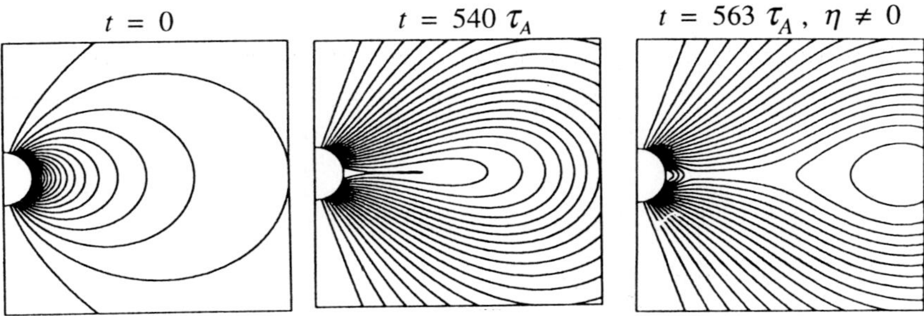

Figure 2. The arcade model of Mikic. and Linker (1994) The evolution is nominally ideal until $t = 520\ \tau_A$ at which time the resistivity is turned on. (*i.e.* greatly enhanced above the numerical value).

In order for the above mechanism to work, the reconnection rate must undergo a sudden transition. Prior to the eruption it must be slow compared to the time scale of the photospheric motions, so that energy can be stored in the coronal currents. After the eruption it must be fast, so that energy can be released rapidly. Thus, a complete model of the eruption process must explain why the reconnection rate suddenly changes at the time of the eruption. There are several possible mechanisms which could lead to a sudden increase in the reconnection rate. For example, if the reconnection in the current sheet is due to the operation of the tearing-mode instability, then reconnection will not occur until the length of the current sheet becomes longer than about 2π times its width (Furth, Killeen, & Rosenbluth, 1963). Alternatively, as the current sheet builds up, its current density may exceed a threshold of a kinetic instability which leads to an anomalous resistivity (*e.g.* Heyvaerts and Priest 1976). The onset of anomalous resistivity triggers rapid reconnection and the ejection of a flux rope as illustrated in Figure 2.

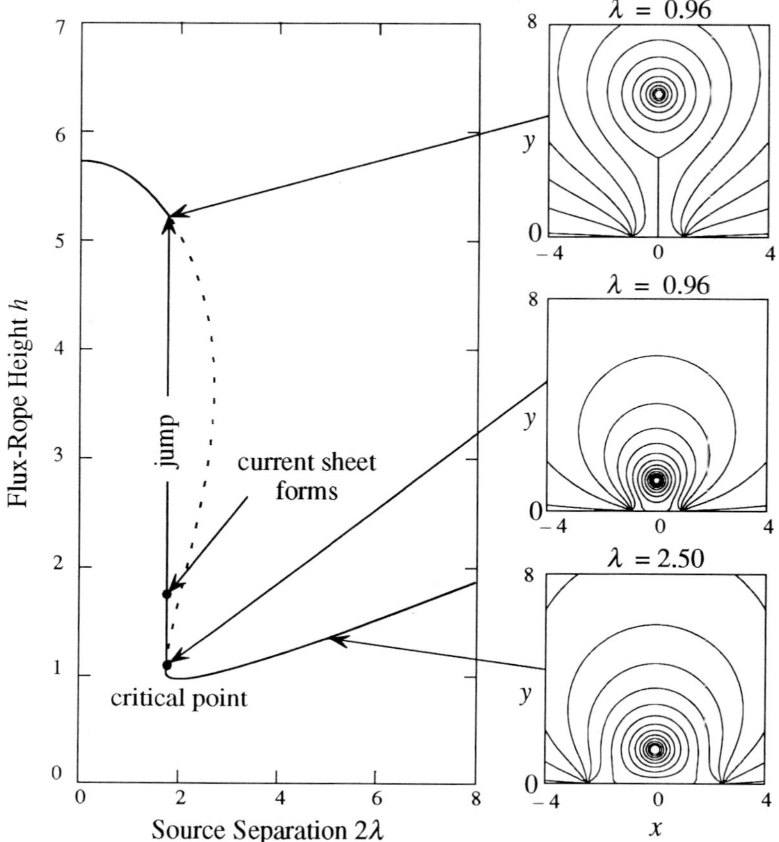

Figure 3. An arcade model containing a flux rope. Eruption occurs when the source regions in the photosphere are pushed together. A current sheet forms only after the eruption begins.

3. Arcades with Flux-Ropes

If arcades are formed in the corona by the emergence of a flux rope from below the photosphere, it is unlikely that this process always leads to the special case shown in Figure 1a where the central axis of the flux rope lies exactly at the photospheric surface. It is more likely that the axis will, at least at some locations, rise up to form a coronal flux rope which is anchored only at its ends (which are out of the plane of Figure 1a). Arcade models which contain such a flux rope provide an attractive mechanism for supporting the prominences that are often associated with eruptions (Kuperus and Raadu 1974). The upwardly concave field lines at the bottom of the flux rope provide an ideal site for supporting cool plasma as it condenses out of the hot corona. More importantly, the presence of the flux rope may make it possible for the arcade to lose equilibrium and erupt even when no current sheet or x-line is present (Van Tend 1979).

Figure 3 shows a two-dimensional flux-rope model which loses equilibrium when the photospheric sources of the field approach one another (Forbes and Priest 1995). When the sources reach a critical distance, which is on the order of the initial height of the flux rope, the flux rope jumps to an new equilibrium at a higher altitude. The new equilibrium has a vertical current sheet under the flux rope, but the total magnetic energy of the configuration is lower than before. The amount of energy which is released by the jump is about 5% of the free magnetic energy stored in the system prior to the jump.

The model shown in Figure 3 is somewhat idealized because all of the shear in the system is contained in a flux rope which does not intercept the photosphere. However, an eruption still occurs even if the radius of the flux rope is much larger (Forbes and Priest 1995). Shearing the arcade fields overlying the flux rope also enhances the energy released in the ideal-MHD jump (Démoulin *et al*. 1991). Although this particular model does not explain how a flux rope is created, it does explain why colliding active regions often lead to eruptive flares (Dezju *et al*. 1980, Kurokawa 1987).

4. Model Predictions

In principle, it should be possible to distinguish observationally between the various mechanisms discussed above for triggering an eruption. If reconnection is operating at a slow rate prior to the eruption, it may reveal itself by low level heating of the plasma. Furthermore, weak flare ribbons, perhaps too weak to observe, would be present before onset and become bright at onset, and they would already be separated as they become visible. By contrast, if the trigger is a loss of magnetic equilibrium, then there could be a delay of a few seconds (*i.e.* an Alfvén scale time) between onset and the appearance of the ribbons. The various possibilities are summarized in Table 1.

Table 1. Arcade Model Predictions

	Pre-existing current sheet	Pre-existing x-line	No current sheet or x-line
Ribbon timing	strengthen at onset	appear at onset	appear after onset
Ribbon location	separated at onset	separated at onset	not separated
Initial shape	elongated	x-type field at onset	arcade / helix

The appearance and location of the ribbons depends, not only upon the trigger mechanism, but also upon the processes which produce the ribbons. Figure 4 shows one proposed scenario for the creation of the ribbons and the X-ray flare loops lying above them (Forbes and Acton 1996). According to this scenario, the outer edges of the flare ribbons map to a neutral line in the corona, while the inner edges map to the inner edge of the soft X-ray and UV flare loops. This inner edge lies at the feet of the cool Hα loops where large downflows are observed. The amount of mass which flows down in the Hα loops during their lifetime is greater than the mass of the entire corona (Kleczek 1964).

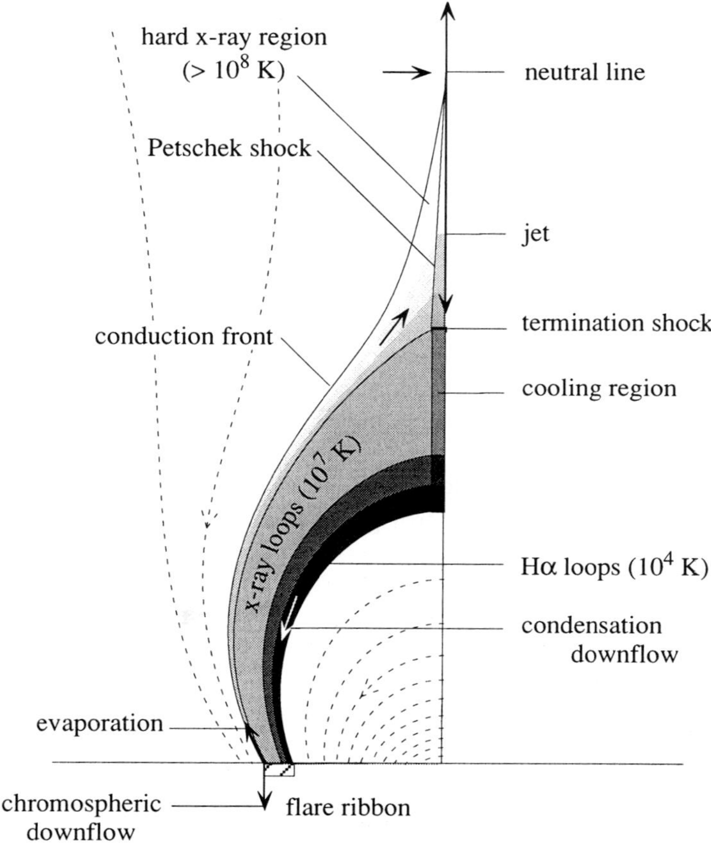

Figure 4. A model for the formation of flare loops and ribbons by reconnection and evaporation.

As reconnection proceeds, the neutral line and flare loops move upward, while the flare ribbons consequently move outwards. An example of the resulting motion is shown in Figure 5 for a long duration flare observed by *Yohkoh* on 22 April 1993. This flare, which occurred on 1993 Apr 22-23 some 33° away from the limb (heliographic coordinates E53 N24), is a weak event.

The characteristic slowing of the ribbon separation velocity with time is a property that is predicted by reconnection dynamics (Forbes and Priest 1984, Lin *et al.* 1995). For the 22 Apr 1993 event, the legs of the X-ray loops have little or no separation at the onset of the event. Initially, the outer ribbon edges must have been less than 3×10^4 km apart and separating at a speed greater than 10 km/s. After 10 hours, they are about 1.2×10^5 km apart and nearly stationary. However, there are other larger events where the ribbons are widely separated at onset (De La Beaujardière *et al.* 1995). These events suggest that there is a neutral line or current sheet in the corona prior to the eruption.

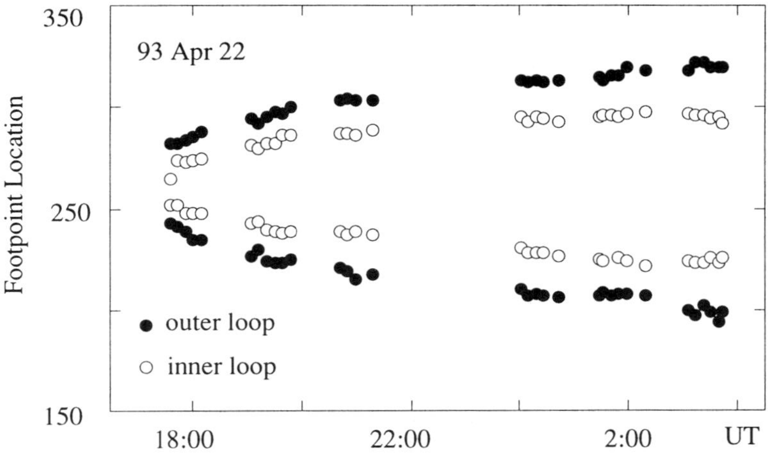

Figure 5. The position of the inner and outer edges of the flare loop legs as a function of time. The locations are expressed in terms of the SXT full resolution half-pixel unit (equal to 1.25" or 890 km).

5. Conclusions

The mechanism of impulsive behavior in eruptive events such as large two-ribbon flares and coronal mass ejections is still a mystery. The rapid onset of most eruptions suggests that a loss of magnetic equilibrium or stability occurs, and there are two-dimensional MHD models of arcades with flux ropes, which show that an eruption can indeed be triggered by small changes in the photospheric field. However, at the present time, no one has been able to construct a realistic three-dimensional model because of the intractability of the MHD equations.

An alternative way, to trigger an eruption is to shear the footpoints of an arcade until a current layer forms which is thin enough to be unstable to tearing. This latter case requires that a current sheet exist prior to the eruption, but at the present time there is no direct evidence that such a sheet exists.

Even if reconnection is not itself the trigger, it is still an essential aspect of the eruption. Without reconnection the magnetic energy release is insufficient to explain large events, and a continuous build-up of magnetic flux in interplanetary space would occur with time.

The work was supported by NASA grants NAGW-3463 (Solar SR&T) and NAG5-1479 (Space Physics Theory Program) to the University of New Hampshire.

References

Aly, J. J. (1991) *Astrophys. J.* **375**, L61.

Aly, J. J. (1994) *Astron. and Astrophys.* **288**, 1012.

De La Beaujardière, J.-F., et al. (1995) *Astrophys. J.* **440**, 386.

Démoulin, P., Priest, E. R. & Ferreira, J. (1991) *Astron. & Astrophys.* **245**, 289.

Dezju, L., Gesztelyi, L., Kondas, L., Kovacs, A. & Rostas, S. (1980) *Solar Phys.* **67**, 317.

Finn, J. M. & Chen, J. (1990) *Astrophys. J.* **349**, 345.

Forbes, T. G. & Acton, L. W. (1996) *Astrophys. J.* **459**, in press.

Forbes, T. G. & Priest, E. R. (1984) in *Solar Terrestrial Physics: Present and Future* (eds. D. M. Butler, K. Papadopoulous) NASA, p. 1.

Forbes, T. G. & Priest, E. R. (1995) *Astrophys. J.* **446**, 377.

Furth, H. P., Killeen & Rosenbluth, M. N. (1963) *Phys. Fluids* **6**, 459.

Heyvaerts, J. & Priest, E. R. (1976) *Solar Phys.* **47**, 223.

Inhester, B., Birn, J. & Hesse, M. (1992) *Solar Phys.* **138**, 257.

Kleczek, J. (1964) In *AAS-NASA Symposium on the Physics of Solar Flares* (ed. W. N. Hess) NASA, SP-50, pp. 77.

Kuperus, M. & Raadu, M. A. (1974) *Astron. Astrophys.* **21**, 189.

Kurokawa, H. (1987) *Solar Phys.* **113**, 259.

Lin, J., Forbes, T. G., Priest, E. R. & Bungey, T. N. (1995) *Solar Phys.* **159**, 275.

Martinell, J. J. (1990) *Astrophys. J.* **365**, 342.

Mikic, Z., Barnes, D. C. & Schnack, D. D. (1988) *Astrophys. J.* **328**, 830.

Mikic, Z. & Linker, J. A. (1994) *Astrophys. J.* **430**, 898.

Steinolfson, R. S. (1991) *Astrophys. J.* **382**, 677.

Sturrock, P. A. (1991) *Astrophys. J.* **380**, 655.

Svestka, Z. & Cliver, E. W. (1992) in *Eruptive Solar Flares* (eds. Z. Svestka, B. V. Jackson, M. E. Machado) Springer-Verlag, New York, p. 1.

Van Tend, (1979) *Solar Phys.* **61**, 89.

PROBLEMS FOR ARCADE AND LOOP FLARE MODELS REVEALED BY YOHKOH

YUTAKA UCHIDA

Science University of Tokyo, Shinjuku-ku, Tokyo 162, Japan

Abstract. Observation from *Yohkoh* have revealed new clues that advance our knowledge about various solar coronal and flare-related problems. Here we concentrate on new clues for arcade and loop flare mechanisms revealed by Yohkoh, and draw the readers' attention to some of the problems which the widely accepted models for respective types of flares may have to face. Alternative possibilities are suggested.

1. Introduction — Observations from YOHKOH

The wide dynamic range and rapid cadence of observations by the Yohkoh soft X-ray telescope (SXT), together with the co-aligned images of hard X-ray telescope (HXT) and the temperature and velocity information derived from the Bragg crystal spectrometers (BCS), have demonstrated the capability of Yohkoh in revealing the clues to the mechanisms of flares that were hidden in the still dark preflare and very early phases of flares. Yohkoh observations have also made it clear that the behavior of fainter X-ray arcade formation events in the regions far from active regions as part of an unexpected dynamicity of the background corona. These arcades are considered to be larger scale, fainter versions of arcade flares, and contribute to the clarification of the mechanism for arcade flaring. In the following, we deal with arcade flares together with their larger scale fainter version, X-ray arcade formation events, as well as another class of flares Yohkoh confirmed, namely, small simple loop flares.

Y. Uchida et al. (eds.),
Magnetodynamic Phenomena in the Solar Atmosphere – Prototypes of Stellar Magnetic Activity, 295–302.

2. Arcade Flares and Arcade Formations outside Active Regions

The most widely accepted model for an arcade flare may be said to be the "opening-up – reclosing simple arcade model" discussed by Sturrock (1966) and others (Hirayama 1974, Kopp and Pneuman 1976, Forbes and Priest 1987, Priest and Forbes 1990). Here, we refer to these scenario as "classical arcade flare model". This model assumes that the dark filament is suspended on the top of a simple magnetic arcade, and the arcade is pulled up and eventually cut open when the dark filament is destabilized and flies away. As the legs of the thus-opened field lines are (somehow) pushed together from the sides, and reclose through magnetic reconnection. The energy in the opened field is thought to be released in the reconnection, producing various processes in the flare.

Our task here is to examine whether this widely accepted "classical arcade flare model" is actually the case or not in the light of new Yohkoh observations, especially, of still faint initial phases of arcade flares.

The first clear *Yohkoh* observation of an arcade flare was for the event of Feb 21, 1992, which showed a candle-flame-like shape with a clear dark tunnel below it (Tsuneta et al.1992). This arcade flare observed axis-on was taken to support the "classical arcade flare model" because the cusped shape resembled what was often depicted in the classical models.

Further scrutiny, however, showed several features not compatible with the classical arcade flare models (Uchida 1996). Those features were found in the re-examination of the dark preflare and very initial phases of flare, and are (a) the presence of larger scale back-connections from the top of the preflare "core" to the photosphere at several times of the width of the "core" on both sides, and (b) the presence of a "partition-like" structure seen inside the dark tunnel, together with a low-lying bright feature near the axis of the tunnel in the initial phase. Similar features, the back-connections from the top of the preflare "core" to the photosphere on both sides, and a bright feature along the axis of the tunnel, were seen also in the flare of Dec 2, 1991 (Tsuneta et al. 1993), and we are sure that the presence of these features are *not* coincidental, but intrinsic (Uchida 1994, 1996, see Fig 1a and 1b therein). These suggest strongly that the magnetic fields involved in arcade flares may *not* be a simple bipolar arcade as assumed in the classicalarcade flare models.

Another set of evidence comes from a detailed examination of the pre-event features for the case of X-ray arcade formation events outside of active regions. We looked into these because the magnetic configuration is larger and simpler, and the processes occur more slowly, and therefore, the details of the processes can better be examined. Before Yohkoh, those fainter sources, especially in the pre-event stage, could not be seen so clearly.

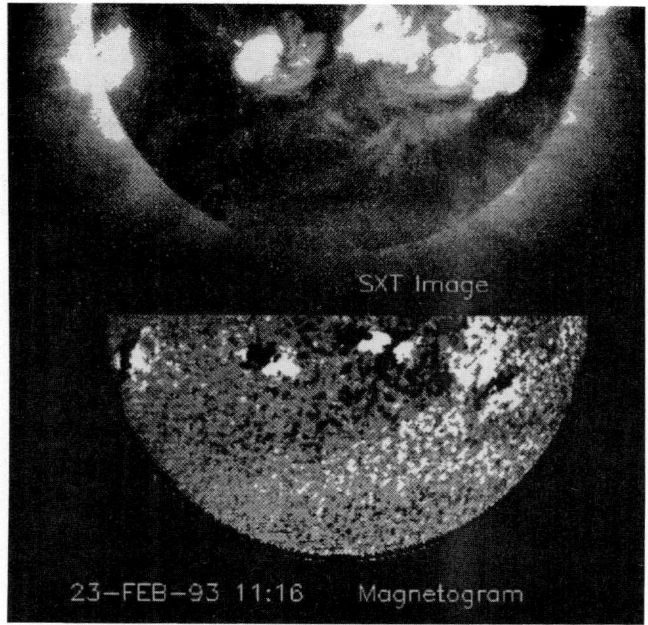

Figure 1. 1a(top) The pre-event corona of the Feb 24-25, 1993 arcade formation event shown in Fig. 2. The corona at the relevant location had a double-arcade structure in which each of them are on each side of the field-polarity reversal line, *but* with their inside footpoints rooted in the nearest parts of the *other* arcade in an intermingled way. 1b(bottom) The Kitt Peak magnetogram rotated to the exact position corresponding to Fig. 1a. It is seen on close examination that the field-polarity reversal line is only for average field, and there is a belt of mixed polarity on both sides of it, explaining the intermingled footpoints of those arcades in this region.

In the Feb 24-25, 1993 X-ray arcade formation event (Fujisaki et al. 1996), it was found that the pre-event coronal structure *had an overlapping pair* of coronal arcades, one on each side of the polarity reversal line above which the X-ray arcade formation occurred with their inner footpoints rooted in the closer part within the domain of the other arcade in an intermingled way (Fig. 1a). A close examination of the Kitt Peak magnetogram rotated precisely to the times shows that the so-called field-polarity reversal line with an Hα dark filament lying above it *is only for the average field*, and there is a belt with mixed polarity in that region (Fig. 1b). This explains the intermingled landing of the inner footpoints forming an overlapping pair of arcades in the pre-event structure, and effectively correspond to the quadruple source configuration (Uchida et al. 1996c).

The X-ray arcade formation event started with the appearance of a bright "spine" rising directly above the polarity reversal line of the "aver-

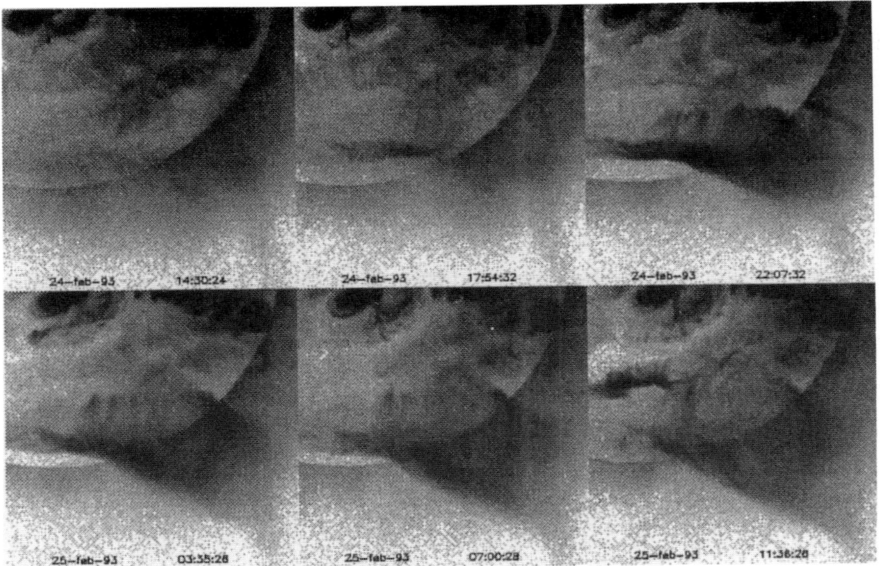

Figure 2. Time development of the Feb. 24-25, 1993 arcade formation event. A bundle of bright threads ("spine") rose above the (average) field-polarity reversal line with loops connecting to the photosphere on the north-west part, and then the western portion of the "spine" started ballooning up to result in a large cusp-like structure just like the Jan. 25, 1992 Giant Cusp event.

age" field some time after the disappearance of the dark filament. It is quite clear that the "spine" was *not* the locus of the reconnecting points of the once opened arcade as expected in the classical arcade flare models, but a bundle of *longitudinal threads*, as already noted in the case of Sep 28, 1991 X-ray arcade formation event (McAllister et al. 1992). Later, the bright "spine" ballooned up at its western part, and this eventually grew into a huge cusp-like structure near the limb (Fig 2). Since this structure was very similar to the well-known "giant cusp" of Jan 25, 1992 (Hiei et al.1995), we examined (Uchida et al. 1996c) the pre-event coronal structure for the latter case closely. We found even clearer characteristics of *non-simple arcade* in the pre-event coronal structure for the "giant cusp" event. Furthermore we found that there was an elongated southern "leg" of the "giant cusp" extended to the west by tens of degrees in longitude. This "leg" clearly corresponded to the elongated eastern part of the "spine" in the case of Feb 24-25, 1993 event. This is difficult to explain in the classical simple arcade models. Both of the above-mentioned features in arcade flares and arcade formation events indicate observational problems for the classical "opening-up – reclosing simple arcade" picture.

Our alternative suggestion is to revisit the proposal we made years ago that the dark filament suspension may be in the neutral sheet of $B_\perp$, the magnetic field component perpendicular to the field-reversal line in the photosphere in a field due to a quadruple photospheric field sources, say, elongated sources A(+), B(-), C(+), and D(-) side by side (Uchida and Jockers 1979)(note that the observed mixed polarity belt along the polarity-reversal line of the average field can be regarded effectively as a quadruple field). The model has sets of field lines connecting (i) the inner pair, B to C, (ii) the outer pair, A to D, (iii) A to B, and (iv) B to D, and thus has a neutral point above thecentral field polarity-reversal line between B and C. If there is a field component parallel to the polarity reversal line, possibly due to the skewness of the outer pair with respect to the inner pair, then this component, even weak, will provide a field component dominating in this locus of the neutral points. This neutral line in $B_\perp$, having $B_\parallel$ in it, will turn into a narrow sheet if the sources A,B,C and D are squeezed toward the polarity-reversal line due to photospheric motion. Mass loaded on $B_\parallel$ in the neutral sheet of $B_\perp$ represent the dark filament. Here $B_\perp$ is already "open" due to the effect of the outer pair of sources, not due to the cutting up of the strong field by the rising filament. When the dark filament flies away, the lines of force connecting A to B, and C to D are squeezed into a state directly facing to each other without a separator, the dark filament, and provide a reconnection site. A flare will be seen in just *the cusped shape* as observed, if the most intensely heated part is considered. The difference from the classical arcade flare models is the presence of the high connections from the reconnection site to the regions A and D, just like what are found by *Yohkoh*, together with the lower-lying part of the dark filament pressed downward, as a low-lying bright S-shaped structure connecting B to C in skew. That may be seen as a partition-like structure and a bright feature near the axis of the tunnel when observed along the axis. In this model, the reconnection between the squeezed connections A to B, and C to D, starts near the neutral line, and the reconnected part will also be seen as the bright cusp above the pressed S-shape structure (Uchida 1980, Uchida et al. 1996b). All those may expand, as observed, because the upper part of the structure has been taken away. The advantages of this model are to be able to avoid the unnatural way of suspension of the dark filament, as well as the energy problem (the dark filament rise should have greater energy than the flare itself in order to be able to cut the field open, and then the flare would be a minor repairing process), which the classical arcade models have. The readers are referred to Uchida et al. (1996b)

3. Simple Loop Flares

The treatments for the loop flares in the last one or two decades have been restricted to the process of how the heated mass is supplied to the flaring loops. Two possible ways for supplying energy to the chromospheric mass to fill the flaring loop have been discussed so far (cf., Antonucci et al. 1986): (i) "Heat conduction – evaporation (ablation) models" (HCE models), in which a production of a superhot heat source is *assumed* to appear at the loop top, and the chromospheric mass "evaporates" into the loop in response to the heat conduction, and (ii) "electron-bombardment – evaporation models" (EBE models), in which high energy electrons are somehow created at the loop top, and as they impinge into the chromosphere, they produce hard X-ray bursts, and at the same time, cause the "evaporation" of the heated material into the soft X-ray emitting loop.

Now, our task is to examine whether what are predicted in these models can actually be seen by *Yohkoh*, or not.

The loop flare of Feb 17, 1992 that occurred near the west limb has been analysed by Uchida et al. (1996a). It showed in its faint preflare stages a quite interesting dynamical behavior. First, two footpoints of the loop brightened, and then a brightening appeared in the loop top region. Subsequently, bright blob emerged from that bright region, moved down to the footpoints and appeared to bounce back to the loop top (somewhat different from what HCE models tell). Eventually, the bright region settled down in the loop top region, *and increased* its brightness about ten times and remained as the brightest region throughout the flare life without further motion. The loop shape deformed into a "question mark" shape in projection by the time the bright blob appeared, and the distortion relaxed before the flare maximum.

In this event, one of the hard X-ray sources appeared at the location of the SXT loop-top source, and moved down with it, and a strong hard X-ray spike burst occurred when the SXT source contacted the northern footpoint (Fig. 3). This may suggest that the high energy particles were contained in a structure moving with a few hundred km/s (most likely an MHD shockfront), and dumped when it came to the footpoint, suggesting a considerably different situation from the free streaming bombarding particle picture in EBE models.

Another problem is the precise timing relation. The rise of the BCS line intensity with blueshift seems to have start 30 sec to 1 min before the non-thermal hard X-ray burst began (Doschek et al. 1983, Tanaka et al. 1983 and confirmed by Yohkoh). This may give rise to a problem for the EBE models. The rise curves of the BCS line intensities are smooth, and it is no likely that the rise part is due to another unrelated small event. It seem

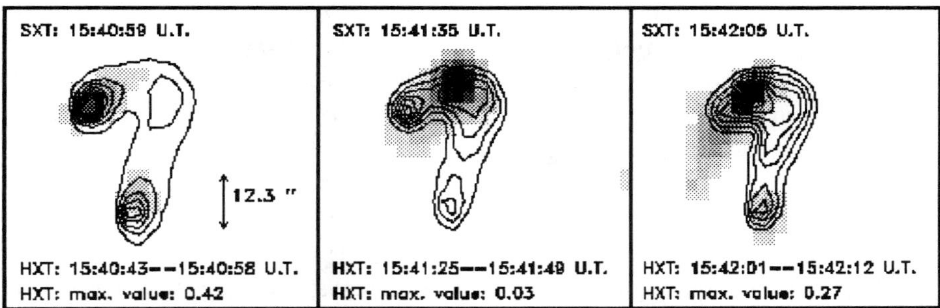

Figure 3. Relative location between the soft X-ray (contours) and the hard X-ray (half-tone) sources as the flare progressed in its initial phases. It is seen that the footpoint hard X-ray sources coincided with those of soft X-rays. A hard X-ray source is seen to move with the soft X-ray bright blob toward the northern footpoint, and the hard X-ray emission bursted up when the soft X-ray source started contacting the northern foot.

that there must be some process other than EBE mechanism in operation for the very early start up phase of loop flares. In this connection, it may be of interest that in the loop flare of Apr 22, 1993, we found a loop top source in the HXT/L-band before strong non-thermal footpoint sources appeared. This loop top source had a characteristic suggestive of superhot thermal source, in favor of HCE models (cf., Masuda 1995). A startling finding, however, was that we found a faint but clear pair of footpoint sources with the same thermal character before that loop top source. Some process other than electron bombardment seem to have started at the footpoints prior to the loop top source.

A loop flare model proposed by Uchida and Shibata (1988) has some properties favorable to the new observations. This model was based on the view that active regions are where the currents (local twists of the flux tubes produced in the convection zone under the photosphere) are disposed of and dissipate. Note that the property of the current considered here is different from the steady current considered in the so-called "current-circuit model". The current considerd here is a locally self-closed current equivalent to a packet of localized twist that can be produced, transferred, and dissipated magnetohydrodynamically. Normally, the current disposed (either due to relaxation along the loop, or due to some injection through magnetic reconnection at the footpoint of a loop) into a loop causes a heated gas injection by the so-called sweeping pinch effect (Uchida and Shibata 1984). The hypersonic (Alfven velocity in low-β corona) heated gas flow along the loop may correspond to the "active region transient brightenings" found in active regions by Yohkoh (Shimizu et al. 1992). If, in contrast, the disposition of such current packets may take place, with certainly much smaller probability, from both ends of one and the same loop, then a very high temperature region will be produced when two of the hypersonic flows collide at a high part of the loop and convert their kinetic energy into heat in shocks, and the magnetic energy of the toroidal component into unwinding motion, and ultimately into heat [Uchida and Shibata 1988, referred to here as DCI (dynamic current injection) model].

A distortion of the loop in the early phase, and relaxation before the flare maximum, may be most naturally explained in this DCI model as due to helical kink instability when the amount of the toroidal flux is maximum before annihilation starts following the collision at the loop top. High energy particles are first accelerated in betatron acceleration in the pinch, and accelerated further to higher energies with Fermi I process between the twist packets approaching to each other with Alfven velocity. The electrons and ions will be released simultaneously from the region between the "pistons" when the magnetic twists decreased in the unwinding through the collision with the opposite helicity packet.

The author acknowledges all the members of the Yohkoh Team for their cooperative efforts in the construction of hardware and software, in the operation of the satellite, and in the data analysis, all those making the wonderful satellite successful.

References

Antonucci, E., et al. (1986) *Adv. Space Res.*, **6**, 151

Doschek, G., et al. (1983) *Astrophys. J.*, **265**, 1103

Forbes, T.G. (1987) *Rev. Geophys.*, **25**, 1583

Fujisaki,K.,et al. (1996) in these Proceedings

Hiei, E., Hundhausen, A., and Sime, D. (1993) *Geophys. Res. Letters*, **20**, 2785

Hirayama, T. (1974) *Solar Phys.*, **34**, 323

Kopp, R.A. and Pneuman, G.W. (1976) *Solar Phys.*, **50**, 85

Masuda, S., et al. (1994) *Nature*, **371**, 495

McAllister, A.H. (1992) *Publ. Astron. Soc. Japan*, **44**, L205

Priest, E., and Forbes, T.G. (1990) *Solar Phys.*, **126**, 319

Shimizu, T. (1992) *Publ. Astron. Soc. Japan*, **44**, L147

Sturrock, P.A. (1966) *Nature*, **221**, 695

Tsuneta, S., et al. (1993) in *The Magnetic and Velocity Fields of Solar Active Regions*, eds. H. Zirin, G-X. Ai, and H. Wang (ASP Conference Series No.46), p239

Tsuneta, S. (1992a) *Publ. Astron. Soc. Japan*, **44**, L63

Uchida, Y., and Jockers, K., 1979, *Max Planck Institute for Astrophysics Preprint*

Uchida, Y. (1980) in *Skylab Workshop, Solar Flares*, ed. P.A. Sturrock (University of Colorado Press), p67, and p110,

Uchida, Y. and Shibata, K. (1988) *Solar Phys.*, **116**, 291

Uchida, Y., et al. (1994) in *X-ray Solar Physics from Yohkoh*, eds. Y.Uchida, T.Watanabe, K.Shibata, and H.Hudson (Universal Academy Press), p161

Uchida, Y. (1996) *Adv. Space Res.*, **17**, (4/5)19-28

Uchida, Y., Khan, J., Doschek, G., Masuda, S., McAllister, A., Hirose, S., Feldman, U., and Cheng, C.C. (1996a) to be published

Uchida, Y., Jockers. K., McAllister, A., Khan, J., and Fujisaki, K. (1996b) to be pubulished

Yashiro et al. (1996) in these Proceedings

BIFURCATION AND STABILITY OF THE CORONAL MAGNETIC FIELD

K.KUSANO, Y.SUZUKI, K.MORIYAMA, K.FUJIE

AND

K.NISHIKAWA

Department of Materials Science, Faculty of Science, Hiroshima University, Higashi-Hiroshima 739, Japan

1. Introduction

It is widely believed that a solar flare is an event in which the free energy accumulated in the coronal magnetic field is catastrophically released. However, current understanding of the question of what triggers flare events is not yet theoretically unified. We investigate the magnetohydrodynamics (MHD) of the solar coronal magnetic field based on the Woltjer-Taylor minimum energy principle (Woltjer, 1958; Taylor, 1974) as well as using the numerical simulations, and propose the new insight that a solar flare is a transition process between bifurcated equilibrium branches.

2. Minimum Energy Coronal Magnetic Field

We obtain a general solution for the Woltjer-Taylor minimum energy state in the solar coronal geometry, which is modeled as a rectangular domain. It is found that the solution bifurcates into two different branches when the magnetic helicity integral or the geometrical factor defined as the ratio of the height to the width of the domain is sufficiently increased (Kusano *et al.*, 1995). One branch is a coupled solution (CS), which connects with the potential field, and another is a mixed solution (MS), which consists of the CS and a component decoupled from the potential field. The topological structure in the magnetic field is different between two branches (CS and MS). Whereas in the CS any magnetic field line connects with the photospheric boundary, the MS possesses an isolated flux tube (plasmoid).

Once the solution bifurcates into the CS and the MS, the CS becomes unstable against global MHD perturbations. Through detailed stability

Y. Uchida et al. (eds.),
Magnetodynamic Phenomena in the Solar Atmosphere – Prototypes of Stellar Magnetic Activity, 303–304.

analyses, we find that there are two different instabilities for a coronal arcade which is in the Woltjer-Taylor state. One is the symmetrical mode instability (SMI), in which the magnetic arcade expands keeping translational symmetry for the direction parallel to the magnetic neutral line. Another is the undulating mode instability (UMI), in which the magnetic arcade periodically swells along the magnetic neutral line.

3. Numerical Simulations

In order to investigate the nonlinear dynamics of the SMI, we carry out two-dimensional MHD numerical simulations. The initial state is given by the CS plus the small perturbation which is the most unstable linear eigenfunction. We can numerically observe that the nonlinear development of the instability causes magnetic reconnection which generates plasmoids on the top of the magnetic arcade. The reconnection process can release about half of the excess energy in the timescale as short as about ten times the characteristic Alfvén transit time. It is confirmed that, as a result of reconnection, the decoupled component is generated and the magnetic configuration is switched from the CS into the MS.

4. Bifurcation–Transition Flare Model

Based on the analyses above, we theoretically propose that a solar flare is a transition process between the bifurcated Woltjer-Taylor states. The theory predicts that the solar flare is triggered when the vertical size of the coronal magnetic loop becomes longer than the horizontal size. Since the topological structure in the coronal magnetic field must be dramatically changed in the transition process, magnetic reconnection must accompany a solar flare in general as observed in the numerical simulations. These results are quite consistent also with the recent observations (Masuda *et al.*, 1994).

References

Masuda, S., Kosugi, T., Hara, H., Tsuneta, S., and Ogawara, Y. (1994) A Loop-top Hard X-ray Source in a Compact Solar Flare as Evidence for Magnetic Reconnection, *Nature* **371** pp. 495–497

Kusano, K., Suzuki, Y., and Nishikawa, K. (1995) A Solar Flare Triggering Mechanism Based on the Woltjer-Taylor Minimum Energy Principle, *Astrophys. J.* **441** pp. 942–951

Taylor, J.B. (1974) Relaxation of Toroidal Plasma and Generation of Reversed Magnetic Fields, *Phys. Rev. Lett.* **33** pp. 1139–1141

Woltjer, L. (1958) A Theorem on Force-Free Magnetic Fields, *Proc. Nat. Acad. Sci. (USA)* **44** pp. 489–492

MHD SIMULATIONS OF CORONAL CURRENT SHEET DYNAMICS

B. KLIEM AND J. SCHUMACHER
Astrophysical Institute Potsdam
An der Sternwarte 16, D-14482 Potsdam, Germany

Dynamical processes in current sheets in the coronae of the Sun and stars are supposed to be the origin of various impulsive energy release phenomena ranging from non-resolvable events ('nanoflares') up to the largest flares. The occurrence of hard X-ray and radio spikes with time scales Δt down to ~ 0.02 s in flares suggests that the energy release is fragmentary, occurring in small parts of the whole energy release region at scales $l \sim V_A \Delta t \sim 10^7$ cm (for coronal parameters $N = 10^{10}$ cm^{-3}, $T = 2.5 \times 10^6$ K, $B = 200$ G).

We study current sheet dynamics initiated by the localized occurrence of anomalous resistivity in several spots scattered along the sheet. This implies that the sheet has a width, $2l_{CS}$, which corresponds to the current density threshold for onset of a kinetic instability, j_{cr}. Taking the LHD instability as an example, we have $l_{CS} < l_{cr} \sim 4\beta^{-1} r_{ci} \sim 7 \times 10^3$ cm (β – plasma-beta). On the other hand, the extent, L_{CS}, of the current sheet is determined by the scale of the driving process and will be of order 10^9 cm or higher. Since $L_{CS}/l_{CS} > 10^5$, one cannot expect that the threshold for the kinetic instability is reached simultaneously along most parts of the sheet, instead it will first be reached at one or a few points scattered along the sheet (which will be distinguished due to fluctuations of plasma parameters with spacings $> \lambda_{mfp} \sim 5 \times 10^6$ cm). Since the kinetic instabilities develop practically instantaneously as compared with MHD time scales, anomalous resistivity will be produced locally before the threshold of kinetic instability is reached in intermediate sections of the sheet during the MHD evolution.

Our model is the Harris current sheet, $\mathbf{B} = B_0 \tanh(y/l_{CS})\, \mathbf{e}_x$, of uniform density at rest, initially disturbed by a few anomalous resistivity spots. We use a 2D Lax-Wendroff scheme with open boundaries. After $t = 4\tau_A$, the resistivity is determined self consistently from the local current density:

$$\eta(x,y,t) = \begin{cases} \eta_0 & \text{for } j < j_{cr} \\ \eta_{an} = 25\eta_0 + 25(j - j_{cr}) & \text{for } j > j_{cr} \end{cases}$$

Y. Uchida et al. (eds.),
Magnetodynamic Phenomena in the Solar Atmosphere – Prototypes of Stellar Magnetic Activity, 305–306.

The occurrence of *several localized areas* of anomalous resistivity has two primary effects, both of which lead to a very dynamical evolution of the sheet: it creates *current filaments* (magnetic islands) and *jets* (localized flows) emanating from the η_{an}-spots. Fig. 1 shows the creation of islands by five initially prescribed η_{an}-spots ("induced tearing") and the coalescence of the islands. Subsequently the coalesced island undergoes repeating phases of expansion and renewed compression by newly formed jets from the outer X-points. Finally the island tears into pair, which is accelerated by a jet from the central X-point to $\sim 0.4V_A$ and leaves the box. At the front side of moving islands the flux is piled up such that localized negative current density enhancements temporally occur, while at the back side of coalescing islands new plasma is pulled into the sheet, which leads to positive current density enhancements. Both effects are present at magnetic Reynolds numbers $R_m = 200$ and increase towards higher R_m: current density enhancements by a factor $\lesssim 10$ occur for $R_m \lesssim 10^4$. The whole evolution comprises a sequence of kinetic energy peaks and electric field peaks suggestive of fragmentary energy release and bursty particle acceleration.

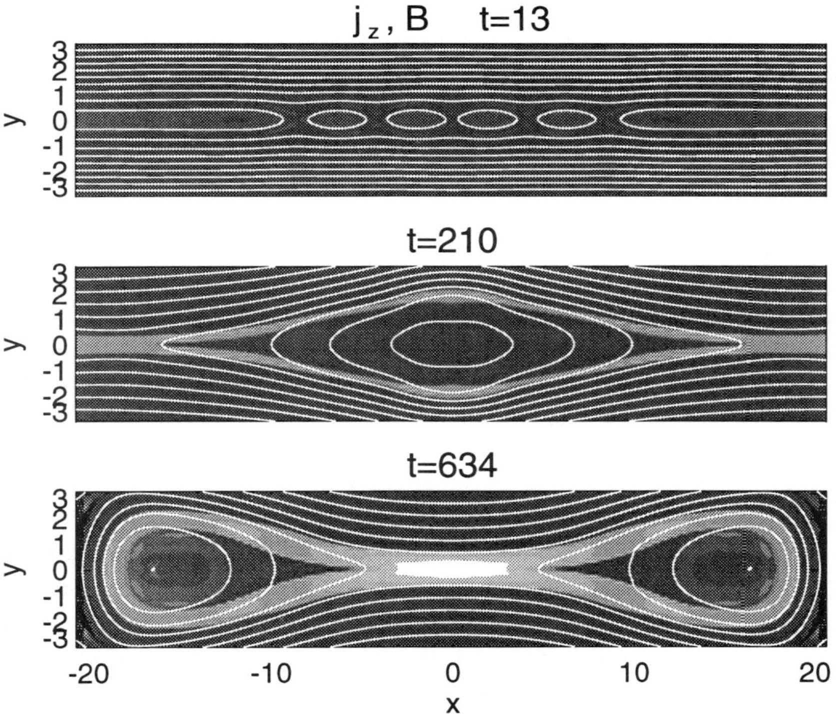

Figure 1. Snapshots of dynamic current sheet evolution: Magnetic field lines and current density (grayscale, ranging from -2.9 [dark] to 3.5 [light]). $R_m = 2000$, $\beta = 0.1$, $j_{cr}/j_0 = 2$; time in units of $\tau_A = l_{CS}/V_A$.

THE ACTIVE ROLE AND MODELS OF RECONNECTING CURRENT SHEETS IN THE SOLAR ATMOSPHERE

B.V. SOMOV
Astronomical Institute of the Moscow State University
Universitetskii Prospect 13, Moscow B–234, Russia 119899

1. Introduction

Before the Yohkoh mission, the process of magnetic reconnection in current sheets as a mechanism, which accumulates the energy of interacting magnetic fluxes and suddenly releases it in the solar atmosphere, was a matter of strong debates and doubts. Here both theoreticians and observers seemed to have strong feelings, often based on belief rather than indisputable theoretical or observational grounds. For example, many years ago, the theoreticians (see review [1]) argued that current sheets can create very fast directed ejections of plasma during the stage of fast reconnection. However, only after the observations of the so–called 'X-ray jets' by SXT on Yohkoh [2], this idea looks well supported by observations and numerical modeling. Moreover, the idea that reconnecting current sheets (RCS) play a key role in conversion of accumulated magnetic energy to kinetic and thermal energies of ejected plasmas [3] becomes obvious.

2. Models of Reconnecting Current Sheets

High–temperature RCS in the solar atmosphere are essentially collisionless. The anomalous coefficients of electric and thermal conductivity have to be used to describe the plasma inside the sheet (see Chapter 3 in [4]). That is why they are called the high–temperature *turbulent* current sheets (HTTCS). However, the idealized model of *neutral* RCS cannot explain the rate of energy release in solar flares. Real RCS are *non–neutral* in two senses. They are *magnetically* non–neutral. It means they have small transverse component of magnetic field, which increases the energy output

Y. Uchida et al. (eds.),
Magnetodynamic Phenomena in the Solar Atmosphere – Prototypes of Stellar Magnetic Activity, 307–308.

of the RCS and its stability, as well as the longitudinal component, which allows to accelerate electrons in RCS more efficiently (for a review see [5]).

As well known, the electric charge separation creates the transverse electric field outside the RCS. In this sence the RCS are also *electrically* non–neutral. The transverse electric field efficiently 'locks' non–thermal ions in the vicinity of the RCS, thus allowing their acceleration by the direct (lomgitudinal) electric field related with the reconnection process [6].

The model, describing in terms of self–similar solution the process of surronding plasma heating by thermal fluxes from a magnetically non–neutral HTTCS [7], is presented. This model demonstrates very high efficiency of such mechanism of coronal heating in active regions and flares; it can be used to compute the differential emission measure, spectral line intensities, and spatial distribution of soft X-ray emission observed by SXT on Yohkoh.

References

1. Somov, B.V. and Syrovatskii, S.I. (1976) Physical processes in the solar atmosphere associated with flares, *Soviet Physics Usp.*, **Vol. 19, no. 10**, pp. 813–835.
2. Shibata, K.,Yokoyama, T., and Shimojo, M. (1996) Coronal X-ray Jets Observed with Yohkoh/SXT *Adv. Space Res.*, **17**, no.4/5 pp. 197-200
3. Syrovatskii S.I. and Somov B.V. (1980) Physical driving forces and models of coronal responces, in M. Dryer and E. Tandberg-Hanssen (eds.), *Solar and Interplanetary Dynamics*, D. Reidel, Dordrecht, pp. 425–441.
4. Somov, B.V. (1992) *Physical Processes in Solar Flares*. Kluwer Academic Publishers, Dordrecht.
5. Somov, B.V. (1994) *Fundamentals of Cosmic Electrodynamics*. Kluwer Academic Publishers, Dordrecht.
6. Litvinenko, Yu.E. and Somov, B.V. (1995) Relativistic acceleration of protons in reconnecting current sheets of solar flares, *Solar Phys.*, **Vol. 158, no. 2**, pp. 317–330.
7. Oreshina, A.V. and Somov B.V. (1995) Reconnecting current sheets in flares as a source of the solar corona heating, *Astronomy Reports*, in press.

THE EVOLUTION OF MAGNETIC ARCADES UNDER FOOTPOINT MOTION

G. S. CHOE AND L. C. LEE
Geophysical Institute, University of Alaska
Fairbanks, AK 99775, USA

A typical sequence of a solar eruptive process consists of distention of the helmet, formation of a CME loop, prominence eruption, opening up of field lines and a flare (e.g., Hundhausen, 1995). The partial opening of magnetic fields is usually regarded as a result of magnetic nonequilibrium, which is an ideal MHD process. As far as force-free magnetic fields are concerned, numerical (Mikić and Linker, 1994; Roumeliotis *et al.*, 1994) and analytical (Lynden-Bell and Boily, 1994) studies have shown that magnetic fields tend to open when the footpoint shear exceeds a certain critical value. However, no field opening has been observed in a numerical simulation with a non-zero β plasma (Mikić and Linker, 1994). In the resistive MHD simulation by Mikić and Linker (1994), magnetic reconnection takes place in a sheared arcade, but the field configuration is far from a partially open field. In this paper, the evolution of solar magnetic arcades under footpoint shearing is studied using MHD simulations and a magnetofrictional method in a 2-D Cartesian geometry. In this geometry, there is no opening up of field lines, which is though not a disadvantage of the study as far as a non-zero β plasma is concerned.

The ideal MHD evolution of magnetic arcades in a low β plasma is found stable for all the shear values we tried. The evolutionary sequence under footpoint shearing is composed of three distinct phases. In the first phase, the overall arcade expansion is very mild, but the growth rates of the toroidal field strength, toroidal field energy and toroidal current density are the highest of all three phases. In the second phase, the toroidal field starts to decrease and there is little variation in the maximum toroidal current density. In the third phase, the most remarkable feature is the formation of a central current layer, where the toroidal current density keeps increasing with increasing shear. The current layer is found to bifurcate a bit above the bottom boundary and the footpoints of the current layer get close to

Y. Uchida et al. (eds.),
Magnetodynamic Phenomena in the Solar Atmosphere – Prototypes of Stellar Magnetic Activity, 309–310.

each other as the shear increases. It can be thus inferred that the open field with a tangential discontinuity is an asymptotic state for an infinite shear in a 2-D Cartesian geometry.

By applying resistivity to pre-sheared arcades, we have found that there exists a critical value of shear, beyond which magnetic reconnection can take place and a magnetic island can be formed. Interestingly, this critical shear approximately coincides with the starting point of the third phase in the ideal MHD evolution. The characteristic of resistive evolutions depends on the spatial pattern of resistivity. With a spatially uniform resistivity, fast reconnection rates cannot be achieved with small shock angles. When resistivity is confined to a very small region, reconnection outflows are so highly collimated to tear the magnetic island into a pair. The island system thus resembles the prominence of a broken shape as observed during a CME. The tip of the island system propagating with a high speed generates either a fast mode structure or a fast shock propagating a distance ahead, which corresponds to a CME loop. After the island system is ejected out of the computational domain, a considerable portion of the flux still remains unreconnected to form a partially open configuration.

When an arcade is sheared under non-zero resistivity from the potential field, a bursty reconnection is repeated at some time intervals creating a new island. This process may explain the phenomena of homologous flares. When a new island is formed, the line-tied field enclosing it as well as the field in it are reconnected very rapidly with the field of the island formed earlier, and two islands quickly merges. The resulting field configuration contains less flux in the island system than before the merging and shows highly stretched line-tied fields. It is thus suggested that more field lines can look open at least in a finite domain by a reconnection process effecting transition to a lower energy state.

From our simulation results, it is inferred that the swelling of the helmet before a CME can be attributed to an ideal MHD process. To the contrary, the main process of a CME, which involves a lot of kinetic energy, is regarded to result from magnetic reconnection.

References

Hundhausen, A. J. (1995) Coronal mass ejections: a summary of SMM observations, in K. Strong, J. Saba and B. Haisch (eds.), *The Many Faces of the Sun*, Springer-Verlag, New York, in press.

Lynden-Bell, D., and C. Boily (1994) Self-similar solutions up to flashpoint in highly wound magnetostatics, *Mon. Not. R. Aston. Soc.*,**267**, 146.

Mikić, Z., and J. A. Linker, (1994) Disruption of coronal magnetic arcades, *Astrophys. J.*,**430**, 898.

Roumeliotis, G., P. A. Sturrock, and S. K. Antiochos, (1994) A numerical study of the sudden eruption of sheared magnetic fields, *Astrophys. J.*,**423**, 847.

HYDRODYNAMIC MODELING OF FLARES WELL-OBSERVED BY YOHKOH/SXT

F. REALE AND G. PERES
Istituto e Osservatorio Astronomico di Palermo, Palazzo dei Normanni, I-90134 Palermo, Italy

AND

H. HUDSON
Institute for Astronomy, University of Hawaii, Honolulu, HI 96822, USA

By comparing X-ray image sequences collected by Yohkoh/SXT and the hydrodynamic calculations made with the Palermo-Harvard code (Peres et al. 1982) we aim at physical insight of soft X-ray solar flares.

SXT flare images, with their simultaneous high spatial and temporal resolution, are very informative, but their physical content is not obvious, since the brightness distribution results from a non linear folding of the emission from a highly structured plasma through the instrument filters. In order to have a deeper interpretation of SXT data, our approach is to fold the results of our hydrodynamic model, providing the evolution of density, temperature and velocity along the loop, with the instrument spectral response in its various filters, to synthesize loop images as they would be detected by SXT, and therefore compare them directly to real images. The power and validity of this approach has already been shown in Peres & Reale (1993) and Reale & Peres (1995).

Here we try to model a specific event occurred in January 13, 1992 at 17:30 UT, selected as appropriate for modeling because it is a limb flare inside a semicircular loop, with very clear and steady geometry, and observed in detail by Yohkoh/SXT. This flare has been already analyzed by Masuda et al. (1994) and Doschek et al. (1995). In our study, we first derive relevant loop characteristics and parameters from the observation: the loop semilength is $\approx 2\ 10^9$ cm and the peak flare temperature ≈ 15 MK. For a first set of simulations, we have assumed a single flaring loop whose

Y. Uchida et al. (eds.),
Magnetodynamic Phenomena in the Solar Atmosphere – Prototypes of Stellar Magnetic Activity, 311–312.

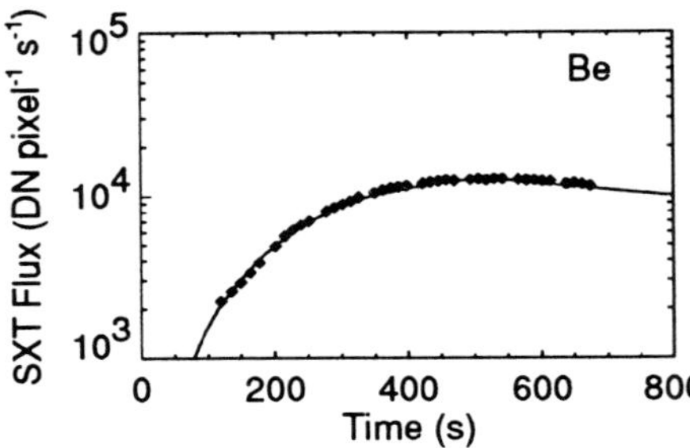

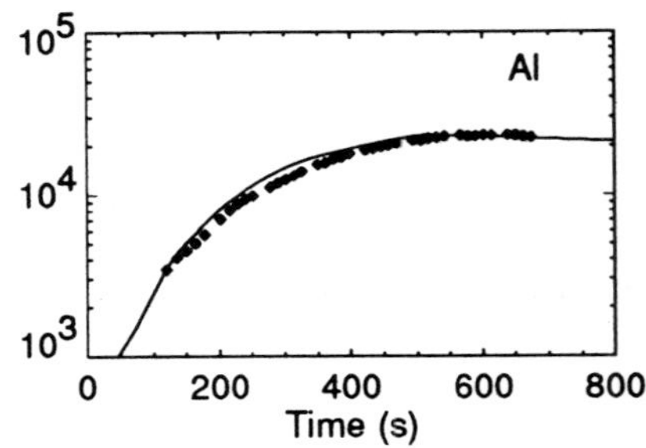

Figure 1. **Model flare light curves in two SXT bands (cf. labels and text) obtained with the Palermo-Harvard model (solid line) and observed data (points). In the model we assume a flare heating constant for 300 s and then decaying with an *e*-folding time $\tau = 800$ s. The heating is deposited at the loop apex, with a characteristic gaussian width of 5 10^8 cm and maximum intensity of 2.2 erg cm^{-3} s^{-1}.**

geometry does not change during the evolution, and a transient heating, located either at the apex or at the footpoints of the loop.

We compare the model results to the data in various SXT bands on two progressively higher levels of detail: i) the whole loop light curve; ii) the evolution of the loop brightness morphology during the flare. One of the first simulations has given a good fit to the light curves observed with the two hardest SXT filters (Al 12 μm and Be 119 μm), under the assumption of transient flare heating deposited at the loop apex (Fig.1). An equivalently good fit could not be obtained with a heating at the loop footpoints, proving that the light curves provide a diagnostics of the heating (Peres et al. 1987).

Fitting the brightness evolution in the images is a more difficult task. The observation shows us that the loop initially has very bright footpoints in both bands, and, after about 100 sec, the apex becomes steadily the brightest loop region. While the model can reproduce the initial brightening of the loop footpoints, a possible signature of the so-called *chromospheric evaporation*, such a long-lasting steady brightening at the loop apex is harder to simulate. Therefore, the interpretation of all the details of the emission during this flare requires further analysis and modeling.

References

Betta, R., Peres, G., Reale, F., and Serio, S. (1995) in preparation.
Doschek, G. A., Strong, K. T., and Tsuneta, S. (1995) *Ap. J.*, **440**, 370.
Masuda, S., Kosugi, T., Hara, H., Tsuneta, S., and Ogawara, Y. (1994), **371**, 495.
Peres, G., and Reale F. (1993), *Astr. Ap. Lett.*, **275**, L13.
Peres, G., Reale, F., Serio, S., and Pallavicini, R. (1987) *Ap. J.*, **312**, 895.
Peres, G., Rosner, R., Serio, S., and Vaiana, G. S. (1982) *Ap. J.*, **252**, 791.
Reale F., Peres, G. (1995) *Astr. Ap.*, **299**, 225.

ON THERMAL HYDRODYNAMICAL MODEL OF SOLAR FLARES

W.Q. GAN
Purple Mountain Observatory
Nanjing 210008, China

The existing hydrodynamical models of solar flares predict a predominantly blueshifted emission of the CaXIX w line at the early phase of the flare. The theoretical blueshift is as large as 3 to 6 mÅ(e.g., Gan et al. 1992). However, observations from previous missions showed that, except for a few flares in which blueshifted Ca XIX emission dominates, the Ca XIX w line exhibits more commonly blue-asymmetric line profile, with the stationary component much stronger than the blueshifted component (e.g., Doschek 1990). Recent observations obtained by the Bragg Crystal Spectrometer (BCS) on Yohkoh, with higher temporal resolution and sensitivity, have shown that most flares exhibit blueshifted Ca XIX and Fe XXV lines before and during the corresponding hard X-ray burst in the rise phase of the flares (e.g., Cheng, Rilee, & Uchida 1994). In addition, there are a few flares that show downward mass motion manifested as redshifted Ca XIX lines (e.g., Mariska, Doschek, & Bentley 1993). To account for the newly observed behavior of Ca XIX and Fe XXV emissions, it is appropriate to make some refinements of the hydrodynamical models, with different parameters, such as loop geometry, heating function, initial atmospheric state, and the effects of two or three dimension transportation. In this paper, we investigate the effects of a high coronal density in the initial model atmosphere for a thermal hydrodynamical model.

The detailed algorithm used here can be found in the papers of Gan & Fang (1990) and Gan et al. (1991). We assume that the flare process is confined in a semicircular loop with a constant cross section, whose two foot points are located in the deep layer of the photosphere. The chromospheric part is properly included besides the coronal part. The thermal energy is assumed to be released around the top of the loop.

We have considered several cases for the initial model atmosphere with a high coronal density ($5 \sim 11 \times 10^{10}$ cm^{-3}). The results of the hydrody-

Y. Uchida et al. (eds.),
Magnetodynamic Phenomena in the Solar Atmosphere – Prototypes of Stellar Magnetic Activity, 313–314.

namic simulation show that, in the initial phase of the flare, there is always a strong downward motion in the corona. This motion with a velocity as large as 700 km s^{-1} can last a few tens of seconds, leading to a pronounced redshift of CaXIX w line. This is not consistent with the observations of majority of flares which show predominantly blue-shifted or blue-asymmetric Ca XIX and Fe XXV lines. However, the results could account for the red-shifted Ca XIX emission observed in a small number of flares, indicating a high initial coronal density in these flares. We conclude that only when initial coronal density is smaller than 10^{10} cm^{-3} can the strong downward motion of the coronal material at the early phase of the flare be suppressed. Taking together the present and previous results of flare simulations, it appears that a model with both thermal and nonthermal energy sources may be more realistic in describing the flare hydrodynamics.

References

Cheng,C.C., Rilee,M.L., & Uchida,Y. (1994), in Proceedings of Kofu Symposium: New Look at the Sun with Emphasis on Advanced Observations of Coronal Dynamics and Flares, eds. S.Enome, & T.Hirayama, 213

Doschek,G.A. (1990), *ApJS*, **73**, 117

Gan,W.Q. and Fang,C. (1990), *ApJ*, **358**, 328

Gan,W.Q., Fang,C., and Zhang,H.Q. (1991), *A&A*, **241**, 618

Gan,W.Q., Rieger,E., Fang,C., & Zhang,H.Q.(1992), *A&A*, **266**, 573

Mariska,J.T., Doschek,G.A., and Bentley,R.O. (1993), *ApJ*, **419**, 418

MHD SIMULATION OF X-RAY JETS BASED ON MAGNETIC RECONNECTION MODEL

YOKOYAMA, T. AND SHIBATA, K.
National Astronomical Observatory
Mitaka 181 Japan

X-ray jets are a new phenomenon discovered by the soft X-ray telescope (SXT) aboard the *Yohkoh* satellite (Shibata *et al.* 1992; Shimojo *et al.* 1995). Shibata *et al.* (1994) proposed a phenomenological model in which X-ray jets are accelerated and heated by *magnetic reconnection* between emerging flux and pre-existing coronal magnetic field. We performed two-dimensional MHD numerical simulations of this model (for detailed results, see Yokoyama and Shibata 1995a,b). When the coronal field is horizontal, a pair of horizontal jets (two-sided-loop type jets; Fig.1a) is ejected. When the coronal filed is vertical or oblique, a vertical or oblique jet (anemone-jet type; Fig.1b) is ejected. One of the interesting results is that reconnection produces not only hot (X-ray) jets but also cool jets which may correspond to Hα surges. This may explain the observed simultaneous co-existence of an Hα surge with an X-ray jet and (small) X-ray flares (Canfield *et al.* 1995; Okubo *et al.* in these proceedings).

References

Canfield, R. C., *et al.* 1995, submitted to *Astrophys. J.*.
Shibata, K., *et al.* 1992, *Pub. Astr. Soc. Japan*, **44**, L173.
Shibata, K., *et al.* 1994, in *Proc. of the International Symp. on the Yohkoh Scientific Results, X-ray Solar Physics from Yohkoh*, eds. Y. Uchida, T. Watanabe, K. Shibata, and H. S. Hudson, (Tokyo: Univ. Academy Press), p29-32.
Shimojo, M., *et al.* 1995, *Pub. Astr. Soc. Japan*, in press.
Yokoyama, T., and Shibata, K. 1995, *Nature*, **375**, 42-44.
Yokoyama, T., and Shibata, K. 1995, submitted to *Pub. Astr. Soc. Japan.*

Y. Uchida et al. (eds.),
Magnetodynamic Phenomena in the Solar Atmosphere – Prototypes of Stellar Magnetic Activity, 315–316.

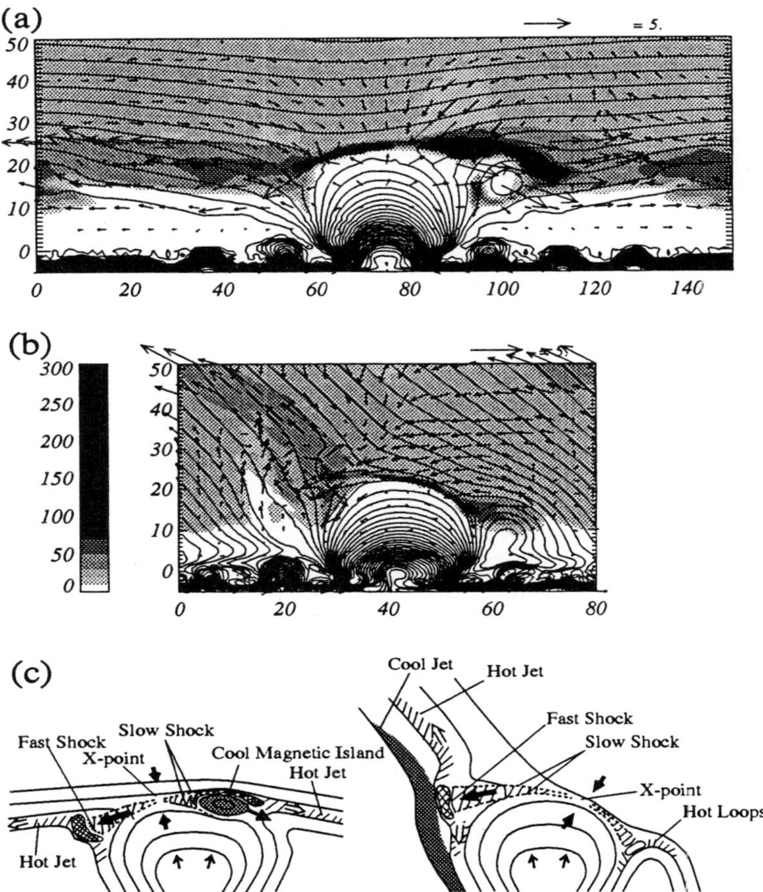

Figure 1. Simulation results of (a) the horizontal coronal field case which models the two-sided-loops(jets) type interaction, and of (b) the oblique coronal field case which models the anemone-jet type interaction. Two-dimensional plot of temperature (gray scale), magnetic field lines (solid lines) and flow field (arrows, whose length indicate velocity) are shown. The scale of length, velocity, temperature are 200km, 10km/sec, 10^4 K, respectively. (c) Schematic picture of results.

ENERGY RELEASE IN THE CURRENT-CARRYING MAGNETIC LOOPS

V.V. ZAITSEV
Institute of Applied Physics, Russian Academy of Science
Ulyanov st., 46, Nizhnii Novgorod, 603600, Russia

In the present talk we consider 1) generation of currents inside a magnetic loop due to photospheric convection; 2) some features of a steady-state Joule dissipation of these currents; 3) possibility of explosive Joule dissipation caused by current disruption in a magnetic loop due to a flute instability. Generation of an electric current in a magnetic loop by a photospheric plasma motion is described by a generalized Ohm's law, and by an equation for a plasma motion. If velocity of convective motion in the photosphere $\mathbf{V} = \{V_r, 0, 0\}$, where $V_r = V_0 r/r_0$ $(r < r_0)$, $V_r = V_0 r_0/r$ $(r > r_0)$ is directed towards the axis of a vertical magnetic loop, we obtain the following set of equations for the components B_z, B_φ of magnetic field in the loop:

$$\frac{\partial B_z}{\partial r} = \frac{4\pi\sigma V_r}{c^2}\frac{B_z}{1+\alpha B_z^2+\alpha B_\varphi^2}, \quad \frac{1}{r}\frac{\partial(rB_\varphi)}{\partial r} = \frac{4\pi\sigma V_r}{c^2}\frac{B_\varphi}{1+\alpha B_z^2+\alpha B_\varphi^2} \tag{1}$$

where $\sigma = \dfrac{e^2 n}{m_e(\nu_{ei}+\nu_{ea})}$, $\alpha = \dfrac{\sigma F}{c^2 n m_i \nu_{ia}}$, ν_{ei}, ν_{ea}, ν_{ia} are respectively effective frequencies of collisions of electrons with ions and neutrals, and ions with neutrals, p_k, is gas kinetic pressure, ρ, plasma density, $F = \rho_a/\rho$, relative density of neutrals, n, number density of electrons.

The characteristic radius of the magnetic tube is determined by the formula (Henoux and Somov, 1991) $r_0 = \dfrac{B_z^2(0)F}{4\pi n m_i \nu_{ia} \mid V_0 \mid (1 - ln2)}$. Total longitudinal electric current I_z generated in the magnetic loop is determined in the following way:

$$I_z = 2\pi \int_0^\infty \frac{1}{r}\frac{\partial(rB_\varphi)}{\partial r} r dr = \frac{cB_z^3(0)F}{8\pi n m_i \nu_{ia} \mid V_0 \mid (1 - ln2)} \tag{2}$$

Y. Uchida et al. (eds.),
Magnetodynamic Phenomena in the Solar Atmosphere – Prototypes of Stellar Magnetic Activity, 317–318.

For upper photosphere ($h = 500km$) this current can be sufficiently large for production of a flare ($I_z \simeq 10^{12}A$ for $| V_0 | \simeq 0.1km \cdot s^{-1}$, $B_z(0) = 2 \cdot 10^3 G$). The arguments given by Melrose and Khan (1989) that all existing photospheric dynamo models are untenable can not be cogent because they did not take into account the Ampere's force in their analysis of the problem.

In the upper photosphere where the inequality $\alpha B^2 \gg 1$ is satisfied the rate of a steady-state Joule dissipation $q_J = \frac{nm_i\nu_{ia}V_r^2(r)}{F} \simeq 1.7 \cdot 10^4\ erg \cdot cm^{-3} \cdot s^{-1}$ may be much more effective than the optical radiation losses (Sen and White, 1972). So this dissipation can cause a temperature increase at a photospheric foot-points of a loop. Moreover, if velocity of convection is maximal near the boundary of a loop, the surface of a loop should be hotter in comparison with its inner parts.

Explosive Joule dissipation is based on a strong increase of a resistance of a loop due to flute instability (Zaitsev and Stepanov, 1992), In this case a rate of the Joule dissipation $q_J = \frac{F^2}{c^2 nm_i\nu_{ia}}(\mathbf{j} \times \mathbf{B})^2$ gives the energy release of the order of one of a solar flare for the most effective case when a current is perpendicular to a magnetic field ($\mathbf{j} \perp \mathbf{B}$). Wheatland and Melrose (1995) noticed that under the coronal conditions a magnetic tube is a force-free one, so a current density $\mathbf{j}$ is practically parallel to a magnetic field and it is difficult to explain the energy release of a flare by explosive Joule dissipation. However, we must take into account that the flux of a prominence or chromospheric plasma penetrating into the current channel due to flute instability changes a structure of a magnetic field in the loop in accordance with the induction equation. And as a result a force-free structure of a loop is disturbed.

References

Henoux, J.C. and Somov, B.V. (1991) The photospheric dynamo, *Astronomy and Astrophys.* **241**, 613.

Melrose, D.B. and Khan, J.I. (1989) Comments on the photospheric dynamo model of Henoux and Somov, *Astronomy and Astrophys.* **219**, 308.

Sen, H.K. and White, M.L. (1972) A physical mechanism for the production of solar flare, *Solar Phys.* **23**, 146.

Wheatland, M.S. and Melrose, D.B. (1995) Energy release in a prominence loaded flaring loop, *Solar Phys.* accepted.

Zaitsev, V.V. and Stepanov, A.V. (1992) Towards the circuit theory of solar flares, *Solar Phys.* **139**, 343.

IV. Magnetic Behavior of the Sun and Stars and Their Activity Cycles

IV.1. Active Zones and Coronal Holes of the Sun and Thier Cycle Variation

MAGNETIC ACTIVITY CYCLE IN THE X-RAY CORONAL STRUCTURES

H. HARA
National Astronomical Observatory, Mitaka, Tokyo 181, Japan

Abstract.
This paper reviews a study of the large-scale coronal activity of the Sun related to the 11-year activity cycle, based upon the *Yohkoh* soft X-ray observations. *Yohkoh* was launched in August 1991, just after the solar maximum of the cycle 22, and continues to observe the Sun in the minimum phase of the magnetic activity cycle. The soft X-ray flux from the whole Sun in the declining phase essentially decreases with the decreasing sizes of active regions, and with the decreasing X-ray surface brightness of background components which consist of quiet regions and coronal holes. The soft X-ray flux does not monotonically decrease, but there are periods of enhancement with about a one-year interval in the whole-Sun X-ray flux. The activity appears as bright clusters in the butterfly diagram of the soft X-ray flux, and corresponds to the emergence of *complexes of activity* in the sunspot zones. At high latitudes we find that the X-ray intensity fluctuates with a time scale of about one year. There is no poleward-drifting pattern extending from the bright low-latitude regions in the time–latitude diagram of X-ray intensity, though such a poleward-drifting pattern is seen in the time–latitude diagram of photospheric magnetic flux.

1. Introduction

We study the evolution of the solar corona in the declining phase of the cycle 22 based upon the *Yohkoh* soft X-ray observations. It is well known that the soft X-ray flux changes with the relative sunspot number and total photospheric magnetic flux in phase. Although the X-ray variability has long been monitored by simple photometric data (*e.g.* GOES soft X-ray

Y. Uchida et al. (eds.),
Magnetodynamic Phenomena in the Solar Atmosphere – Prototypes of Stellar Magnetic Activity, 321–328.

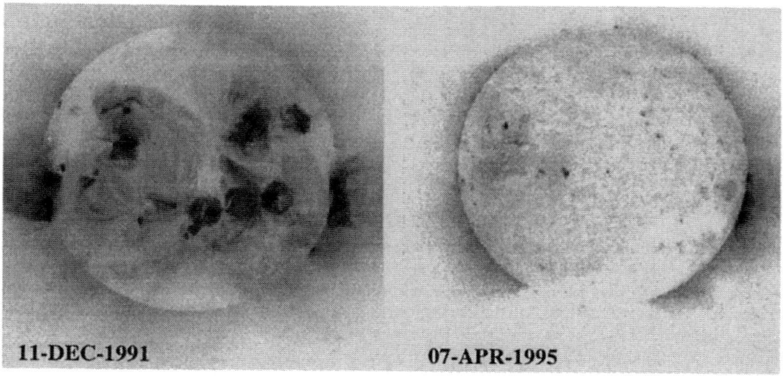

Figure 1. Soft X-ray images near the solar maximum (left) and minimum (right) phase.

data), the soft X-ray telescope (SXT; Tsuneta *et al.* 1991) aboard *Yohkoh* (Ogawara *et al.* 1991) allows us for the first time to study the long-term soft X-ray variability of individual coronal structures in detail. *Yohkoh* has been observing the Sun since September 1991 during a period in which the solar activity was drastically changing (figure 1).

Yohkoh soft X-ray full-disk images taken with thin-Al filter are used in the present study. The pixel size of the images is about $5''$. We use soft X-ray intensity histograms which are made from the X-ray images for the purpose of investigating the long-term variability of the solar corona. Coronal synoptic charts are made from the X-ray images to see a relation between structure in the temporal variation of the soft X-rays and the corresponding spatial structures. By using all X-ray synoptic charts, we show zonal structures at both low and high latitudes. The high-latitude zonal structures are discusssed in association with the *extended solar activity cycle* whose concept is recently proposed by Wilson *et al.* (1988).

2. Long-Term Soft X-ray Variability of the Sun

Figure 2 shows an intensity histogram $[-dN(>I)/dI]$ of a soft X-ray image as a function of X-ray intensity I, where $N(>I)$ is the number of pixels with X-ray intensities larger than I. The unit of soft X-ray intensity I is DN/s/$5''$pixel, where 1 DN is equivalent to an energy deposition of approximately 365 eV onto one pixel of the CCD. The ordinate of the histogram is proportional to the projected area in a given intensity range. At an X-ray intensity of 100 DN/s/$5''$pixel, the histogram slope typically changes. This X-ray intensity threshold discriminates active regions from other darker regions as seen in figure 2. Active regions correspond to a power-law portion

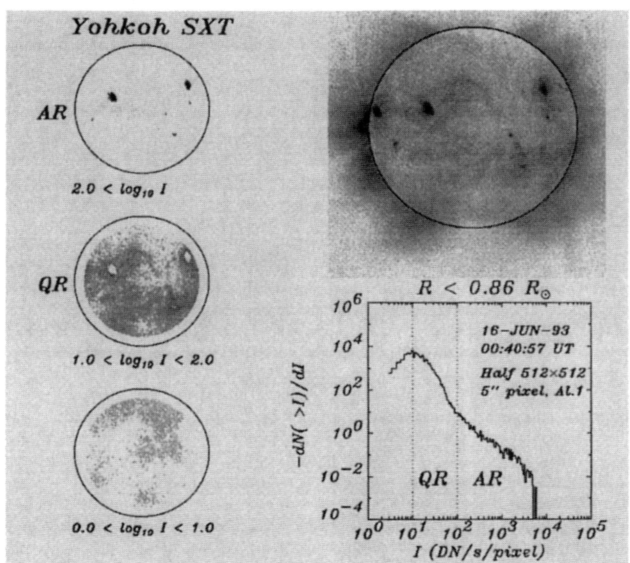

Figure 2. A soft X-ray histogram (bottom–right) on 1993 June 16 made from an SXT composite image (top–right) within 0.86 $R_\odot$. AR and QR indicate active and quiet regions, respectively. Left figures show the locations which correspond to three portions separated by the intensity thresholds given in the bottom-right figure. Circles in the X-ray images show the limb of the Sun.

in the X-ray intensity histogram, and darker regions, which consist of quiet regions and coronal holes, show a different slope of intensity distribution in the histogram (Hara 1994).

Figure 3 shows the temporal variation of the soft X-ray flux. The sinusoidal curve in the bottom–left of the figure reflects the change of the area due to the seasonal variation of the distance between the Sun and Earth. The total X-ray flux (top–left in figure 3) gradually decreases with time, oscillating with time scales of ~27 days and ~1 year. The period of ~27 days is due to the solar rotation and implies the nonuniformity of occurrence of the bright regions in longitude. The modulation of one-year period in the X-ray flux can be also seen in the temporal variations of the total magnetic flux (Harvey 1993). In the previous cycle, 10.7 cm radio flux, Ca K plage index, and Lα flux also show a modulation of one-year period in the declining phase (Pap *et al.* 1991).

When we define the mean surface brightness of the active regions B as

$$B \equiv \int_{I_{th}}^{\infty} I \cdot \left(-\frac{dN(>I)}{dI}\right) \, dI \cdot \left[\int_{I_{th}}^{\infty} \left(-\frac{dN(>I)}{dI}\right) \, dI\right]^{-1}, \qquad (1)$$

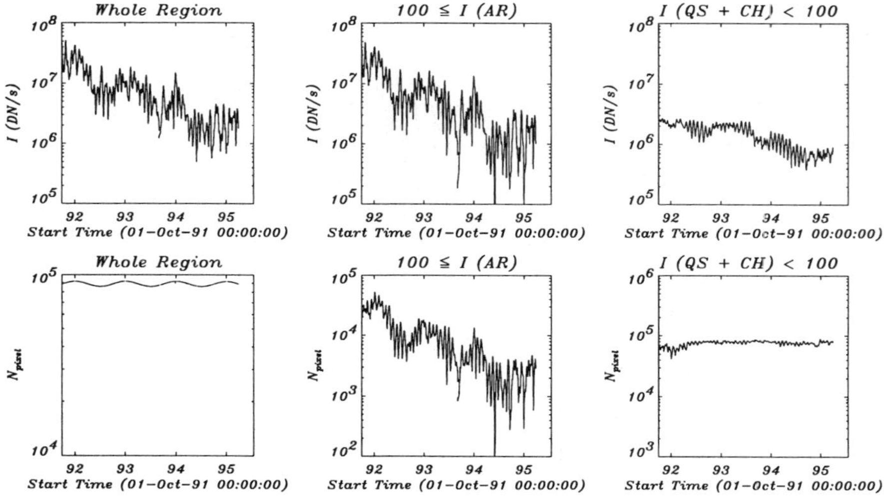

Figure 3. Temporal variation of the soft X-rays in the region within 0.86 $R_\odot$. The 10-days' running mean of the integrated X-ray intensity I (top) and the number of image pixels N_{pixel} (bottom) are shown. Integrations over the whole region (left), active regions (middle), and dark regions that consist of the quiet regions and coronal holes (right) are indicated. The annual variation of distance between the Sun and Earth is corrected in the top figures.

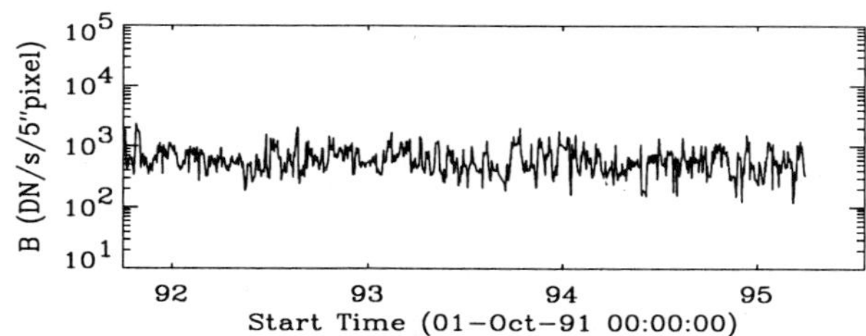

Figure 4. Temporal variation of the surface brightness of active-region corona defined by equation (1).

where $I_{th} = 100$ DN/s/5$''$pixel, it does not systematically decrease for about three years during the declining phase of the cycle 22, though there is a large fluctuation around the mean value in a short time scale less than one month (figure 4). The fluctuation is associated with the different X-ray intensity distributions of active regions appearing at certain periods (Hara 1994).

We call a region where $I < 100$ DN/s/5$''$pixel “background corona” here. So-called quiet regions and coronal holes are constituents of this background corona. Although an X-ray flux monitor such as those aboard the GOES satellites has been mostly monitoring the variation of bright regions, *i.e.* active regions, such an observation cannot discriminate between the background corona and the active regions because of lack of spatial resolution. We show the variation of the X-ray flux from the background corona based upon full-disk observations in the top-right portion of figure 3. Since the area occupied by the background corona is almost constant over the observing period as seen in the bottom-right of figure 3, the background corona has changed its brightness in the declining phase of the current cycle. The X-ray intensity from the background corona around 1995 March is smaller than that around 1991 October by a factor of three. The decrease of X-ray intensity from the background corona is well seen as a shift of the peak to the darker side in the X-ray intensity histogram (Hara 1996).

3. Low-Latitude Activities

The time–latitude diagram of the soft X-ray intensity is presented in figure 5. This figure is made from the soft X-ray synoptic charts by averaging the X-ray intensity in the direction of the Carrington longitude. Therefore structures in the longitudinal direction are smeared out. In low-latitude regions ($\theta < 30°$) it is clear that bright structures intermittently appear, and an active phase roughly comes almost every one-year interval, though there are also cases of shorter intervals. These bright structures correspond to the bumps in the soft X-ray temporal variation as seen in the figure 3. The bright structures do not consist of a single active region with a long lifetime about one year, but they comprise many active regions because they have several local maxima inside of their own structures in both the latitudinal and temporal directions. The bright regions in the low latitudes of the time–latitude diagram consist of clusters of active regions, that is, *complexes of activity* found in the magnetic-field observations (Bumba & Howard 1965, 1969; Gaizauskas *et al.* 1983).

4. High-Latitude Activities

We can also see activity at high latitudes in figure 5. The X-ray intensity in high-latitude regions fluctuates with time. Periods when high-latitude regions become bright are Carrington rotations 1862–1865, 1877–1879, and 1889–1891 in the northern hemisphere, and 1863–1869, 1877–1882, and 1895–1899 in the southern hemisphere. Although these phases roughly coincide with times when the distance between the Sun and Earth is minimum, it is a mere coincidence because the seasonal variation of the distance has

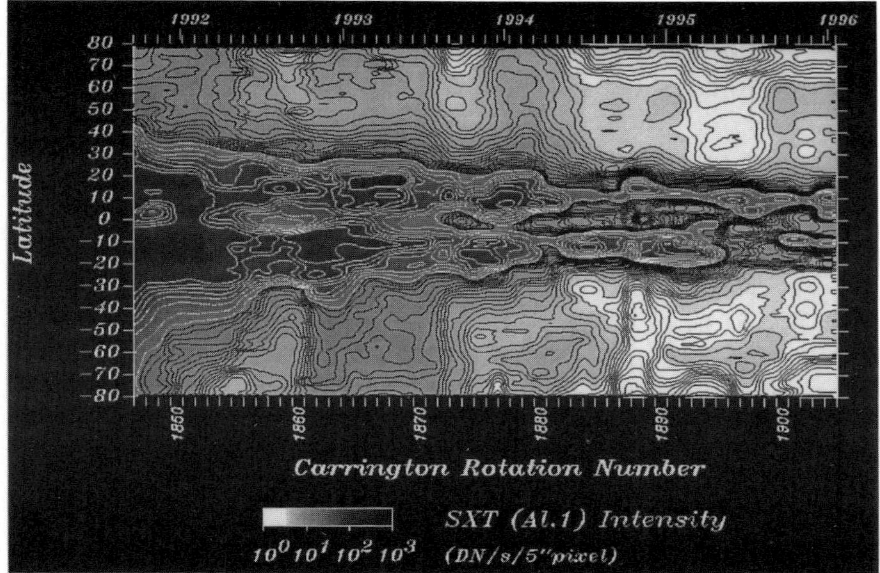

Figure 5. Time–latitude diagram of soft X-ray intensity through the thin Al filter, which covers a period between Oct 1991 and Jan 1996. The annual change of the distance between the Sun and Earth is corrected. Iso-intensity contours are indicated by white (black) lines for low-latitude (high-latitude) regions. The step of contour levels is constant in logarithmic scale, and the intensity at a certain contour level is 1.31 (1.11) times brighter or darker than intensities of the neighboring contours in white (black) lines.

been corrected, and because there is a north–south asymmetry when the high-latitude X-ray intensity locally becomes a maximum. The existence of the high-latitude activity zones in the corona has been already known (Waldmeier 1957; Trellis 1957; Bretz & Billings 1959; Hansen *et al.* 1969; Leroy & Noens 1983; Altrock 1988, 1992). The high-latitude activity zones which were observed with *Yohkoh* SXT are the equatorward-migrating high-latitude activity zones (EHAZ) in Altrock's nomenclature (Altrock 1992). Intensity fluctuations of the EHAZ have been found for the first time in the present study. From the *Yohkoh* soft X-ray observations between 1991 and 1994, Hara (1996) identifies the structures correseponding to the EHAZ with the polar-side legs of large-scale coronal loops, which are quite different from the active regions at low latitudes. This suggests that the coronal extended structures ($\theta \sim 60°$) with respect to the *extended solar activity cycle* (Wilson *et al.* 1988) have no association with low-latitude ($\theta < 30°$) active regions appeared in the following cycle.

5. Origin of High-Latitude Magnetic Field

It is well known that the time–latitude diagram of the net magnetic flux at the photospheric surface shows apparent poleward streams (Snodgrass 1992). This observational fact has been thought to be evidence for the poleward meridional flow which transports magnetic flux from the low latitudes toward the poles. If the magnetic field transported from active-region sites to the high-latitude regions is the origin of the high-latitude magnetic field which forms high-latitude coronal structures, it is reasonable to consider that a poleward drifting pattern would be also seen in the time–latitude diagram made from *Yohkoh* soft X-ray images. There is, however, no drifting pattern from low latitudes toward the pole in the corona. This suggests that the high-latitude magnetic fields do not originate from low-latitude decaying active regions due to the transport by the meridional flow. This may suggest that the high-latitude magnetic flux also come just below the photosphere at high latitudes. This idea with respect to the origin of high-latitude magnetic field is close to Stenflo's (1992).

6. Large-Scale Structures in the Corona

Figure 6 shows the soft X-ray synoptic chart for Carrington rotation number 1894. The X-ray intensity of dark regions is emphasized so that a sinusoidal structure is clearly seen. Two bright regions are seen well in high latitudes; one is located at longitude 330–150° and latitude –60°, and the other at longitude 120–240° and latitude 60–70°. These zones are located at the outer edge of the large-scale structures, and separate polar coronal holes. These correspond to EHAZ. The sinusoidal structure in the synoptic chart (figure 6) reminds us of the coronal streamer zones observed in the white-light K-corona (Hundhausen 1977). Although the pattern of the sinusoidal structure has a similar shape to the calculated large-scale magnetic neutral line at the source surface (*Solar-Geophysical Data*), the pattern of X-ray synoptic chart is much more extended toward the polar regions than that of the magnetic-field data.

References

Altrock, R. C. 1988, *Solar and Stellar Coronal Structure and Dynamics*: in Proc. of Sacramento Peak 9th Summer Workshop, ed. R. C. Altrock, p.414.
Altrock, R. C. 1992, *Bull. Am. Astron. Soc.*, **24**, 746.
Bretz, M. C., and Billings, D. E. 1959, *Astrophys. J.*, **129**, 134.
Bumba, V., and Howard, R. 1965, *Astrophys. J.*, **141**, 1502.
Bumba, V., and Howard, R. 1969, *Solar Phys.*, **7**, 28.
Gaizauskas, V., Harvey, K. L., Harvey, J. W., and Zwaan, C. 1983, *Astrophys. J.*, **265**, 1056.
Hansen, R. T., Garcia, C. J., Hansen, S. F., and Loomis, H. G. 1969, *Solar Phys.*, **7**, 417.

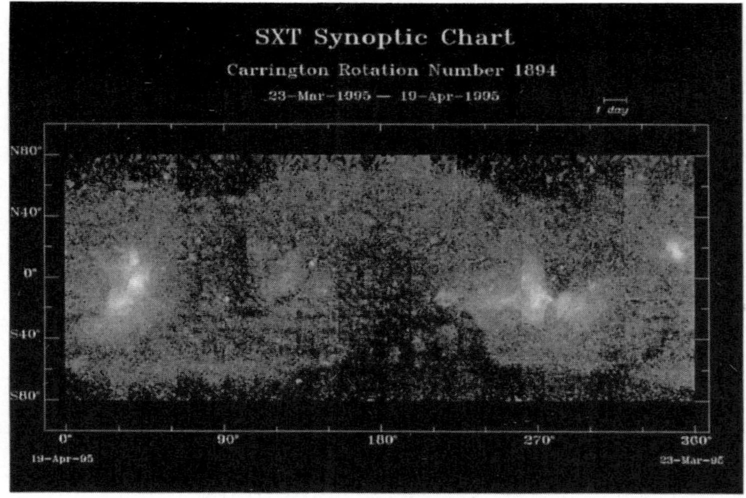

Figure 6. Soft X-ray synoptic chart for Carrington number 1894. The dark-intensity level is exaggerated to show a large-scale sinusoidal structure in the corona. The high-latitude activity zones are seen in both hemispheres as bright zones which elongate in the longitudinal direction.

Hara, H. 1994, in *New Look at the Sun with Emphasis on Advanced Observations of Coronal Dynamics and Flares, NRO report*, **360**, eds. S. Enome and T. Hirayama, p.57.

Hara, H. 1996, Ph.D. thesis, The University of Tokyo.

Harvey, K. L. 1993, Ph.D. thesis, University of Utrecht.

Hundhausen, A. J. 1977, in *Coronal Holes and High Speed Wind Streams*, ed. J. B. Zirker, p.225.

Leroy, J. L., and Noens, J. C. 1983, *Astron. Astrophys.*, **120**, L1.

Ogawara, Y., Takano, T., Kato, T., Kosugi, T., Tsuneta, S., Watanabe, T., Kondo, I., and Uchida, Y. 1991, *Solar Phys.*, **136**, 1.

Pap, J. M., London, J., and Rottmann, G. J. 1991, *Astron. Astrophys.*, **245**, 648.

Snodgrass, H. B. 1992, *The Solar Cycle*: in Proc. of Sacramento Peak 12th Summer Workshop, ed. K. L. Harvey, San Francisco, California, p.205.

Stenflo, J. O. 1992, *The Solar Cycle*: in Proc. of Sacramento Peak 12th Summer Workshop, ed. K. L. Harvey, San Francisco, California, p.83.

Trellis, M. 1957, *Ann. d'Astrophys. Suppl.*, **5**.

Tsuneta, S., Acton, L., Bruner, M., Lemen, J., Brown, W., Caravalho, R., Catura, R., Freeland, S., Jurcevich, B., Morrison, M., Ogawara, Y., Hirayama, T., and Owens, J. 1991, *Solar Phys.*, **136**, 37.

Waldmeier, M. 1957, *Die Sonnenkorona, Vol. 2*, (Verlag Birkhauser, Basel).

Wilson, P. R., Altrock, R. C., Harvey, K. L., Martin, S. F., and Snodgrass, H. B. 1988, *Nature*, **333**, 748.

THE DYNAMICS OF MAGNETIC FLUX TUBES IN THE SOLAR CONVECTION ZONE

A Study of Active Region Formation

G. H. FISHER
Space Sciences Laboratory
University of California
Berkeley, CA 94720-7450

Y. FAN
National Solar Observatory
950 North Cherry Avenue
Tucson, AZ 85719

AND

D. W. LONGCOPE AND M. G. LINTON
Space Sciences Laboratory
University of California
Berkeley, CA 94720-7450

Abstract. We describe some of our recent work on the dynamics of flux tubes in the solar convection zone. We focus on two topics, the first being the orientation ("tilt") of active regions, and how comparisons between observations of tilt and tilts computed from numerical simulations of flux tubes can be used to infer properties of magnetic fields deep in the solar interior. The second topic is an investigation of the kink instability of twisted flux tubes in the solar interior, and its possible relationship with flare-productive δ spot regions.

1. Introduction

Much of our research has focused on studying how magnetic flux emerges from the base of the solar convection zone, where the solar cycle dynamo is believed to operate, to the surface of the sun, where the effects of magnetic activity are observed. Our approach has been to develop, refine and study

Y. Uchida et al. (eds.),
Magnetodynamic Phenomena in the Solar Atmosphere – Prototypes of Stellar Magnetic Activity, 329–336.

theoretical models for the motion of magnetic flux tubes. Such models are strongly motivated by detailed observations of magnetic flux emergence, which indicate that flux emerges in the form of discrete tubes surrounded by field-free gas. These models have been very successful in explaining many observed properties of active regions. Here, we review a few of our recent accomplishments.

2. Description of the Thin Flux Tube Numerical Model

A model for the dynamics of an isolated magnetic flux fiber, termed a *thin flux tube*, was proposed by Spruit (1981). The tube is "thin" in that all of its properties vary along its axis over scales much larger than the diameter of its cross section. This fact can be used to greatly simplify the equations of compressible magnetohydrodynamics (MHD). In the simplified equations, the tube itself is modeled as a three dimensional space-curve (its axis) moving within the interior of a star. Mass density, temperature, pressure, and magnetic field strength all vary along the tube's length and evolve according to dynamical equations. The axis itself moves under the actions of physical forces including: magnetic buoyancy, magnetic tension, aerodynamic drag (coupling the tube to the external plasma), and the Coriolis effect (the equations of motion are posed in a co-rotating reference frame). All of these forces are derived from the application of MHD equations to the case of a thin flux tube. There is also an effect of enhanced inertia due to the motion of the external medium around the moving tube. We have employed these principles to develop a sophisticated numerical model of flux tubes moving within the solar interior. The equations of motion used in our thin flux tube model are derived in Appendix A of Fan, Fisher and McClymont (1994, henceforth FFM), and the numerical techniques for their solution are described in the Appendix of Fan, Fisher and DeLuca (1993, henceforth FFD). Similar numerical models have been developed and are being used by several other research groups (see e.g. Choudhuri 1989; Caligari, Moreno-Insertis and Schüssler 1995).

3. Results of Model Calculations

3.1. JOY'S LAW

Perhaps the best known success of all of the numerical thin flux tube models has been their quantitative reproduction of "Joy's Law". Joy's Law is an empirical relation between the latitude of a bipolar magnetic region or a sunspot group and the observed angle (relative to E-W) made by its poles (Zirin 1988, Hale *et al.* 1919). This angle, termed its *tilt angle*, α, is defined to be positive when the leading pole is closer to the solar equator than the following pole. Binned active region data, from spot groups (Hale *et al.* 1919, Howard 1991b, Fisher, Fan and Howard 1995) and magnetogram measurements (Wang and Sheeley 1989), show a systematic increase of the tilt angles of active regions with the absolute value of latitude. For latitudes near 30° mean tilt angles are in the 8°–10° range. This effect may be extremely important; in several global models of the solar magnetic field (Wang and Sheeley 1991, Leighton 1964, 1969) the eleven year cycle is a direct consequence of this preferential orientation in newly emerged flux.

Tilt angles arise in a simple and compelling manner in thin flux tube models. As the apex of the loop rises, the plasma inside it expands. The Coriolis force on a parcel of expanding fluid will rotate it in the sense opposite to rotation (i.e. in an inertial frame, its rotation rate decreases). Furthermore, the amplitude of this rotation varies with latitude, θ, in the same manner as the Coriolis effect itself: $\sin(\theta)$. Thus the sense of the tilt angles and their dependence on latitude can be qualitatively explained in terms of a thin flux tube.

Numerical thin flux tube calculations by D'Silva and Choudhuri (1993), Schüssler *et al.* (1994), and ourselves (FFM) have all reproduced Joy's Law to a convincing level of accuracy. Consistency with observation can be achieved with plausible choices for physical parameters such as the magnetic flux Φ in active region tubes, field strength B_0 at the base of the convection zone, and initial latitude θ_0 of the toroidal tube. In fact the latter two parameters are unobservable attributes of the solar dynamo. The application of thin flux tube models to match observations provides one of the few constraints available for these quantities.

3.2. ASYMMETRIC MOTIONS

The asymmetry in the shape of the emerging flux loop, clearly visible in Fig. 1, is also a consequence of the Coriolis effect, as described by Moreno-Insertis (1994). This shape could explain an observed asymmetry in the proper motion of spots during flux emergence (Moreno-Insertis 1994, Caligari *et al.* 1995): The shallower slope of the leading leg makes the leading spot appear to move faster than the following spot, in general agreement with observation. The Coriolis effect also generally results in a greater field strength in the leading leg of the flux loop, which we argue (FFD) is the origin for the more compact morphology of the leading side of active regions when compared to the following side.

3.3. A SCALING LAW FOR ACTIVE REGION TILTS

A more detailed analysis of the tilt angle variation was given in FFM. By considering the scaling of forces with Φ, B_0, and θ, we found that the tilt

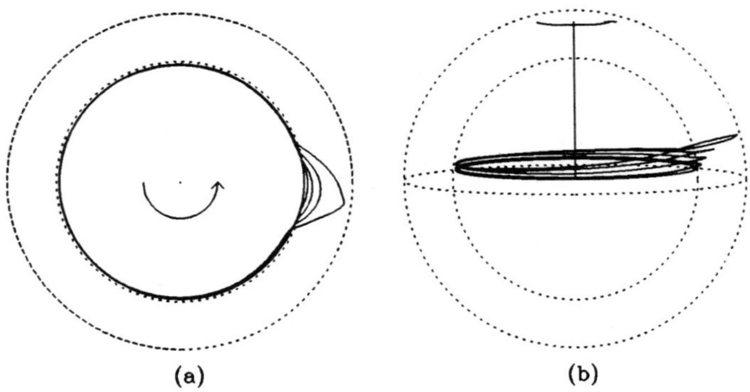

Figure 1. The shape of an evolving flux ring with flux $\Phi = 10^{22}$ Mx and the initial field strength $B_0 = 3 \times 10^4$ G. The flux ring is viewed from the north pole and from 5 degrees above the equator in panels (a) and (b). 5 instants of the flux ring's evolution are shown as the solid curves; the inner and outer dotted circles mark the base of the convection zone and the photosphere.

angles should follow

$$\alpha \propto \sin(\theta)\ \Phi^{1/4}\ B_0^{-5/4} \tag{1}$$

This simple relation agrees reasonably well with the results of our full numerical simulations provided $B_0 \gtrsim 2 \times 10^4$G (Figure 11, FFM). Thus the theoretical calculations not only reproduce the observed Joy's Law behavior, but also make testable predictions of how active-region tilts should vary with B_0 and Φ.

3.4. TESTS OF THE PREDICTED TILT-ANGLE SCALING LAW WITH SPOT-GROUP DATA

The tilt angle scaling law, Eq. (1), contains a dependence on latitude (Joy's Law) as well as a previously untested dependence on net flux, Φ. Motivated by this theoretical prediction, we sought to ascertain whether such a dependence exists in observed tilt angles. To do this we studied tilt angles in a dataset of 24,701 sunspot groups observed in white light between 1917 and 1985 at the Mt. Wilson Observatory (Howard, Gilman and Gilman 1984). Tilt angles were determined from the spot groups with an algorithm described in §2 of Fisher, Fan and Howard (1995, henceforth FFH). The size of this dataset permits statistically meaningful tests to be made on many subsets of the data. White light observations contain no direct magnetic information, but previously established proxy relationships allow this to be inferred. Specifically, Howard (1992) has established a proportionality between the net (unsigned) flux in an active region (from magnetograms) and the distance separating the centroids of its leading and following polarities

This enabled us to use the polarity separation d of a spot group as a proxy for the magnetic flux Φ.

We found in FFH that the *mean* tilt behavior was consistent with the θ and Φ variation predicted in equation (1). (For reasons discussed in FFH, the quantity B_0 is not directly measurable, but its range of values appears to be quite restricted — see FFH for a more complete discussion of this issue). A more intriguing result of FFH was the large amplitude of the *fluctuations* $\Delta\alpha$ of individual tilts away from the mean behavior. These fluctuations are very significant, are much larger than estimated measurement errors, are not a function of latitude (unlike the "Joy's Law" behavior of the mean tilts), but are strongly decreasing functions of d. In FFH we argue that the fluctuations are solar in origin, and most likely from perturbations of the rising tubes by convective motions.

3.5. CONVECTION ZONE TURBULENCE IN MODELS OF RISING FLUX TUBES

In Longcope and Fisher (1995, henceforth LF) we show that turbulent motions in the convection zone do lead to tilt angle fluctuations $\Delta\alpha$ of the magnitude observed in FFH. We further find that a simple model explains the observed relation between tilt angle fluctuation and footpoint separation. To do this, we have developed a new technique for adding statistical perturbations to the thin flux tube model.

The first part of this technique defines an algorithm for generating realizations of the turbulent velocity field in the convection zone. A given turbulent realization is a function of space and time; the behavior of an ensemble of realizations gives statistical moments consistent with those assumed under mixing length theory. Although mixing-length convection models are highly idealized, they probably mimic the qualitative depth dependence of the velocity field that a rising flux tube encounters. For each individual realization of the velocity field, one can then compute the resultant perturbations to the motion of a rising flux tube.

To calculate the effect of the turbulent velocity field, we consider the limit of small deflections to a flux tube which has risen through a stationary atmosphere (i.e. we introduce the forces driven by convection zone turbulence as a first order perturbation to the general flux tube equation of motion). To further simplify our calculation, we use, for the unperturbed rise, a straight horizontal tube. After Fourier-transforming the perturbed equation of motion, we derive forced-damped harmonic oscillator equations for the evolution of each Fourier mode in the tube. This results in a spectrum of deflections which is linearly related to the spectrum of the turbulence. Using 250 numerically generated realizations of the turbulent velocity field, it is possible to calculate the fluctuations, $\Delta\alpha$, in the tilt angle of the emergent tube. Results of these Monte-Carlo calculations compare favorably to the observed values found in FFH (see Fig. 2).

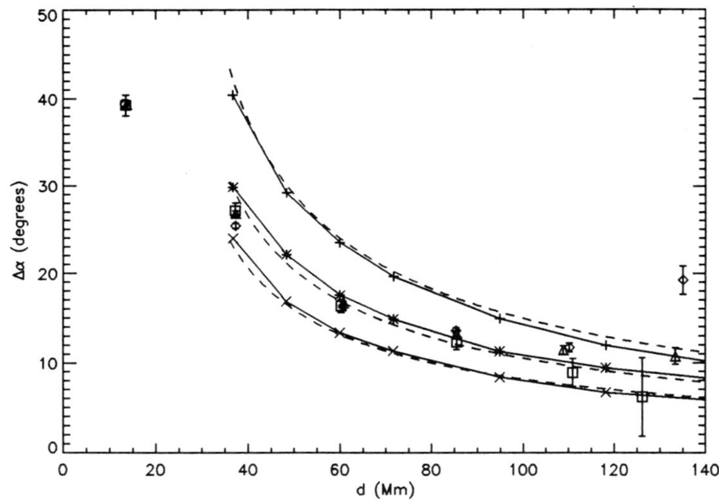

Figure 2. The RMS tilt angle, $\Delta\alpha$ as a function of d (Mm). The solid lines pass through the results of Monte-Carlo calculations with initial magnetic field strengths B_0, of 20 kG (+), 30 kG ($*$), 40 kG ($\times$). Observed values of $\Delta\alpha$ from FFH are shown for high (□), mid (△) and low (◇) latitudes.

4. The Helical Kink Instability of Twisted Flux Tubes in the Interior

There is a small but important class of active regions which appear to be twisted and possibly kinked or knotted when they emerge. The structure and dynamics of the magnetic field in these active regions is especially important because they are strongly linked to the occurrence of large solar flares. The link between solar flares and emerging flux loops which are kinked can be inferred from the fact that the "island δ" spot configuration is typically the site of the largest solar flares (see *e.g.* Leka *et al* 1995; Kurokawa 1991; Tanaka 1991; Zirin 1988, p. 337). Here, two spot umbræ of opposite magnetic polarity emerge within one spot penumbra; furthermore, the orientation of the two polarities is frequently reversed from the usual Hale configuration. One interpretation of the δ spot configuration is that a rising loop of a twisted magnetic flux tube has kinked into a braided structure, similar to that which results when twist is applied to the loop of a rubber band. Such an interpretation would account qualitatively for the reversed polarity configuration, the close proximity of the opposite polarity spots, and the occurrence of flares as the oppositely directed magnetic fields in the intertwined loop legs undergo reconnection.

Motivated by a desire to understand the dynamics of twisted active region flux tubes below the solar photosphere, we have investigated the kink stability of these tubes (Linton *et al* 1995). Following previous studies of the kink mode, we apply linearized equations of MHD to a cylindrical magnetic equilibrium (screw pinch), but with significant differences from the earlier work. In our case, the magnetic field vanishes outside some radius, $r = R$, where it is confined by the higher pressure of the unmagnetized plasma. This outside boundary of the tube is free to move, displacing the unmagnetized plasma as it does so. We consider equilibria in which all field lines have the same helical pitch: $B_\theta / r B_z = q =$ const. Our main results are as follows:

(1) These equilibria are stable to kinking, provided that the field line pitch does not exceed a threshold; $q \leq q_{\rm cr}$ for stability. The threshold is $q_{\rm cr} = \sqrt{\alpha}$, where α is the r^2 coefficient in the Taylor series expansion of the equilibrium axial magnetic field (B_z) about the tube axis ($r = 0$): $B_z(r) = B_0(1 - \alpha r^2 + ...)$. When this criterion is violated, there are unstable eigenmodes, $\xi \propto e^{i(\theta + kz)}$. The most unstable of these have a helical pitch k which is near (but not equal to) the field line pitch.

(2) For weakly twisted tubes ($qR \ll 1$) we are able to derive growth rates and unstable eigenfunctions analytically. For strongly twisted tubes ($qR \gtrsim 1$) we find growth rates and unstable eigenfunctions numerically.

(3) The maximum growth rate and range of unstable wavenumbers for a strongly twisted tube can be predicted qualitatively by using the analytical results from the weakly twisted case. The maximum growth rate in that case is given by $\omega_{max} = v_A R(q^2 - q_{\rm cr}^2)/3.83$, where v_A is the axial Alfvén speed. The range of unstable wavenumbers is given by $(-q - \Delta k/2) < k < (-q + \Delta k/2)$, where $\Delta k = 4qR\sqrt{q^2 - q_{\rm cr}^2}/3.83$.

(4) The kink instability we find corresponds primarily to internal motion. Helical translations of the entire tube are found to be stable.

(5) We argue that an emerging, twisted magnetic flux loop will tend to have a uniform q along its length. The increase in the tube radius R as it rises results in a decreasing value of $q_{\rm cr}$. This means that the apex of the flux loop will become kink unstable before the rest of the tube.

(6) Our results lead us to believe that most twisted flux tubes rising through the convection zone will be stable to kinking. Those few tubes which are kink unstable, and which presumably become knotted or kinked active regions upon emergence, only become kink unstable some time after they have begun rising through the convection zone.

This work was supported by NASA, NSF, ONR, the California Space Institute, and the Miller Foundation for Basic Research in Science.

References

Caligari, P., Moreno-Insertis, F. & Schüssler, M. 1995 "Emerging flux tubes in the solar convection zone. I. Asymmetry, tilt and emergence latitude", *Ap. J* **441**, 886.

Choudhuri, A.R. 1989, "The evolution of loop structures in the flux rings within the solar convection zone", *Sol. Phys.* **123**, 217.

D'Silva, S. & Choudhuri, A.R. 1993, "A theoretical model for tilts of bipolar magnetic regions", *A & A* **272**, 621.

Fan, Y., Fisher, G.H. & DeLuca, E.E. 1993 (FFD), "The origin of morphological asymmetries in bipolar active regions", *Ap. J* **405**, 390.

Fan, Y., Fisher, G.H. & McClymont, A.N. 1994 (FFM), "Dynamics of emerging active region flux loops", *Ap. J* **436**, 907.

Fisher, G.H., Fan, Y. & Howard, R.F. 1995 (FFH), "Comparisons between theory and observations of active region tilts", *Ap. J* **438**, 463.

Hale, G.E., Ellerman, S., Nicholson, S.B. & Joy, A.H. 1919, "The magnetic polarity of sunspots", *Ap. J* **49**, 153.

Howard, R., Gilman, P.A. & Gilman, P.I. 1984, "Rotation of the Sun measured from Mount Wilson white-light images", *Ap. J* **283**, 373.

Howard, R.F. 1991b, "Axial tilt angles of sunspot groups", *Sol. Phys.* **136**, 251.

Howard, R.F. 1992 "The rotation of active regions with differing magnetic polarity separation", *Sol. Phys.* **142**, 233.

Kurokawa, H. 1991, "Optical Observations of Flare Productive Flux Emergence", *Lecture Notes in Physics*, **387**, 39 (Springer: Berlin).

Leighton, R.B. 1964, "Transport of magnetic fields on the Sun", *Ap. J* **140**, 1547.

Leighton, R.B. 1969, "A magneto-kinematic model of the solar cycle", *Ap. J* **156**, 1.

Leka, K.D., Canfield, R.C., McClymont, A.N., & van Driel-Gesztelyi, L. 1995, "Evidence for Current Carrying Emerging Flux", *Ap. J.*, submitted.

Linton, M.G., Longcope, D.W., & Fisher, G.H. 1995, "The Helical Kink Instability of Isolated, Twisted Magnetic Flux Tubes", *Ap. J* , submitted.

Longcope, D.W. & Fisher, G.H. 1995 (LF), "The effect of convection zone turbulence on a rising flux tube", *Ap. J* accepted for publication.

Moreno-Insertis, F., Schüssler, M. & Ferriz-Mas, A. 1992, "Storage of magnetic flux tubes in a convective overshoot region", *A & A* **264**, 686.

Moreno-Insertis, F. 1994, "The magnetic field in the convection zone as a link between the active regions on the surface and the field in the solar interior", p. 117 in *Solar Magnetic Fields*, eds. Schüssler, M. & Schmidt, W., Cambridge Univ. Press.

Schüssler, M., Calgari, P., Ferriz-Maz, A. & Moeron-Insertis, F. 1994, "Instability and eruption of magnetic flux tubes in the solar convection zone", *A & A* **281**, L69.

Spruit, H.C. 1981, "Motion of magnetic flux tubes in the solar convection zone and chromosphere", *A & A* **98**, 155.

Tanaka, K. 1991, "Studies on a very flare-active δ group: Peculiar δ spot evolution and inferred subsurface magnetic rope structure", *Sol. Phys.* **136**, 133.

Wang, Y.-M. & Sheeley, N.R., Jr. 1989, "Average properties of bipolar magnetic regions during sunspot cycle 21", *Sol. Phys.* **124**, 81.

Wang, Y.-M. & Sheeley, N.R., Jr. 1991, "Magnetic flux transport and the Sun's dipole moment: new twists to the Babcock-Leighton model", *Ap. J* **375**, 761.

Zirin, H. 1988, *Astrophysics of the Sun*, Cambridge Univ. Press.

HOT SPOTS FOR SOLAR ACTIVITY

T. BAI
HEPL, Stanford University
Stanford, CA, 94305, USA

Abstract. Studies on hot spots and active longitudes are briefly reviewed in this paper. Although the concept of active longitudes are old, interesting discoveries were made by recent studies, which applied new analysis methods to data covering several solar cycles. The main characteristics of hot spots are as follows: (1) Hot spots, initially recognized as areas where major flares erupt preferentially, are also preferred areas for emergence of big sunspot groups. (2) Double hot spots appear in pairs that rotate at the same rates, separated by about 180° in longitude. Single hot spots have no such companions. (3) The northern and southern hemispheres behave differently in organizing solar activity in longitude. Because of this north-south asymmetry, the areas for preferred flare eruptions are called hot spots (Bai 1987). (4) Lifetimes of hot spots range from one to several solar cycles. (5) A hot spot is not always active throughout its lifetime but goes through dormant periods. (6) Hot spots with different rotation periods can co-exist in the same hemisphere during the same solar cycle.

1. Introduction

The concept of "active longitudes" has been around for decades. This term refers to longitude intervals where flare activity is for durations of years much higher than elsewhere. Implicit in this term is that the same longitude intervals are active in the both northern and southern hemispheres. Most early studies on this subject were not quantitative (e.g., Warwick 1965; Svestka and Simon 1969; Dodson and Hedeman 1975a), although they stimulated interest in the subject. Exceptionally, Wilcox and Schatten (1967), Haurwitz (1968), and Fung *et al.* (1971) performed quantitative studies early on, by analyzing longitude distributions of flares taking the rotation period as a free parameter. These papers analyzed flares observed

Y. Uchida et al. (eds.),
Magnetodynamic Phenomena in the Solar Atmosphere – Prototypes of Stellar Magnetic Activity, 337–344.

mostly during cycle 19, without separating southern-hemisphere flares from northern-hemisphere flares.

Recent interest on active longitudes was aroused by the discovery of the 154-day periodicity (Rieger *et al.* 1984). In order to test the idea that two active longitudes or hot spots with slightly different rotation periods interact once every 154 days to enhance flare activity periodically (Bai and Sturrock 1987), it was necessary to investigate first their existence. Bai (1987) performed a quantitative analysis of longitude distributions of major flares observed from 1980 through 1985 by the Hard X-ray Burst Spectrometer (HXRBS) aboard the *SMM (Solar Maximum Mission)*. By separating northern and southern hemisphere flares in his analysis, he found that flares are organized differently in longitudes in the two hemispheres. He found a prominent hot spot with a synodic rotation period of 26.75 days in the northern hemisphere. (In this paper, rotation periods are synodic unless specified otherwise.) Bai (1988) extended his analysis back to cycle 19 by including major flares selected from compilations made by Dodson and Hedeman (1971, 1975b, 1981). He found that the prominent hot spot discovered in his previous paper was one of the double hot spots that had persisted for two solar cycles (cycles 20 and 21) in the same longitude intervals. The refined rotation period of these hot spots was 26.72 days. For the first time, it was established by quantitative analysis that lifetimes of hot spots could be longer than one solar cycle.

The north-south asymmetry in longitude organization, first noted by Bai (1987), was confirmed by his later papers (Bai 1988, 1990). It was also seen by an analysis of coronal magnetic field rotation (Hoeksema and Scherrer 1987) and by analysis of surface magnetic fields (Antonucci *et al.* 1990). Because of the north-south asymmetry, Bai (1987) coined the term "hot spot" to eliminate the north-south symmetry implied by the term "active longitude."

It is a challenge to understand how hot spots can maintain their identity for many years in the presence of differential rotation of the Sun. Two points at 10° and 20° latitude initially at the same longitude would drift apart more than 90° in longitude in only a year if they rotated at average rotation rates corresponding to their latitudes. Therefore, hot spots must be due to coherent patterns (or structures) that withstand differential rotation. Insights on hot spots may help us understand the underlying causes of solar activity, solar internal structure, and solar convection and improve space-weather forecasts. In this paper, we summarize recent studies by Bai (1987, 1988, 1990), Bai *et al.* (1996), and Akioka *et al.* (1992).

The most comprehensive study is by Bai *et al.* (1996), who analyzed major flares observed from 1955 to 1995 (cycles 19 through 22), big sunspot groups observed during the same period, and large-scale surface magnetic field strengths observed by the Wilcox Solar Observatory (WSO) at Stanford University. For cycles 19 and 20, they use major flares selected on the basis of comprehensive flare index (CFI; cf. Dodson and Hedeman (1979 for its definition) from the compilations made by Dodson and Hedeman (1971, 1975b, 1981). For cycles 21 and 22, they select flares with GOES soft X-ray class $\geq$M3.0.

2. Spatial Organization of Superactive Regions

By studying flare distributions for cycles 19 through 22, Bai *et al.* (1996) found more hot spots in addition to the ones found earlier. Some of them were active only during one solar cycle, but several hot spots were active in the same locations for longer than one solar cycles. Double hot spots with a synodic rotation period of 26.73 days were active in the northern hemisphere during cycles 20 through 22. Double hot spots with a synodic rotation period of 27.41 days were active in the northern hemisphere through all four solar cycles. A single hot spot with a synodic rotation period 28.00 days were active in the southern hemisphere through all four solar cycles, although its statistical significance is not very high.

A superactive region is defined as an active region that produced 4 or more major flares. Major flares are defined as flares with CFI>5 for cycle 19, as flares with CFI>6 for cycle 20, and as flares with GOES soft X-ray class $\geq$M3.0 for cycles 21 and 22. Bai (1988, 1990) and Bai *et al.* (1996) adopted a lower threshold for cycle 19 than cycle 20 because data coverage for cycle 19 was less complete. The reason for adopting different criteria is that there is no uniform observations of solar flares through the last four solar cycles. Major flares selected on the basis of CFIs are available for the rising part of cycle 21 (from 1976 through 1979), and major flare selected on the basis of *SMM* HXRBS peak fluxes are available for the period from 1980 through 1989. GOES soft X-ray detector observed solar flares for the entire solar cycle 21. Hence, for some periods lists of major flares selected by more than one criterion are available. The longitude distributions of major flares for these periods are found to be similar regardless of the selection criteria (Bai 1990).

Bai (1988) found that superactive regions preferentially emerge in hot spots. Therefore, central meridian passage times of superactive regions must have preferred values instead of being random. Figure 1 is the Rayleigh power spectrum of the central meridian passage times of northern-hemisphere superactive regions for cycles 20 through 22. We find a prominent peak at 13.37 days. The probability of this peak is due to random chance is $165\exp(-14.76) = 6.4 \times 10^{-5}$ (where 165 is the number of independent frequencies in the period interval from 12.5 to 15 days and 14.76 is the

is a site where flares preferentially occur but not a site where big sunspot groups preferentially emerge. In this case, sunspot group emergence is more or less random, but there are localized agents which enhance flare activity in hot spots. (b) A hot spot is a location where big sunspot groups emerge preferentially, and the high flare activity is consequential. Our analysis shows that hot spots are also sites for preferential emergence of big sunspot groups. (See also Akioka *et al.* 1992.) Therefore, the second possibility is valid, except for a few hot spots for which an additional mechanism for enhancing flare activity might be necessary.

TABLE 1. Hot-Spot Systems in the Northern Hemisphere[a]

Number of Hot Spots	Period (days)	Solar Cycle	Norm. Power[b]	Chance Probability	Amp. of Mod.[c]	Power[d] (Spots)	Amp.[e] (Spots)
1	28.88	19	9.9	1.4×10^{-3}	0.71	5.2	0.37
1	27.00	21	7.6	1.8×10^{-2}	0.79	7.9	0.68
2	26.73	20 & 21	13.0	3.0×10^{-4}	0.69	11.2	0.55
2	27.40	19–21	13.7	2.1×10^{-4}	0.55	12.2	0.40

Notes:
a. This table from Bai *et al.* (1996).
b. See Bai *et al.* (1996) for an explanation on how to get normalized powers.
c. The amplitude of modulation, A, is obtained by fitting the distribution of major flares to a sinusoidal function $f(l) = A\{1 + A\cos(nl + l_0)\}$, where l is the longitude, $n = 1$ for single hot spot systems and $n = 2$ for double hot spot systems, and l_0 is a phase angle. See Bai (1992) for the relationship between A and the Rayleigh power.
d. Normalized power for sunspot area distributions.
e. Amplitude of modulation for sunspot area distributions.

(2) Double hot spots appear in pairs that rotate at the same rates, separated by about 180° in longitude. Single hot spots have no such companion.

Tables 1 and 2 summarize properties of hot spots found from analysis of major flares and large suspot groups. The last two columns are the results of big sunspot group anaysis.

(3) The northern and southern hemispheres behave differently in organizing solar activity in longitude.

This is evident from the fact that the power spectra for northern hemisphere flares and southern hemisphere flares have different peaks. Because of this north-south asymmetry, the areas of enhanced flare activity is called hot spots (Bai 1987).

(4) The lifetimes of hot spots range from one to several solar cycles.

TABLE 2. Hot-Spot Systems in the Southern Hemisphere[a]

Number of Hot Spots	Period (days)	Solar Cycle	Norm. Power[b]	Chance Probability	Amp. of Mod.[c]	Power[d] (Spots)	Amp.[e] (Spots)
1	24.98	20	9.0	1.1×10^{-2}	1.13	4.7	0.55
1	25.50	22	5.4	9.0×10^{-2}	0.64	5.5	0.63
1	28.00	19–22	8.6	2.1×10^{-2}	0.43	6.3	0.30
2	25.08	21	6.0	5.0×10^{-2}	0.71	2.2	0.35

Notes: See notes for Table 1.

For example, a double hot-spot system with a rotation period of 26.73 days persisted in the same locations in the northern hemisphere from during cycles 20 through 22. Several other examples are found in Tables 1 and 2.

(5) A hot spot is not always active throughout its lifetime but goes through dormant periods.

Additionally, in double hot-spot systems, the activity levels of two hot spots seem to be anti-correlated. When one hot spot is active, its companion is inactive, and vice versa. This suggests that two companions in a double hot-spot systems are globally linked. This needs further investigation.

(6) Hot spots with different rotation periods co-exist in the same hemisphere during the same solar cycle.

For example, double hot spots with a rotation period of 26.73 days and a single hot spot with a rotation period of 27.00 days co-existed in the northern hemisphere during cycle 21. There is evidence that a single hot spot with a rotation period of 28.00 days persisting through four solar cycles (19 through 22) in the southern hemisphere. During cycle 20, a prominent hot spot with a rotation period of 24.98 days operated in the southern hemisphere.

We also notice that the rotation periods of hot spots range from 25 days to more than 29 days. This range exceeds the range of rotation periods of the activity belts at the surface and in the convection zone. It seems reasonable to interpret that they are rotation periods of wave patterns in the convection zone.

In this regard, it is interesting to note that "at the edge of chaos" self-organizing patterns emerge (Langton 1990). If the solar convection is at the edge of chaos, it may produce and maintain regular patterns in the convective flow. If hot spots are due to large-scale up-drafting plumes, the surface temperatures in hot spots will be higher than elsewhere. Because one can determine the rotation periods of persistent hot spots, folding future temperature measurements of the surface in a coordinate system rotating with the period of a hot-spot system could reveal minute temperature differences. It is also interesting to see whether the conjectured convective plumes can be detected by helioseismology.

This research was funded by NSF grants ATM 9312424 and ATM 9400298, NASA grant NAGW 2502, and an Office of Naval Research contract N00014-89.

References

Akioka, M., Kubota, J., Suzuki, M., and Tohmura, I. (1992) *Solar Phys.*, **139**, 177.
Antonucci, E., Hoeksema, J.T., and Scherrer, P.H. (1990) *Astrophys. J.*, **360**, 296.
Bai, T. (1987), *Astrorphys. J.*, **314**, 795.
Bai, T. (1988), *Astrophys. J.*, **328**, 860.
Bai, T. (1990), *Astrophys. J. Lett.*, **364**, L17.
Bai, T. (1992), *Astrophys. J.*, **397**, 84.
Bai, T., Hoeksema, J.T., and Scherrer, P.H. (1996), *Astrophys. J*, to be published.
Bai, T., and Sturrock, P.A. (1987),. *Nature*, **327**, 601.
Dodson, H.W., and Hedeman, E.R. (1971) Experimental, Comprehensive Solar Flare Indices and Its Derivation for "Major" Flares, 1955–1969, *World Data Center for Solar Terrestrial Physics Report* **UAG-14**. NOAA, Boulder.
Dodson, H.W., and Hedeman, E.R. (1975a) *Solar Phys.*, **42**, 121.
Dodson, H.W., and Hedeman, E.R. (1975b) Experimental, Comprehensive Solar Flare Indices for Certain Flares, 1970–1974, *World Data Center for Solar Terrestrial Physics Report* **UAG-52**. NOAA, Boulder.
Dodson, H.W., and Hedeman, E.R. (1981) Experimental, Comprehensive Solar Flare Indices and Its Derivation for "Major" and Certain Lesser Flares, 1975–1979, *World Data Center for Solar Terrestrial Physics Report* **UAG-80**. NOAA, Boulder.
Fung, P.C., Sturrock, P.A., Switzer, P.S., and van Hoven, G. (1971), *Solar Phys.*, **18**, 90.
Haurwitz, M.W. (1968) *Astrophys. J.*, **151**, 351.
Hoeksema, J.T., and Scherrer, P.H. (1987) *Astrophys. J.*, **318**, 428.
Langton, C. (1990) *Artificial Life II*, eds. Taylor, C., Farmer, J.D., Rassmussen, S., Addison Wesley, Redwood City, CA.
Mardia, K.V. (1972) *Statistics of Directional Data* Academic Press, New York.
Rieger, R. *et. al.* (1984) *Nature* **312**, 623.
Svestka, Z., and Simon, P. (1969) *Solar Phys.*, **10**, 3.
Warwick, C.S. (1965) *Astrophys. J.*, **141**, 500.
Wilcox, J.M., and Schatten, K.H. (1967) *Astrophys. J.*, **147**, 364.

SOLAR DYNAMO DRIVEN BY GLOBAL CONVECTION AND DIFFERENTIAL ROTATION AND THE MAGNETO-THERMAL PULSATION OF THE SUN

HIROKAZU YOSHIMURA
Department of Astronomy, University of Tokyo
Tokyo 113, Japan

1. Introduction

Since 1972 when we proposed a mechanism of the solar dynamo driven by mass flows of global convection and differential rotation in the solar convection zone, each process that constitutes the complex machinery of the solar dynamo has been clarified step by step (Yoshimura 1972, 1975a,b, 1978a,b, 1979, 1981, 1983a, 1985, 1993b). We now know that the basic process of the dynamo can be visualized by topological deformation of stretching, winding, and folding of magnetic field lines in the solar convection zone by combined flows of the differential rotation and the global convection under influence of the rotation. In this paper, we briefly review the basic part of the dynamo process in terms of the topological deformation of the field lines and proceed to the second major step of advancement of understanding of the dynamo machinery achieved from analysis of observed evidences of nonlinear interaction of mass and heat flows and magnetic field in form of magnetic flux tubes of multiple spacial and temporal scales, which is a new category of solar phenomena called the magneto-thermal pulsation of the Sun.

2. The Solar Dynamo Machinery

The basic part of the solar dynamo machinery can best be understood by the fundamental concept of magnetohydrodynamics (MHD) of frozen-in magnetic field lines in a fluid system in the limiting case of infinite electric conductivity. The field lines in this case move along with fluid particles of mass motions. If the streamlines of the flows are smooth and simple, the

Y. Uchida et al. (eds.),
Magnetodynamic Phenomena in the Solar Atmosphere – Prototypes of Stellar Magnetic Activity, 345–354.

magnetic field lines become also smooth and simple. If the streamlines are complex, the field lines become complex and entangled.

The basic classical question of cosmic dynamo problem in general has been whether or not there are any fluid motions in a continuous media that amplify a magnetic field from an infinitesimal level to a finite level and maintain it.

The problem of the solar dynamo asks further whether or not these fluid motions can reverse the magnetic field lines in every 11 years. Hence the problem of the cosmic dynamo and the solar dynamo is whether or not a kind of fluid motions can move around the magnetic field lines in the fluid and amplify the magnetic field energy from an infinitesimal level to a finite level and reverse the direction of the field lines. This is a question of topology of the field lines in three-dimensional space. The amplification of magnetic field energy is possible by stretching of the field lines. The question is then what kinds of flows can continue the stretching of the field lines and can reverse their direction. Since electric conductivity in a real cosmical system is finite, diffusion process enters into the scene. This process destroys the amplified magnetic field against the dynamo process. Particularly when small scale turbulence is present in the system, spacial scale of diffusion process can become so small that even a very small magnetic diffusivity or equivalently a very large electric conductivity makes the diffusion process of magnetic field very effective. The diffusion process that involves reconnection of field lines is a necessary process of the solar dynamo machinery to organize field lines that are entangled by the deformation process due to the dynamo driving flows.

The basic classical question of cosmic dynamo problem and of the solar dynamo machinery can be understood by a linear concept of the MHD. Given a kinematical structure of a kind of mass flows, the MHD process is linear in the sense that the magnetic field as a solution of the MHD process either exponentially grows or decays depending on whether the dynamo action of the flows is stronger or weaker than the diffusion action of small scale turbulence.

The actual process of the solar dynamo machinery can be understood by combination of theoretical considerations of fluid dynamics of rotating system and observations of the flows. Observed flows of the differential rotation have long been known to be vital to the operation of the solar dynamo by its stretching action on the magnetic field lines in longitudinal direction. At the same time, however, it also has long been known that the axisymmetric flows of the differential rotation alone are not sufficient to maintain the solar dynamo. The component of the magnetic field in the meridional plane perpendicular to the direction of rotation cannot be created by the flows of the differential rotation and thus eventually dies away. The stretching process of the field lines from that component into the component in the direction of rotation will also cease.

As a possible candidate of the much needed flows which are, together with the flows of the differential rotation, are sufficient to drive and maintain the solar dynamo, we have proposed flows of the global scale convection which we named the global convection (Yoshimura 1972). Theoretical considerations of the nature of the flows of the global convection in a rotating spherical shell show that two characteristics of the flows, both of which are caused by Coriolis force due to the existence of rotation, are vital to the operation of the solar dynamo. One characteristics is that flow pattern of the convection propagates around the rotational axis of the Sun by the Coriolis force. Rossby waves in a rotating spherical shell also propagate around the rotational axis due to the Coriolis force but whose flows are driven by horizontal temperature difference while the convective flows are driven by radial forcing of gravity (Yoshimura 1974). The other characteristics is that streamlines of the convective flows are twisted by the Coriolis force under the influence of the same rotation and that, as a result of the twisting process, the flow pattern has a helical structure. Because of the propagation of the flow pattern, the streamlines look like wave flows of water of a river without being rolled up around the convective cell when the flows are seen from a framework which moves with the same speed as the flow pattern of the convective cell. In the case of supergranulation of the Sun whose flow pattern does not propagate, the field lines are wound up around the convective cells and are concentrated at the boundaries of the cells. Because of this nature of the global convection, the pattern of the global scale convection does not look like a cell. This is the reason why we call the convection the global convection. We distinguish it from another concept of global scale giant cell convection.

The existence of this global convection was examined by analysing various characteristics of global scale magnetic field as well as active region scale magnetic field (Yoshimura 1971, 1973). Various efforts were done to directly detect the velocity field of this global convection. We think that we found a significant signatures of existence of the velocity field in the Stanford dopplergram data. But the duration of the continuous data that showed these signatures at the time of analysis was two separate intervals of one month length for each interval (Yoshimura 1987).

There are several kinds of approaches to the proof of the solar dynamo machinery model. We must prove that the combined flows of the differential rotation and the global convection can actually work as the solar dynamo. One is to formulate a dynamo equation from the original MHD equation by averaging the MHD equation over whole longitudes which governs magnetic field evolution in the rotating solar convective shell (Yoshimura 1972). The velocity field of the differential rotation is arbitrarily given. Since the nature of magnetic field solutions of the formulated dynamo equation is sensitive to the structure of the differential rotation, the structure of the differential rotation inside the convection zone can inversely be inferred by comparing behavior of the magnetic field solutions of the formulated dynamo equation with behavior of the observed magnetic field (Yoshimura 1975a,b, 1976a,b, 1977a,b). This becomes possible due to the fact that the magnetic field solutions of the dynamo equation propagate along iso-rotation surface inside the convection zone as dynamo waves (Yoshimura 1975b).

The velocity field of the global convection, however, is not arbitrarily given. It is created by modifying a solution of linearized Navier-Stokes equation which governs fluid motions inside a rotating spherical shell with Boussinesq approximation. The modification is done so that the structure of the velocity field has a freedom to be topologically deformed so that the structure can simulate the real velocity field of the global convection in the strongly stratified solar convection zone. The structure of the velocity field is explicitly given so that effects of rotation and geometry of the sphere are accurately treated. As a result of this treatment, the propagation of the convective flow pattern and its helical characteristics are found to be vital to the operation of the solar dynamo machinery.

Solutions of the formulated dynamo equation show that the magnetic field can be amplified exponentially from an infinitesimal level to a finite level and thus the dynamo by the combined flows of the differential rotation and the global convection works. The solutions also show that the combined flows can reverse the direction of the magnetic field lines in every 11 years. The time scale of 11 years is a function of speed of deformation of field lines and thus is a function of the velocity field. The solutions are not sensitive to any particular mode of the global convection as long as the convection has the two important characteristics of the propagation of the flow pattern and of the helical structure.

Almost all the basic characteristics of the solar cycle can be simulated by the solutions. This is due to the fact that the solutions of the dynamo equation propagates along iso-rotation surface inside the convection zone as dynamo waves and that the surface distribution of the various characteristics of the observed solar cycle reflects the surface section of the propagating dynamo waves (Yoshimura 1975a,b, 1976a,b). Hence the direction of propagation of the dynamo waves is important. While detailed structure of the global convection does not influence much the nature of the solutions, the depth of the global convection flows where the horizontal component of the convective flows changes their direction is a critical factor for the solutions and can affect the nature of the solutions. The direction of helical twists of the flows and of the magnetic field is different for the upper part above and the lower part below the depth. The direction of propagation of the dynamo waves, as a result of the difference of direction of helical twists, becomes different for the two parts. This aspect influences much the nature of solutions of the whole system (Yoshimura 1972, 1975a,b, 1993b). In other words, the nature of solutions depends strongly on whether the upper part or the lower part plays a dominant role in the whole dynamo machinery.

A keen criticism was raised to the this formalism, particularly to the approximation which was used in the process of formulation of the dynamo equation. The formulation is essentially a simplification of description of the MHD process that follows time evolution of magnetic field by using a linearized non-axisymmetric magnetic field solution of the MHD equation in formulating the longitude-averaged dynamo equation. The linearized non-axisymmetric magnetic field solution represents infinitesimal deformation of field lines and is a linear function of infinitesimal non-axisymmetric velocity field component of the flows. Hence the resulting dynamo equation becomes bilinear with regards to the non-axisymmetric velocity field component of the flows. This approximation is called quasi-linear approximation.

To answer the criticism, the original MHD equation in three-dimensional space without using the quasi-linear approximation was solved numerically with the same formalism and philosophy as in the case of the dynamo equation with a given velocity structure of the differential rotation and of the global convection (Yoshimura 1983a). Since we fully followed time development of the deformation process of the field lines and did not treat the field development as an infinitesimal deformation, we needed not use the quasi-linear approximation. When the reaction of the magnetic field on the flows in form of the Lorentz force was ignored, the MHD equation became linear with respect to the magnetic field. The solutions of the MHD equation in three-dimensional space had the same characteristics as those of the solutions of the dynamo equation and hence answered definitively and positively to the question of the cosmic dynamo in general and of the solar dynamo machinery. The combined flows of the differential rotation and the global convection could work as a dynamo and generate and maintain the magnetic field and reverse the direction of the field lines in every 11 years.

Since the MHD equation in three-dimensional space was solved by a numerical method, it was not straightforward to perceive time evolution of the spacial structure of the magnetic field and its relation with the structure of the velocity field. In an effort to visualize the dynamo process, a graphic representation was done so that time evolution of the spacial structure of the magnetic field and its association with the structure of the velocity field could clearly be understood (Yoshimura 1993b). As a result of this effort, it became straightforward to visually understand the basic features of the dynamo process driven by the combined flows of the differential rotation and of the global convection without following details of calculation of the MHD equation. It became clear that the basic process of the solar dynamo machinery is stretching and winding and folding of magnetic field lines by the combined flows of the differential rotation and of the global convection with helical twists and propagation characteristics. Stretching of the field lines creates magnetic field energy. Folding of the field lines reverses the magnetic field lines beneath the existing field lines. The repetition of the reversal process forces the magnetic field system to propagate as dynamo waves along iso-rotation surface. The whole process also creates giant flux tubes encircling the Sun, each flux tube of which causes solar activity of one solar cycle. We call these flux tubes the main flux tubes of the Sun.

3. The Nonlinear Solar Cycle Dynamo

The second part of the solar dynamo results from nonlinear interaction of mass and associated heat flows and magnetic flux tubes created by the basic part of the dynamo process. In this part, we have two basic problems of cosmic magnetism. The first problem is why and how magnetic field in cosmos is almost always in form of magnetic flux tubes, why and how the tubes are formed and interact nonlinearly with ambient plasma. The second problem is how the nonlinear reaction of the dynamo-generated magnetic field takes place to modify the dynamo-driving flows and weaken the dynamo so that the field-generating dynamo and the field-destroying diffusion due to small scale turbulence can be in balance. The second problem is equivalent to the question of the mechanism that determines the level of solar activity and the amplitude of solar cycle and hence is related to the problem of understanding of long-term behavior of the solar cycle. Since, as we have discussed above, the basic part of the solar dynamo machinery creates huge magnetic flux tubes encircling the Sun, the second problem can be considered as a part of the first problem.

An orthodox theoretical way to solve these problems is to formulate the MHD equation together with Navier-Stokes equation, which includes a term of Lorentz force which describes process of reaction of magnetic field on mass flows, and an equation of ionization and an equation of state. We then solve the system of equations to show that the system has solutions that are capable to reproduce the nonlinear phenomena described above. This approach to the nonlinear dynamo is often called dynamical dynamo problem. Such a theoretical effort is not fully successful yet.

The other way to approach to the solutions of the basic problems is to obtain information from various observed data to build up concepts to understand the nonlinear dynamo mechanism. Recently in a series of analysis of various data, we have found evidences that confirm concepts that were developed by a theoretical study of the nonlinear dynamo. The theoretical study we have performed which we found useful to understand the nonlinear solar dynamo machinery and possibly to understand any cosmic nonlinear dynamo mechanism is the study of the long-term behavior of the solar cycle over at least 1000 years. We built a simplified model to investigate the nonlinear process by representing two basic aspects of the nonlinear dynamo using two kinds of parameters in a formula that represents the strength of the dynamo in the dynamo equation which describes the longitude-averaged MHD processes due to the flows of the differential rotation and of the global convection (Yoshimura 1978a,b, 1979). The formula is a simplified replacement of the role that Lorentz force must play in the nonlinear dynamo. The replaced formula represents two basic aspects of the nonlinear process. One is that the Lorentz force of the dynamo-generated magnetic field modify flows of the dynamo and weaken the dynamo. The other aspect is that it needs a certain time for this modification to take place. The first aspect determines the level of solar activity and the level of the solar cycle. The second aspect of delay time is determined by the strength of magnetic field inside the Sun where the dynamo is working since the delay time is the time needed for the force to modify the velocity field. Conversely, from the delay time, in principle, the strength of the field can be determined. One of the most important results of this formulation of the nonlinear dynamo is that time delay of about 20-year is necessary for a solution of the formulated nonlinear dynamo equation to reproduce the observed 55-year grand cycle modulations of the 11-year solar cycle (Yoshimura 1978b, 1979). This delay time of 20 years was found in the 110-year modulation of the solar differential rotation calculated from displacement of positions of sunspot groups (Yoshimura and Kambry 1993a,b,c,d; Yoshimura 1993a). This finding indicates that the Lorentz force of the solar cycle magnetic field is strong enough to modify flows in the solar convection zone and hence the same Lorentz force of the solar magnetic field must work on the flows of the convection. A purpose of next step of this series of efforts, therefore, became to find an evidence of time delayed phenomenon of the nonlinear reaction of the magnetic field on the global convection.

4. The Magnetic Flux Tubes

One major part of the nonlinear interaction processes between flows and magnetic field is the dynamics of magnetic flux tubes. Among other problems related to the flux tubes, the concept of magnetic buoyancy is the most important one for the solar dynamo machinery. If temperature distribution of a magnetic flux tube is such that inside temperature and outside temperature are the same, the magnetic pressure inside of the flux tube makes inside gas pressure be lower than that of outside, which makes inside mass density be lower than surroundings and hence the flux tubes become lighter than surroundings. Because of this, it has been argued that every magnetic flux tube inside the solar convection zone, once it is formed, becomes buoyant and erupts to the surface with time scale much smaller than the time scale of solar cycle. Hence, it has been argued that magnetic flux tubes we observe on the surface of the Sun cannot be generated in middle of the solar convection zone. If it is generated, it would erupt onto the surface of the Sun too soon that there would be no time for the solar cycle dynamo to amplify the magnetic field. These arguments lead to the conclusion that the solar dynamo must take place at the base of the convection zone.

In our context of the solar dynamo machinery theory driven by flows of the differential rotation and of the global convection, the dynamo should take place in middle of the solar convection zone (Yoshimura 1983b). If inside temperature of a magnetic flux tube is lower than outside temperature, then inside mass density of a flux tube could be equal to or larger than that of outside. Then the magnetic buoyancy cannot work and thus the magnetic flux tubes can stay inside the convection zone (Yoshimura 1985). Temperature difference which is necessary for the flux tube to stay in the convection zone was found to be extremely small. This theoretical conjecture shows that cooling of magnetic flux tubes can take place inside the convection zone.

5. The Magneto-Thermal Pulsation of the Sun

This conjecture was unexpectedly confirmed by an analysis of various kinds of observed data sets. One was the data of total solar irradiance (TSI) observed from space by the Active Cavity Radiometer Irradiance Monitor (ACRIM) I experiment on board of Solar Maximum Mission (SMM) and by the Earth Radiation Budget (ERB) experiment of Nimbus 7. The other was the data observed from ground of the sunspot relative number, total magnetic field flux, Ca II K index, and He I 10830 Å, all of which turned out to be good indices of the magnetic field of the Sun. We found that the luminosity calculated from the TSI data and the magnetic field of the Sun were pulsating in unison with at least two kinds of time scales. One was 200 days. The other was 11 years. The former was associated with formation of clusters of magnetic flux tubes, each tube of which, we thought, was responsible for formation of a solar active region. The latter was associated with the basic dynamo process of the solar cycle. We called the former phenomenon, the 200-day magneto-thermal pulsation and the latter, the 11-year magneto-thermal pulsation. The 200-day magneto-thermal pulsation was superposed on the relatively slowly changing 11-year magneto-thermal pulsation. In the 200-day magneto-thermal pulsation, the TSI increased about 50 days after an increase of the magnetic field indices of the Sun and decreased about 20 days before the increase of the magnetic field indices.

We have argued that these evidences mean that blocking of heat flow and cooling of magnetic flux tubes take place inside the convection zone. The existence of the surface is not needed for the cooling and the convection zone works as a temporary heat reservoir. This mechanism makes the flux tubes heavy enough to hinder expulsion of flux tubes from the solar convection zone and thus the dynamo can operate in the middle of the solar convection zone. We also found that, in the 11-year magneto-thermal pulsation, the time profiles of the ERB and ACRIM TSI and of the ACRIM TSI were more similar to that of the sunspot relative number when the time profiles of the ERB and ACRIM TSI were displaced toward the past by an amount of about 10.3 years, which is one solar cycle prior to the observed time interval of the ERB and ACRIM data. This is the delay time we have been looking for as an evidence of the modulation of the convective heat flow by the solar cycle magnetic field. In this way, each concept that constitutes the nonlinear dynamo machinery is being verified one by one. More detailed review of this series of work on the magneto-thermal pulsation is to be found in Yoshimura (1995b).

References

Yoshimura, H. (1971) Complexes of Activity of the Solar Cycle and Very Large Scale Convection, *Solar Phys.* **18**, 417–433.

Yoshimura, H. (1972) On the Dynamo Action of the Global Convection in the Solar Convection Zone, *Astrophys. J.* **178**, 863–886.

Yoshimura, H. (1973) On the Characteristics of the Basic Framework of Solar Active Regions and the Magnetohydrodynamical Structure of the Convection Zone, *Solar Phys.* **33**, 131–143.

Yoshimura, H. (1974) Global Fluid Motions in a Uniformly Rotating Spherical Shell Studied as a Model of the Solar Convection Zone by a Variational Method, *Publ. Astron. Soc. Japan* **26**, 9–51.

Yoshimura, H. (1975a) A Model of the Solar Cycle Driven by the Dynamo Action of the Global Convection in the Solar Convection Zone, *Astrophys. J. Suppl.* **29**, 467–494.

Yoshimura, H. (1975b) Solar-Cycle Dynamo Wave Propagation, *Astrophys. J.* **201**, 740–748.

Yoshimura, H. (1976a) Solar Cycle General Magnetic Fields of 1959-1974 and Dynamical Structure of the Solar Convection Zone, *Solar Phys.* **47**, 581–600.

Yoshimura, H. (1976a) Phase Relation between the Toroidal Solar-Cycle General Magnetic Fields and Location of the Origin of the Surface Magnetic Fields, *Solar Phys.* **50**, 3–23.

Yoshimura, H. (1977a) Coronal General Magnetic Field Evolution as a New Parameter of the Solar Cycle, *Solar Phys.* **52**, 41-52.

Yoshimura, H. (1977a) Solar-Cycle Evolution of the Coronal General Magnetic Field of 1959-1974 and the Synchronous Variation of High-Speed Solar Wind Streams and Galactic Cosmic Rays, *Solar Phys.* **54**, 229-258.

Yoshimura, H. (1978a) Nonlinear Astrophysical Dynamos: The Solar Cycle as a Nonlinear

Oscillation of the General Magnetic Field Driven by the Nonlinear Dynamo and the Associated Modulation of the Differential-Rotation-Global-Convection System, *Astrophys. J.* **220**, 692–711.

Yoshimura, H. (1978b) Nonlinear Astrophysical Dynamos: Multiple-Period Dynamo Wave Oscillation and Long-Term Modulations of the 22 Year Solar Cycle, *Astrophys. J.* **226**, 706–719.

Yoshimura, H. (1979) The Solar-Cycle Period-Amplitude Relation as Evidence of Hysteresis of the Solar Cycle Nonlinear Magnetic Oscillation and the Long-Term (55 Year) Cyclic Modulation, *Astrophys. J.* **227**, 1047–1058.

Yoshimura, H. (1981) Solar Cycle Lorentz Force Waves and the Torsional Oscillation of the Sun, *Astrophys. J.* **247**, 1102–1112.

Yoshimura, H. (1983a) Dynamo Generation of Magnetic Fields in Three-Dimensional Space: Solar Cycle Main Flux Tube Formation and Reversals, *Astrophys. J. Suppl.* **52**, 363–385.

Yoshimura, H. (1983b) Solar Cycle Dynamo Wave Origin of Sunspot Intensity and X-ray Bright Point Number Variation, *Solar Phys.* **87**, 251–259.

Yoshimura, H. (1985) Cooling of Magnetic Flux Tubes as a Mechanism for Suppression of Magnetic Buoyant Escape of the Flux Tubes from the Sun and Stars, *Publ. Astron. Soc. Japan* **37**, 171–181.

Yoshimura, H. (1987) The Detection of Global Convection on the Sun by an Analysis of Line Shift Data of the John M. Wilcox Solar Observatory at Stanford University, in B. R. Durney and S. Sofia (eds.), *The Internal Solar Angular Velocity*, D. Reidel Publishing Company, Dodrecht, pp. 89–95.

Yoshimura, H. (1993a) Nonlinear Coupling between the 110-Year Periodic Modulations of Solar Differential Rotation and Solar Cycle, in F. Krause, K. -H. Rädler, and G. Rüdiger (eds.), *The Cosmic Dynamo*, Kluwer Academic Publishers, Dodrecht, pp. 63–69.

Yoshimura, H. (1993b) The Solar Dynamo and Planetary Dynamo, in F. Krause, K. -H. Rädler, and G. Rüdiger (eds.), *The Cosmic Dynamo*, Kluwer Academic Publishers, Dodrecht, pp. 463–479.

Yoshimura, H. (1994a) The Time-Delayed Solar Cycle Luminosity Modulations by Sub-Surface Magnetic Flux Tubes, *Astron. Nachr.* **315**, 189–203.

Yoshimura, H. (1994b) Darkening of the Sun Prior to Surface Appearance of Sunspot Flux Tubes and Magneto-Thermal Pulsation of the Sun, *Astron. Nachr.* **315**, 371–390.

Yoshimura, H. (1995a) The 10-Year Time Delay of the Solar Cycle Luminosity Modulation of the Sun behind the Solar Cycle Magnetic Oscillation, *Astron. Nachr.* **316**, 175–186.

Yoshimura, H. (1995b) Coupling of Total Solar Irradiance and Solar Magnetic Field Variations with Time Lags: Magneto-Thermal Pulsation of the Sun, in *Solar Drivers of Interplanetary and Terrestrial Disturbances*, Proceedings of the 16th National Solar Observatory / Sacramento Peak International Workshop, Astronomical Society of the Pacific Conference Series, in press.

Yoshimura, H. and Kambry, M. A. (1993a) Secular Acceleration of Solar Rotation from 1943 to 1986, *Solar Phys.* **143**, 205–214.

Yoshimura, H. and Kambry, M. A. (1993b) The Secular Modulation from 1943 to 1992 and its Time-Delayed Correlation with the 55-Year Grand Cycle of the 11-Year Solar Cycle, *Solar Phys.* **148**, 11–26.

Yoshimura, H. and Kambry, M. A. (1993c) The 100-Year Periodic Modulation of Solar Rotation, *Astron. Nachr.* **314**, 9–19.

Yoshimura, H. and Kambry, M. A. (1993d) The Time-Delayed Correlation of the 100-Year Modulation of the Solar Rotation with the Long-Term Modulations of the Solar Cycle,*Astron. Nachr.* **314**, 21–30.

FORMATION OF A KINKED ALIGNMENT OF SOLAR ACTIVE REGION

R. MATSUMOTO
Department of Physics, Faculty of Science, Chiba University, Inage-ku, Chiba 263, Japan and ASRC, JAERI, Naka, Japan

T. TAJIMA, W. CHOU
University of Texas at Austin, Austin, TX 78712, USA

AND

K. SHIBATA
National Astronomical Observatory, Mitaka, Tokyo 181, Japan

Images of the solar corona as observed by the soft X-ray telescope aboard the *Yohkoh* satellite sometimes show a sequence of S-shaped X-ray emitting regions in the low latitudes of the northern or southern hemisphere (Matsumoto et al. 1996). Such a structure suggests that they are emergent portions of a submerged large scale helical flux tube in the convection zone. When the magnetic twist exceeds a threshold value, the flux tube deforms itself into a helical structure through the kink instability. We have carried out three-dimensional MHD simulations of such a kink-unstable flux tube in a convectively unstable gas layer. Here we briefly describe our model and numerical results (see Matsumoto et al. 1996 for details).

We use Cartesian coordinates with z in the vertical direction and assume an ideal gas with adiabatic index $\gamma = 1.4$. The initial state consists of the hydrostatic gas layer with (1) a convection zone in $-2.5H < z < 2H$ with temperature gradient $dT/dz = -(\gamma - 1)T_0/H$, where H and T_0 are the scale height and temperature at $z = 0$, respectively, (2) a cold, isothermal ($T = 0.2T_0$) layer in $2H < z < 4H$, and (3) a hot, isothermal ($T = 5T_0$) corona in $4H < z$. A twisted flux tube with the Gold-Hoyle force-free magnetic field distribution is imbedded at $z = 0$ in the convection zone. We cut off the flux tube at $r = R$ and assume pressure balance between the flux tube and the external medium. Periodic boundary conditions are applied at $x = 0$ and $x = L$. Small v_z perturbations are imposed inside the flux tube at $t = 0$.

Y. Uchida et al. (eds.),
Magnetodynamic Phenomena in the Solar Atmosphere – Prototypes of Stellar Magnetic Activity, 355–356.

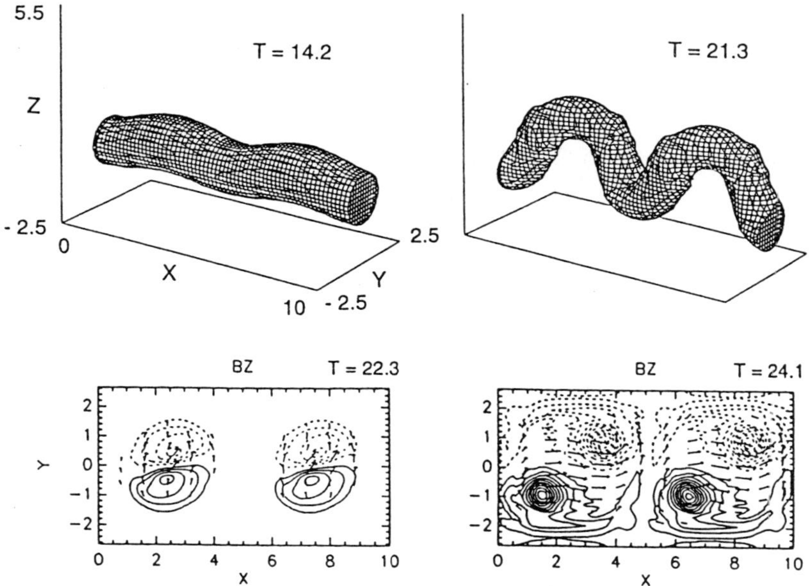

Figure 1. Evolution of a kink-unstable flux tube. Upper panels show the $B^2 = 0.36\rho_0 C_s^2$ isosurface. Lower panels show isocontours of vertical magnetic fields and horizontal magnetic vectors at photospheric level ($z = 2.55H$). Solid curves and dotted curves correspond to different polarities.

In figure 1, we show typical results of the simulation. The model parameters are $v_A^2 = 2C_s^2$, $R = H$, and $L/R = 10$, where C_s is the sound speed at $z = 0$ and v_A is the initial Alfvén speed at the center of the flux tube. The number of grid points is $64 \times 64 \times 150$. At $t = 21.3H/C_s$, the flux tube has deformed itself into a helical structure due to the kink instability. The top portions of the helical flux tube continue to rise in the upper convection zone. When they finally emerge into the corona, the emergent portions form a sequence of S-shaped regions similar to what we see in X-ray images. The lower panels show photospheric magnetic fields. The regions with most intense vertical magnetic fields correspond to sunspots. As the main part of the flux tube emerges, the f- and p-spots move approximately in the east-west direction ; their axial tilt to the equator gradually decreases. These numerical results qualitatively reproduce many characteristics of observed solar active regions such as their kinked alignment, rising motion of magnetic loops, formation of sunspots, and their drifting motions.

References

Matsumoto, R., Tajima, T., Chou, W., Okubo, A., and Shibata, K. (1996) to be submitted to *Astrophysical Journal Letters*

IV. Magnetic Behavior of the Sun and Stars and Their Activity Cycles

IV.2. Observed Domain of Activity on Stars

STELLAR MAGNETIC ACTIVITY, ACTIVITY CYCLES, AND DYNAMOS

R. PALLAVICINI
Osservatorio Astrofisico di Arcetri
Largo Fermi 5, I-50125 Firenze, Italy

Abstract. Observational aspects of magnetic activity in stars are discussed with emphasis on their implications for dynamo models. The dependence of chromospheric and coronal emission on rotation in convective stars is a strong, albeit indirect evidence for the occurrence of some kind of dynamo action in all late-type stars. The available data, including those on stellar activity cycles, are as yet insufficient to constrain dynamo models in an effective way. The stellar data, anyway, seem to suggest a much more complex picture of the dynamo process than the one indicated by solar observations.

1. Introduction

The purpose of this paper is to discuss observational evidences of magnetic activity on stars and their relevance for constraining dynamo models. This is not an easy task. In fact, while there is ample evidence at present that stellar activity is of magnetic origin and likey due to some sort of dynamo action, there is no compelling evidence to make a selection between alternative models or to constrain them in an effective way. Probably, the most significant result emerging from the available stellar data is that dynamo action in stars can be much more complex that could be inferred from observations of the solar activity cycle alone. While many stars appear to behave similarly to the Sun, many others show a much more chaotic behaviour or may be deprived at all (either temporarily or permanently) of a significant dynamo action. Understanding how these different behaviours depend on basic stellar parameters like rotation, convection, or age is one

Y. Uchida et al. (eds.),
Magnetodynamic Phenomena in the Solar Atmosphere – Prototypes of Stellar Magnetic Activity, 359–366.

of the great challenges in modern stellar astrophysics, one that will require a major effort on both the observational and theoretical sides.

This is an area in which the solar-stellar connection can play a crucial role. Observations of the Sun have the highest spatial, spectral and temporal resolution, and span a much larger time interval than stellar observations (the latter point is particularly important in view of the long time scales –years and even centuries– typically associated with activity cycles). Solar observations have been, and still are, the benchmark against which to test dynamo models: as a matter of fact, these models are calibrated on the Sun, which makes their predictive power for other stars rather limited. By comparing the Sun with late-type stars with different masses, rotation rates and evolutionary states one hopes to understand how the dynamo process depends on basic stellar parameters and hence obtain a better understanding of the dynamo itself. Unfortunately, while the observable quantities refer to the surface layers of a star, the dynamo action operates in the deep interior, probably at the bottom of the convection zone, and the connection between surface activity and dynamo generated magnetic fields is far from being a direct one.

In addition to this fundamental limitation, there are other practical, but not less important difficulties in using stellar observations to constrain dynamo models. All observable quantities are in fact affected, at least to some extent, by selection effects, temporal variability, and background problems (e.g. the possible presence of a non-magnetic "basal flux" in Ca II H+K observations). In many cases, the data samples used so far have been limited to small numbers of objects, and one should worry whether such samples are statistically significant especially in the presence of short-term variability that may obscure or significantly alter the long-term trends associated with activity cycles. In the absence of theoretical guidelines, it is difficult, if not impossible, to decide which are the most meaningful observables to be used to constrain models, i.e. whether luminosities, surface fluxes, or luminosities normalised to bolometric luminosities should be used. Also unclear is whether one should use the rotation period or a combination (like the Rossby number) of rotation and convective properties in studying the dependence of radiative losses upon dynamo action. Different correlations have been found using different parameters (and even by different authors using the same parameters!), thus it is questionable whether one can learn much on stellar dynamos from this kind of approach.

With these caveats in mind, I will discuss in the following some of the observational constraints put on stellar dynamo mechanisms by ground based and space data. This paper is not for the specialist and provides only some general background mainly for the benefit of readers from the solar physics community. First, I will discuss basic results about the dependence of stellar magnetic activity upon rotation, convection, and age. Then I will report on the extensive Mt. Wilson program for detecting activity cycles in other stars, and I will discuss a few attempts to use these data to infer dynamo properties. The main conclusion will be that there are no simple observational tests to discriminate between different dynamo models: while the available data are certainly consistent with dynamo action, they are by and large insufficient to effectively constrain the dynamo mechanism in stars other than the Sun.

2. Dependence upon Rotation and Convection

Early observations made in the sixties showed that both rotation and chromospheric activity decline with age in solar-type stars (see review by Hartmann and Noyes 1987). This was summarised by Skumanich (1972) in his famous $t^{-1/2}$ law, which is probably the first and most convincing attempt to relate stellar activity to dynamo action. The decline of rotation with age is attributed to the braking action of magnetised stellar winds, while the associated decline of chromospheric activity with age is believed to be a consequence of the decreased rotation rate and dynamo efficiency. It is interesting to note that the Skumanich's law was derived by using only three data points (the Pleiades, the Hyades, and the Sun) and by attributing a unique average value to the rotation rate and Ca II emission of solar-type stars in the Pleiades and the Hyades. Subsequent observations of chromospheric and coronal radiative losses in clusters of different ages have substantially confirmed the decline of stellar activity with age, but have also revealed a much more complex picture.

Recent results, which also include pre-main sequence (PMS) stars in star forming regions as well as field stars of different age populations, show that on average chromospheric and coronal emission do decline with age, but not with a $t^{-1/2}$ law; an exponential decline is more consistent with the data, while there is also evidence for an additional dependence on spectral type. Even more interestingly, cluster data (particularly young clusters) show a large spread in rotational velocities and chromospheric/coronal radiative losses, so it is questionable whether one can attribute a single data point to a given cluster. Particularly relevant in this respect are the results of Bouvier (1994) on the evolution of the stellar rotation rate as a function of age, from the PMS (T-Tauri) phase to old stars like-the Sun. Not only the average rotation rate increases during PMS contraction on the radiative tracks (as expected from conservation of angular momentum), but also a large spread is produced in rotation rates, probably as a consequence of the different degrees of coupling between the star and the circumstellar disk. The spread is particularly large for α Persei (age $\sim$ 50 Myr) whose late-type stars have rotation rates ranging from the minimum detectable value ($\sim$ 10 Km/s) to almost 200 Km/s. The spread is still large for the Pleiades (age $\sim$ 70 Myr), but then rapidly decreases for older stars and has virtually disappeared at the age of the Hyades ($\sim$ 700 My), presumably because the braking action of magnetised winds is stronger for more rapidly rotating stars.

These results have two important implications. First, that rotation, rather than age, is the controlling factor in determining the efficiency of dynamo action and hence the level of chromospheric and coronal activity (a conclusion enforced by the high activity of tidally locked but relatively old evolved stars like RS CVn binaries); and second, that the various processes of acceleration and deceleration which occur during stellar evolution are likely to act differentially throughout the stellar body, causing significant radial gradients in the stellar angular velocity. While the dependence of chromospheric and coronal emission upon rotation has been demonstrated in many studies (albeit finding often different functional dependences), the effects of different rotational profiles in the stellar interior, and in general the influence of different convection zone properties, are much more difficult to evidentiate from an observational point of view and have received therefore much less attention.

According to mean-field dynamo theory, the mean field of a rotating convective star is sustained by the fluid motions against magnetic diffusion losses and periodically regenerated by a mechanism which involves differential rotation and cyclonic convection (see, e.g., Rosner and Weiss 1992). In particular, while both differential rotation and cyclonic convection can contribute to the formation of a toroidal component from an originally poloidal field, cyclonic convection (i.e. the effect of Coriolis forces upon the convective motions) regenerate a poloidal field of opposite sign from the toroidal field. For an $\alpha\omega$ dynamo (in which the contribution of cyclonic convection to the toroidal component is negligible), the efficiency of dynamo action is expressed by a dynamo number N_D which is the product of two magnetic Reynolds numbers $R_m = vl/\eta_t$, one for fluid motions due to differential rotation (the ω-effect) and the other for the action of cyclonic convection (the α-effect). The dynamo number N_D can be expressed as:

$$N_D = \frac{\alpha \, (\nabla_r \Omega) \, l^4}{\eta_t^2} \tag{1}$$

where $\alpha \sim < v \cdot \nabla \times v > \tau_c$ is the product of the mean helicity times the convective overturn time, $\nabla_r \Omega$ is the radial gradient of angular velocity, and $\eta_t \sim < v^2 > \tau_c$ is the turbulent magnetic diffusivity.

Assuming, as order of magnitude estimates, that $\alpha \sim \Omega l$, $\nabla_r \Omega \sim \Omega/l$ and $\eta_t \sim l^2/\tau_c$, the dynamo number is given by

$$N_D \sim \Omega^2 \tau_c^2 \sim R_o^{-2} \tag{2}$$

where $R_o = P_{rot}/\tau_c$ is the so-called Rossby number, i.e. the ratio of the rotation period of the star to the convective overturn time. According to dynamo theory, therefore, the efficiency of dynamo action should depend not only on the rotation period of the star, but also on the convection zone properties, in particular the convection zone depth. For main sequence stars, τ_c is a rapidly increasing function of the colour index $B - V$ up to $B - V \sim 0.8 - 0.9$ and almost constant for cooler temperatures (cf. Noyes et al. 1984a). Do we have evidence for such a dependence of stellar activity upon both rotation *and* convection ?

Noyes et al. (1984a) were the first to point out the importance of the Rossby number for a correct interpretation of stellar activity in terms of the dynamo mechanism. They found that a plot of Ca II H+K surface fluxes normalised to the stellar bolometric flux σT_{eff}^4 vs. the Rossby number R_o has significant less scatter than a plot of the Ca II H+K fluxes versus rotation period P_{rot}, although the latter plot also shows an acceptable correlation. They also found that the improvement could be obtained only for a specific value (i.e. 2.0) of the ratio of the mixing length to the pressure scale height which enters in the calculation of τ_c. Moreover, if the Ca II H+K fluxes were not normalised to the bolometric flux, the correlation was in fact worse. Although this result, as well as subsequent similar results by many other authors (cf. Simon 1992), is certainly attractive, it is not a proof than an $\alpha\omega$ dynamo is really responsible for the observed stellar activity. First, the improvement in the correlation is only marginal and depends on the use of a poorly determined theoretical quantity. Secondly, both the normalised flux and the Rossby number depend on spectral type and it may be that this conspires to produce somewhat less scatter than a plot of surface fluxes vs. rotation period (see also Hartmann and Noyes 1987).

A recent reanalysis of the same question by Hempelmann et al. (1995a) using X-ray data from the ROSAT all-sky survey shows that the ratio of X-ray luminosities to bolometric luminosities for a sample of field and cluster stars correlate well with both the rotation period and the Rossby number. There is no strong reason from an observational point of view to chose one formulation rather than the other. Thus, while there is ample evidence for a connection between stellar activity and rotation, in qualitative agreement with the predictions of simple dynamo models, it appears unlikely that correlations of the type we have discussed can be used to constrain dynamo mechanisms for stars of different spectral types and convective zone properties. Naively, one should expect that stars with the same rotation period and different spectral types should have different chromospheric and coronal fluxes, with the stars of later spectral type being systematically more active at least up to a certain $B - V$ colour (see Eq. 2 above). It does not seem that this simple expectation is confirmed by the available data, although temporal variability, small data samples and possibly other (unkown) effects may effectively mask the expected trend.

3. Activity Cycles

Somewhat better constraints on dynamo mechanisms could be provided by observations of stellar activity cycles, particularly the long term program of Ca II observations carried out at Mt. Wilson (Wilson 1978, Baliunas and Vaughan 1985, Baliunas et al. 1995a). These data, which now span a quarter of a century, have shown a variety of behaviours which can basically be reduced to three different types. About one third of the observed stars, mostly old quiet stars like the Sun, show a cyclic behaviour, with periods ranging from about 3 to 15 years. These are the stars which behave more similarly to the Sun. Another group of stars, usually young active stars, show a chaotic behaviour with no obvious periodicities. Finally, the remaining stars appear to be constant, with no indication at all of an activity cycle. These stars could be in a dormient state of magnetic activity, similarly to the seventeenth century solar Maunder minimum, or could be stars in which the dynamo action is permanently suppressed.

Within the group of stars which show solar-type activity cycles, there is no significant dependence of the cycle period on other stellar parameters (Saar and Baliunas 1992). The cycle period does not seem to depend on spectral type (i.e. depth of the convection zone), nor on rotation (neither rotation period nor Rossby number), nor on the star activity level. It is interesting to note that some dynamo models have made specific predictions about the dependence of the cycle period on, e.g., spectral type, although these predictions were often contradictory (both cycle periods increasing and decreasing with spectral type were predicted; see also Noyes et al. 1984b for a reported dependence of the cycle period on the rotation period, based on a smaller data sample). Things are not much better with regard to the cycle amplitude, although here some trends start to appear (Saar and Baliunas 1992). The cycle amplitudes tend to be longer for later spectral types and for more slowly rotating (i.e. less active) stars, but the scatter is large, and selection effects in the data sample can bias the results. There is also no apparent correlation between cycle amplitude and period.

A more evident correlation was found by Saar and Baliunas (1992) between the normalised cycle frequency $\Omega^*_{cyc} = \Omega_{cyc} R^2/\eta_t$ and the dynamo

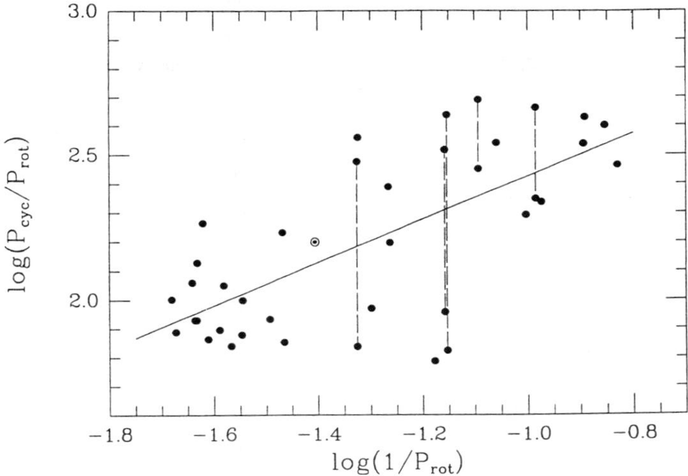

Figure 1. **The ratio P_{cyc}/P_{rot} vs. the inverse of P_{rot} for solar-type stars with cyclic activity (from Baliunas et al. 1995b).**

number N_D. Not only Ω^*_{cyc} increases with N_D, but now we have a clear separation between stars of different activity levels. Quiet slowly rotating stars like the Sun define an upper locus in the plane Ω^*_{cyc} vs. N_D, while more active stars define a separate locus systematically shifted to lower Ω^*_{cyc} and higher N_D than for quiet stars (see Fig. 8 in Saar and Baliunas 1992). This suggests that stars in different regimes of rotation may have different kinds of dynamos, as predicted by some theoretical models (see discussion by Weiss in this volume). The problem with this kind of correlations (see also Tuominen et al. 1988) is that they combine purely observational quantities with theoretical computed quantities like the convective overturn time or the turbulent magnetic diffusion time, with all the uncertainties asssociated with the mixing length theory of convection (this applies also to the use of the Rossby number by Noyes et al. 1984a, as discussed above).

Recently, Soon et al. (1993) and Baliunas et al. (1995b) have suggested the use of a purely observational quantity, the ratio P_{cyc}/P_{rot} of the cycle period to the rotation period, to characterise magnetic activity in stars with well defined cycles. They argue that this ratio is a measure of the dynamo number and could be used as an observational test of the predictions of the dynamo theory. A plot of this ratio vs. the colour index $B-V$ shows clearly a separation between active rapidly rotating stars (with larger P_{cyc}/P_{rot}) and quiet slowly rotating stars, in agreement with the interpretation of this ratio as a measure of dynamo efficiency (Soon et al. 1993). Different dynamo models predict that this ratio should scale as $\sim N_D^\beta$ with β in the range $\sim 0.3 - 0.9$; the observational data on activity cycles are consistent with this trend, provided one uses the inverse of the rotation period as a measure of the dynamo number and considers only stars that have similar internal structure and convective zone properties (Fig. 1). Within this framework, the cycle amplitude shows an inverse correlation with the dynamo number, but only for the old slowly rotating stars (Baliunas et al. 1995b).

Finally, Hempelmann et al. (1995b) have investigated the X-ray emission of stars with different types of Ca II activity cycles using data from the *ROSAT* all sky-survey. They find that stars with a chaotic behaviour have typically higher X-ray emission that stars with solar-type regular cycles or no cycles at all. This was expected, since stars with chaotic behaviour and irregular cycles were known to have typically higher levels of Ca II emission. More interestingly, stars with regular cycles or no cycles are also those with the largest Rossby numbers (i.e. smaller dynamo numbers), suggesting again a different dynamo pattern for rapidly and slowly rotating stars.

To go beyond the general qualitative statements above seems premature at this stage. There are many factors that enter the complex hydromagnetic process of field generation by dynamo action. Trying to reduce this complex process to correlations between a few parameters is certainly an oversimplification. The data themselves, which are based on a limited time span, also contain many uncertainties (particularly in the determination of P_{cyc}). Conclusions based on simple correlations and scaling arguments, although attractive, should always be considered with caution.

References

Baliunas, S.L., et al. (1995a) *Ap. J.* **438**, 269

Baliunas, S.L., Nesme-Ribes, E., Sokoloff, D., and Soon, W.H. (1995b) *Ap. J.*, in press

Baliunas, S.L., and Vaughan, A.H. (1985) *Ann. Rev. Astron. Ap.* **23**, 379

Bouvier, J., 1995, in J.-P. Caillault (ed.) *Eight Cambridge Workshop on Cool Stars, Stellar Systems, and the Sun*, ASP Conference Series **24**, 151

Hartmann, L.W., and Noyes, R.W. (1987) *Ann. Rev. Astron. Ap.* **25**, 271

Hempelmann, A., Schmitt, J.H.M.M., Schultz, M., Rüdiger, G., and Stępień, K. (1995a) *A&A* 294, 515

Hempelmann, A., Schmitt, J.H.M.M., and Stępień, K. (1995b) *A&A*, in press

Noyes, R.W., Hartmann, L.W., Baliunas, S.L., Duncan, D.K., and Vaughan, A.H. (1984a) *Ap. J.* **279**, 763

Noyes, R.W., Weiss, N.O., and Vaughan, A.H. (1984b) *Ap. J.* bf 287, 769

Rosner, R., and Weiss, N.O. (1992) in K.L. Harvey (ed.) *The Solar Cycle*, ASP Conference Series **27**, 511

Saar, S.H., and Baliunas, S.L. (1992) in K.L. Harvey (ed.) *The Solar Cycle*, ASP Conference Series **27**, 150

Simon, T., 1992, in M.S. Giampapa and J.A. Bookbider (eds.) *Seventh Cambridge Workshop on Cool Stars, Stellar Systems, and the Sun*, ASP Conference Series **26**, 3

Skumanich, A., 1972, *Ap. J.* **171**, 565

Soon, W.H., Baliunas, S.L., and Zhang, Q. (1993) *Ap. J. Letters* **414**, L33

Tuominen, I., Rüdiger, G., and Brandenburg, A. (1988) in O. Havnes et al (eds.) *Activity in Cool Star Envelopes*, Kluwer, p. 13

Wilson, O.C., 1978 *Ap. J.* **226**, 379

RECENT MEASUREMENTS OF STELLAR MAGNETIC FIELDS

S.H. SAAR
Smithsonian Astrophysical Observatory
MS 58, 60 Garden St. Cambridge, MA 02138, USA

1. Introduction

Since Robinson's (1980) groundbreaking paper, the magnetic properties of cool stars (i.e., the surface filling factor of active regions, f, and their mean field strength, B) have been studied by several groups. A vaguely disquieting trend has developed over the last decade, however: measured magnetic fluxes have generally decreased with time. Values for ϵ Eri (a moderately active K2 dwarf), for example, have progressed (see Saar 1990; Valenti et al. 1995; and refs. therein) from $fB = 0.57 - 0.78$ kG in 1984 to $fB = 0.30 - 0.35$ in 1989, with only a solitary upper limit ($fB \leq 0.11$ kG, if $B \equiv 1.5$ kG; Saar 1988) significantly deviating from the trend.

Much of the change can be attributed to improvements in the analyses. As more physical effects were included in the modeling (treatment of radiative transfer, blends, disk-integration, etc.), more sources of line broadening were treated and the magnetic parameters (especially f) decreased. But one must ask: will $fB \rightarrow 0$ eventually? In other words, are the measurements real, or the product of still not-understood or overlooked systematic effects?

2. Results of Recent Observations

Several analyses of new, primarily IR, data, indicate that the measurements *are* indeed real. The first evidence comes from a careful study of infrared (IR) spectra by Valenti et al. (1995), who find $B = 1.44$ kG and $f = 8.8\%$ for ϵ Eri by modeling the $g_{eff} = 3$ 1.56μm Fe I and other features; lower limits were determined for two inactive stars. Thus, $fB(\epsilon$ Eri$) \approx 0.12$ kG, clearly non-zero, but significantly smaller than results from optical data.

The advent of cryogenic echelles and IR array detectors promises an explosion of new and improved magnetic field measurements. The first look at data from these devices (the CSHELL on NASA's IRTF; Greene et al.

Y. Uchida et al. (eds.),
Magnetodynamic Phenomena in the Solar Atmosphere – Prototypes of Stellar Magnetic Activity, 367–374.

1993) is presented here and in Saar (1996). We obtained spectra of four members of the 2.22μm Ti I multiplet (Saar & Linsky 1985) at a resolution $\lambda/\Delta\lambda \approx 35{,}000$ and $S/N \geq 100$ in January 1995. These were sky-subtracted, flat-fielded, telluric lines removed (using scaled spectra of A and B stars), and residual fringing (from the circular variable filter) removed. So far, only one pair of Ti I lines has been modeled. Because of this and the fringing problems, the results given here should be regarded as preliminary.

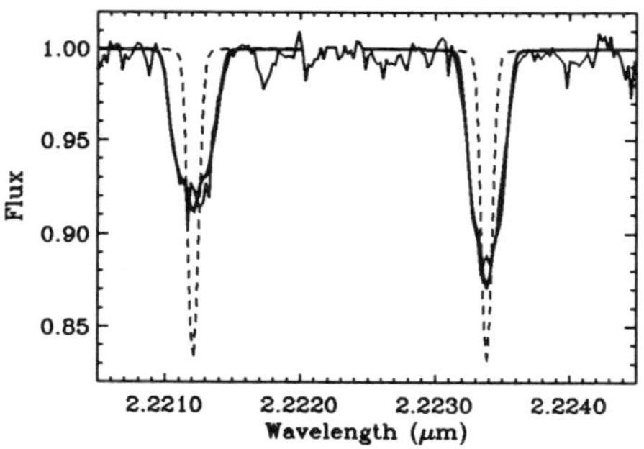

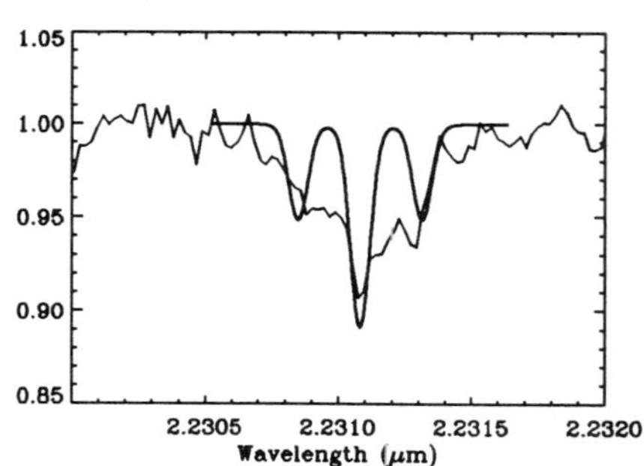

Figure 1. Left: IRTF CSHELL spectrum of GL 171.2A (thin solid line, $\lambda/\Delta\lambda = 35,000$), showing Ti I lines at 2.2211 μm ($g_{\rm eff} = 2.00$) and 2.2233 μm ($g_{\rm eff} = 1.67$) and line models with B= 2.8 kG and $f = 50\%$ (heavy solid) and $B = 0$ (dashed), showing clear magnetic broadening. Right: CSHELL spectrum of AD Leo showing the Ti I line at 2.2310 μm ($g_{\rm eff} = 2.50$) and a model with B= 4.0 kG and $f = 60\%$ (heavy solid). This and other lines clearly show that a single component model is inadequate; a distribution of B values, either across the surface or vertically or both, is needed.

Among the dozen or so K and M stars we observed was GL 171.2A (= BD +26°730), a nearly pole-on K5Ve flare star known to exhibit magnetic line broadening (Saar et al. 1990; Basri & Marcy 1994). We fit the Ti I line profiles with a two component model: the observed flux $F_{\rm obs} = fF_{\rm mag}(B) + (1-f)F_{\rm quiet}(B=0)$ where $F_{\rm mag}$ and $F_{\rm quiet}$ are the disk-integrated fluxes from a completely magnetic and a $B = 0$ star, respectively. We adopt identical Milne-Eddington atmospheres for both components, and include full Zeeman patterns, magneto-optical effects, and radial-tangential macroturbulence. We find the Ti I lines are much broader than implied by $v \sin i$ alone, and are best modeled if $B \approx 2.8$ kG and $f \approx 50\%$ (Fig. 1, left). These values are quite consistent with previous optical results, indicating that at least *some* of the older magnetic measurements are accurate.

Other CSHELL results (Table 1) confirm strong B on dMe stars (Saar 1994a), and reveal details of horizontal and vertical magnetic structure on stars. Since the Ti I features increase in strength with decreasing $T_{\rm eff}$ and are quite weak in normal K stars (e.g., ϵ Eri; Saar 1988), their anomalous strength in LQ Hya (K2Ve) and II Peg (K2IVe) together with widths $\gg$

$v \sin i$ imply *starspot* magnetic fields are being observed for the first time. Optical Zeeman Doppler images of LQ Hya (Donati 1996), which map the local $\sum(f\vec{B})$ distribution across the stellar surface, show a much smaller net flux. This suggests that much of the magnetic flux on LQ Hya (or at least the plage-like regions which dominate the optical signal) is in the form of bipoles with separations smaller than a surface resolution element. Late M dwarfs (AD Leo and YZ CMi) clearly show unusually broad Zeeman σ components (Fig. 1, right) which might arise either from a spread in B in different active regions on the surface, or a significant vertical gradient in B over the height of formation of the Ti I lines, or some combination of both (see also Saar 1992, Johns–Krull & Valenti 1995). These results are described in more detail in Saar (1996).

3. Analysis

3.1. THE DATA SET

With these new, improved magnetic field measurements in hand, it is clearly useful to revisit the relationships between f and B, stellar properties, and magnetic activity indicies such as X–ray and Ca II HK emission. Caution is required, however, since the latest results (e.g., Valenti et al. 1995) imply some of the older measurements are faulty. I have therefore compiled a carefully selected sample of magnetic measurements from analyses which treat radiative transfer effects and use disk-integration in their models (see Landolfi et al. 1989). In addition, I (ruthlessly!) neglect results from low S/N IR data, measurements using Fe I 8468Å in K dwarfs (they disagree with IR data for ϵ Eri and 61 Cyg A), Zeeman/magnetic Doppler imaging results (due to problems detecting all the flux), and curve–of–growth analyses (which cannot separate f and B, measuring only their product at lower accuracy). This leaves the new CSHELL results presented here, and the data compiled in Table 1. In what follows, I focus primarily on dwarf stars (and leave the RS CVn II Peg out of the fits). It is important to realize that uncertainties in the physical properties of stellar fluxtube atmospheres introduce significant, incompletely understood, and difficult to quantify systematic effects in the magnetic parameters (especially f); these will dominate over random errors in f and B (Basri et al. 1990; Saar et al. 1994; Saar & Solanki 1996).

3.2. CORRELATIONS WITH STELLAR PROPERTIES

Previous studies (e.g., Saar 1990, 1994a) have indicated that B is confined by photospheric gas pressure: $B \leq B_{\rm eq} = (8\pi P_{\rm gas})^{0.5}$ (where $P_{\rm gas}$ is the gas pressure). Using newer model atmospheres (Kurucz 1991) and taking

TABLE 1. Selected Magnetic Parameter Measurements and Stellar Data

Star/ HD #	sp. type	P_{rot} (d)	f (%)	B (kG)	ref	$\log F_X$	$\log F_{CIV}$ (ergs cm^{-2} s^{-1})
sun	G2V	25.4	1.5	1.5	1	4.8	4.0
115383	G0V	4.9	19	1.0	2	6.2	5.1
20630	G5V	9.4	20	1.8	3	5.9	4.8
131156A	G8V	6.2	18	1.9	2,9	6.4	4.9
131511	K1V	9	6	1.7	2	5.8	4.7
26965	K1V	37	≤2.7	1.71	4*	5.3	3.7
185114	K1V	27.2	≤1.9	1.36	4*	5.0	4.0
22049	K2V	11.3	8.8	1.44	4*	5.7	4.6
17925	K2V	6.76	35	1.5	5	6.1	5.0
LQ Hya†	K2Ve	1.60	70	3.5	6*	7.3	5.6
GL 171.2A	K5Ve	1.85	50	2.8	6*,10	7.1	5.7
EQ Vir	K5Ve	3.9	55	2.5	7	7.0	5.3
DT Vir	M2Ve	1.54	50	3.0	6*	6.9	...
AD Leo	M3.5Ve	2.6	60	4.0	6*,11*	6.6	5.5
YZ CMi	M4.5Ve	2.78	67	4.2	6*	6.8	5.3
EV Lac	M4.5Ve	4.38	50	3.8	8	6.7	...
GL 729	M4.5Ve	...	50	2.6	8	6.2	...
II Peg†	K2IVe	6.7	60	3.0	6*	7.4	5.9

References: [1]Montesinos & Jordan (1993); [2]Linsky et al. (1994); [3]Saar & Baliunas (1992); [4]Valenti et al. (1995); [5]Saar (1991); [6]prelim. result, this paper; [7]Saar (1992; unpubl.); [8]Johns–Krull & Valenti (1995); [9]Marcy & Basri (1989); [10]Saar et al. (1990); [11]Saar (1994a). *IR data; †significant starspot contribution.

$P_{gas}(T = T_{eff})$, Saar (1996) shows that the situation may be more complex (Fig. 2, left). While $B \leq B_{eq}$ is generally valid in most GK stars and at least one less active M dwarf (GL 729), $B > B_{eq}$ in the heavily spotted K dwarf LQ Hya and in more active dKe and dMe stars.

One possible explanation is that we are beginning to see contributions of starspots on the most active stars. This is clearly true for LQ Hya (and II Peg), and may also be true for the dKe/dMe stars. It is also consistent with the increased photometric variability in stars with "saturated" magnetic activity (O'Dell et al. 1995). Thus, for stars with angular velocities $\Omega > \Omega_{sat}$, the magnetic flux continues to increase (Solanki 1994), appearing in the form of spots (with $B_{spot} > B_{plage} \sim B_{eq}$) which are more easily detected in the IR. Alternately, $B_{plage} > B_{eq}$ in "saturated" stars.

If these ideas are correct, $\Omega \sim \Omega_{sat}$ will mark a change in f and B. For $\Omega < \Omega_{sat}$, $B \leq B_{eq}$ and f will be a strong function of rotation. Such

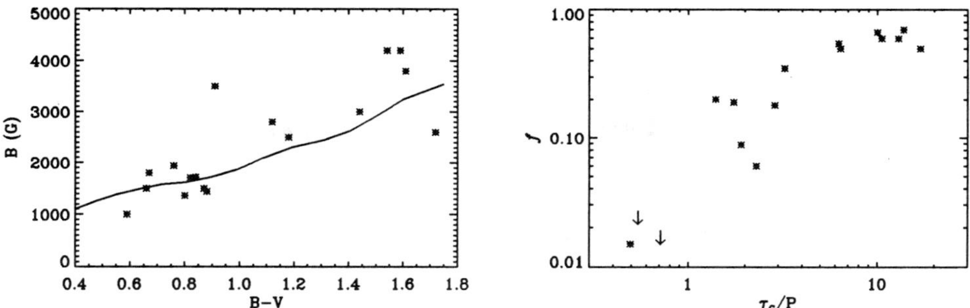

Figure 2. Left: Measured B vs. $B - V$ color; the pressure equipartition field strength B_{eq} is also shown (solid line). Right: Measured f vs. inverse Rossby number, τ_C/P_{rot}, showing a non-linear increase with saturation (see text).

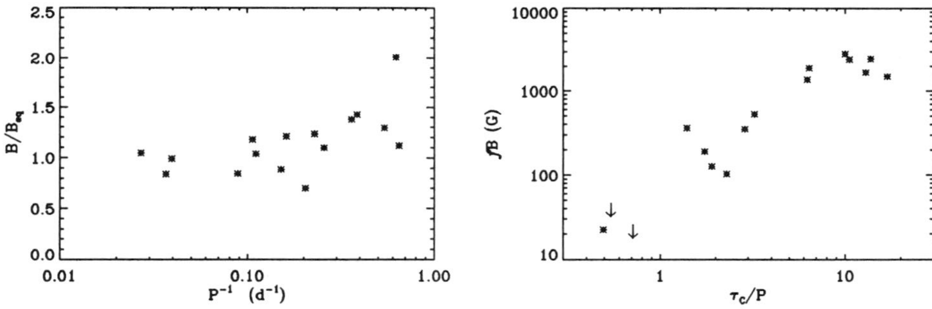

Figure 3. Left: B/B_{eq} vs. inverse rotation period. An increase is suggested for $P <$ 3 days. Right: Measured fB vs. inverse Rossby number, τ_C/P_{rot}, showing a non-linear increase with no clear saturation ($fB \propto Ro^{-1.3}$; see text).

relationships have been noted before (e.g., Saar 1990; 1994b), and the new data confirm and refine them: Saar (1996) shows $f \propto P_{rot}^{-1.8}$ for $P_{rot} > 3$ days, and $f \approx 60\%$ for $P_{rot} < 3$ days. Similar correlations can also be constructed with the Rossby number $Ro = P_{rot}/\tau_C$, where τ_C is the convective turnover time. The f – Ro relation (Fig. 2, right) can be fit with either a power law plus saturated state ($f \propto Ro^{-1.3}$ for $Ro^{-1} < 8$, $f \approx 60\%$ for $Ro^{-1} > 8$) or with $\log f = -0.26 - 0.85Ro$, a result in remarkable agreement with the theoretical estimate of Montesinos & Jordan (1993). This magnetic area saturation for $Ro^{-1} > 8$ ($P_{rot} \sim 3$ days for a mid K dwarf) is consistent with the activity saturation at Ω_{sat} (e.g., Vilhu 1984). There is also some evidence for an increase in B with rotation for $\Omega > \Omega_{sat}$ (Fig. 3, left). Finally, there is *not* clear evidence for saturation in magnetic flux density fB versus Ro (Fig. 3, right; see also Saar 1996), where $fB \propto Ro^{-1.3}$

for all Ro ($\log fB \propto -1.0Ro$ also works). This result lends further support for the idea that while f reaches a maximum for $\Omega > \Omega_{\rm sat}$, B (and thus fB) continues to increases with Ω (i.e., $B \geq B_{\rm eq}$, perhaps due to increased $f_{\rm spot}/f$; Saar 1994b). (I note in passing that while correlations between f or fB and Ro are not significantly better than with $P_{\rm rot}$ for dwarf stars, use of the Rossby number *does* permit II Peg to be included in the same relationship.)

4. Correlations with Magnetic Activity Diagnostics

It is particularly relevant in the context of this meeting to also take a fresh look at relationships between magnetic parameters and emission from stellar outer atmospheres. Table 1 also lists surface fluxes of coronal X-rays, and transition region C IV emission (1550Å). These values have been taken primarily from Rutten et al. (1989, 1991), Hempelmann et al. (1995), Ayres et al. (1995), Wood et al. (1994), and results of several IUE and ROSAT campaigns.

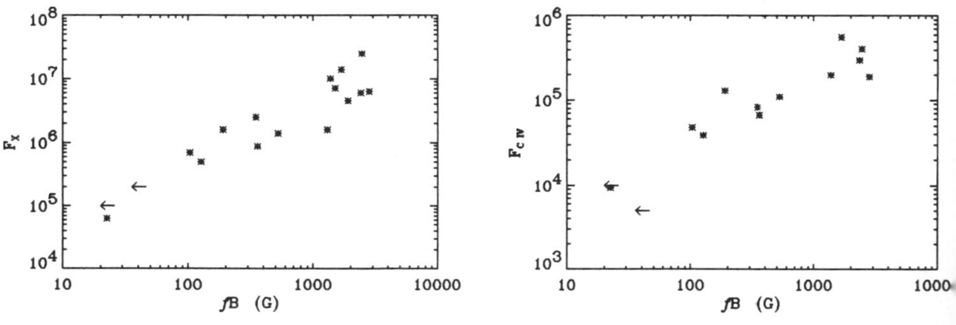

Figure 4. X-ray (left) and C IV (right) surface flux densities (ergs cm^{-1} s^{-1}) vs. magnetic flux density fB; the data can be fit with $F_X \propto (fB)^{1.0}$ and $F_{\rm CIV} \propto (fB)^{0.8}$.

I find $F_X \propto (fB)^{1.0}$ (Figure 4, left), in agreement with previous results (Schrijver et al. 1989; Schrijver 1991). The relation proposed by Stępień (1994; $F_X \propto T_{\rm eff}^{8.3}(fB)^{1.9}$) does not fit the data as well. The C IV data shows $F_{\rm CIV} \propto (fB)^{0.8}$ (Fig. 4, right), again consistent with previous results (Schrijver 1990, 1991). Note however, that almost equally good fits can be found for $F_X \propto f^{1.2}$ and $F_{\rm CIV} \propto f^{1.0}$. The near linear relationship between fB and F_X suggests not all the photospheric magnetic flux reaches (and heats) the corona (Stępień 1988; Montesinos & Jordan 1993). Schrijver (1990) proposes a combined heating mechanism for the transition region.

One can also explore the relationship between magnetic and chromospheric flux. Here, I use the Ca II HK flux excess ($\Delta F_{\rm HK}$) as a proxy for chromospheric emission, computed mostly from values in Rutten et al. (1991). I have ignored the M dwarfs, since the Ca II HK emission lines in these stars are a much less important part of the chromospheric energy budget (H emission dominates) and (uncertain) correction factors are needed to correct for this (Rutten et al. 1989). The result is $\Delta F_{\rm HK} \propto (fB)^{0.5}$ for $fB <$ 0.4 kG, above which chromospheric heating in Ca II appears to saturate. This result, which appears to be due to the topology of merging flux tubes (Solanki et al. 1991; Schrijver 1993), is roughly in agreement with Schrijver et al. (1989), but is based only 11 stars, and should be viewed with caution. Nevertheless, the emission flux – magnetic flux correlations found here are overall consistent with the analogous, spatially resolved relationships seen on the sun (Schrijver 1993), and with stellar flux–flux correlations.

Further accurate measurements of f and B, in particular for both inactive and very active stars, are needed to confirm the tentative results discussed above. Fortunately, new IR spectrographs and arrays, together with improved analysis methods, are now making this feasible.

This paper made use of data obtained at the NASA IRTF, operated by the Univ. of Hawaii under contract with NASA, and the NSO McMath–Pierce telescope, operated by AURA under contract with the NSF. I would like to thank J. Bookbinder for help observing at the IRTF, T. Greene and J. Rayner for deft technical support, and D. Jaksha, M. Giampapa and all involved with the late, great McMath–Pierce Solar–Stellar program for all their help. Thanks also to those referenced below for many enlightening discussions. This work was supported by a Smithsonian ROF grant, and NASA grants NAG5-2173, NAG5-1536, and NAG5-1975.

References

Ayres, T.R., et al. (1995), ApJS **96**, 223
Basri, G.S., Marcy, G.W. (1994) ApJ, **431**, 844
Basri, G.S., Marcy, G.W., Valenti, J. (1990), ApJ **360**, 650
Donati, J.-F. (1996) in *Stellar Surface Structure* (ed. K.G. Strassmeier), Kluwer, in press.
Greene, T., Tokunaga, A., Toomey, D.W., Carr, J. (1993), *Proc. SPIE*, **1946**, 313
Hempelmann, A., Schmitt, J.H.M.M., Schultz, M., Rüdiger, G., Stepień, K. (1995), A&A **294**, 515
Johns–Krull, C.M., Valenti, J.A. (1995) in *Cool Stars, Stellar Systems, and the Sun*, (eds. R. Pallavicini, A. Dupree), ASP Conf. Ser., in press.
Kurucz, R.L. (1991) in *Precision Photometry: Astrophysics of the Galaxy* (eds. P. Davis, A. Upgren, & K. Janes), L. Davis., 27
Landolfi, M., Landi Degl'Innocenti, M., Landi Degl'Innocenti, E. (1989) A&A **216**, 113
Linsky, J.L., Andrulis, C., Saar, S.H., Ayres, T.R., Giampapa, M.S. (1994) in *Cool Stars, Stellar Systems, and the Sun*, (ed. J.-P. Caillault) ASP Conf. Ser. 64, p. 438
Marcy, G.W., Basri, G.S. (1989) ApJ **345**, 480
Montesinos, B., Jordan, C. (1993) MNRAS **269**, 900
O'Dell, M.A., Panagi, P., Hendry, M.A., Collier Cameron, A. (1995), A&A, **294**, 715

Robinson, R.D. (1980), ApJ, **239**, 961

Rutten, R.G.M., Schrijver, C., Zwaan, C., Duncan, D., Mewe, R. (1989), A&A **219**, 239

Rutten, R.G.M., Schrijver, C.J., Lemmens, A.F.P., Zwaan, C. (1991), A&A **252**, 203

Saar, S.H. (1988) ApJ, **324**, 441

Saar, S.H. (1990) in *IAU Symposium 138, The Solar Photosphere: Structure, Convection, and Magnetic Fields* (ed. J. O. Stenflo), Kluwer, p. 427

Saar, S.H. (1991) 1991, in *Mechanisms of Chromospheric and Coronal Heating*, (eds. P. Ulmschneider, E. Priest, R. Rosner), Springer, p. 273

Saar, S.H. (1992) in *Cool Stars, Stellar Systems, and the Sun*, (eds. M.S. Giampapa, J.A. Bookbinder) ASP Conf. Ser. 26, p. 252

Saar, S.H. (1994a) in *Infrared Solar Physics*, (eds. D. Rabin, J. Jeffries), Kluwer, p. 493

Saar, S.H. (1994b) in *Cool Stars, Stellar Systems, and the Sun*, (ed. J.-P. Caillault), ASP Conf. Ser. 64, p. 319

Saar, S.H. (1996) in *Stellar Surface Structure* (ed. K.G. Strassmeier), Kluwer, in press.

Saar, S.H., Baliunas, S.L. (1992) in *The Fourth Solar Cycle Workshop*, (ed. K. L. Harvey), ASP Conf. Ser. 27, p. 197

Saar, S.H., Bünte, M., Solanki, S.K. (1994), in *Cool Stars, Stellar Systems, and the Sun*, (ed. J.-P. Caillault), ASP Conf. Ser. 64, p. 474

Saar, S.H., Golub, L., Bopp, B.W., Herbst, W. & Huovelin, J. (1990) in *Evolution in Astrophysics* (ed. E. Rolfe). ESA SP–310, p. 431

Saar, S.H., Linsky, J.L. (1985) ApJ, **299**, L47

Saar, S.H., Solanki, S.K. (1996), A&A, in press

Schrijver, C.J. (1990) A&A **234**, 315

Schrijver, C.J. (1991) in *Mechanisms of Chromospheric and Coronal Heating*, (eds. P. Ulmschneider, E. Priest, R. Rosner), Springer, p. 257

Schrijver, C.J. (1993) A&A **269**, 395

Schrijver, C.J., Coté, J., Zwaan, C. & Saar, S.H. (1989) ApJ **337**, p. 964

Solanki, S.K. (1994) in *Cool Stars, Stellar Systems, and the Sun*, (ed. J.-P. Caillault), ASP Conf. Ser. 64, p. 477

Solanki, S.K., Steiner, O., Uitenbroek, H. (1991) A&A **250**, 220

Stepień, K. (1988), ApJ **335**, 892

Stepień, K. (1994), A&A **292**, 191

Valenti, J.A., Marcy, G.W., Basri, G.S. (1995) ApJ **439**, 939

Vilhu, O. (1984) A&A, **133**, 117

Wood, B.E., Brown, A., Linsky, J.L., Kellet, B.J., Bromage, G.E., Hodgkin, S.T., Pye, J.P. (1994), ApJS **93**, 287

ACTIVE LONGITUDES AND ROTATION OF ACTIVE STARS

M. RODONÒ[1,2] AND A. F. LANZA[2]
[1] *Istituto di Astronomia dell'Università di Catania*
[2] *Osservatorio Astrofisico di Catania*
Viale A. Doria, 6 - I 95125 Catania, Italy

Abstract. The interpretation of light curves and line profiles of active stars indicates that surface features exist at different atmospheric heights and these features, assumed as tracers of stellar rotation, suggest the persistence of active longitudes for timescales up to decades, as well as a differential rotation regime. The current results and some newly proposed methods to investigate activity and rotation are reviewed in the framework of recently proposed stellar dynamo models.

1. Introduction

Activity phenomena related to magnetohydrodynamics processes occurring in stellar interiors and atmospheres have been identified in a number of single and binary stars, among which are chromospherically active main sequence single stars (Baliunas & Vaughan 1985), T Tauri stars (Appenzeller & Dearbon 1984, Appenzeller & Mundt 1989, Montmerle et al. 1991), FK Comae giants (Bopp 1983), and close binaries belonging to the RS Canum Venaticorum class (Bopp & Fekel 1977, Catalano 1983, Rodonò 1992, 1994), UV Ceti flare stars (Rodonò 1990), Algols (Olson 1984) and cataclismic binaries (Bianchini 1990). These variables show a solar-like activity, characterized by magnetic fields which control the transfer of energy and momentum into the atmosphere producing cool spots in the photosphere and non-radiatively heated structures in the upper levels (chromospheric and coronal plages and flares).

The observations indicate that the basic ingredients for the development of solar-like activity are a deep convective envelope and a fast ro-

Y. Uchida et al. (eds.),
Magnetodynamic Phenomena in the Solar Atmosphere – Prototypes of Stellar Magnetic Activity, 375–386.

tation (usually $> 5 - 10$ km s^{-1} at the equator), in agreement with the magnetohydrodynamics dynamo model (Moffatt 1977, Krause & Rädler 1980, Zeldovich et al. 1983, Weiss 1994). However, a satisfactory comprehension of the physics of magnetic fields in a turbulent stellar envelope at very high magnetic Reynolds numbers is still lacking, essentially because of our inadequate description of the non-linear processes governing the turbulence itself. Therefore a leading role is played by the observational study of magnetic activity, which allows us to put some constraints on the basic parameters of the dynamo mechanism, such as the differential rotation amplitude and turbulent diffusivity, and to test its basic results concerning activity cycle lengths, large scale field structures, and surface migration of the activity centers.

In the present review, basic observational methods are reviewed in the framework of the so-called *solar-stellar connection* and their capabilities and results are briefly discussed, addressing in particular the problem of the surface distribution of activity and its relationship with stellar rotation.

2. Active Longitudes

Solar active regions show a marked tendency to appear in close proximity both in time and space, the probability of formation of a new active region being significantly higher near a preexistent active region (Zwaan 1992, Harvey & Zwaan 1993). Sunspot *activity complexes* or *activity nests* have lifetimes of the order of six months (e.g., Gaizauskas et al. 1983, Castenmiller et al. 1986, Harvey & Zwaan 1993). Moreover, the longitude distribution of major solar flares is not random, showing the existence of *hot spots* with lifetimes longer than the 22-year magnetic cycle and rotation rates usually slower than that of the activity belts (Bai 1988, 1990).

In the case of the active stars, the tendency for a clustering of the photospheric spots around certain longitudes is very pronounced, as shown, e.g., by the existence and the long-time phase coherence of a photometric wave in the light curves of the RS CVn binaries (Catalano 1983, Rodonò et al. 1995). Modelling the wide-band flux modulation in terms of discrete starspots (e.g., Strassmeier & Bopp 1992, Henry et al. 1995) or a continuous distribution of spots (e.g., Eaton & Hall 1979, Rodonò et al. 1995), the distribution of the spotted area versus longitude can be derived. Ground-based photometry of eclipsing binaries allows us to reach a resolution of $10^\circ - 30^\circ$ in longitude during eclipses, when the occulting star acts as a moving screen progressively scanning the spotted companion, and of $50^\circ - 90^\circ$ outside eclipses.

The actual resolution in latitude is poor when wide-band photometry alone is used, but a significant improvement can be achieved by spectroscopy using the Doppler Imaging technique, provided that the star has a projected equatorial rotation velocity $v \sin i$ in the range $20 - 90$ kms^{-1} (Vogt et al. 1987, Piskunov et al. 1990, Cameron 1992).

The main result from multi-band photometry and Doppler imaging techniques is the detection of cool spots with areas $10^3 - 10^4$ times that of sunspots, but with a lower temperature deficit, usually between $800 - 1200$ K (Eaton 1992) and lifetimes from a few months up to several years (Hall & Busby 1990). It is not known whether such regions are large coherent single spots or a loose cluster of many small and short-lived solar-size spots. Only in the latter case the analogy with the solar activity complexes should be fully appropriate. However, we shall denote the locations of starspots as *stellar active longitudes* and characterize them by their motion and lifetime.

Short-period RS CVn active components show evidence of two active longitudes near the quadrature points with one longitude being usually more active than the other at any given time (Zeilik 1991, Zeilik et al. 1994 and references therein). A similar behaviour has been detected in the apparently single giant star FK Comae (Jetsu et al. 1993, 1994).

In longer period RS CVn binaries ($P_{orb} > 1 - 2$ days) one active longitude is usually present. It drifts slowly with respect to the coordinate system co-rotating with the orbital motion and the drift rate can be variable in amplitude and sign (Catalano 1983), but the moving active longitude seems to be a long-lived structure with a lifetime of the order of the decade. Sometimes two (or more) active longitudes may be present as found by Rodonò et al. (1995) for RS CVn (see Figure 1).

In the case of single stars, there is some indication that the spot distribution in longitude tends to become more uniform as the level of activity decreases, thus approaching the solar behaviour (Baliunas & Vaughan 1985, Hall 1991, Radick et al. 1995).

The distribution of spots in latitude, as inferred by the Doppler imaging technique, usually shows high-latitude or polar-cap spots, whose reliability is strongly debated. Recently, new arguments in favour of high-latitude spots have been proposed (Hatzes et al. 1995), but accurate line profile models are needed for a conclusive interpretation of the observations (Byrne 1992, Unruh & Cameron 1995).

3. Differential Rotation

In principle, latitudinal differential rotation can be determined from the Fourier transform of the rotationally broadening function of the photospheric line profiles (Bruning 1981). This technique is suited for stars with a uniform photospheric temperature distribution and has indeed been ap-

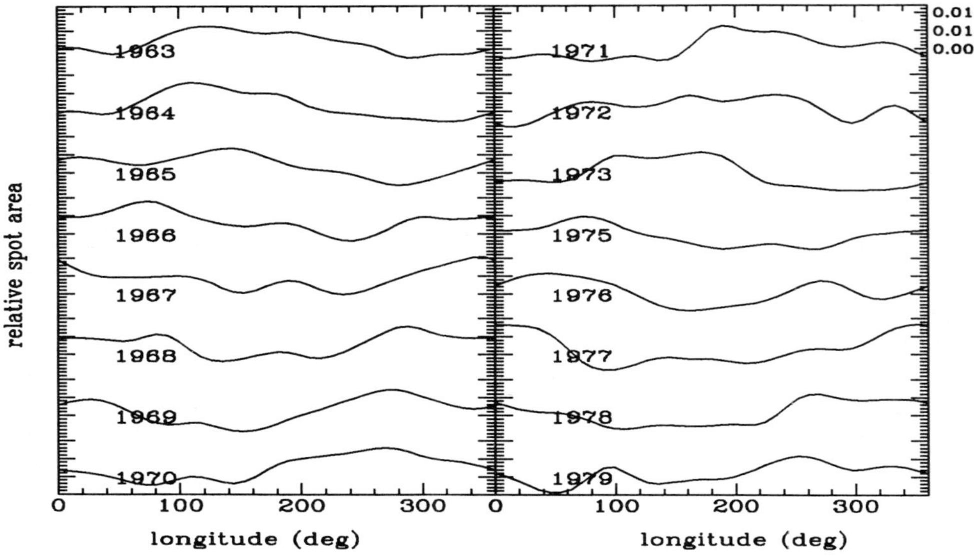

Figure 1. The relative spotted area (total active star photosphere = 1) per 9° longitude bin vs. longitude derived from the observed yearly light curves of RS CVn from 1963 to 1979 (Rodonò et al. 1995).

($D_r = \frac{\Omega_e - \Omega_p}{\Omega_e}$), the main result has been that $D_r = 0.0 \pm 0.3$ is consistent with the observations for both groups of stars (for the Sun $D_r = 0.2$). Active solar-like stars have surface inhomogeneities which can be used as tracers of the stellar rotation. The first method proposed to measure latitudinal differential rotation is based on the change of the drift rate of the photometric wave in RS CVn systems (Rodonò 1981, Catalano 1983). It assumes that different drift rates are due to spots at different latitudes, so that an estimate of the amplitude of stellar differential rotation can be obtained from a long time sequence of light curves. As an example we present in Figure 2a the drift of the longitude of maximum spottedness on the active component of RS CVn during a period of about 30 years. The drift is measured with respect to a coordinate system co-rotating with the orbital motion and its variation gives a D_r between 5% and 20% of the solar value, in the hypothesis that the main activity belt migrates in latitude between 25° and 10°. A particularly interesting and intriguing phenomenon is the possible correlation between the apparent acceleration of the stellar rotation and the decrease of the orbital period near the end of 80's (see Figure 2b).

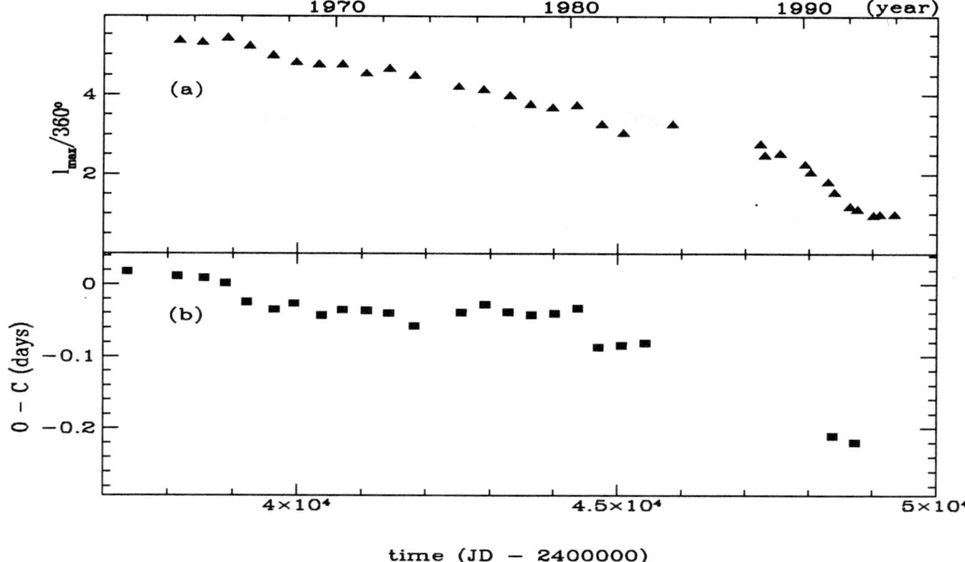

Figure 2. The longitude of the spotted area maximum l_{max} from spot modelling of the 1963-1993 yearly light curves **(a)**, and the orbital period variation **(b)**, vs. time.

Hall (1991) has reviewed the available data for the migration of spots' longitudes in a sample of 85 stars. In the hypothesis that the active longitude drift reflects latitudinal differential rotation, a relationship between rotation and differential rotation amplitude can be derived. It turns out that D_r is of the order of $10^{-1} - 10^{-3}$ of the solar value and decreases with increasing Ω, ($D_r \sim \Omega^{-0.85}$) contrary to the intuitive expectation that the amplitude of differential rotation should increase with the increasing effect of the Coriolis force on convective motions. The fact that a large fraction of the stars in the sample are binaries does not seem to affect the value of D_r so that tidal forces can not be invoked to explain such a result.
The evaluation of the sign of the difference $\Omega_e - \Omega_p$ can be obtained from spectroscopic data. Vogt & Hatzes (1991) have deduced a polar acceleration with $D_r = -0.02 \pm 0.002$ for the active component of UX Arietis from a sequence of Doppler images of the star.
In favour of such low values of differential rotation, Hall & Busby (1990) report also the long observed life-times of spots. In fact such large surface features should be smeared out by the shear due to a latitudinal differential rotation of solar amplitude within a few rotations. However, the situation

sequence of Doppler images of the star.

In favour of such low values of differential rotation, Hall & Busby (1990) report also the long observed life-times of spots. In fact such large surface features should be smeared out by the shear due to a latitudinal differential rotation of solar amplitude within a few rotations. However, the situation can be more complicated if starspots consists of several small size spots which emerge and decay at random within a larger activity complex (see, e.g., Eaton et al. 1995).

Another method to estimate stellar differential rotation in active stars is the analysis of the time series of the spectroscopic index S which measures the strength ofthe chromospheric emission (Baliunas et al. 1983, 1985, Donahue & Baliunas 1992). The value of S sometimes shows a quasi-periodic modulation with the rotation phase and the analysis of an enough long and continuous time series allows us to determine the rotation period P_{rot}. Usually P_{rot} is not constant but shows variations between 1% and 10% among observing seasons. As in the previous case, such changes are attributed to changes in the latitude of the chromospheric plages along an activity cycle, so that the range of the observed rotation periods can be used to estimate the differential rotation amplitude. In the case of the Sun the expected change of P_{rot} along the 11-year cycle has indeed been detected in disk-integrated data (Donahue 1993). From the analysis of 37 main sequence single stars, Donahue (1993) has deduced that D_r increases with the rotation rate Ω ($D_r \sim \Omega^{0.7}$). Moreover, for several stars, the variation of the rotation period versus the activity level along the cycle (measured by the average value of the S index itself) could allow us to infer also the sign of the difference $\Omega_e - \Omega_p$, as described in detail by Donahue & Baliunas (1992) and Donahue (1993).

The above described methods are based on the migration in latitude of the spots or plages, but in the case of the RS CVn variables no established evidence exists of a regular migration along the activity cycle (Hall 1991, Rodonò et al. 1995). In order to overcome this problem and obtain an estimate of D_r in a time scale short with respect to the length of the activity cycle, Lanza et al. (1993, 1994) have proposed a method based on the Fourier analysis of a continuous sequence of wide-band light curves. If starspots consist of many solar-size spots, each rotating with the angular velocity of the surrounding plasma at its latitude, as suggested by the observation of sunspot groups (Gilman & Howard 1986), then the spot pattern will be progressively distorted by the latitudinal differential rotation and eventually each individual latitude belt will drift with respect to the others. The disk-integrated photometric modulation produced by the spot pattern will show the rotation frequencies of the various small-size spots and the ensemble of the frequencies can be resolved by a Fourier analysis

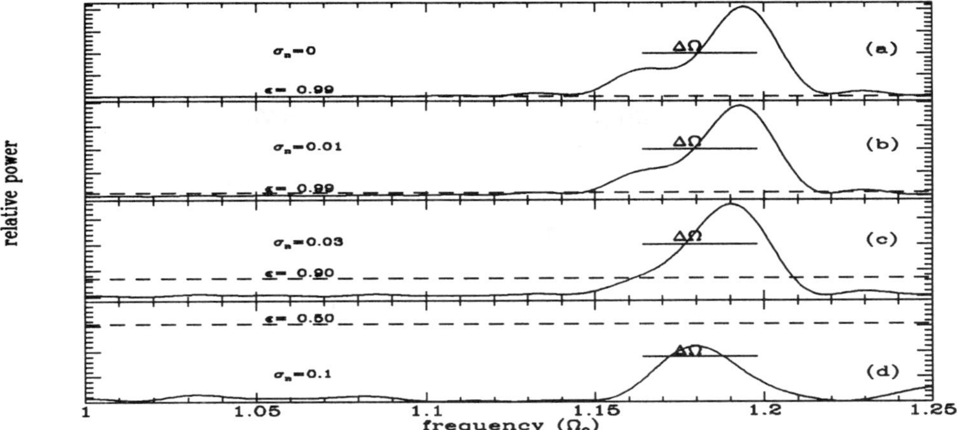

Figure 3. Periodograms obtained from simulated light curve sequences of a differentially rotating spotted star with a varying level of noise added. The noise is Gaussian with standard deviation σ_n measured in unit of the flux of the unspotted star. The horizontal bar indicates the interval ($\Delta\Omega$) covered by the rotation frequencies of the spot pattern. The confidence level for the periodogram main peak is indicated by the dashed line. The multiperiodic nature of the modulation and the amplitude of the differential rotation can be reliably detected up to $\sigma_n = 0.01$.

are not easily met using ground-based telescopes, however a possible multiperiodicity has been detected by Baliunas et al. (1985) in some of their most extended time series.

It is also possible to obtain information on the sign of the latitudinal differential rotation, without reconstructing a sequence of surface maps, by applying the same technique to spectroscopic data (see Lanza & Rodonò 1996). The method can be applied to any star suited for Doppler Imaging and does not suffer from problems related to line profile modelling or regularization criteria, because it directly exploits the correspondence between latitude belts and line profile intervals along a rotationally broadened line profile. It requires a sequence of line profiles with high signal-to-noise ratio ($\frac{S}{N} \geq 300$) and a coverage without gaps longer than ~ 0.2 in rotation phase. In order to measure the amplitude and sign of a differential rotation of order $\Delta\Omega$, the extension of the sequence should be at least of $\frac{2\pi}{\Delta\Omega}$. An example of the results obtained from simulated data is shown in Figure 4. The application of such a technique to real stars requires a coordinated whole-Earth network of telescopes or a dedicated space-borne observatory

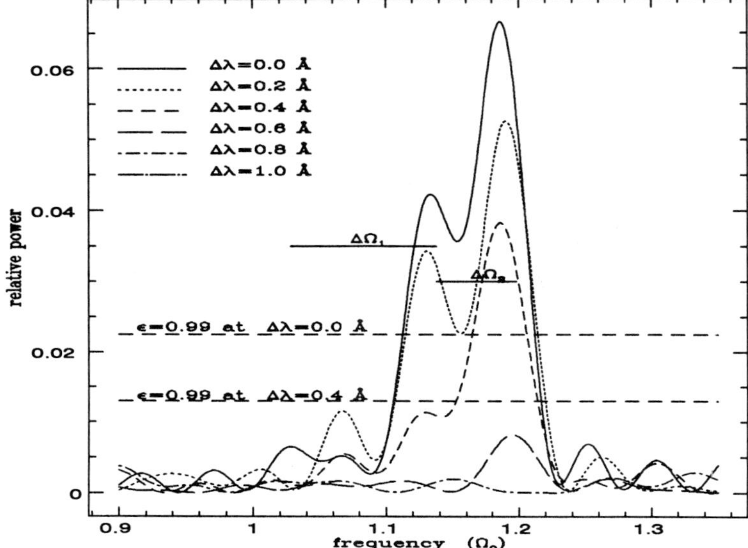

Figure 4. Periodograms obtained from a sequence of simulated line profiles with $\frac{S}{N} = 450$, labelled according to the wavelength difference $\Delta\lambda$ with respect to the line center at $\lambda_0 = 6000.0$ Å . The two continuous lines indicate the rotation frequency intervals of the assumed high-latitude ($\Delta\Omega_1$) and low-latitude ($\Delta\Omega_2$) starspots, respectively (Ω_0 is the angular velocity at the poles). The dashed lines indicate the 99% confidence levels of the main periodogram peaks at $\Delta\lambda = 0.0$ Å and $\Delta\lambda = 0.4$ Å, respectively. The weakening of the high frequency peak toward line wings gives evidence of an equatorial acceleration ($\Omega_e > \Omega_p$), as explained in Lanza & Rodonò (1996).

on a high-altitude orbit, in order to avoid gaps as much as possible in the time series.

The lifetime of the adopted tracers is of primary importance for the above discussed methods. If the lifetime of the individual spots or plages is shorter than the time interval needed to resolve the differential rotation, the multi-periodic nature of the modulation can not be detected (Lanza et al. 1994). Moreover, Donahue & Baliunas (1992) and Donahue (1993) have pointed out that the best rotation period determination with a given kind of surface tracer is achieved for an optimal length of the time series. For instance, in the solar case, a time length of about six months appears to be the most appropriate when the chromospheric emission is used. It is worthy noting that such a time scale corresponds to the lifetime of the activity complexes mentioned above. Tracers showing a short-term evolution, such as sunspots, are not adequate for an accurate measurement of the rotation period and of its variations along the 11-year cycle because their areal growth and decay induce a strong noise in the data, thus severely limiting the frequency reso-

of surface tracer is achieved for an optimal length of the time series. For instance, in the solar case, a time length of about six months appears to be the most appropriate when the chromospheric emission is used. It is worthy noting that such a time scale corresponds to the lifetime of the activity complexes mentioned above. Tracers showing a short-term evolution, such as sunspots, are not adequate for an accurate measurement of the rotation period and of its variations along the 11-year cycle because their areal growth and decay induce a strong noise in the data, thus severely limiting the frequency resolution (LaBonte 1982, 1984). In the stellar case, such effects could prevent the measurement of differential rotation for the most active stars showing short-term changes in their active regions and frequent flaring.

4. A Comparison with Recent Dynamo Models

The mean-field dynamo models have proved quite successful in explaining the basic properties of the solar activity cycle, so that their extension to the active stars has been natural.

Rädler et al. (1990), Moss et al. (1991), Rüdiger & Elstner (1994) and Moss et al. (1995) have proposed models in which large-scale non-axisymmetric magnetic fields are excited at values of the dynamo number comparable or even lower than those corresponding to the excitation of the usual axisymmetric fields. A common results of such analyses is that the non-axisymmetric modes are stable only if the differential rotation is low ($D_r < 0.1 - 0.2$). As a matter of fact, a large differential rotation, as predicted by, e.g., Kitchatinov & Rüdiger (1993) and Küker et al. (1993), tends to eliminate any asymmetry in the large scale field by means of the induced shear within a few rotations. However, given the observed low values of surface differential rotation inspotted stars, these models should be able to explain the existence of active longitudes due to a non-axisymmetric mean field, but are not capable of explaining the high-latitude spots appearing in Doppler images.

An explanation of the high-latitude activity has been suggested by Schüssler & Solanki (1992) and Schüssler (1996) in the framework of the flux tube model proposed for the solar subsurface field dynamics (Caligari et al. 1995). They assume that the magnetic field of a star is stored in the form of toroidal slender flux tubes in the overshoot region below the base of the convective zone. When the field strength of a flux tube is intensified beyond a certain critical value between $10^5 - 10^7$ G, it becomes unstable, and one or two crests develop, entering in the superadiabatic part of the convective zone, where the buoyancy force drives them toward the photosphere. In a star which has a convective zone deeper and a rotation rate faster than the Sun, the Coriolis force plays an important role in determing the path followed by the emerging flux tube. Inextreme cases the flux tube is forced to emerge parallel to the rotation axis and high-latitude activity is produced. In general, the latitude of emergence of a flux tube will depend mainly on its initial field strength, because it determines the relative balance between the radial buoyancy and the Coriolis force.

It is interesting to note that the first unstable mode for a toroidal slender flux tube is characterized by an azimuthal wave number $m = 1$ or $m = 2$ so that one or two activity complexes are expected to form in photosphere separated by $\sim 180°$ as observed in some short-period RS CVn binaries (Zeilik 1991, Zeilik et al. 1994). However, it remains to be explained why the perturbations are preferentially excited at the quadrature longitudes with one longitude being more active at any given time.

The presence of a system of toroidal flux tubes in the overshoot layer might modify the distribution of the angular momentum in a convective envelope because, in order to maintain mechanical equilibrium, the plasma inside the flux tubes should rotate faster than that outside (Moreno-Insertis et al. 1992). Some preliminary calculations suggest that such an effect might support the mechanism proposed by Applegate (1992) to explain the orbital period variations in active binaries (see also Rodonò et al. 1995).In conclusion, the model proposed by Schüssler (1996) seems promising, also because the predicted flux tube dynamics is not strongly dependent on the profile of the differential rotation nor on the presence of large scale flows. However, the process responsible for the generation of the assumed super-equipartition toroidal flux tubes in the overshoot region remains to be specified (Ferriz-Mas et al. 1994, Moreno-Insertis et al. 1995) and in this context the development of non-linear models, which try to include the interaction between differential rotation and dynamo, is of fundamental importance (e.g., Kitchatinov & Rüdiger 1994, Kitchatinov et al. 1994).

References

Appenzeller, I., Dearbon, D. P. S. (1984) *ApJ* **278**, 689
Appenzeller, I., Mundt, R. (1989) *A&AR* **1**, 291
Applegate, J. H. (1992) *ApJ* **385**, 621
Bai, T. (1988) *ApJ* **328**, 864
Bai, T. (1990) *ApJ* **364**, L17
Baliunas, S. L. et al. (1983) *ApJ* **275**, 752
Baliunas, S. L. et al. (1985) *ApJ* **294**, 310
Baliunas, S. L., Vaughan, A. H. (1985) *ARA&A* **23**, 379
Bianchini, A. (1990) *AJ* **99**, 1941
Bopp, B. W. (1983) in *Activity in Red Dwarf Stars, IAU Coll. 71*, P. B. Byrne & M. Rodonò (Eds.), D. Reidel Publ. Co., Dordrecht, p. 363
Bopp, B. W., Fekel, F. C. (1977) *AJ* **82**, 490
Bruning, D. H. (1981) *ApJ* **248**, 274

Byrne, P. B. (1992) in *Surface Inhomogeneities on Late-Type Stars*, P. B. Byrne & D. J. Mullan (Eds.), Springer-Verlag, Berlin, p. 3

Caligari, P., Moreno-Insertis, F., Schüssler, M. (1995) *ApJ* **441**, 886

Cameron, A. C. (1992) in *Surface Inhomogeneities on Late-Type Stars*, P. B. Byrne & D. J. Mullan (Eds.), Springer-Verlag, Berlin, p. 33

Castenmiller, M. J. M., Zwaan, C., van der Zalm, E. B. J. (1986) *Sol. Phys.* **105**, 237

Catalano, S. (1983) in *Activity in Red Dwarf Stars*, P. B. Byrne & M. Rodonò (Eds.), D. Reidel Publ. Co., Dordrecht, p. 343

Donahue, R. A. (1993) *Ph.D. Thesis*, New Mexico State University, Las Cruces, New Mexico

Donahue, R. A., Baliunas, S. L. (1992) *ApJ* **393**, L63

Eaton, J. A., 1992, in Surface Inhomogeneities on Late-Type Stars, P. B. Byrne & D. J. Mullan (Eds.), Springer-Verlag, Berlin, p. 15

Eaton, J. A., Hall, D. S. (1979) *ApJ* **227**, 907

Eaton, J. A., Henry, G. W., Fekel, F. C. (1995) *ApJ* submitted

Ferriz-Mas, A., Schmitt, D., Schüssler, M. (1994) *A&A* **289**, 949

Gaizauskas, V. et al. (1983) *ApJ* **265**, 1065

Gilman, P. A., Howard, R. (1986) *ApJ* **303**, 480

Gray, D. F. (1982) *ApJ* **258**, 201

Hall, D. S. (1991) in *The Sun and Cool Stars: activity, magnetism and dynamos*, I. Tuominen et al. (Eds.), Springer-Verlag, Berlin, p. 353

Hall, D. S., Busby, M. R. (1990) in *Active Close Binaries*, C. Ibanoglu (Ed.), Kluwer Ac. Publ., Dordrecht, p. 377

Harvey, K. L., Zwaan, C., 1993, *Sol. Phys.* **148**, 85

Hatzes, A. P. et al. (1995) in *Stellar Surface Structures, IAU Symp. 176*, K. G. Strassmeier (Ed.), Univ. of Vienna, Vienna, p. 9

Henry, G. W. et al. (1995) *ApJS* **97**, 513

Jetsu, L., Pelt, J., Tuominen, I. (1993) *A&A* **278**, 449

Jetsu, L., et al. (1994) *A&A* **282**, L9

Johns-Krull, C. M. (1996) *A&A* in press

Kitchatinov, L. L., Rüdiger, G. (1993) *A&A* **276**, 96

Kitchatinov, L. L., Rüdiger, G. (1994) in *Cool Stars, Stellar Systems and The Sun*, J.-P. Caillault (Ed.), A.S.P. Conf. Series vol.**64**, p. 196

Kitchatinov, L. L., Rüdiger, G., Küker, M. (1994) *A&A* **292**, 125

Krause, F., Rädler, K.-H. (1980) *Mean-Field Magnetohydrodynamics and Dynamo Theory*, Pergamon Press, Oxford

Küker, M., Rüdiger, G., Kitchatinov, L. L. (1993) *A&A* **279**, L1

LaBonte, B. J. (1982) *ApJ* **260**, 647

LaBonte, B. J. (1984) *ApJ* **276**, 335

Lanza, A. F., Rodonò, M. (1996) *A&A* in press

Lanza, A. F., Rodonò, M., Zappalà, R. A. (1993) *A&A* **269**, 351

Lanza, A. F., Rodonò, M., Zappalà, R. A. (1994) *A&A* **290**, 861

Moffatt, H. K. (1977) *Magnetic Field Generation in Electrically Conducting Fluids*, Cambridge Univ. Press, Cambridge

Montmerle, T. et al. (1991) in *Protostars and Planets*, E. H. Levy & J. I. Lunine (Eds.), Univ. of Arizona Press

Moreno-Insertis, F., Schüssler, M., Ferriz-Mas, A. (1992) *A&A* **264**, 686

Moreno-Insertis, F., Caligari, P., Schüssler, M. (1995) *ApJ* **452**, 894

Moss, D., Tuominen, I., Brandenburg, A. (1991) *A&A* **245**, 129

Moss, D. et al. (1995) *A&A* **294**, 155

Olson, E. C. (1984) *Advances in Photoelectric Photometry 2*, 15

Piskunov, N. E., Tuominen, I., Vilhu, O. (1990) *A&A* **230**, 363

Radick, R. R. et al. (1995) *ApJ* **452**, 332

Rädler, K.-H. et al. (1990) *A&A* **239**, 413

Rodonò, M. (1981) in *Photometric and Spectroscopic Binary Systems*, E. B. Carling &

Z. Kopal (Eds.), D. Reidel Publ. Co., Dordrecht, p. 285

Rodonò, M. (1990) in *Flares in Star Clusters, Associations and the Solar Vicinity, IAU Symp. 137*, L. V. Mirzoyan et al. (Eds.), D. Reidel Publ. Co., Dordrecht, p. 371

Rodonò, M. (1992) in *Evolutionary Processes in Active Close Binaries, IAU Symp. 151*, Y. Kondo et al. (Eds.), Kluwer Ac. Publ., Dordrecht, p. 71

Rodonò, M. (1994) in *The Fourth MU.SI.CO.S. Workshop*, C. Catala et al. (Eds.), Beijing Astronomical Observatory, p. 69

Rodonò, M., Lanza, A. F., Catalano, S. (1995) *A&A* **301**, 75

Rüdiger, G., Elstner, D. (1994) *A&A* **281**, 46

Schüssler, M. (1996) in *Stellar Surface Structures, IAU Symp. 176*, K. G. Strassmeier & J. L. Linsky (Eds.), Kluwer Ac. Publ., Dordrecht, in press

Schüssler, M. Solanki, S. K. (1992) *A&A* **264**, L13

Strassmeier, K. G., Bopp, B. W. (1992) *A&A* **259**, 183

Unruh, Y. C., Cameron, A. C. (1995) *MNRAS* **273**, 1

Vogt, S. S., Hatzes, A. P. (1991) in *The Sun and Cool Stars: activity, magnetism, dynamos*, I. Tuominen et al. (Eds.), Springer-Verlag, Berlin, p. 297

Vogt, S. S., Penrod, G. D., Hatzes, A. P. (1987) *ApJ* **321**, 496

Weiss, N. O. (1994) in *Lectures on Solar and Planetary Dynamos*, M. R. E. Proctor & A. D. Gilbert (Eds.), Cambridge Univ. Press, Cambridge, p. 59

Zeilik, M. (1991) in *The Sun and Cool Stars: activity, magnetism and dynamos*, I. Tuominen et al. (Eds.), Springer-Verlag, Berlin, p. 370

Zeilik, M. et al. (1994) *ApJ* **421**, 303

Zeldovich, Ya. B., Ruzmaikin, A. A., Sokoloff, D. D. (1983) *Magnetic Fields in Astrophysics*, Gordon & Breach Sci. Publ., New York

Zwaan, C. (1992) in *Sunspot: Theory and Observations*, J. H. Thomas & N. O. Weiss (Eds.), Kluwer Ac. Publ., Dordrecht, p. 75

DYNAMOS IN DIFFERENT TYPES OF STARS

N. O. WEISS
Department of Applied Mathematics and Theoretical Physics
University of Cambridge, Cambridge CB3 9EW, U.K.

1. Introduction

It is generally accepted that magnetic fields in late-type stars are produced by dynamo action in their convection zones. In this review I shall discuss the impact on dynamo models of the observations that have been described by Pallavicini, Byrne and Giampapa *et al.* in these Proceedings. The magnetic behaviour of these stars is governed by a few control parameters: the mass and composition determine stellar structure and, in particular, the depth of the convection zone, while the rotation rate, Ω, controls the degree of magnetic activity, which in turn affects magnetic braking. I shall focus on G-stars like the Sun, whose rotational history is well established.

There is an important distinction between turbulent amplification, which only yields transient magnetic fields, and dynamo action, which can maintain a stationary or oscillating field. The effect of rotation on convection is to produce cyclonic eddies with helicity, represented by the α-effect. These eddies are a key ingredient of stellar dynamos. Then there is a further contrast between "magnetic weather" and "magnetic climate". Small-scale turbulent dynamos typically provide disordered fields, while systematic dynamo action, which can generate magnetic cycles, results from a combination of helicity and differential rotation (the ω-effect).

Although stars spin down owing to magnetic braking, growing less active as they age, it is more appropriate in a theoretical treatment to reverse stellar evolution by considering the effects of gradually increasing Ω. So I shall start by discussing slow rotators with magnetic cycles, like the Sun, and then go on to consider rapidly rotating and more active stars. Finally, I shall comment on the consequences of generating strong magnetic fields and of the deep convection zones in low-mass M-stars. Many of these topics have been dealt with recently at greater length (Weiss 1994).

Y. Uchida et al. (eds.),
Magnetodynamic Phenomena in the Solar Atmosphere – Prototypes of Stellar Magnetic Activity, 387–394.

2. Slow Rotators with Magnetic Cycles

2.1. THE SOLAR DYNAMO

The whole theory of stellar dynamos is based on detailed observations of the Sun. Its internal rotation can be deduced from helioseismological measurements: although the results obtained by inverting the frequency-splitting of p-modes are still ambiguous (Sekii, Gough & Kosovichev 1995), it seems clear that Ω is predominantly a function of latitude in the convection zone, with an abrupt radial shear (the tachocline) in a layer that includes a region of weak convective overshoot. There is a general consensus that the solar dynamo operates in a shell at the base of the convection zone, with strong toroidal fields (10^4–10^5 G) generated by the radial shear. Direct observations of the solar cycle show that the irregular 11-yr activity cycle (corresponding to a 22-yr magnetic cycle) was interrupted by the Maunder minimum in the 17th century; proxy measurements, derived from variations in the production rates of radioisotopes such as ^{14}C and ^{10}Be allow us to infer that the pattern of cyclic activity modulated by grand minima has persisted for at least 10^4 yr (Stuiver 1994; Beer *et al.* 1994). Moreover, other slowly rotating stars appear to share this pattern of behaviour. Dynamo theory has to explain the basic cycle, its modulation on a longer timescale and the irregular behaviour, which is apparently chaotic.

It has been shown numerically, in self-consistent but simplified models, that a nonlinear dynamo can produce magnetic cycles. Most studies rely, however, on mean field dynamo theory, with the effects of turbulent cyclonic convection parametrized by α (proportional to the mean helicity) and ω (an angular velocity gradient). Then the crucial parameter is the dynamo number $D \propto \alpha\omega \propto \Omega^2$. If $|D|$ is too small, magnetic fields will decay but as $|D|$ passes through a critical value D_c the field-free state becomes unstable to oscillatory perturbations whose amplitude is limited by nonlinear effects. Here there are several possibilities (Brandenburg 1994). At one extreme, the mean field $\mathbf{B}$ may exert a macrodynamic couple that alters ω. Since the Lorentz force is quadratic in $\mathbf{B}$, that leads to variations in Ω with an 11-yr period – corresponding to the "torsional oscillations" that have been observed. At the other extreme, the Lorentz force may act through microdynamics, as small-scale fluctuations quench the α-effect or alter the angular velocity. In a self-consistent treatment, valid provided there is a separation of scales in the convection zone, turbulent Reynolds stresses are modified by rotation (the Λ-effect) so as to allow large-scale differential rotation, and the Λ-effect is itself altered by the fluctuating magnetic fields. This approach has been applied to model a solar dynamo in the region of convective overshoot and to compute the "torsional oscillations" (Kitchatinov, Rüdiger & Küker 1994; Küker, Rüdiger & Pipin 1995).

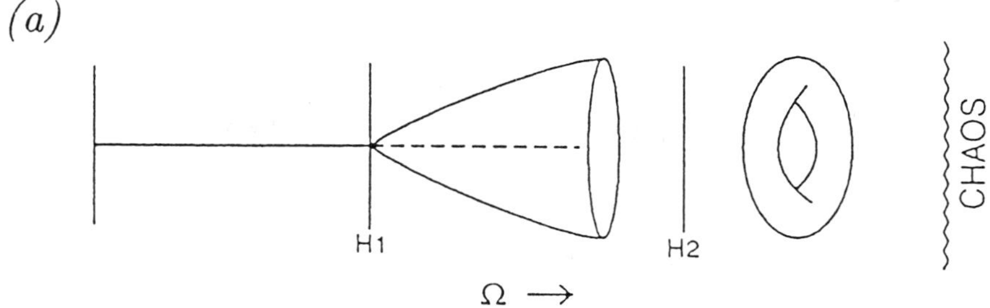

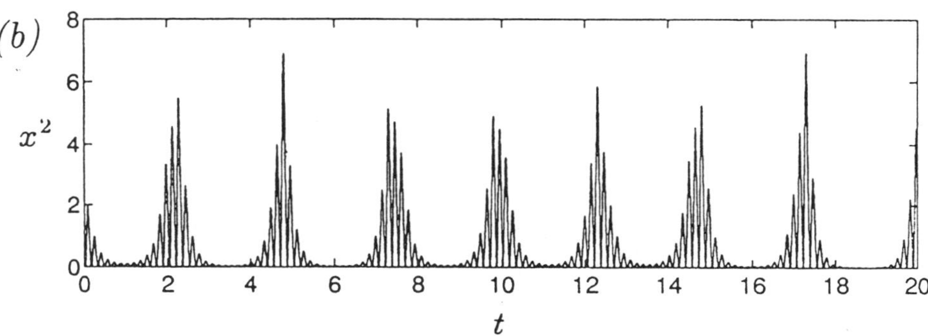

Figure 1. (*a*) Schematic bifurcation diagram, showing two successive Hopf bifurcations (H1, H2) followed by a transition to chaos as Ω is increased. (*b*) Chaotically modulated activity cycles: x^2 as a function of time t in a toy model (after Tobias *et al.* 1995).

2.2. ORIGIN OF CHAOTIC MODULATION

Figure 1*a* illustrates the essential bifurcation sequence in a nonlinear stellar dynamo. As D (or Ω) is increased, there is first an oscillatory (Hopf) bifurcation at D_c that is followed by strictly periodic dynamo action. The second Hopf bifurcation leads to regularly modulated cycles, with trajectories in phase space that lie on two-tori instead of forming limit cycles. Subsequent bifurcations then lead to the chaotic modulation that is observed in the Sun. The underlying bifurcation sequence can be demonstrated in a simple model. All that is needed is a suitable third-order system of nonlinear ordinary differential equations (which is necessary to describe motion on a two-torus and sufficient to allow chaos).

Tobias, Weiss & Kirk (1995) consider such a toy system; in the three-dimensional phase space, the variables x and y represent the toroidal and poloidal fields, while the entire hydrodynamic state, including convection and differential rotation, is collapsed onto the z-axis. The coupled nonlinear differential equations are in fact the normal form equations for a saddle-node/Hopf bifurcation, so the behaviour exhibited by their solutions is expected to be robust. As a single parameter, corresponding to D or Ω, is increased, there is indeed a Hopf bifurcation followed by a secondary Hopf bifurcation that leads to doubly-periodic modulated cycles. The final transition to aperiodic modulation is associated with a global bifurcation in which the torus is destroyed. Figure 1*b* shows the behaviour of x^2 (which represents the sunspot number) as a function of time in the chaotic regime. In the plot the irregular cycles are modulated on a longer timescale, so that there are intervals with enhanced activity, separated by episodes of reduced activity. In this model, these episodes are associated with grand minima, which have a characteristic timescale of about 200 yr. A variant, in which two timescales of modulation are involved, is discussed by Tobias elsewhere in these Proceedings. Although such low-order models lack predictive power, they do display *generic* patterns of behaviour that help to explain the origin of cycles with grand minima in actual stars.

These toy systems reproduce complicated temporal behaviour without describing spatial structure. The simplest spatiotemporal patterns are one-dimensional travelling waves, whose nonlinear development also displays aperiodic modulation. Two-dimensional structure is more realistic and Parker (1993) devised an attractive linear model in cartesian geometry, with an $\alpha\omega$-dynamo operating at the interface between a layer (the convection zone) where the α-effect is concentrated and one (the radiative zone) in which the shear is located. Tobias (1995) has extended the partial differential equations to include nonlinear effects. The most significant results are obtained when saturation occurs through the action of the macrodynamic Lorentz force on the velocity shear. Figure 2*a* shows a nonlinear travelling wave solution; the shaded contours indicate the toroidal field and the velocity perturbation generated by the Lorentz force. As expected, the latter has half the period of the former. As $|D|$ is increased, the travelling wave gives way to a modulated wave, which is illustrated in Figure 2*b*. For yet larger values of $|D|$, chaotically modulated waves appear.

If we include lateral boundaries (at the poles) and allow for equatorial symmetries, the initial bifurcations give rise to two families of oscillatory solutions, with dipole and quadrupole symmetry about the equator, and subsequent symmetry-breaking bifurcations then produce mixed-mode solutions. Jennings & Weiss (1991) explored the bifurcation structure for a one-dimensional cartesian model and Tobias (1995) has investigated corresponding behaviour for the two-dimensional interface dynamo described above. Similar bifurcations also appear in spherical geometry.

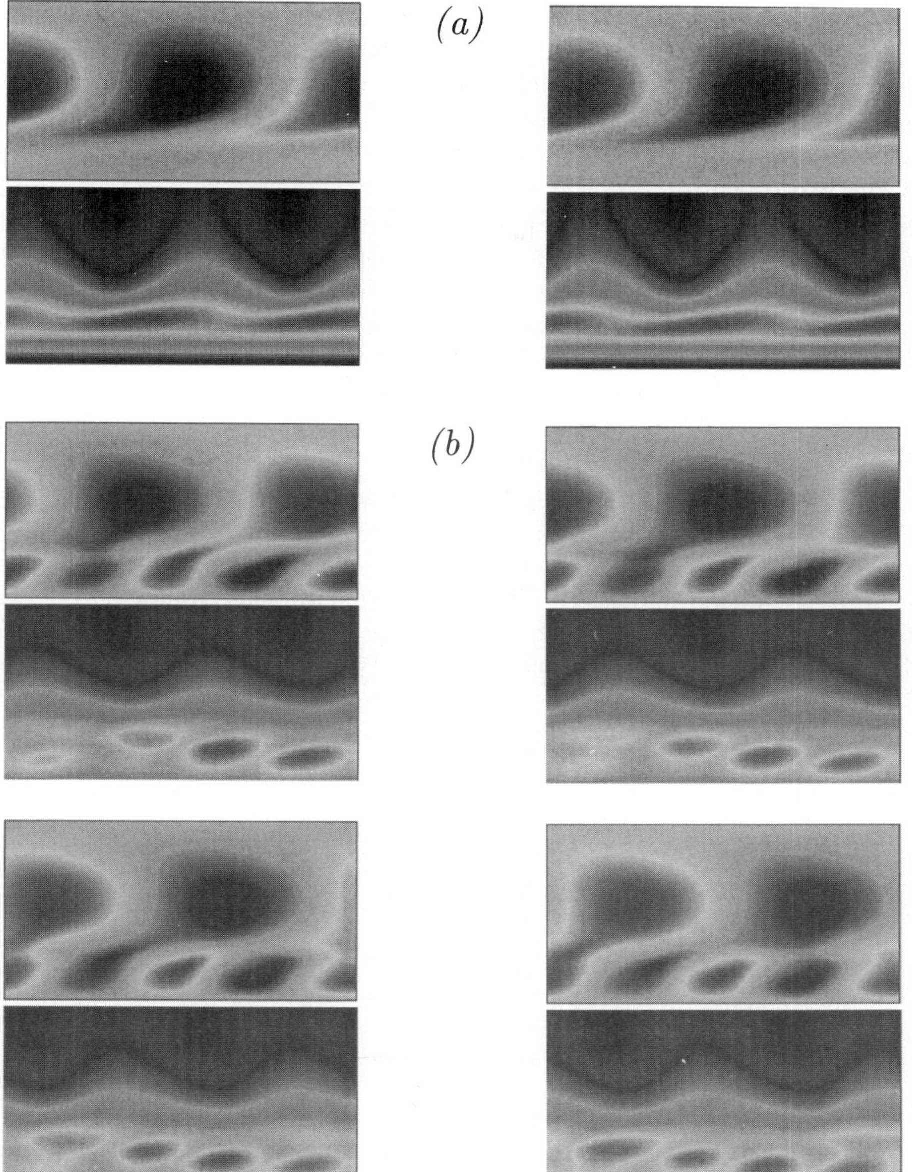

Figure 2. Modulation in a two-dimensional cartesian dynamo (*a*) Rightward travelling waves at two successive times. (*b*) Modulated waves at four successive times. Upper panels show the toroidal field and lower panels the velocity perturbation. (After Tobias 1995.)

3. Rapidly Rotating Active Stars

3.1. EFFECTS OF INCREASING Ω

Rotation affects convection through the Coriolis force and convective eddies are then responsible for differential rotation. The mechanism that leads to the rotation profile in the Sun is not yet fully understood. From the Proudman-Taylor theorem, it is clear, however, that in a rapidly rotating star large-scale convection will occur in banana cells, elongated parallel to the rotation axis, and that the angular velocity will tend to be constant on cylindrical surfaces; this has been confirmed by computation and experiment. The change from a profile similar to that revealed by helioseismology to one with the angular velocity more nearly constant on cylinders has been modelled successfully in terms of the Λ-effect (Küker, Rüdiger & Kitchatinov 1993; Kitchatinov & Rüdiger 1995).

Increasing Ω alters the dynamo in two ways. From §2 we expect magnetic fields to increase in strength and to exhibit richer time-dependence, while developing more complicated spatial structure. In addition, the new pattern of rotation reduces the shear in angular velocity, removing the need for a shell dynamo to generate toroidal flux. Indeed, the contrast between slow rotators, with recognizable magnetic cycles, and more rapid rotators, with a different relation between rotation periods and "periods" derived by analysing records of activity (Saar & Baliunas 1993), suggests that the dynamo mechanism itself has changed. Irregular cycles apparently give way to unstable dynamos with sporadic (but frequent) reversals – a transition from quasi-AC to quasi-DC behaviour. Idealized numerical models of stellar dynamos also reveal an altered pattern of behaviour in rapidly rotating systems (Gilman 1983). We should expect the total magnetic flux to continue to increase with increasing Ω although the toroidal field becomes less dominant compared with the poloidal field. Conservation of angular momentum also ensures that any rising flux tubes drift away from the equator to accumulate near the poles.

3.2. EFFECTS OF INCREASING $\langle B^2 \rangle$

Rapid rotation leads to stronger fields which themselves alter the nature of the dynamo. Chromospheric emission saturates for stars rotating with 5–10 times the Sun's angular velocity, $\Omega_\odot$. Thereafter more and more magnetic flux accumulates in starspots, giving rise to stronger photometric variability. The dynamo process itself saturates only for $\Omega > 20\ \Omega_\odot$. The strong fields can then lead to enhanced magnetic braking, sufficient to explain the reduction of Ω from 80 $\Omega_\odot$ to 20 $\Omega_\odot$ during the gap of 2×10^7 yr that separates young stars in the α Persei cluster from those in the Pleiades, *without* decoupling the convection zone from the radiative core (Collier Cameron & Li 1994).

The presence of polar spots in rapidly rotating stars suggests that there is a dominant poloidal field. For example, about 20% of the surface area of the K0 dwarf AB Doradus, rotating with $\Omega \approx 50\Omega_\odot$, is covered by spots, which are concentrated near one pole. In addition, there are features at lower latitudes which show little variation over an interval of a year (Collier Cameron 1995). The strong polar fields are bound to influence the pattern of convection and might be expected to drive a dipolar meridional circulation. Indeed, overall convective transport may be favoured if magnetic flux is confined to the neighbourhood of a pole. The intense field will then interact with local convective processes, which are likely to favour a group of isolated spots rather than a single huge spot, if the largest spot groups on the Sun can be taken as a guide.

3.3. LOW MASS STARS

In going from G through K to M-stars, the stellar mass decreases and the convection zone grows deeper, until M5-stars are fully convective. So there is no possibility of a shell dynamo in an M-star. Some differential rotation may remain but it seems more likely that such a star only retains a turbulent (α^2) dynamo, giving rise to strong but disordered fields. There may still be a tendency to polar activity, though observations favour spots at lower latitudes (suggesting that the dynamo is essentially non-axisymmetric). In the absence of a strong large-scale field with a long lever-arm, magnetic braking will be less efficient; the consequent reduction in the rate of spin-down is in agreement with observations (Durney, De Young & Roxburgh 1993).

4. Conclusion

I have tried to clarify various current approaches to modelling nonlinear stellar dynamos, whether through toy systems, or treatments based on Reynolds stresses, or by creating $\alpha\omega$-models that illuminate the physics. It would be premature to attempt the massive computation required for a fully consistent model until the cause of differential rotation is properly understood. So far, all detailed calculations have been guided by solar observations. As I have indicated, it is dangerous to extrapolate from the Sun to rapidly rotating and hyperactive stars, which may have qualitatively different types of dynamo. At the moment we are not equipped to describe them more precisely.

In the immediate future we can, however, look forward to advances in helioseismology associated with SOHO and GONG, and to the accurate determination of the variation of Ω with radius and latitude in the Sun. So we must focus on hydrodynamics and produce a self-consistent model of convection and differential rotation in the Sun. Then we can construct a convincing nonlinear model of the solar dynamo. Finally, we shall be ready to tackle rapidly rotating stars. I suspect that this programme will be enough to occupy not only the present generation of research students but also their successors.

References

Beer, J., Joos, F., Lukasczyk, C., Mende, W., Rodriguez, J., Siegenthaler, U. & Stellmacher, R. (1994) ^{10}Be as an indicator of solar variability and climate, in E. Nesme-Ribes (ed.), *The Solar Engine and its Influence on Terrestrial Atmosphere and Climate*, pp. 221–233. Springer, Berlin.

Brandenburg, A. (1994) Solar dynamos: computational background, in M.R.E. Proctor & A.D. Gilbert (eds) *Lectures on Solar and Planetary Dynamos*, pp. 117–159. Cambridge University Press.

Collier Cameron, A. (1995) New limits on starspot lifetimes for AB Doradus, *Mon. Not. Roy. Astr. Soc.* **275**, 534–544.

Collier Cameron, A. & Li, J. (1994) Magnetic braking of G and K dwarfs without core-envelope decoupling, *Mon. Not. Roy. Astr. Soc.* **269**, 1099–1111.

Durney, B.R., De Young, D.S. & Roxburgh, I.W. (1993) On the generation of the large-scale and turbulent magnetic fields in solar-type stars. *Solar Phys.* **145**, 207–225.

Jennings, R.L. & Weiss, N.O. (1991) Symmetry breaking in stellar dynamos. *Mon. Not. Roy. Astr. Soc.* **252**, 249–260.

Kitchatinov, L.L. & Rüdiger, G. (1995) Differential rotation in solar-type stars: revisiting the Taylor-number puzzle. *Astr. Astrophys.*, submitted.

Kitchatinov, L.L., Rüdiger, G. & Küker, M. (1994) Λ-quenching as the nonlinearity in stellar-turbulence dynamos, *Astr. Astrophys.* **292**, 125–132.

Küker, M., Rüdiger, G. & Kitchatinov, L.L. (1993) An $\alpha\Omega$-model of the solar differential rotation, *Astr. Astrophys.* **279**, L1–L4.

Küker, M., Rüdiger, G. & Pipin, V.V. (1995) Solar torsional oscillations as due to magnetic quenching of the Reynolds stress, *Astr. Astrophys.* submitted.

Parker, E.N. (1993) A solar dynamo surface-wave at the interface between convection and non-uniform rotation, *Astrophys. J.* **408**, 707–719.

Saar, S.H. & Baliunas, S.L. (1992) Recent advances in stellar cycle research, in K. Harvey (ed.), *The Solar Cycle*, pp. 150–167. Astron. Soc. Pac., San Francisco.

Sekii, T., Gough, D.O. & Kosovichev, A.G. (1995) Inversions of BBSO rotational splitting data, in R.K. Ulrich, E.J. Rhodes & W. Däppen (eds), *Gong 94: Helio and Astero-Seismology from the Earth and Space*, pp. 59–62. Astron. Soc. Pac., San Francisco.

Stuiver, M. (1994) Atmospheric ^{14}C as a proxy of solar and climatic change, in E. Nesme-Ribes (ed.), *The Solar Engine and its Influence on Terrestrial Atmosphere and Climate*, pp. 203–220. Springer, Berlin.

Tobias, S.M. (1995) *Nonlinear Solar and Stellar Dynamos.* Ph.D. dissertation, University of Cambridge.

Tobias, S.M., Weiss, N.O. & Kirk, V. (1995) Chaotically modulated stellar dynamos. *Mon. Not. Roy. Astr. Soc.* **273**, 1150–1166.

Weiss, N.O. (1994) Solar and stellar dynamos, in M.R.E. Proctor & A.D. Gilbert (eds) *Lectures on Solar and Planetary Dynamos*, pp. 59–95. Cambridge University Press.

CORONAL MAGNETIC ACTIVITY IN LOW MASS PRE-MAIN SEQUENCE STARS

FREDERICK M. WALTER
State University of New York
ESS. Dept, Z=2100, Stony Brook NY 11794-2100 USA

Today's Sun is a mere shadow of its former self. Four and one-half billion years ago the Sun was a fully convective, rapidly-rotating K star, with about 3 times its present radius. By all indications, it was magnetically active, with profound implications for the evolution of planetary atmospheres and the establishment of a an environment suitable for life on Earth. The Sun today retains no information on this stage of its evolution: we must explore it vicariously by observing low mass pre-main sequence (PMS) stars 4.5 billion years younger than our Sun. These studies afford the opportunity to explore what is possibly a distinctly non-solar-like dynamo.

Two distinct groups comprise the low mass PMS stars: the classical (cTTS) and the naked (nTTS) T Tauri stars. The difference between the two classes rests in the presence or absence of detectable amounts of circumstellar dust radiating in the near-IR. The accretion of circumstellar material results in the prominent Hα emission line characteristic of the cTTS, as well as emission in other chromospheric lines. This strong emission is not diagnostic of stellar magnetic activity: it is a consequence of accretional heating of the accreting gas, and will not be discussed here. The nTTS, on the other hand, show every indication of possessing true "solar-like" magnetic activity.

As more nTTS are identified, it is possible to study the characteristics of a statistically significant sample of stars. Here I discuss analysis of the X-ray properties about 60 stars in the Tau-Aur, upper Sco, and R CrA associations for which I have complete data.

Based on EINSTEIN and ROSAT observations, the nTTS have X-ray luminosities between about 10^{29} and 10^{31} erg s^{-1} and characteristic temperatures of order 1 keV. Spectral fits suggest the emission is thermal with two temperature components, similar to those of active main sequence (MS) and post-MS stars. S. Skinner and I have recently obtained ASCA spectra

Y. Uchida et al. (eds.),
Magnetodynamic Phenomena in the Solar Atmosphere – Prototypes of Stellar Magnetic Activity, 395–396.

of two bright nTTS. Fits to the spectrum of HD 142361, a $2M_\odot$, 1 million year (Myr) old G2 IV star in the upper Sco association (Walter *et al.* 1994), require ~0.7 and 2 keV components, with twice the emission measure in the hot component. The fits give acceptable χ^2_ν for solar abundances. Aside from the abundances, the spectral parameters are quantitatively similar to those of RS CVn systems, rapidly rotating post-MS G-K subgiants.

Two coronal decay laws have been published for solar-mass stars: $L_X \sim t^{-0.6}$ (Feigelson *et al.* 1993) and $F_X \sim \exp(-\sqrt{\frac{t}{400Myr}})$ (Walter & Barry 1991). Both relations fit the data for stars younger than about 10^9 years. The surface flux F_X does not decay among PMS stars. On PMS Hayashi tracks, the stellar radius shrinks as $t^{-\frac{1}{3}}$, so that, for constant F_X, $L_X \sim R^2 \sim t^{-\frac{2}{3}}$. Hence the decrease in L_X among PMS stars appears attributable mostly to the shrinking of the stellar radius.

The nTTS span a factor of 10 in mass, from 0.2 to 2 $M_\odot$. In the upper Sco association, which is coeval at an age of about 1 Myr, I find that $F_X \sim \sqrt{M}$ and $L_X \sim M^{\frac{3}{4}}$. The flux ratio $\frac{F_X}{F_{bol}}$ is independent of mass, at a value of -3.37±0.23. As the association ages and the higher mass stars evolve off their Hayashi tracks, these relations will change.

Grankin (1993) noted a correlation between L_X and rotation period P_{rot} in a small sample of PMS stars. I note a similar result: F_X does decrease as the stellar rotation period lengthens. However, this is due to the relation $F_X \sim \sqrt{M}$. The more massive stars have shorter P_{rot} than the lower mass stars: there is no evidence for a rotation-activity relation among the low mass PMS stars. I note that Majer *et al.* (1986) have argued that this is also true of the RS CVn systems.

I conclude that the coronae of the nTTS are RS CVn-like. The nTTS and RS CVns have similar P_{rot} and surface gravities, suggesting that these parameters, and not the base of the convection zone, determine the coronal morphology. Whether or not the activity is truly solar-like is debatable.

References

Feigelson, E.D., Casanova, S., Montmerle, T., and Guibert, J. (1993) ROSAT X-ray Study of the Chamaeleon I Dark Cloud I. The Stellar Population, *ApJ*, **416**, pp. 623–632.

Grankin, K.N. (1993) Periodic Light Variations in Ten Weak Emission T Tauri Stars in Taurus-Auriga Complex, *IBVS*, **3823** pp. 1-3.

Majer, P., Schmitt, J.H.M.M., Golub, L., Harnden, F.R. Jr., and Rosner, R. (1986) X-Ray Spectra and the Rotation-Activity Connection of RS Canum Venaticorum Binaries, *ApJ*, **300**, pp. 360-373.

Walter, F.M., and Barry, D.C. (1991) Pre- and Main Sequence Evolution of Solar Activity, in C.P. Sonett, M.S.Giampapa, and M.S. Matthews (eds.) *The Sun in Time*, The University of Arizona Press, Tucson, pp. 633–657.

Walter, F.M., Vrba, F.J., Mathieu, R.D., Brown, A., and Myers, P.C. (1994) X-ray Sources in Regions of Star Formation. V. The Low Mass Stars of the Sco OB2 Association, *AJ*, **107**, pp. 692–719.

V. Governing Factors of Solar/Stellar Activity

THE SUN AS THE ULTIMATE CHALLENGE TO ASTROPHYSICS: THE VITAL PHASE OF SOLAR PHYSICS

EUGENE N. PARKER
The University of Chicago Enrico Fermi Institute
and Depts. of Physics and Astronomy
933 East 56th Street, Chicago, Illinois 60637, U.S.A.

Abstract. Observations indicate that the various active phenomena observed on the Sun are the principal constituents of the activity of other solitary late main sequence stars although in different proportions and prominence. But our ability to provide quantitative descriptions of the activity of other stars is limited observationally by our inability to resolve their disks and theoretically by the absence of clear ideas and quantitative theory for the phenomena in the Sun. Thus, for instance, it is not clear why the Sun is obliged to form spots, so it is not clear how to interpret the enormous spot areas that appear on some M-dwarfs. Similary, it is not clear why some M-stars produce flares 10^3 times more energetic than on the Sun. The same holds for the structure of the X-ray coronas of other stars, the strength of their stellar winds, the nature of the stellar dynamos, and the luminosity variations in step with their general level of activity. None of these phenomena are properly understood for the Sun, yet there is reason to expect that the crucial observational studies of the Sun (solar microscopy and spectrometry in visible, UV, and X-rays, helioseismology) as well as critical attention to theoretical possibilities, promise progress with several aspects of the problem. The essential point is that the magnetohydrodynamics of the high Reynolds and Lundquist numbers characterizing the convective zone of the Sun is qualitatively different from the familiar concepts of hydrodynamics and the plasma physics laboratory. The fibril state of the magnetic field at the visible surface of the Sun is the direct indication of that fact, and it appears that the dynamo and the several consequences of the dynamo fields can be understood only in terms of the complicated dynamics of magnetic fibrils. We shall be able to estimate the

Y. Uchida et al. (eds.),
Magnetodynamic Phenomena in the Solar Atmosphere – Prototypes of Stellar Magnetic Activity, 399–410.

properties of the fibril fields of other stars only when we better understand the physics of fibrils in the Sun.

1. The Challenge

The outstanding scientific challenge to the field of astrophysics is both the variability and the suprathermal activity of the ordinary star like the Sun. The suprathermal activity is itself highly variable, from the individual flare to the X-ray corona, the coronal mass ejection, and the solar wind. Then there is the observed variation by 0.2 percent in the total brightness of the Sun in step with the cyclic 11-year variation of the suprathermal activity. Observations indicate similar behavior of other solitary late main sequence stars (Baliunas, et al., 1995; Zhang, et al., 1994). It is not surprising to find, then, that the northern temperate zone of Earth shows a mean temperature closely tracking the level of activity of the Sun and, by inference, the varying brightness of the Sun (Eddy, 1977a,b, 1983; Friis-Christensen and Lassen, 1991; Ribes and Nesmes-Ribes, 1993; Hameed and Gang, 1994 and Parker, 1994a). The challenge is to understand the physics of the activity and the variability. It is important that the variability of the Sun now, today, and the possibilities for future variation, be understood as soon as scientifically feasible. The direct effect of the changing brightness of the Sun on the climate at all levels in the terrestrial atmosphere, from the ozone layer to the mean annual air temperature and the ocean surface water temperature, is of vital concern for human society.

The study of solar variability and the terrestrial consequences began with the *indirect* phase of space science when we studied the fluctuations in conditions at ground level and inferred that there was activity in the ionosphere and magnetosphere above. From the correlation with solar activity we inferred that the Sun was the prime mover. Then came the *direct* phase of space science when we placed instruments in the magnetosphere and in interplanetary space to see directly what was happening there. We have learned a lot and we are still in the direct phase. But we have learned enough now to see that our living conditions here on Earth are driven up and down by the activity of the Sun. It is vital, therefore, that we address the question of what has happened with the Sun in the past, what is happening at the present time, and what are the possibilities in the future. We are in the *vital* phase of space science and solar physics, and it is imperative that we all understand and appreciate that fact. Needless to say, there is no lack of ideas on the nature of the activity of the Sun. It is widely believed that the cyclical magnetic field of the Sun is produced by an $\alpha\omega$-

dynamo in the convective zone, the X-ray corona is created by the heat from dissipating Alfven waves, sunspots are explained by the magnetic inhibition of convective heat transport, the solar wind arises from the matter injected by spicules into the corona, most stars are better understood than the Sun, etc. Many of these ideas are only guesses at best, and several of them are demonstrably inadequate, if not absurd. So if we are to address the physics of the Sun it is essential to view the problems with a critical eye, to see the problems in perspective and to appreciate what is and is not solid physical understanding. The widespread astrophysical practice of declaring the nature of active unresolved celestial objects is more entertainment than science. And that process of sequential conjecture is obsolete for the Sun where we have enough observational information to recognize the unanticipated mechanisms of the suprathermal activity. Nature is simply too complex to be explained by our meager repertoire of apriori guesses. Close observational scrutiny invariably reveals effects outside the realm of familiar physics.

Returning to the observations, it is curious that in its first 10^8 years the total luminosity of a star anti-correlates with the general level of activity, while later in life the two are in step, as with the Sun today. The studies of Foukal and Lean (1986, 1988; Foukal, 1990) show that a large part of the variation in total brightness of the Sun can be understood from the competition between dark sunspots and bright faculae and plages. The faculae dominate the sunspots because of their greater areas and longer life, so that the brightness increases with the general level of magnetic activity. One can imagine, then, that early in life a star like the Sun produces more spots in preference to faculae and plages, so that the brightness varies inversely with the activity.

This remarkable situation serves to point up the fact that we do not understand the basic physics of plages, faculae, or sunspots. We know a lot about them, but we do not understand why a star is obliged by the laws of nature to produce them. So we do not understand why the proportions of spots and faculae should be different when a star is young. In fact, there is more to the variation of solar brightness than faculae and spots (Willson and Hudson, 1991; Kuhn and Libbrecht, 1991; Foukal, Harvey, and Hill 1991) and presumably, therefore, more to the variation of the luminosity of other stars. Foukal suggests that the brightness variation arises primarily from the deep roots of sunspots and faculae (Foukal, Fowler, and Livshits, 1983; Foukal and Fowler, 1984). The assumption is that the sunspots and faculae interact with the upward heat flow at depths so great that 10 or 10^2 year changes in solar surface brightness have negligible effect on the massive thermal capacity. The characteristic thermal relaxation time for the convective zone is estimated by the total thermal energy divided by the heat flux, giving approximately 10^5 years. However, it must be remembered that if the temperature at the base of the convective zone is fixed, then the thermal energy available is not more than the height integrated difference $\Delta dT/dz$ between the actual and the adiabatic temperature gradient. With the evidence now that the brightness of the Sun may have varied by two or more parts in 10^3 for periods of the order of 10^2 years (e.g. the Maunder Minimum (17th century) and the Medieval Maximum (12th and 13th century)), it is not clear that this alone is sufficient. For instance, if one integrates $\Delta dT/dz$ a distance 10^{10} cm upward from the bottom of the convective zone to a position half way to the surface, the result is 1K. The thermal capacity of the middle half of the convective zone is approximately 10^{10} cm times the specific heat $\rho C_p \sim 3 \times 10^7$ ergs/cm^3 of matter at the half way point (where $\rho \sim 0.1$ gm/cm^3, $T \cong 0.9 \times 10^6$K). So a temperature change of 1K involves 3×10^{17} ergs/cm^2. The energy flux is approximately 10^{11} ergs/cm^2 sec. Therefore a change of two parts in 10^3, or 2×10^8 ergs/cm^2 sec would carry away the available 3×10^{17} ergs/cm^2 in a time 1.5×10^9 sec or 50 years. That is to say, $\Delta dT/dz$ would double in 50 years providing much greater convective heat transport if the surface brightness were decreased, and $\Delta dT/dz$ would fall to zero in 50 years if the surface brightness increased. So it is not entirely clear that the thermal resevoir in the deep convective zone is sufficient. And in any case, plages may be much more superficial, enhancing the heat transport only at depths of $1 - 2 \times 10^9$ cm or less.

It should be noted that the brightness increase with activity may be simply an unavoidable consequence of the appearance of magnetic fields at the surface of the Sun. For the magnetic activity of a star like the Sun is presumably the result of the formation of Ω-loops in the azimuthal flux bundles below the bottom of the convective zone. Each rising Ω-loop creates a wake in the form of an updraft. The convective forces drive the updraft and the essential point is that the successive rising Ω-loops extend from the bottom to the top of the convective zone, providing a vertical coherence length within the updraft that is substantially in excess of the coherence length of the ambient convection. The net result is an increased convective delivery of heat to the surface (Parker, 1994b,c, 1995a,b). The different modes by which the extra energy is radiated from the surface is then the next consideration, involving the unknown physics of faculae, spots, etc.

Now if the brightness enhancement is primarily a consequence of the emerging Ω-loops, then it is curious that early in life the net delivery of heat to the surface of a star varies inversely rather than directly with the 6 - 10 year cyclic level of activity. So one might imagine the accumulation of vast areas of long lived spots on the surface of the young star, with only a limited emergence of fresh Ω-loops and relativity few faculae. Presumably, the long-lived magnetic roots of the spots extend down through the convective zone in sufficient numbers to inhibit the convective transport of heat to the surface. But if this is the case, why is the internal hydrodynamics and magnetohydrodynamics so different in the first 10^8 years from the hydrodynamics and magnetohydrodynamics after 10^9 years? The difference seems to be qualitative. Unfortunately we understand so little of the physics of the internal circulation in a star, the dynamics of Ω-loops, and the formation of sunspots that we can do nothing beyond guess work.

Indeed, there is very little about the activity of the Sun that is established on a solid scientific foundation. For instance, the flare appears to be a consequence of rapid reconnection of oppositely directed magnetic field components (Sweet, 1958a,b; Parker, 1957a,b), but from there on the detailed particle acceleration shows endless variety. One obvious question is why are flares limited to about 10^{32} ergs on the Sun and yet reach 10^{35} ergs on some *dM* dwarf stars? Is it merely a matter of scaling up the size and intensity of the field in the dwarf? The X-ray corona of the Sun is created by magnetic field dissipation of about 10^7ergs/cm^{2}sec (Withbroe and Noyes, 1977). We suggest that the heat input arises from the dissipation of free magnetic field energy at the many spontaneous surfaces of discontinuity that form within the deformed and internally interwoven bipolar magnetic fields above the surface of the Sun (Parker, 1981, 1983a,b; 1994d). But it has yet to be established by observation of the intermixing of the fibril footpoints that the necessary interweaving of the field lines exists. It has been suggested (Porter and Moore, 1986; Martin, 1986) that the near coronal hole is heated principally by microflaring in the network fields. But it has not yet been shown that there is enough energy in the microflaring. On the other hand, there is no evident alternative. The far coronal hole, beyond several solar radii is conjectured to be heated by Alfven waves generated by the granules. But again there is no observational confirmation (cf. Parker, 1991a,b).

The astonishing fibril state of the magnetic field at the photosphere has been known for twenty five years and the causes of this peculiar state maybe understood to some degree (cf. Spruit, 1979, Parker, 1984a) in terms of the convective and radiative heat transport. It has been speculated for many years now that the magnetic field is in a fibril state throughout the entire convective zone (Zwaan, 1978, 1985; Piddington, 1978: Parker, 1979a,b,c, 1984a,b,c; Tsinganos, 1979). This idea has come into sharper focus in the last few years, beginning with the result from helioseismology that the principal gradient in angular velocity of the Sun is concentrated immediately below the bottom of the convective zone, from which one infers that the azimuthal magnetic field lies principally in that same layer below the bottom of the convective zone. Recent studies of the hydrodynamics of flux bundles of azimuthal field (Spruit and van Ballegooijen, 1982a,b; Fisher, McClymont, and Chou, 1991; D'Silva, 1993; D'Silva and Choudhuri, 1993; Fan, Fisher, and DeLuca, 1993; Fan, Fisher and McClymont, 1994; Schüssler, et al, 1994) indicate that the fibrils of azimuthal field have intensities of the order of $0.5 - 1.2 \times 10^5$ gauss if magnetic buoyancy, tension, and Coriolis force is to provide the emerging Ω-loops at the observed latitudes with the observed modest tilt to the east-west direction. It is conjectured that the extreme intensity of 10^5 gauss may be a direct consequence of the continuing formation of Ω-loops by the magnetic buoyancy, and, as already noted, the continual emergence of successive Ω-loops enhances convective heat transport from the bottom of the convective zone to the visible surface Parker (1994a,b,c).

When one looks at the 22-year oscillatory behavior of the magnetic fields in the Sun, there seems to be no alternative to some form of $\alpha\omega$-dynamo, except that there is a fundamental difficulty with the usual formulation of the theory. The problem arises from the fact that an MHD dynamo is intrinsically irreversible. In particular, the generation of the poloidal field component, expressed in terms of an azimuthal vector potential, by interaction of cyclonic convection with the azimuthal field–the α-effect–involves the formation of rotated local loops of field with nonvanishing projection on meridional planes. It is essential that dissipation merge these loops into a diffuse general poloidal field, for otherwise their considerable Maxwell stress would untwist them and restore the initial azimuthal field. The formation of the poloidal field, then, is necessarily irreversible. The azimuthal field is formed by shearing the general poloidal field in the nonuniform rotation of the star, and that too must be irreversible as a consequence of diffusion if the field pattern is to be cyclic, as in the Sun.

The traditional formulation of the problem calls upon turbulent diffusion to provide an effective resistive diffusion coefficient $\eta_t \sim 10^{11} - 10^{12}$ cm^2/sec. It is necessary that η_t be large enough that the fields diffuse across a characteristic distance ℓ in a characteristic time τ where $\ell^2 \sim 4\eta\tau$ in order of magnitude. With $\ell \sim 10^{10}$cm, representing the depth of the lower convective zone and $\tau \sim$ 5-years, the result is $\eta_t = 2 \times 10^{11}$cm^2/sec. Mixing length models (cf. Sprüit, 1974) of the convective zone are based on a characteristic eddy velocity v and eddy size λ at each level suggesting a dimensional diffusion coefficient $\eta_t \cong 0.1\lambda v$. A few hundred km below the visible surface it is estimated that $v \sim 0.4$ km/sec with $\lambda \sim 0.5 \times 10^3$ km, for which $\eta_t \cong 2 \times 10^{12}$ cm^2/sec. At a depth of 10^5 km, one obtains $v \sim 3 \times 10^3$ cm/sec, $\lambda \sim 3 \times 10^9$cm, and $\eta_t \sim 10^{12}$ cm^2/sec, etc. So as far as one could tell, there was sufficient diffusion. Detailed quantitative kinematic magnetohydrodynamical dynamo models do very well in duplicating the observed cyclic behavior of the magnetic fields at the surface of the Sun (cf. the Proceedings edited by Krause, Rädler and Rüdiger, 1993).

The difficulty with these kinematical dynamo models is that observations of the magnetic flux appearing at the surface of the Sun (Gaizauskis, et al., 1983) indicates a total azimuthal flux of at least 10^{23} Maxwell's in each hemisphere (Parker,1987), from which it follows that the mean azimuthal field intensity is 10^3 gauss or more. The mixing length models of the convection zone of the Sun (cf. Sprüit, 1974) provide theoretical equipartition fields with a maximum of about 3×10^3 gauss, falling to zero at the top and bottom of the convective zone. The result is that the mean azimuthal field is not far from the equipartition field, and such fields are too strong to be stretched into random flux bundles of sufficiently small thickness ϵ to be subject to resistive diffusion in the lower convective zone where $\eta \sim 10^3 - 10^4$ cm^2/sec.

The kinematical theory of turbulent diffusion of magnetic fields necessarily involves the development of random filaments of thickness ϵ so small that the characteristic diffusion time ϵ^2/η is comparable to the turnover time λ/v of the dominant eddies. The result is $\epsilon = \lambda/N_m^{\frac{1}{2}}$ where $N_m = \lambda v/\eta$ is the magnetic Reynolds number for the eddies. The reduction in thickness ϵ comes about because the eddies breakup the initial mean field B on the scale λ. The length of each bundle increases exponentially with time, more or less as $exp(vt/\lambda)$ because of the local elongation rate v/λ all along the bundle. The bundle is drawn out into a coiled ribbon with a width that remains of the same order as λ while the thickness $\epsilon \cong \lambda exp(-vt/\lambda)$ in order to conserve volume. The cross sectional area of the bundle declines approximately as $exp(-vt/\lambda)$, and conservation of magnetic flux in the bundle means a field larger than the mean field by the factor $exp(vt/\lambda) = \lambda/\epsilon = N_m^{\frac{1}{2}}$.

Appealing to rapid reconnection between neighboring filaments leads to a smaller factor, with $\lambda/\epsilon = N_m^{\frac{1}{3}}$ (Parker, 1963, 1979d). But the difficulty is evident in either case. The magnetic Reynolds number N_L is of the order of 10^9 at the midlevel of the convective zone so that $N_L^{\frac{1}{2}} = 3 \times 10^4$ and $N_L^{\frac{1}{3}} \sim 10^3$. But the mean field is already comparable to the equipartition field and the Reynolds stresses could not possibly force the stretching of filaments of the desired small thickness and large field intensity.

The problem is discussed at some length in the literature (Parker, 1993, 1994a,e) where it pointed out that recent results of helioseismology, that the nonuniform rotation of the Sun is concentrated in a relatively thin layer (thickness $\sim 3 \times 10^9$ cm) below the bottom of the convective zone, may relieve the problem to some degree. The point is that the azimuthal field would be generated in the shear layer below the convective zone, leaving the convective zone relatively clear ($\sim 10^2$ gauss) for the formation of the weak (10 gauss) poloidal field by the cyclonic convection. This does not entirely solve the problem, because it still calls for small-scale fields greatly in excess of the equipartition field.

Therefore we have taken up the possibility that the field is everywhere in a fibril form (Parker, 1982) pointing out that the interdiffusion of two nonparallel fibril fields is then merely a matter of rapid reconnection across the fibril diameters δ. As noted above, there is reason to think that the azimuthal fibrils are 10^5 gauss with a filling factor of the order of 3×10^{-2}, providing a mean of 3×10^3 gauss over the shear layer below the convective zone. The individual azimuthal fibrils below the convective zone move with the general flow, occasionally forming buoyant Ω-loops that extend upward through the convective zone to the surface. The small scales necessary for dissipation of opposite azimuthal fields during each 11-year half cycle are provided by the dynamics of the reconnection of interacting fibrils, of course, so that there is not the extensive stretching and mixing of ever more intense and thinner ribbons of field that is part of the traditional picture of kinematical turbulence.

The upward extension of Ω-loops to form the active regions observed at the surface does not stifle the formation of a mean poloidal field of 10 gauss or so over a significant fraction of the volume. It appears that the solar $\alpha\omega$-dynamo becomes plausible on this basis. A crude model was developed by Vainshtein, Parker, and Rosner (1993) to illustrate how the fibril dynamo might operate. Vishniac (1995) carries the ideas on the dynamics of magnetic fibrils into much greater detail.

The point of this sojourn in the realm of dynamo theory is to provide an illustration of the depth of the qualitative and quantitative uncertainty in our understanding of the magnetic activity cycle of the Sun. We really are only guessing at the origins of what is observed at the surface. Therefore, our ideas on the nature of the activity on solar-type stars, and particularly stars that are not close in age, mass, and composition to the Sun, can not go much beyond a phenomenological description of the quirks observed in the spectrum of their integrated light.

This statement is not a cry of despair. It is a challenge – a call to arms. Because there are observations to be made and theory to be explored that can greatly advance our understanding of the physics of a star and help us to understand the variations of the Sun that have such impact on our terrestrial climate. This is the vital phase of solar physics!

2. The Task Ahead

First of all we need to define the uncertainties in the conventional retinue of explanations. Then we need to sort out those uncertainties that may reasonably be expected to yield to a determined observational and theoretical assault. On the observational side there are several fundamental studies that need to be carried out to determine the path for theoretical development. With the advent of the GONG observing system there is the hope that in some way helioseismology can detect some of the effects of subsurface magnetic fields and subsurface flows, providing guidance for theoretical inquiry. The fundamental step is a solar "microscope" to resolve the individual magnetic fibrils, estimated to have diameters of the order to 100 km. The solar microscope would consist of a diffraction limited 120 cm mirror, achieving a resolution of $0.1''$, or 75 km at the Sun. Ideally the optical microscope would be operated with co-aligned UV and X-ray telescopes mounted on the same orbiting platform and operated over the years of both minimum and maximum solar activity. However several flights of a balloon borne optical system complemented by appropriately coordinated UV and X-ray observations from sounding rockets would be a giant step forward at a fraction the cost of the orbiting system. For we must not lose sight of the fact that there are several fundamental issues that cannot be settled without direct information on the internal structure of the individual fibril, the rate of intermixing of the fibrils, the details of the activity of fibrils in the network regions, the fine-scale structure of the mysterious facula and plage, the fine-scale structure of sunspots, etc. to mention the more outstanding mysteries of the moment. Then we may be sure that the solar microscope will open up a new world of hitherto unknown phenomena that currently lie outside our realization. What will be seen in the fine structure of flares, prominences, spicules, photospheric wave motions. And what is the cause of the facula? Until we know the nature of the faculae, it will not be possible to determine from what depth below the surface it picks up its energy, which is important if we wish to work out the physics of the brightness variations of the Sun.

The solar microscope was proposed over twenty years ago. It is well within technical capability but does not fit into the political agenda of the space agency. So the solar microscope has been stalled for two decades and a couple of years ago was openly rejected by NASA as either an orbiting observatory or as the much less expensive balloon-borne observatory. The solar microscope is the next fundamental step in probing the many mysteries of the Sun. The Sun is the most mysterious star in the sky, for the simple reason that we resolve the solar disk in the telescope. Other stars are just as complicated but their mysteries are not conspicuous. One might wonder why other stars emit X-rays, and 10 year cycles of activity. But there is not much that can be learned of the physics of the activity cycles and the X-ray emission by studying the distant stars alone. The Sun is the key, and unfortunately many essential aspects of the activity of the Sun are on such small-scales that a special solar microscope above the terrestrial atmosphere is needed to study them. But there is no reason why that fascinating project cannot be undertaken immediately. Until the solar microscope is put into operation the physics of the surface activity remains guess work, from which it is not possible to determine the nature of the activity that is so manipulative of terrestrial climate. Nor is it possible to say much about the causes of the activity on other solitary stars.

In the absence of the microscope, the obvious theoretical tasks are the development of the dynamics of fibril magnetic fields. First of all, why is the field in a fibril state in the first place? Evidently the convection in the interstices remains relatively field-free because it continually turns over. It sees the magnetic fields as continually circling the boundaries of the field-free region and not penetrating more than the appropriate skin-depth into the convection. Second, what is the rate at which the individual fibrils reconnect with other fibrils with different orientation? Unfortunately the answer to this question depends upon the unknown detailed structure of the fibrils. Then is it possible to summarize the dynamics of a fibril field with mean-field equations? Can one really construct a self-consistent theoretical solar dynamo based on sound physical principles? What is the large-scale hydrodynamics of the convective zone? How far can one go to construct a deductive theoretical model of the rotating stratified convective zone, with a minimum of arbitrary parametrization? Can such a model account for the Maunder Minimum (17th century) or the protracted Medieval Maximum (12th and 13th centuries)? Can we understand, then, why the brightness of a young star varies inversely with the magnetic activity and late in life varies directly? These are some of the obvious general questions that should one day be answered if we are to understand the solar manipulation of the terrestrial climate and if we are to have more than a superficial understanding of the star phenomenon in general.

References

Baliunas, S.L. et al, (1995) Chromospheric variations in main sequence stars, *Astrophys. J.* **438**, 269.

D'Silva, S. (1993) Can equipartition fields produce the tilts of bipolar magnetic regions, *Astrophys. J.* **407**, 385.

D'Silva, S. and Choudhuri, A.R. (1993) A theoretical model for tilts of bipolar magnetic regions, *Astron. Astrophys.* **272**, 621.

Eddy, J.A. (1977a) Climate and the changing Sun, *Clim. Change* **1**, 173.

Eddy, J.A. (1977b) Historical evidence for the existence of the solar cycle, in The Solar

Output and its Variation, Boulder, Colorado Associated University Press, ed. O.R. White, p. 51.

Eddy, J.A. (1983) The Maunder Minimum, a reappraisal, *Solar Phys.* **89**, 195.

Fan, Y., Fisher, G.H., and DeLuca, E.E. (1993) The origin of morphological asymmetries in bipolar active regions, *Astrophys. J.* **405**, 390.

Fan, Y., Fisher, G.H., and McClymont, A.N. (1994) Dynamics of emerging active region flux loops, *Astrophys. J.* **436**, 907.

Fan, G.H., McClymont, A.N., and Chow, D.Y. (1991) The stretching of magnetic flux tubes in the convective overshoot region, *Astrophys. J.* **374**, 766.

Foukal, P. and Lean, J. (1986) The influence of faculae on total solar irradiance and luminosity, *Astrophys. J.* **302**, 826.

Friis-Christensen, E., and Lassen, K. (1991) Length of the solar cycle: An indicator of solar activity closely associated with climate, *Science* **254**, 698.

Gaizauskas, V., Harvey, K.L., Harvey, J.W., and Zwaan, C. (1983) Large-scale patterns formed by solar active regions during the ascending phase of cycle 21, *Astrophys. J.* **265**, 1056.

Hameed, S.,and Gong, G. (1994) Variation of spring climate in lower-middle Yangtse River Valley and its relation with solar-cycle length, *Geophys. Res. Lett.* **21**, 2693.

Krause, F., Rädler, K.H., and Rüdiger, G. (1993) The Cosmic Dynamo, Dordrecht, Kluwer Academic Publishers.

Kuhn, J.R., and Libbrecht, K.G. (1991) Nonfacular solar luminosity variations, *Astrophys. J. Lett.* **381**, L35.

Martin, S. (1988) The indentification and interaction of network, intranetwork, and ephemeral-region magnetic fields, *Solar Phys.* **117**, 243.

Parker, E.N. (1963) A kinematical model of turbulent hydromagnetic fields, *Astrophys. J.* **138**, 226.

Parker, E.N. (1979a) Sunspots and the physics of magnetic flux tubes. II. Aerodynamic drag, *Astrophys. J.* **230**, 914.

Parker, E.N. (1979b) Sunspots and the physics of magnetic flux tubes. III. Aerodynamic lift, *Astrophys. J.* **231**, 250.

Parker, E.N. (1979c) Sunspots and the physics of magnetic flux tubes. VI. Convective propulsion, *Astrophys. J.* **232**, 282.

Parker, E.N. (1979d) Cosmical Magnetic Fields, Oxford, Clarendon Press, §17.6.

Parker, E.N. (1981) The dissipation of inhomogeneous magnetic fields and the problem of coronae. II. The dynamics of dislocated flux tubes, *Astrophys. J.* **244**, 644.

Parker, E.N. (1982) The dynamics of fibril magnetic fields. II. The mean field equations, *Astrophys. J.* **256**, 302.

Parker, E.N. (1983a) Magnetic neutral sheets in evolving fields. I. General theory, *Astrophys. J.* **264**, 635.

Parker, E.N. (1983b) Magnetic neutral sheets in evolving fields. II. Formation of the solar corona, *Astrophys. J.* **264**, 642.

Parker, (1984a) Stellar fibril magnetic fields I. Reduced energy state, *Astrophys. J.* **283**, 343.

Parker, E.N. (1984b) Stellar fibril magnetic fields. II. Two dimensional magnetohydrodynamic equations, *Astrophys. J.* **249**, 47.

Parker, E.N. (1984c) Stellar fibril magnetic fields. III. Convective counterflow, *Astrophys. J.* **294**, 57.

Parker, E.N. (1987) The dynamical oscillation and propulsion of magnetic fields in the convective zone of the Sun. I. General considerations, *Astrophys. J.* **312**, 868.

Parker, E.N. (1991a) Heating solar coronal holes, *Astrophys. J.* **372**, 719.

Parker, E.N. (1991b) The phase mixing of Alfven waves, coordinated modes, and coronal heating, *Astrophys. J.* **376**, 355.

Parker, E.N (1993) A solar dynamo surface wave at the interface between convection and nonuniform rotation, *Astrophys. J.* **408**, 707.

Parker, E.N. (1994a) Summary comments, in The Solar Engine and its Influence on

Terrestrial Atmosphere and Climate, Berlin, Springer-Verlag, ed. E. Nesme-Ribes, p. 527.

Parker, E.N. (1994b) Theoretical interpretation of magnetic activity, in The Sun as a Variable Star, Solar and Stellar Irradiance Variations, Proc. IAU Symp. No. 143, Boulder, Colorado 1993, Cambridge, Cambridge University Press, ed. J.M. Pap, C. Fröhlich, H.S. Hudson, and S.K. Solanki, p. 264.

Parker, E.N. (1994c) Theoretical properties of Ω-loops in the convective zone of the Sun. I. Emerging bipolar magnetic regions, *Astrophys. J.* **433**, 867.

Parker, E.N. (1994d) Spontaneous Current Sheets in Magnetic Fields, New York, Oxford University Press.

Parker, E.N. (1994e) Origins of the solar magnetic field, in Solar Magnetic Fields, Cambridge, Cambridge University Press, ed. M. Schüssler and W. Schmidt, p. 94.

Parker, E.N. (1995a) Theoretical properties of Ω-loops in the convective zone of the Sun. II. Origin of enhanced solar irradiance, *Astrophys. J.* **440**, 415.

Parker, E.N. (1995b) Theoretical properties of Ω-loops in the convective zone of the Sun, *Astrophys. J.* **442**, 405.

Piddington, J.H. (1978) The flux-rope-fiber theory of solar magnetic fields, *Astrophys. Space Sci.* **55**, 401.

Porter, J.G. and Moore, R.L. (1988) in Proc. 9th Sacramento Peak Summer Symposium, 1987, Sunspot. N.M.: NSO/
Sacramento Peak, ed. R.C. Altrock, p. 30.

Ribes, J.C., and Nesmes-Ribes, E. (1993) The solar sunspot cycle in the Maunder Minimum AD 1645 to AD 1715, *Astron. Astrophys.* **276**, 549.

Schüssler, M., Caligar, P., Feriz-Mas, A., and Moreno-Insertis, F. (1994) Instability and eruption of magnetic flux tubes in the solar convective zone, *Astron. Astrophys.* **281**, L69.

Sprüit, H.C. (1974) A model of the solar convection zone, *Solar Phys.* **34**, 277.

Sprüit, H.C. (1979) Convective collapse of flux tubes, *Solar Phys.* **61**, 363.

Sprüit, H.C., and van Ballegooijen, A.A. (1982a) Stability of teroidal flux tubes in stars, *Astron. Astrophys.* **106**, 58.

Sprüit, H.C., and van Ballegooijen, A.A. (1982b) Stability of teroidal flux tubes in stars, erratum, *Astron. Astrophys.* **113**, 350.

Tsnganos, K.C. (1979) Sunspots and the physics of magnetic flux tubes. IV. Aerodynamic lift on a thin cylinder in convective flows, *Astrophys. J.* **231**, 260.

Vainshtein, S.I., Parker, E.N., and Rosner, R. (1993) On the generation of strong magnetic fields, *Astrophys. J.* **404**, 773.

Vishniac, E.T. (1995) The dynamics of flux tubes in high β plasma, *Astrophys. J.* (in press).

Willson, R.C., and Hudson, H.S. (1991) The Sun's lumnosity over a complete cycle, *Nature* **351**, 42.

Withbroe, G.L., and Noyes, R.W. (1977) Mass and energy flow in the solar chromosphere and corona, *Ann. Rev. Astron. Astrophys.* **15**, 363.

Zhang, Q., Soon, W.H., Baliunas, S.L., Lockwood, G.W., Skiff, B.A., and Radick, R.R. (1994) A method for determing possible brightness variations of the Sun in past centuries from observations of solar-type stars, *Astrophys. J. Lett.* **427**, L111.

Zwaan, C. (1978) On the appearance of magnetic flux in the solar photosphere, *Solar Phys.* **60**, 213.

Zwaan, C. (1985) The emergence of magnetic flux, *Solar Phys.* **100**, 397.

Poster Papers

Posters for Session I

MEASUREMENT OF THE CORONAL ELECTRON TEMPERATURE AT THE TOTAL SOLAR ECLIPSE ON 3RD NOV. 1994

K. ICHIMOTO, K. KUMAGAI, I. SANO
T. KOBIKI AND T. SAKURAI
National Astronomical Observatory
Mitaka, Tokyo 181, Japan

AND

A. MUÑOZ
Asociacion Chilena de Astronomia y Astronautica,
Marcoleta 485 - Of. H. Santiago, Chile

The continuous coronal spectrum is created by Thomson scattering of the photospheric radiation by free electrons in the corona. Because of the large thermal motions of the coronal electrons, most absorption features in the photospheric spectrum are washed out to form the nearly continuous spectrum of the K-corona. However, weak depressions are still expected in the K-corona spectrum at wavelengths where the strong absorption lines are concentrated in the photospheric spectrum, i.e. $\sim$ 3900Å including the CaII lines and the G band at $\sim$ 4300Å (Cram 1976). The shape of the continuous coronal spectrum, therefore, can offer a direct measure of the coronal electron temperature.

A spectroscopic observation was carried out at the total solar eclipse on 3rd Nov.1994 in Putre, Chile. Spectral images covering the wavelength range of 3500Å $\sim$4720Å were obtained for 3 different locations of the slit; two heights in the eastern large streamer and the southern coronal hole. After subtraction of the F and E-components from the spectra and correction of the wavelength-dependent detector sen-

Y. Uchida et al. (eds.),
Magnetodynamic Phenomena in the Solar Atmosphere – Prototypes of Stellar Magnetic Activity, 413–414.

sitivity, the obtained continuous coronal spectra are compared with the theoretical curves of the electron-scattering coronal spectra. The best fitting between the observed and the calculated curves gives the electron temperature at each location on the slit.

The results are summarized as following;

- Electron temperature (T_e) of the eastern streamer is estimated to be about 1.7-2.0 MK with the symmetric coronal model. If we take the direction of the streamer extension into account (15° behind the limb), the estimated temperature is reduced to ~1.6 MK.
- A variation of the electron temperature with height is not found in the streamer between $1.1R_\odot$ and $2.0R_\odot$ within the error of the measurment.
- The base of the southern coronal hole obviously shows a lower temperature (0.9-1.1 MK) than the coronal streamer.
- Though T_e in the coronal hole shows a rapid increase with height, this can be partly attributed to the scattered light from the sky.

We also tried to deduce a possible accelelation of the outward flow in the corona from the overall wavelength shift of the coutinuous spectra. A hint of an velocity difference (acceleration) by ~76km/s was found between $1.1R_\odot$ and 2.0 $R_\odot$ in the streamer.

The method for diagnosing the coronal electron temperature from the shape of the continuous spectra is superior to various techniques so far used for measuring the coronal electron temperature in the senses followings; 1) fewer assumptions are required to derive the temperature because the spectral shape contains direct information on the electron thermal motions; the only critical assumption is the geometry of the coronal structure; 2) we may also study the temperature distribution in the extended corona (up to $R \geq 2R_\odot$); 3) only a simple instrument is required to achieve the measurement. The continuous coronal spectra also offer a possibility for measuring the outflow velocity at the acceleration region of the solar wind though fairly accurate correction of stray light and instrumental sensitivity are required.

References

Cram,L.E., 1976, *Solar Phys.*, **48**, 3-19

Ichimoto,K., Kumagai,K., Sano,I., Kobiki,T., Sakurai,T., Muños,A., 1996, submitted to *Publ.Astron.Soc.Japan*.

RESULTS FROM 3 NOVEMBER 1994 SOLAR ECLIPSE

B.H. FOING, J.E. WIIK, N. HENRICH, F. DAVID, T. BEAUFORT
ESA Solar System Division
ESTEC/SO, Postbus 299, 2200 AG Noordwijk, The Netherlands
B. ALTIERI, R. OROSEI, R. LAUREIJS, L. METCALFE
ESA Space Science Department
E. MAURICE
Marseille Observatory, CNRS, France
AND
P. REUTER, D. BOIS
Azimut 360, Chile

Abstract.
We report on our observations of the 3 November 1994 solar eclipse from the North Chile altiplano, using the Solar System Division CCD and transportable telescope experiment, as well as 4 imaging experiments. Spectrophotometry of the quiet corona, prominences and polar plumes was successfully achieved. A summary of scientific results on magnetic coronal structures, coronal densities and temperatures is presented.

1. Eclipse Campaign Objectives and Operations

Before permanent coronographic and UV observations with SOHO and other future solar missions, the 3 November 1994 total solar eclipse was the last good opportunity of the century to observe the solar corona from the ground. The objectives of this campaign were to obtain information on magnetic coronal structures, and on temperature and density coronal diagnostics, as well as preparation for the SOHO data analysis. We operated our instruments from the base of Putre at 3500 m (Foing et al 1995, 1996):

- a Photometrics CCD camera, with an MPP 1284x1024 pixel CCD chip, operating safely under the ambient conditions. The camera was adapted to a portable 25 cm Meade telescope with automatic pointing. This experiment

Y. Uchida et al. (eds.),
Magnetodynamic Phenomena in the Solar Atmosphere – Prototypes of Stellar Magnetic Activity, 415–416.

was used with an objective grating, providing simultaneously (in the zero order) images of the inner and middle corona, and (in the first order) spectra of the inner corona and prominences at the limb.

- a camera attached to a Celestron 5 (12 cm, f/10) recorded photographic images of the inner corona with different exposure times

- a photographic 6x6 Hasselblad camera with 300 mm focal lens was used for imaging the middle and outer corona

In collaboration with the Eclipse 6000 expedition team, we obtained the highest solar eclipse images (at 5300m) with a Cassegrain 500 mm f/8 catadioptric telescope and a 6x6 camera with a 280 mm lens. Finally, at the very summit of the Parinacota volcano (6300m), the Moon shadow running along the eclipse path, and the light scattered from the horizon out of totality could be observed.

2. Results on Coronal Magnetic structures and temperature

We summarise some scientific results from the analysis of these data. The photographic data have been digitised, as well as calibration steps measured on the same film. The digitised images allowed the delineation of a number of fine plumes over the poles, and equatorial streamers. A grid of isophotes was obtained allowing us to find the radial gradient of intensity around the corona to be different for polar and equatorial structures. The geometry of related magnetic structures was determined, and it was found that some extended streamers depart significantly from radial orientation.

The CCD data had much higher dynamic range and signal-to-noise. High-pass filtering allowed us to trace numerous fine polar plumes, and to show a mixture of complex closed loops over the equatorial regions. The scattered light (about 10 per cent at limb) could be corrected accurately. The intensity ratio between equator and coronal holes ranges from 7 to 11, giving the ratio of electronic densities. An intensity scale height of 90 Mm was measured leading to an equivalent hydrostatic temperature of $1.8\ 10^6$ K at the base of the equatorial corona (Henrich 1995).

High-resolution images of prominences in white light (tracing the density) and in lines of Ca II, He I and Balmer, show different fine structures due to optical thickness and NLTE multidimensional transfer effects.

Extended results will be published elsewhere (Foing et al 1996).

References

Foing, B.H., Wiik, J. et al, 1995, Ann. Geophys. Part III, Vol 13, p.C664
Foing, B.H. et al , 1996, in preparation
Henrich N., Master Physics report, sept. 1995 ESTEC/Univ. Paris XI

INTERPRETATION OF SXT DATA CONCERNING THE DIFFUSE CORONA

P. A. STURROCK and M.S. WHEATLAND
Stanford University, Stanford, CA 94305, USA
L.W. ACTON
Montana State University, Bozeman, MT 59717, USA

During the interval 3-15 May 1992, an extended region of diffuse, stable corona crossed the north-east limb of the Sun. During this passage, the soft X-ray telescope on Yohkoh obtained a number of high-quality pairs of images, closely spaced in time, through the two thinnest analysis filters. Our first step was to average the measurements at each radial position over an azimuthal sector about 25 degrees wide. These averaged measurements were represented, at each radial position, by an emission measure and a temperature (derived from the ratio of the signals from the two filters).

We found that the emission measure, so determined, was a good fit to a spherically symmetric isothermal model with a temperature of $10^{6.2}$ K. However the temperature, so determined, was found to be far from isothermal. The temperature profile is a good fit to the trend that one would expect for conserved inward radial heat flux in a spherically symmetric atmosphere, if the heat flux at the surface is about $10^{5.6}$ erg cm^{-2} sec^{-1}.

Since it is clearly unacceptable to adopt an isothermal model, we have considered a barometric spherically symmetric atmosphere with conserved inward heat flux. We have compared this model with the observational data in the following way. We first assume that the region is effectively spherically symmetric, and use the Abel integral transform to estimate the temperature and density as functions of radius. We then compare these functions with the forms that are to be expected on the basis of a barometric atmosphere with inward heat flow. We find that the best least-squares fit is obtained for the following values of the temperature T_0, heat flux F_0 and density n_0 at the base of the model: $T_0 = 10^{6.08}$ K, $F_0 = 10^{5.62}$ erg cm^{-2} s^{-1}, $n_0 = 10^{8.85}$ cm^{-3}.

The downward coronal heat flux will be converted into radiation in the low transition region and upper chromosphere. Moore and Fung (1972) showed that this equivalence leads to a relationship between the downward heat flux and the pressure in the transition region. Their theory leads us to expect that, for a heat flux of $10^{5.62}$ erg cm^{-2} s^{1} and a temperature of $10^{6.08}$ K at the base of the corona, the electron density should be $10^{8.87}$ cm^{-3}, in good agreement with the value derived from our analysis of the observational data.

Although we have adopted a spherically symmetric model in this analysis, it should be emphasized that the region under discussion is not a coronal hole: it appears to be composed of loops that extend to quite large radii. It appears that below $x = 1.5$, where $x = r/R_0$, the atmosphere is approximately spherically symmetric. More important, it appears from our analysis that, for this case, nonthermal heating occurs primarily outside the radius $x = 1.5$, with little direct nonthermal heating below $x = 1.5$. The legs of the loops are heated primarily by heat flux originating in or near the tops of the loops.

Y. Uchida et al. (eds.),
Magnetodynamic Phenomena in the Solar Atmosphere – Prototypes of Stellar Magnetic Activity, 417–418.

If analysis of other regions shows that this is a general result, that nonthermal energy deposition in a coronal loop is typically localized close to the top of the loop, this fact may be a useful discrimator between the two main theories of coronal heating: (a) that coronal heating is a steady process, due to wave propagation and dissipation, and (b) that coronal heating is an episodic process, probably due to flare-like energy release. (For a recent review of coronal heating see, for instance, Ulmschneider, Priest and Rosner 1991). It will be necessary to determine, for each case (steady heating or episodic heating), whether or not the process will deposit energy preferentially near the tops of coronal loops.

This work was supported by NASA grants NAS 8-37334 and NAGW-2265, and by Air Force grant F49620-95-1-0008.

References

Moore, R.L., and Fung, P.C.W. (1972) Structure of the Chromosphere-Corona Transition Region, *Solar Phys.*, **23**, 78-102.

Ulmschneider, P., Priest, E. R., and Rosner, R. (eds.) (1991) *Mechanisms of Chromospheric and Coronal Heating* , Springer Publishing, Berlin.

TEMPERATURE STRUCTURE OF THE DIFFUSE CORONA

C. R. FOLEY AND J. L. CULHANE
Mullard Space Science Laboratory, University College London, Surrey RH5 6NT, UK

L. W. ACTON
Montana State University, Bozeman, MT 59717, US

AND

J. R. LEMEN
Lockheed Palo Alto Research Laboratory, Palo Alto, CA 94304, US

Yohkoh has observed diffuse emission from different regions of the Corona These large diffuse regions are observed with Yohkoh to extend up to 0.8 solar radii above the limb. Although partial loop structures can sometimes be discerned, systematic structures such as streamers or active region loops are not observed. Carefully de-scattered, (Hara *et al.*, 1978), Soft X-ray Telescope (SXT) images recorded though the thin X-ray analysis filters, Al1265 and AlMgMn, have been used to obtain the averaged radial temperature and emission measure structure for the diffuse regions. The temperature is observed to increase, from ~1.8MK to ~2.4MK for an extended diffuse observation made on the 26^{th}August 1992, figure 1. This is typical of other observations made on the 3^{rd}October 1992, and the 8^{th}May 1992, which similarly possess radially increasing temperatures. Small differences are noted between the observations, which may be accounted for in terms of the orientation of the structures to the Line of Sight, LS, which is modulated by the solar rotation.

The results are found to be similar for observations of regions with and without discern-able structure, which is taken as evidence for the closed field nature of the diffuse corona as observed by Yohkoh SXT. The magnetograms and Hα data for all observations are consistent with all the observations being of closed field regions. The response of the SXT for images recorded through the thin x-ray analysis filters peaks in the region of MK, where the signal is ~ 300× that which would be recorded at 1MK. The corresponding response at 2MK is approximately 100×. This has the cor

Y. Uchida et al. (eds.),
Magnetodynamic Phenomena in the Solar Atmosphere – Prototypes of Stellar Magnetic Activity, 419–240.

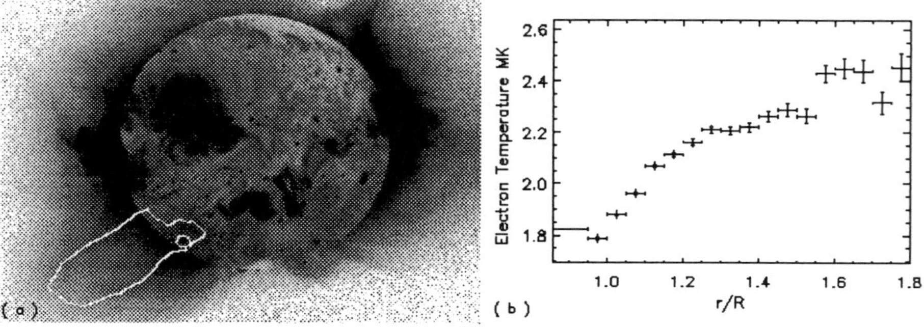

Figure 1. (a) The 26th August 1992 observation with the region of interest bounded. (b) The average radial LS temperature structure of region.

sequence that the higher temperature components located in the fore and background parts of the region in the LS dominate the signal of the cooler plasma, located at lower heliocentric heights above the limb. This results in a radial LS temperature profile which originates from higher temperatures and is thus shallower. Modelling of limb brightening observations may be used to demonstrate the heliocentric radial temperature dependence as opposed to the LS radial temperature dependence (work in progress).

The EUV observations made with the ATM on Skylab which had a peak sensitivity in the region of 1MK, yielded radial temperature characteristics consistent with the results discussed here when the differences in the instrument responses are considered. The results of those observations were found to be well represented by a static closed field model of RTV origin, (Withbroe, 1988), (Rosner *et al.*, 1978). Sturrock *et al.*, 1996, extracts the inward heat flux, for a diffuse observation, and interprets the result with respect to coronal heating.

Recent comparisons of the extended diffuse regions included within this study, with the Mauna Loa solar observatories MK3 Coronameter data have demonstrated that the diffuse regions are often associated with the base region of streamers. A letter is in the final stages of preparation for the Astrophysical Journal regarding this work.

References

Rosner R. and Tucker W. H. and Vaiana G. S. (1978), *ApJ* **220**, 643–665

Withbroe G. L. (1988), *ApJ* **325**, 442–467

Hara H. and Tsuneta T. and Acton L. W. and Bruner M. E. and Lemen J. R. and Ogawara Y.(1994) Temperatures of coronal holes observed with the Yohkoh SXT, *PASJ* **46**, 493

Sturrock P. A. and Acton L. W. and Wheatland M. S.(1996), *These proceedings*

MEASUREMENTS OF ACTIVE AND QUIET SUN CORONAL PLASMA PROPERTIES WITH SERTS EUV SPECTRA

J.W. BROSIUS
Hughes STX Corporation
Code 682.1, NASA/Goddard, Greenbelt, MD 20771 USA

J.M. DAVILA AND R.J. THOMAS
Laboratory for Astronomy and Solar Physics
Code 682.1, NASA/Goddard, Greenbelt, MD 20771 USA

B.C. MONSIGNORI-FOSSI
Arcetri Astrophysical Observatory
Largo Enrico Fermi 5, 50125 Firenze ITALY

AND

J.L.R. SABA
Lockheed Palo Alto Research Laboratory
Code 682.2, NASA/Goddard, Greenbelt, MD 20771 USA

The Solar EUV Rocket Telescope and Spectrograph (SERTS) obtained spectra and images of active regions, quiet Sun areas, and other solar features during successful flights in 1989, 1991, 1993, and 1995. The availability of multiple emission lines from eight ionization stages of iron (1) furnishes line ratio temperature diagnostic tools which are independent of elemental abundance uncertainties, (2) enables verification of the instrumental photometric calibration and atomic physics parameters through density- and temperature-insensitive ratios, (3) provides a means of deriving plasma densities, and (4) supplies input for DEM analyses. Temperatures derived from seven different line ratios are shown in Figure 1. For ratios among the higher ionization stages of iron, the derived active region temperature is greater than that of the quiet Sun; for ratios among the lower ionization stages, the active region and quiet Sun temperatures converge. Differential emission measure distributions were derived using the procedure developed by Monsignori-Fossi & Landini for SOHO. An active region DEM is shown in Fig 2. Details are provided in a paper submitted to *Astrophysical Journal.*

Y. Uchida et al. (eds.),
Magnetodynamic Phenomena in the Solar Atmosphere – Prototypes of Stellar Magnetic Activity, 421–422.

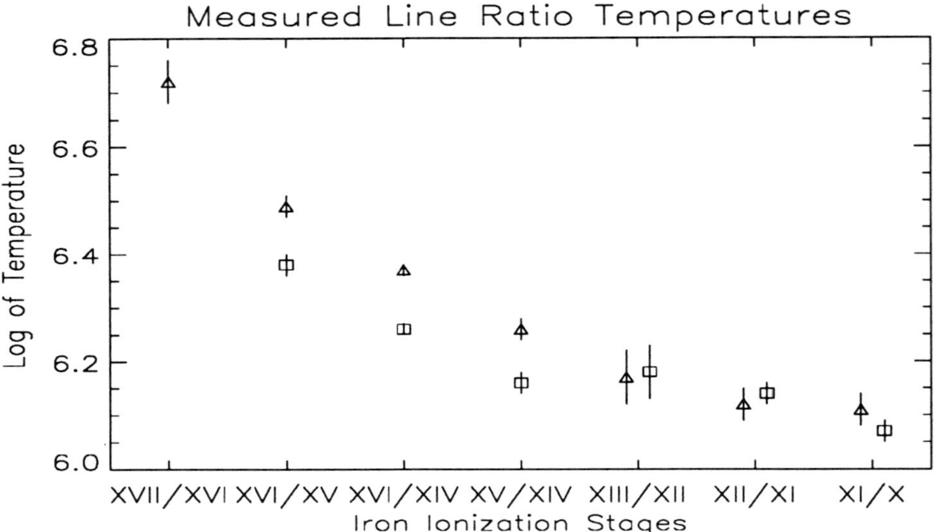

Figure 1. Temperatures derived from line intensity ratios for NOAA region 6615 (1991; triangles) and for quiet Sun areas (1993; squares). The x-axis gives the ionization stages of the two iron lines used for each ratio, and the y-axis gives log T. Error bars include uncertainties in the intensity measurements and in the plasma density. Temperatures derived from ratios involving only Fe X – Fe XIII have been displaced from each other horizontally to improve clarity.

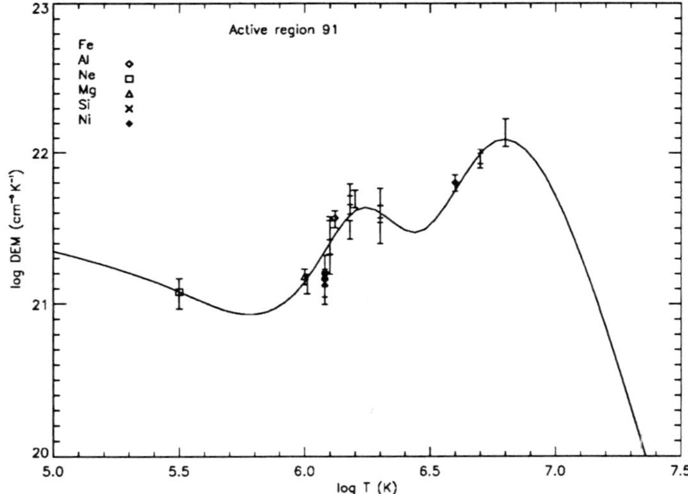

Figure 2. NOAA region 6615 differential emission measure distribution, in $cm^{-5}\ K^{-1}$. The curve was derived by fitting available lines of Fe, Ne, Mg, Si, Al, and Ni. Different symbols designate different elements. The x-coordinate of each symbol indicates the temperature of the maximum contribution to the line emission; the y-coordinate is the product of the DEM and the ratio of the observed to the predicted intensities. Error bars represent uncertainties attributed to intensity measurement only. The largest uncertainties in the DEM curve occur where the fewest emission lines are available.

CALCULATIONS OF SOFT X-RAY IMAGES FROM MHD SIMULATIONS OF HEATING OF CORONAL LOOPS

A.J.C. BELIËN, S. POEDTS AND J.P. GOEDBLOED
FOM-Institute for Plasma Physics 'Rijnhuizen',
P.O. Box 1207, 3430 BE Nieuwegein, The Netherlands

Spatial and time resolutions of soft X-ray cameras have reached the arcsec and second ranges [1, 2], respectively. With these resolutions, in principle it should be possible to do magnetohydrodynamics (MHD) spectroscopy [3], *i.e.* reconstruction of the basic features of the loop from measured MHD frequencies. This can provide a link between observations and theoretical simulations. This paper is devoted to linking computer simulations with soft X-ray observations by comparing features on the image level. This provides another means of linking observations and simulations.

To determine the soft X-ray intensities observed by the SXT aboard Yohkoh, we calculated $\int_0^\infty n_e^2(z)F(T(z))dz$ for each image pixel. F is the thermal response function of SXT for a specific filter. The electron density, n_e, and temperature , T, are obtained from the MHD calculations.

Extensive nonlinear evolution simulations of resonant absorption in solar coronal loops have been carried out [4]. The associated simulated images of soft X-rays intensities, as would be observed by SXT through the Al.1 filter, are shown in the series of snapshots of Figure 1. The development of a resonant layer, its nonlinear broadening, and the heating of the loop are easily seen in the simulated SXT images. This proves that resonant absorption can be seen in soft X-ray images. However, we have not yet found SXT loop observations that closely resemble the simulated ones for resonant absorption. This does not rule out resonant absorption as a possible means of heating coronal loops, but probably, it indicates that the theoretical model is too restrictive. Calculations of resonant absorption, for example, are mainly done for monoperiodic excitation, while it is known that a range of frequencies is available for excitation. Therefore, excitation with a range of frequencies and its implications on SXT observations is currently being investigated.

Another promising aspect of the creation of series of SXT images for a large set of parameter values is that it indicates the dependence of SXT

Y. Uchida et al. (eds.),
Magnetodynamic Phenomena in the Solar Atmosphere – Prototypes of Stellar Magnetic Activity, 423–424.

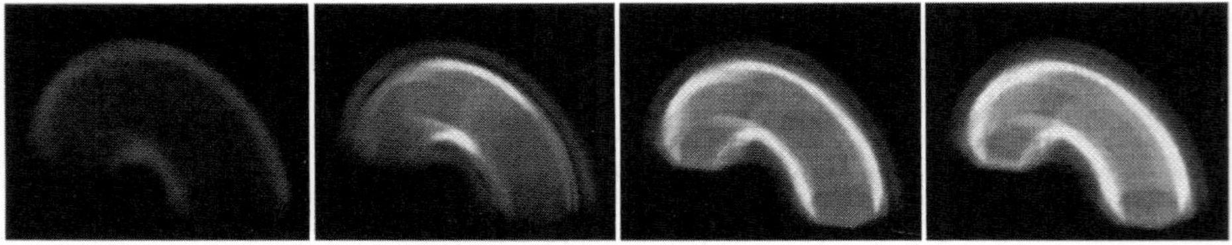

Figure 1. Four simulated Al.1 SXT snapshots showing the evolution of a coronal loop that is heated by means of resonant absorption of Alfvén waves.

images on the parameters. This knowledge might be used to estimate the corresponding parameters attached to a real observation in soft X-rays, premised that the observation under consideration share common characteristics of the underlying theoretical model, for example, resonant layers. Moreover, we have shown that the estimation can be automated using image processing tools [5]. This is, then, another way of doing MHD spectroscopy: reconstruction is based on image information instead of frequencies.

It is our opinion that the creation of the X-ray images is a valuable tool which can also be used to visualize other heating scenarios. Ultimately, this may lead to a choice or rejection of particular scenarios based on the real observations and the simulated observations.

Acknowledgements

The authors wish to thank Jim Klimchuk (NRL Washington) for stimulating discussions and support on SXT modelling.

References

1. L. Golub, M. Herant, K. Kalata, I. Lovas, G. Nystrom, F. Pardo, E. Spiller, and J. Wilczynski. Sub-arcsecond observations of the solar x-ray corona. *Nature*, 344:842–843, April 26 1990.
2. S. Tsuneta, L. Acton, M. Bruner, J. Lemen, W. Brown, R. Caravalho, R Catura, S. Freeland, B. Jurcevich, M. Morrison, Y. Ogawara an T. Hirayama, and J. Owens. *Solar Phys.*, 137:37, 1991.
3. J.P. Goedbloed, S. Poedts, G.T.A. Huysmans, G. Halberstadt, H.A. Holties, and A.J.C. Beliën. Magnetohydrodynamic spectroscopy, large scale computation of the spectrum of waves in plasmas. *Future Generation Computer Systems*, 10:339–343, 1994.
4. S.M. Poedts and G.C. Boynton. Non-linear magnetohydrodynamics of footpoint driven coronal loops. *Accepted for publication in Astron. Astrophys.*, 1995.
5. A.J.C. Beliën, H.J.W. Spoelder, R. Leenders, S. Poedts, and J.P. Goedbloed. Calculation of soft x-ray images from solar coronal MHD heating simulations. *Submitted.*

2D AND 3D NONLINEAR MHD SIMULATIONS OF CORONAL LOOP HEATING BY ALFVÉN WAVES

S. POEDTS AND J.P. GOEDBLOED
FOM-Institute for Plasmaphysics 'Rijnhuizen'
P.O.Box 1207, 3430 BE Nieuwegein, The Netherlands

The resonant absorption of Alfvén waves provides a mechanism to transfer energy from large to very small length scales at which it can be thermalized on relatively short time scales. The role this mechanism might play in the heating of the magnetic loops observed in the solar corona has been studied extensively in the context of linear magnetohydrodynamics (MHD) (see [1] and references therein). However, due to the extremely high (magnetic) Reynolds numbers in the hot coronal plasmas, the linear approximation is directly challenged by the results of these studies and, in particular, by the development of large amplitude fields in the resonant layers. Here, the *nonlinear temporal evolution* of externally excited coronal loops is simulated numerically in *2D and 3D dissipative MHD*. Both sideways excitation by incident magnetosonic waves and footpoint excitation by photospheric convective motions are considered. It turns out that the dynamics in the resonant layers is dominated by nonlinear effects.

First, consider a cylindrical, axisymmetric (2D) coronal loop model with a homogeneous longitudinal magnetic field, a radially stratified density and driven periodically at one end and fixed at the other end, modelling the driving by overshooting convection on the one hand and the field line anchoring ('line-tying') in the dense photosphere on the other hand. The generated torsional Alfvén wave is reflected at the line-tied boundary resulting in a standing Alfvén wave which phase-mixes and which develops a resonant behaviour at those flux surfaces where the local Alfvén frequency matches the frequency of the external driver. In these layers, the dynamics is nonlinear and the driven system *does not evolve to a steady state* in contrast to linear MHD simulations. The nonlinear simulations show much broader resonant layers and the heating turns out to be very efficient with high absorption coefficients ($> 90\%$) and coupling factors and acceptable time scales (much

Y. Uchida et al. (eds.),
Magnetodynamic Phenomena in the Solar Atmosphere – Prototypes of Stellar Magnetic Activity, 425–426.

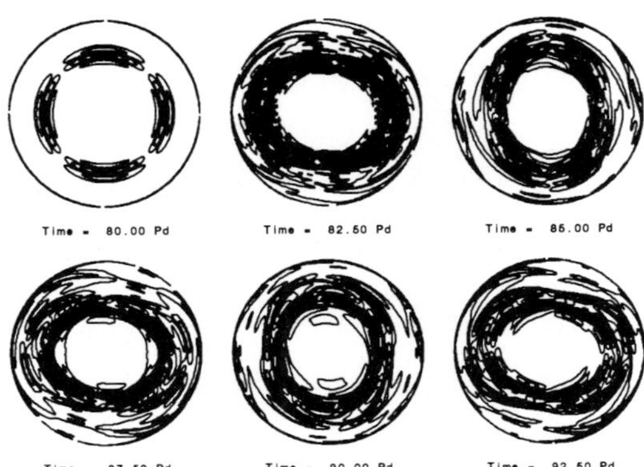

Figure 1. Snapshots of the z-component of the vorticity for a nonlinear continuation of a linear run after 80 driving periods (when a steady state is reached).

shorter than the life time of the loops). Moreover, the footpoint excitation does not require the presence of global modes to be efficient [2].

Extending the model to a 3D cylinder with a helical field reveals more nonlinear effects. Simulations of the resonant absorption of side-ways incident waves show again the development of resonant layers. However, the system makes use of the additional degree of freedom in the third (azimuthal) spatial direction. The background fields, and thus the resonance positions, vary in time and substantial nonlinear mode coupling transfers energy to 'overtones' with smaller wavelengths. For large enough amplitudes and large enough magnetic Reynolds numbers the gradients in the resonant layers become so high that the shear flow becomes Kelvin-Helmholtz unstable. As a result, even smaller length scales are developed and the resonant layers are deformed substantially so that the heat is deposited over the entire loop volume rather than in narrow resonance layers (see Fig. 1).

The nonlinear dynamics of the periodically varying shear flows that occur in the resonant layers is much more complex than predicted by linear MHD simulations. As a matter of fact, in *nonlinear* MHD the background is allowed to vary in response to the heat deposition and *nonlinear mode coupling* occurs. Moreover, for large enough driving amplitudes and magnetic Reynolds numbers, the shear flows in the resonant layers become Kelvin-Helholtz unstable. The effect of this instability on the heating efficiency is yet to be quantified under typical coronal loop circumstances.

References

1. Poedts, S., Beliën, A.J.C., Goedbloed, J.P., *Solar Phys.*, **151**, 271 – 304 (1994).
2. Poedts, S. and Boynton, G.C., *Astron. & Astrophys.*, 1995, in press.

GENERATION OF LINEAR AND NONLINEAR MAGNETIC TUBE WAVES IN THE SOLAR ATMOSPHERE

Z. E. MUSIELAK
University of Alabama, Huntsville, AL 35812, USA
R. ROSNER
University of Chicago, Chicago IL 60637, USA
AND
P. ULMSCHNEIDER
University of Heidelberg, D-6900 Heidelberg, Germany

Abstract. The fact that solar magnetic fields outside sunspots have a "flux tube" structure has been-well known for a number of years. These magnetic flux tubes may become "windows" through which the wave energy generated in the solar convective zone is carried by various types of waves to the overlying chromosphere and corona. From a theoretical point of view it is important to incorporate these highly intermittent spatial magnetic structures into the theory of wave generation. The first calculations of this sort have been reported by Musielak, Rosner and Ulmschneider (1989) who developed a general theory describing the interaction of magnetic flux tubes with the turbulent convection. Recently, this theory has been modified to incorporate an improved description of the spatial and temporal spectrum of the turbulent convection and used to calculate wave energy fluxes carried by linear longitudinal (Musielak *et al.* 1995a) and linear transverse (Musielak *et al.* 1995b) magnetic tube waves in the solar atmosphere (see Table 1).

The described treatment of the generation of magnetic tube waves is purely analytical which means that it is restricted to linear waves. Some numerical simulations (e.g., Malagoli, Cattaneo and Brummell 1990) and recent solar observations (e.g., Muller *et al.* 1994) suggest that nonlinear effects are important in the generation of these waves. We have recently adopted a 1-D, time-dependent, MHD code originally developed by Ulmschneider, Zähringer and Musielak (1991) to study the excitation of nonlinear transverse (Huang, Musielak and Ulmschneider 1995) and nonlinear

Y. Uchida et al. (eds.),
Magnetodynamic Phenomena in the Solar Atmosphere – Prototypes of Stellar Magnetic Activity, 427–428.

longitudinal (Ulmschneider and Musielak 1995) magnetic tube waves and found a significant increase in the wave energy fluxes carried by these waves (see Table 1). In light of these results, the magnetic wave energy fluxes calculated by using the analytical methods must be regarded as only lower bounds on the magnetic wave generation.

It is the purpose of this paper to find out whether the wave energy fluxes generated by turbulent motions in the solar convection zone are sufficient to sustain the mean level of radiative losses observed from active regions in the solar chromosphere (i.e. the chromospheric network) where the enhanced heating is observed. Taking the total radiative losses from the chromospheric network to be 1×10^7 erg/cm^{2}s and the filling factor to be 0.2, it is seen from Table 1 that magnetic flux tubes might be candidates for the heating in magnetically active regions of the solar chromosphere. Note, however, that in order to make this estimate more realistic energy losses due to the wave propagation must also be accounted for.

TABLE 1. Total wave energy fluxes carried by magnetic tube waves in the solar atmosphere

Type of Waves	Magnetic Field [G]	Flux [erg/cm^{2}s]
Linear longitudinal	1500	1×10^7
Linear transverse	1500	8×10^7
Nonlinear longitudinal	1500	5×10^8
Nonlinear transverse	1500	1×10^9
Linear longitudinal	1000	3×10^8
Linear transverse	1000	7×10^8
Nonlinear longitudinal	1000	5×10^8
Nonlinear transverse	1000	1×10^9

References

Huang, P., Musielak, Z.E., & Ulmschneider, P., 1995, *Astron. Astrophys.*, **297**, 579.

Malagoli, A., Cattaneo, F., & Brummell, N. H., 1990, *Astrophys. J.*, **361**, L33.

Muller, R., Roudier, T., Vigneau, J., & Auffret, H., 1994, *Astron. Astrophys.*, **283**, 232.

Musielak, Z.E., Rosner, R., Gail, H.-P., & Ulmschneider, P. 1995a, *Astrophys. J.*, **448**, 865.

Musielak, Z.E., Rosner, R., Gail, H.-P., & Ulmschneider, P. 1995b, *Astrophys. J.*, submitted.

Musielak, Z.E., Rosner, R., & Ulmschneider, P. 1989, *Astrophys. J.*, **337**, 470.

Ulmschneider, P., & Musielak, Z. E., 1995, *Astron. Astrophys.*, submitted.

Ulmschneider, P., Zähringer, K., & Musielak, Z. E., 1991, *Astron. Astrophys.*, **241**, 625.

MAGNETIC ROOTS OF ENHANCED HIGH CORONAL LOOPS

J.G. PORTER, D.A. FALCONER (NAS/NRC) AND R.L. MOORE
NASA/MSFC, Huntsville, AL

K.L. HARVEY (SPRC) AND D.M. RABIN
NSO, Tucson, AZ

AND

T. SHIMIZU
Inst. of Astronomy/University of Tokyo, Tokyo, Japan

We report results from an extension of a previous investigation of the magnetic roots of high-arching bright coronal loops (Porter et al 1994, in Proceedings of Kofu Symposium "New Look at the Sun," ed. S. Enome, NRO Report No. 360, p. 65). In the previous work, the magnetic locations and magnetic structure of the brightest coronal features in a selected active region were determined by registering Yohkoh SXT images with a MSFC vector magnetogram via registration of the sunspots. The active region (AR 6982 on 26 Dec 91) was selected for study because it had a large delta sunspot with a core of strong magnetic shear along the polarity inversion; it was expected that such extemely nonpotential magnetic fields would foster exceptionally strong coronal heating and hence be exceptionally bright in coronal images. It was found that the coronal heating in this active region indeed was markedly more intense in the low-lying sheared core field than in the bulk of the field that arched over the sheared core and spanned the whole bipolar region. In addition, the coronal images showed something that was not anticipated in the selection of this region: a section of the high-arching envelope field was much brighter than the rest, showing that it received much more coronal heating than the rest of the envelope field. These enhanced high coronal loops stemmed from around an embedded island of opposite-polarity flux that was the site of microflaring and enhanced coronal heating. It was therefore surmised that the high bright loops somehow received their enhanced coronal heating from this foot-point activity. In the present work, by registering a full-disk Kitt Peak magnetogram with full-disk Yohkoh SXT images, we have found many more examples of large

Y. Uchida et al. (eds.),
Magnetodynamic Phenomena in the Solar Atmosphere – Prototypes of Stellar Magnetic Activity, 429–430.

enhanced coronal loops having one foot rooted at a site of mixed polarity within an active region. The figure shows part of a Yohkoh SXT full disk image taken at 19:45 UT on Dec 26, 1991 (a composite of a 78 ms and a 2668 ms exposure). AR 6982 is on the right and AR 6985 is on the left. Nine large loops (1-9) and 3 bright "source" areas (A-C) are identified. Note how each of the extended loops has one (narrow) end rooted near a source area: loops 1-6 stem from near source A, loop 7 from near source B, and loops 8 and 9 from near source C. The white/black contours superposed on the figure correspond to field strengths of +/- 150 G in the Kitt Peak magnetogram taken at 16:08 UT. It can be seen that sources A, B, and C mark islands of included magnetic polarity. The extended loops' narrow ends are rooted in the dominant polarity surrounding these islands. The remote ends of loops 1 and 2 lie within the same active region as the source ends. For the other loops, the remote ends fan out in weak-field regions of opposing polarity. Rapid cadence partial frame observations (Porter et al 1994) show frequent microflaring at source A. Less-frequent full disk images suggest that sources B and C are sites of similar behavior. These additional examples serve as further evidence that the enhanced coronal heating in large bright coronal loops is a consequence of microflaring and/or related activity in mixed-polarity fields at one of the loop footpoints.

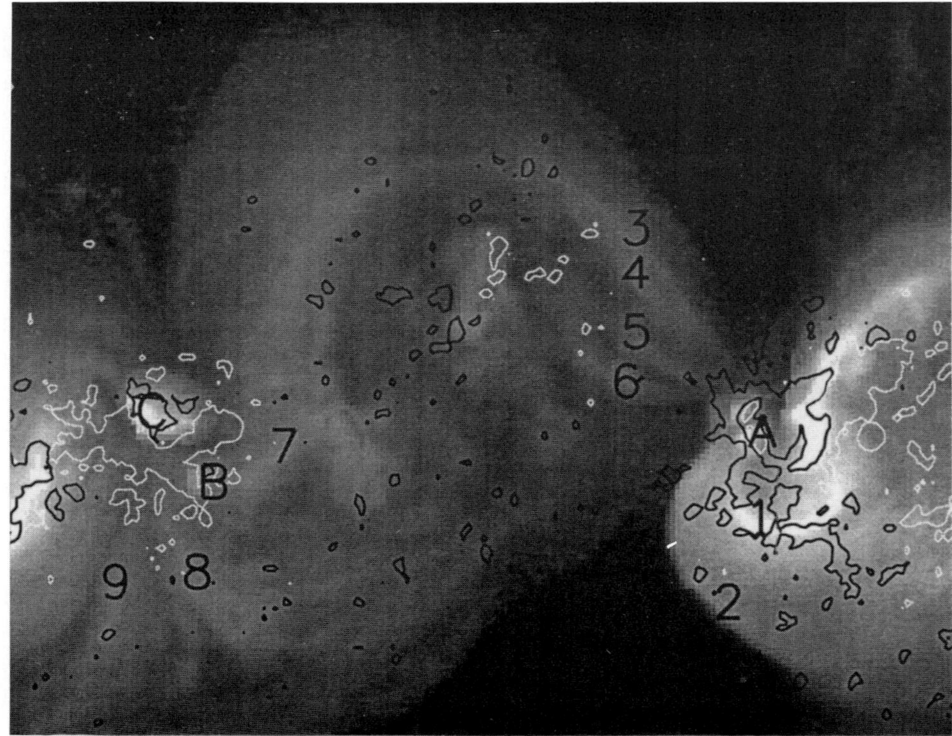

DYNAMIC BEHAVIOR OF THE CORONAL STREAMER BELT

X.P. ZHAO AND J.T. HOEKSEMA
Center for Space Science and Astrophysics
HEPL B204, Stanford University, Stanford, California, U.S.A.

Extended Abstract

In addition to its solar cycle evolution, the coronal streamer belt that lies at the base of the heliospheric current sheet sometimes undergoes drastic changes in its configuration. What causes drastic changes of the global structure?

We first examine the effect of the $\sim$ 22 UT, 3 November 1994 coronal mass ejection (CME) that originated above the southwestern limb in a helmet streamer that was a part of the coronal streamer belt. The *Yohkoh* Soft X-ray and MLSO/HAO white-light solar images observed near the time of the 3 November 1994 solar eclipse and one day, 13 days and 27 days after the CME suggest that the helmet streamer disrupted by the CME reforms in a time interval of 12 hours.

We search for the effect of helmet-streamer CMEs on the global structure of the coronal streamer belt using SMM coronagraph data in 1985. The location of the streamer belt is calculated using the WSO synoptic magnetic field and the source surface model. Figure 1 indicates that helmet-streamer CMEs generally do not significantly change the overall configuration of the belt.

Using ISEE-3 data we identify the temporal and local effect of CMEs on the internal structure of the heliospheric current sheet or coronal streamer belt. Counterstreaming halo electron events are believed to be the manifestation of the helical CME fields and are one of the most reliable indicators of CMEs in interplanetary space. The variations of the IMF azimuthal angle during the passage of CMEs suggest that CMEs often alter the internal structure and increase the thickness of the coronal streamer belt only temporarily and locally (For the details see the fuller paper accepted by *J. Geophys. Res.* in 1995).

Over scales of the order of a solar radius, the corona is controlled by a competition between the tendency of the million degree corona to expand

Y. Uchida et al. (eds.),
Magnetodynamic Phenomena in the Solar Atmosphere – Prototypes of Stellar Magnetic Activity, *431–432.*

into the solar wind and the opposing tendency of bipolar magnetic fields in a highly conducting plasma to seek a closed configuration and resist the solar-wind flow. Depending on the balance the equilibrium corona has two distinct states, typified by the coronal holes and the arcades of helmet-streamers separating these holes. Drastic changes of the global structure may be caused by the physical processes that produce new coronal holes.

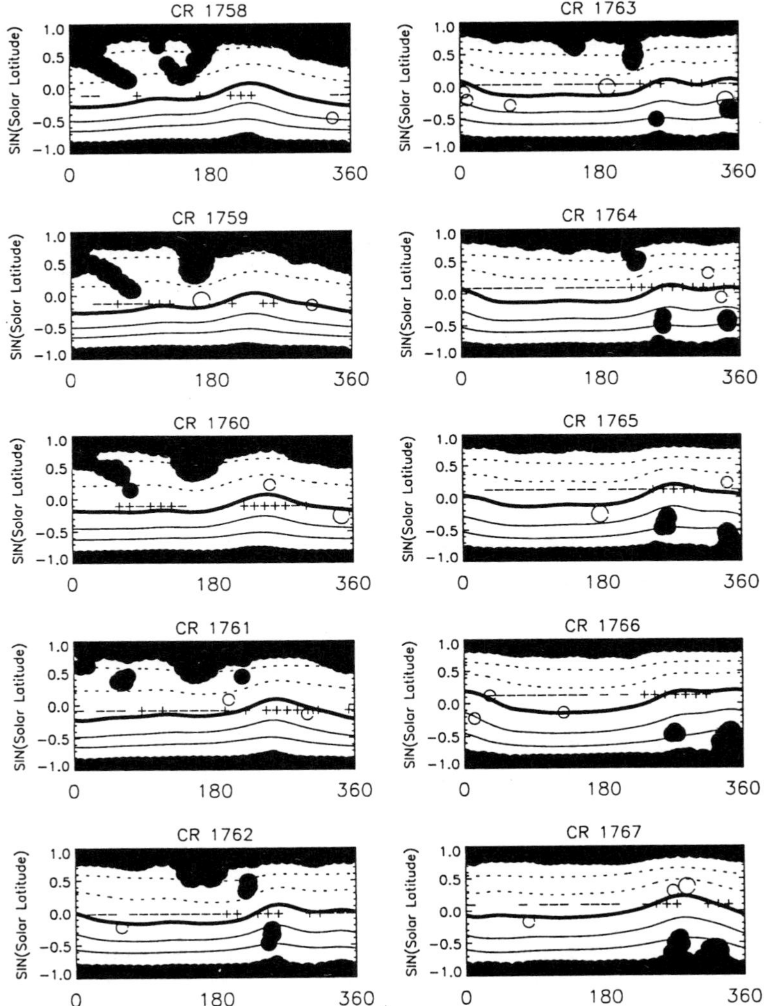

Fig. 1 The effect of blowout-CMEs (large open circles) and disruption-CMEs on the heliospheric current sheet (thick lines). The '+' and '−' symbols denote observed daily-averaged outward and inward IMF polarity. The filled circles denote the foot points of open field lines (coronal holes).

MINOR PHOTOSPHERIC AND CHROMOSPHERIC MAGNETIC ACTIVITY AND RELATED CORONAL SIGNATURES

G. CAUZZI
Institute for Astronomy - Univ. of Hawaii at Manoa
2680 Woodlawn Dr. - Honolulu HI 96822 (USA)

A. FALCHI
Osservatorio Astrofisico di Arcetri
Largo E. Fermi 5 - I-50125 Firenze (Italy)

R. FALCIANI
Dept. of Astronomy and Space Science - Univ. of Florence
Largo E. Fermi 5 - I-50125 Firenze (Italy)

E. HIEI
Meisei University Hodokubo
Hino - Tokyo 191 (Japan)

AND

L.A. SMALDONE
Dept. of Physical Sciences - Univ. "Federico II"
Pad. 19 Mostra d'Oltremare - I-80125 Napoli (Italy)

Solar activity is associated with the appearance, temporal evolution and spatial modifications of the magnetic field at the solar surface and with its interactions with plasma mass flows.

If all the various manifestations of solar activity can be attributed to the same basic physical mechanism (viz., the currently fashionable magnetic reconnection), the study of the small spatial scale and short time-constant phenomena will be particularly valuable for a better understanding of the physical processes since hopefully only the basic mechanisms will be at work.

In this context minor transient activity phenomena (as Ellerman bombs (EB), persistent bright points, microflares, etc.) represent very interesting targets to be investigated for their importance in clarifying the basic mechanisms for energy storage, release and transfer occurring in an apparently simple structure.

Y. Uchida et al. (eds.),
Magnetodynamic Phenomena in the Solar Atmosphere – Prototypes of Stellar Magnetic Activity, 433–434.

During an international coordinated observing campaign in Sept.-Oct., 1992 we got simultaneous observations of minor activity phenomena with the PFI (partial frame image) mode of YOHKOH soft X-ray telescope and the ground-based (GBO) facilities of the National Solar Observatory (both Sacramento Peak and Kitt Peak sites). The overlay accuracy between the various GBO images is of the order of $\pm$ 0.45″. The co-alignment with the soft X-ray (SXR) data has been obtained with an accuracy of $\pm$ 1″.

On Oct. 05, 1992 we observed an EB from its build-up up to the decay phase. In the following we summarize the main results of our analysis.

- In our observations we found spectral signatures of the EB in radiations originated in the chromosphere ($H\gamma \rightarrow H_9$ and Si I - 3905 lines; $H\alpha$ + 1.5 Å and Na - D_2 0.0).
- The EB has a size of about 3″ and occurs just above the photospheric magnetic neutral line.
- We obtained the light curves of the EB $H\alpha$ + 1.5 Å emission (time resolution 2.7 s) and Na-D_2 - 0.0 emission (time resolution 30 s). The maximum contrast is 1.13 of the average nearby quiet intensity. The total lifetime is estimated about 11.5 min. and intensity variations occur on a typical time scale of about one minute.
- The asymmetry of the line emission profiles in the center of the EB area indicates the presence of a line-of-sight velocity field, which changes from upwards ($\approx$ -4.0 km/s) at the onset to downwards later on ($\approx$ +1.5 km/s 2.5 min. after the onset and around +5.0 km/s near the time of maximum emission). This behaviour has never been observed on an EB, but it is in agreement with the results we found for a small solar flare (Cauzzi *et al.*, 1995 A&A, in press).
- No evident counterpart of the chromospheric EB is found in SXR.
- The SXR structures outline a nearby loop system that began to become unstable at least 40 min. before the EB development.
- The SXR isocontour evolution indicates that roughly 20 min. before the EB onset a sudden increase in the SXR emission occurs along the pre-existing SXR loop structure. This seems indicate that an emerging (or sinking) magnetic loop is interacting with the pre-existing SXR loop system and modifies its structural pattern.
- The EB occurs nearby a footpoint of this modified SXR loop (described on the previous item) when its emission begins to decrease.

The final and complete version of this paper will be submitted to *Solar Physics* for publication.

HEI 10830Å IN SOLAR ACTIVE REGIONS AND ITS CORRELAT WITH CHROMOSPHERIC LINES

R. KITAI
Kwasan and Hida Obseratories, Kyoto University
Kamitakara, Gifu 506-13, JAPAN

I. TOHMURA
Osaka Prefectural College of Technology
Saiwai 26-12, Neyagawa, Osaka 572, JAPAN

Y. SUEMATSU
National Astronomical Observatory of Japan
2-21-1 Osawa, Mitaka, Tokyo 181, JAPAN

M. AKIOKA
Hiraiso Solar Terrestrial Research Center, C.R.L.
Ibaraki 311-12, JAPAN

AND

D. SOLTAU
Kiepenheuer Institut für Sonnenphysik
Schoeneckstr. 6, D-79104 Freiburg, GERMANY

EXTENDED ABSTRACT: During the August 1993 International EFR Campaign, five active regions, which were at different phases of development, were observed at the German Vacuum Tower Telescope in Tenerife. We took filtergrams in Hα, in CaII K_{232}, and in green continuum, simultaneously with HeI 10830Å spectra over the target active regions. Spectroscopic quantities, such as equivalent widths, line widths, Doppler shifts, residual intensities etc. of HeI 10830Å, were derived and mutual correlations among them were studied. Our main results are as follows.

(1) The HeI 10830Å absorption line is strong in plage areas and in Hα filaments. In fibril area of active regions, HeI 10830Å shows only weak absorption, not different from that in quiet regions.

(2) Except in active region filament areas, the equivalent width of HeI 10830Å is well correlated with the brightness of K_{232}. Plage and enhanced

Y. Uchida et al. (eds.),
Magnetodynamic Phenomena in the Solar Atmosphere – Prototypes of Stellar Magnetic Activity, 435–436.

network seen in CaII K_{232} are associated with enhanced absorption in HeI 10830Å. On the other hand, Hα on-band brightnesses show less correlation with HeI 10830Å than CaII K_{232}. This result is new and gives a clue to the helium excitation mechanism in the solar atmosphere.
(3) Doppler-shifted streaks were frequently found in HeI 10830Å spectra. They were found both in association with the arch filaments of emerging flux regions and in plage area. Correlation analysis showed that, in plage areas, downflows are dominant in the regions where HeI 10830Å absorption is strong. This finding in HeI 10830Å is consistent with the characteristics of the CIV line (Klimchuk, 1987).

Correlation between HeI 10830Å equivalent widths and CaII K_{232} brightness found in this work may cast some light on the excitation mechanism of helium on the Sun. One of the drawbacks of the "coronal back-irradiation" theory is the difficulty of explaining the existence of fine spatial structures of HeI 10830Å absorption by diffuse coronal radiation. As Ca II K plage areas are expected to be the regions where chromospheric layers are much more strongly heated, collisional processes may contribute much more than expected thus far for the helium excitation. Another line of thought, a modified "coronal back-irradiaton" theory, emerges. As was found, upper chromospheric layers of plage areas are dynamic. There exists plenty of downward-moving material. This material may consist of finer threads or blobs, the outer parts of which are excited by coronal back-radiation. Superposition of these finer scale structures may result in larger optical thickness of the HeI 10830Å line and explain the strong absorption in plage areas. The strong HeI 10830Å absorption in Hα dark filaments can also be interpreted in a similar way, if we assume that dark filaments are composed of fine threads.

Prevailing downward flows in solar active regions, found in this work, may play an important role in the solar coronal mass circulation process. The flows suggests the existence of returning mass flux from the corona to lower layers, probably cooled condensation gases formed in the magnetic corona, flowing down along magnetic field lines.

References

Athay, G., 1976, "The solar chromosphere and corona: Quiet Sun", D. Reidel Publishing Company, p118

Ichimoto, K., Fang, G., and Hiei, E., 1993, in Ai, G. et al. (eds.) "Proceedings of the First China-Japan Seminar on Solar Physics", p158

Kitai, R., Tohmura, I., Suematsu, Y., Akioka, M., and Soltau, D., 1995, in "Proceedings of the Third China-Japan Seminar on Solar Physics", in press

Klimchuk, J.A., 1987, Astrophys. J., 323, p368

Rust, D., and Bridge III, C.A., 1975, Solar Phys., 43, p129.

Venkatakrishnan, P., Jain, S.K., Singh J., Recely, F., and Livingston, W.C., 1992, Solar Phys., 138, p107

X-RAY AND MAGNETIC FEATURES OF Hα SURGES

A. OKUBO, R. MATSUMOTO AND S. MIYAJI
Dept. of Physics, Fac. of Science, Chiba Univ.
1-33 Yayoicho Chiba 263, Japan

M. AKIOKA
Hiraiso Solar Terrestrial Research Center, C.R.L.,
3601 Isozaki Hitatinaka Ibaraki 311-12, Japan

H. ZHANG
Beijing Astronomical Observatory, Chinese Academy of Science,
Beijing 100080, China

M. SHIMOJO
Dept. of Physics, Fac. of Science, Tokai Univ.
Hiratsuka Kanagawa 259-12, Japan

AND

Y. NISHINO, K. ICHIMOTO, K. SHIBATA AND T. SAKURAI
National Astronomical Observatory, Mitaka, Tokyo 181, Japan

The soft X-ray telescope (SXT) aboard the *Yohkoh* satellite has revealed many X-ray jet-like features, *i.e.* transitory X-ray enhancements with apparent collimated motions (Shibata *et al.*, 1992). Many of X-ray jets are associated with flares in X-ray bright points, emerging flux regions, or active regions (Shibata *et al.*, 1992, Shimojo *et al.*, 1995).

Shibata *et. al.* (1992) found an X-ray jet associated with an Hα surge in NOAA7070. They suggested that both the X-ray jet and the Hα surge originated in the same physical mechanism.

In order to elucidate the physical nature of this event (25 Feb. 1992, 00:00:00-04:00:00 UT), we combined partial frame images (PFI) taken by SXT, Hα images (only line center) by the Flare Telescope at Mitaka, and magnetogram data from Huairou (Beijin Astro. Obs.). We coaligned all the data and studied the time variations (Fig. 1).

We found: (1) The X-ray jet started from 01:45:50 and the Hα surge started from 01:47:00. The footpoints of the X-ray jet and the Hα surge coincided (within 5 arcsec) and started to flash simultaneously (within 2 min-

Y. Uchida et al. (eds.),
Magnetodynamic Phenomena in the Solar Atmosphere – Prototypes of Stellar Magnetic Activity, 437–438.

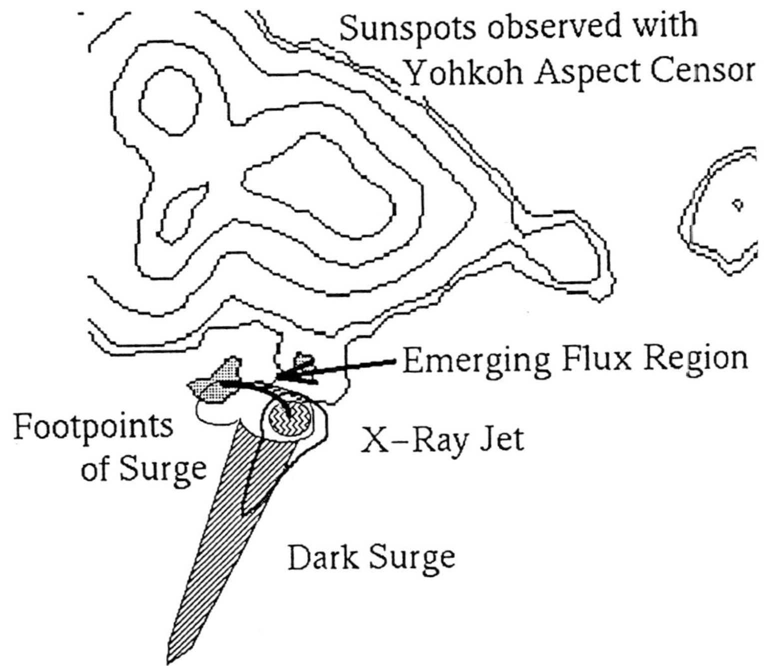

Figure 1. An overlay of SXT partial frame images of an X-ray jet, Hα images of a surge, and a magnetogram image of an emerging flux region in NOAA7070.

utes). (2) Both the X-ray jet and the Hα surge grew towards the southeast. (3) An emerging flux region (recognized by magnetogram data) appeared at the footpoint of the X-ray jet and the Hα surge.

These results are consistent with a model in which magnetic reconnection between the emerging loop and the overlaying magnetic field produces both the X-ray jet and the Hα surge (Yokoyama and Shibata, 1995).

References

Shibata, K., Ishido, Y., Acton, L., Strong, K., Hirayama, T., Uchida, Y., McAllister, A., Matsumoto, R., Tsuneta, S., Shimizu, T., Hara, H., Sakurai, T., Ichimoto, K., Nishino, Y., and Ogawara, Y., (1992) Observations of X-Ray Jets with the *Yohkoh* Soft X-Ray Telescope. *PASJ* 44, L173-L179

Shimojo, M., Hashimoto, S., Shibata, K., Hirayama, T., Hudson, H. S., Acton, L., (1995) Statistical Study of Solar X-Ray Jets Observed with The *Yohkoh* Soft X-Ray Telescope., accepted to PASJ

Yokoyama, T. and Shibata, K., (1995) Magnetic Reconnection as the Origin of X-Ray Jets and Hα Surges on the Sun. *Nature*, **375**, L42-L44

HIGH RESOLUTION OBSERVATION OF SOLAR SPICULES AND THEIR KINEMATIC MODELLING

Y.SUEMATSU
National Astronomical Observatory
Mitaka, Tokyo 181, Japan

Abstract

Solar spicules are known as one of the most prevalent small-scale dynamical phenomena in the solar chromosphere, and are likely to make a considerable contribution to coronal mass supply. In order to understand the dynamical behavior of spicules, we analyzed high resolution filtergrams in Hα line center and wings taken at an enhanced network region near disk center (Suematsu et al. 1995), and found: (1) About 80% of the spicules, many of which were multiple events, could be traced through up and down phase; there were some spicules which were seen in the upward phase but barely visible in the downward phase, and vice versa. (2) Both proper motions and line-of-sight Doppler signals indicate that the spicules are phenomena of true material motion; the motions of the tops of some spicules are well represented by an inclined ballistic trajectory with an initial ejection velocity of about 40 km/sec. (3) Most of bright points in the Hα blue wing tend to appear at the base of spicules at their peak extension and in the falling phase; they are barely seen in the Hα red wing.

In order to understand the observational behavior of spicules, we constructed a simple kinematic model, in which spicule material consists of chromospheric material impulsively ejected into the corona along inclined rigid magnetic flux tubes. The model spicule follows a modified ballistic motion where the deceleration is different from solar gravity and is a function of initial position of the spicule material; the initial velocity distribution of the spicule is given as a function of axial and radial distances.

The initial condition of the spicule is hydrostatic equilibrium, as in the VAL chromosphere model, and the density of the ejected spicule is derived from the conditions of mass conservation and isothermal motion. In

Y. Uchida et al. (eds.),
Magnetodynamic Phenomena in the Solar Atmosphere – Prototypes of Stellar Magnetic Activity, 439–440.

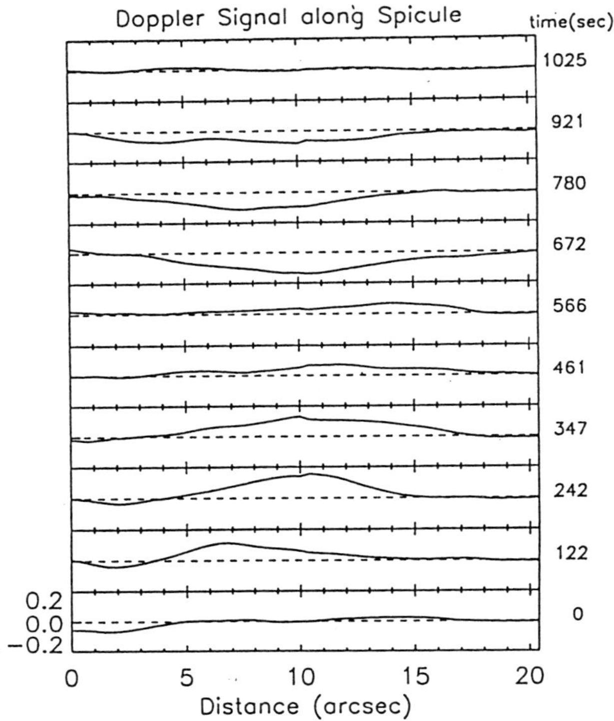

Figure 1. Temporal variation of Doppler signal profile along the trajectory of a spicule. The Doppler signal is defined as $(I_r - I_b)/(I_r + I_b)$, where the I_r and I_b are the intensity at H$\alpha + 0.65$Å and -0.65Å, respectively. It is noted that the Doppler signal always takes a maximum in the middle of a spicule in the extension phase and a minimum in the middle in the receding phase.

calculating the visibility of a spicule ejected into the corona in the Hα line, we used a cloud model (Beckers 1964), assuming that the source function and Doppler width of the line are constant and that the optical thickness of the element is proportional to the density.

The kinetic model of a spicule with a particular initial velocity distribution can explain the observed temporal behavior of visibility in the Hα line and the Doppler signals shown in Figure 1, and would be useful to investigate the energy release mechanism responsible for the spicule formation.

References

Beckers,J.M. 1964, Ph.D. Thesis, Utrecht.
Suematsu,Y., Wang,H, and Zirin,H., 1995, *Astrophys. J.*, **450**, 411.

LOCATION OF TYPE I RADIO CONTINUUM AND BURSTS ON YOHKOH SOFT X-RAY MAPS

S. KRUCKER AND A. O. BENZ
Institute of Astronomy, ETH Zürich, Switzerland
M. J. ASCHWANDEN
Department of Astronomy, University of Maryland, USA
AND
T. S. BASTIAN
National Radio Astronomy Observatory, Socorro, USA

Abstract. The data is presented of a solar type I noise storm observed on 92/07/30 with the radio spectrometer Phoenix of ETH Zürich, the Very Large Array, VLA (333 MHz) and the soft X-ray telescope on board the Yohkoh satellite. The continuum and the bursts of the noise storm are spatially separated by about 100" and apparently lay on different loops as outlined by the SXR. Between periods of strong burst activity, burst-like emissions are also superimposed on the continuum source. These observations contradict the predictions of existing type I theories.

1. Introduction

Solar type I noise storms were the first solar radio events discovered in the metric range, and are also the most common phenomenon at these wavelengths. Type I events last for several hours and sometimes for days. They are associated with active regions, but are not related to solar flares. Two different components of type I emission can be distinguished: Type I emission is composed of short lasting (0.1-1 s), narrowband (several % of the mean frequency) emissions (type I bursts) and a continuous, slowly varying, broadband (50-400 MHz) background emission (type I continuum).

2. Discussion and conclusions

A clear separation of the continuum source and the bursts has been noted. The separation is not only large (up to 10^{10} cm), but transverse to the field lines as outlined by dense X-ray loops (cf. Fig. 1). This seems to contradict

Y. Uchida et al. (eds.),
Magnetodynamic Phenomena in the Solar Atmosphere – Prototypes of Stellar Magnetic Activity, 441–442.

the expectations of both the 'reconnection' (Benz & Wentzel 1981) and the 'shock' (Spicer et al. 1981, Wentzel 1981) models: In these two models, the continuum source and the burst source are both related to the location of trapped electrons, i.e. loops connected to the acceleration region.

If one only considers the location of the observed continuum and burst sources, and looks for a possible magnetic geometry in which a connection by field lines between these two sources is possible, a 'helmet streamer' would be a possible configuration. The continuum sources would be situated below the vertical current sheet, and the burst sources would be located in the neutral sheet above, where reconnection takes place.

On the other hand, the continuum source is not free of burst activity at 333 MHz. This may support a model in which the observations are explained by two independent noise storms, one at the position of the continuum source and the other one at the position of the burst sources. In this model the lack of a continuum emission at the location of the burst sources may be interpreted as an effect of age or loop size.

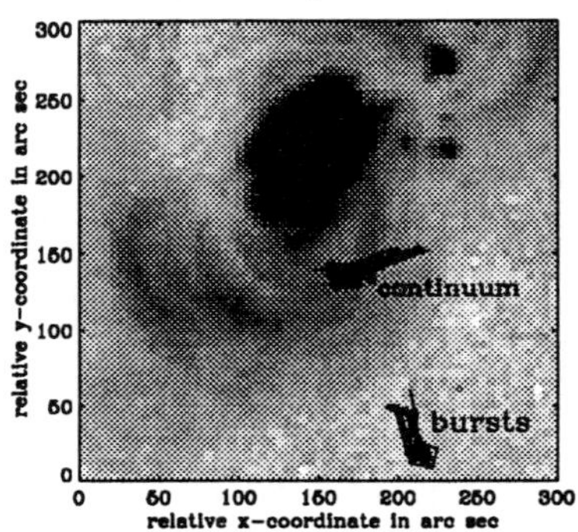

Figure 1. Enlargements of the active region (S20 W10) related to the type I event. On the SXR image taken at 14:56:43 UT the coordinates of the centroids of the continuum source and the burst source are plotted. The approximated density of the type I continuum source ($n_e \approx (1.5 \pm 0.4) \cdot 10^9$ cm^{-3}) is in agreement with fundamental plasma emission.

References

Benz, A. O. and Wentzel, D.G. (1981), *A&A* **94**, 100.

Krucker, S., Benz, A.O., Aschwanden, M.J. and Bastian, T.S. (1995), *Solar Phys.* **160**, 151

Spicer, D. S., Benz, A. O., and Huba, J. D. (1981), *A&A* **105**, 221.

Wentzel, D. G. (1981), *A&A* **100**, 20.

CORONAL MAGNETIC FIELDS FROM POLARIZATION OBSERVATIONS AT MICROWAVES

F. CHIUDERI DRAGO AND F. BORGIOLI
University of Florence, Italy
C.E. ALISSANDRAKIS
University of Ioannina, Greece
M. HAGYARD
Marshall Space Flight Center, Alabama, USA
AND
K. SHIBASAKI
Nobeyama Radio Observatory, N.A.O., Japan

Abstract. We compare high resolution observations of a Solar Active Region (A.R. 7530) performed with the Westerbork Synthesis Radio Telescope, at 5000 MHz ($\lambda = 6\,\mathrm{cm}$), with images obtained with the SXT onboard the Yohkoh satellite and with the coronal magnetic field, extrapolated from Marshall Space Flight Center (MSFC) photospheric magnetograms.

1. Comparison with Sunspots and Soft X-rays

The I and V maps of the active region are shown in the paper by Alissandrakis and Chiuderi Drago (1994, paper I). Their comparison with the SXT and the white light images of the same region shows the coincidence of the I map maxima with the two main sunspots of the active region and of the x-ray loops with the less intense part of the radio map at its border.

The brightness temperature above the two main spots ($T_b \simeq 2.6 \cdot 10^6$ K and $T_b \simeq 3.1 \cdot 10^5$ K respectively) is very well reproduced by the gyroresonance mechanism, using potential extrapolation (Alissandrakis, 1981) of the MSFC photospheric field, provided that different plasma parameters are assumed above the two sunspots.

The shape of the low intensity A.R. contours is qualitatively reproduced by the free-free emission, computed using the electron temperature T_e and

Y. Uchida et al. (eds.),
Magnetodynamic Phenomena in the Solar Atmosphere – Prototypes of Stellar Magnetic Activity, 443–444.

Emission Measure EM derived from the Al12/Al01 ratio of SXT observations from the Yohkoh satellite. The computed T_b is, however, much lower than the observed, since the derived T_e is very high thus giving a very small radio optical depth. We believe that the Al12/Al01 ratio, known to be sensible only to $T_e > 5.\cdot 10^6$ K, *is not suitable for the temperature determination in active regions.*

2. Comparison with the Extrapolated Magnetic Field

The comparison of the V map with the photospheric magnetic field, also shown in paper I, shows that, while the positive circular polarization region is located entirely above a region of positive magnetic field (as expected from the magneto-ionic theory), the negative polarization overlays regions of opposite magnetic polarity, indicating that polarization reversal along the ray path must have occurred.

The observed sign of the circular polarization (CP) depends in fact on the value of the coupling coefficient, $C(QT) = \frac{4.77 \cdot 10^{-18} \nu^4}{N_e B^3} |d\theta/ds|$ at the point where the angle θ between the line of sight s and the magnetic field $\mathbf{B}$ is $\theta \simeq \pi/2$ (Cohen, 1960).

If $C \ll 1$, (Week coupling) *the CP is reversed.*

If $C \gg 1$, (Strong coupling) *the CP is not reversed.*

If $C = 1$, *the CP is cancelled or strongly suppressed.* In this region, called *depolarization strip*, the expected polarization is linear (see paper I).

Using the extrapolation of the MSFC magnetic field mentioned above, we computed the quantity $N_e \cdot C(QT)$ at several points above the A.R. We found that at all points where the CP is reversed the following inequalities hold: $N_e \cdot C(QT) \ll (N_e \cdot C(QT))_0$ and $h_P < h_0$, where h_P is the height of the point where $s \perp \mathbf{B}$ and the suffix 0 refers to the points on the depolarization strip where $C(QT) = 1$. This happens also in the region where is $V < 0$ above the negative field: here, however, all ray path s crosses two times $B_\perp$, reversing the CP twice. Finally at all points located in the region of positive field, where CP is not reversed, we have instead $N_e \cdot C(QT) \gg (N_e \cdot C(QT))_0$ and $h_T > h_0$.

These results indicate that, with any density model such that $dN_e/dh \leq 0$, the requirements of the mode coupling theory are fully satisfied all over the A.R. by the extrapolated potential magnetic field. This is the first time that the extrapolated magnetic field at relatively high coronal levels ($h_0 \simeq 10^5$ Km) is observationally checked.

References

Alissandrakis, C.E. (1981) *Astron. Astrophys.* **100**, 197

Alissandrakis, C.E. and Chiuderi Drago, F. (1994) *Astrophys. J. Letters* **426**, L72

Cohen, M.H. (1960) *Astrophys. J.* **329**, 991

DETECTION OF NONTHERMAL RADIO EMISSION FROM CORONAL X-RAY JETS

M. R. KUNDU AND J. P. RAULIN
Dept. of Astronomy, Univ. of Maryland, College Park, MD 20742, USA

N. NITTA
Lockheed Palo Alto Research Lab., Palo Alto, CA 94304, USA

H. S. HUDSON
Institute for Astronomy, Univ. of Hawaii, Honolulu, HI 96822, USA

A. RAOULT
Observatoire de Meudon, 5 Place Jules Janssen, F-92195, France

K. SHIBATA
National Astronomical Observatory, Mitaka, Tokyo 181, Japan

AND

M. SHIMOJO
Dept. of Physics, Tokai Univ., Hiratsuka, 259-12, Japan

Abstract. Type III bursts are detected in association with dynamic coronal X-ray jets observed by Yohkoh/SXT. The type III bursts are spatially and temporally coincident with the X-ray jets.
The association of type III bursts with X-ray jets implies the acceleration of electrons to several tens of keV, along with the heating responsible for the production of soft X-rays.

Keyword; Radio, X-ray, Jet, Bursts, Nonthermal, Plasma

1. Observations and Results

The Yohkoh X-ray jets of the type discussed here consist of highly collimated plasma structures originating in active regions or X-ray bright

Y. Uchida et al. (eds.),
Magnetodynamic Phenomena in the Solar Atmosphere – Prototypes of Stellar Magnetic Activity, 445–447.

points. They evolve dynamically on time scales of the observations (typical sample intervals on the order of minutes) and they are visible in soft X-rays because of density and temperature enhancements relative to the surrounding coronal material. Most jets are associated with small flares.

It is known that type III emitting electrons propagate along radio structures at the limb corresponding to coronal streamers of electron density higher than ambient density by factors of 4 - 10 (Wild et al. 1959; Kundu et al. 1983). The association of type III-like bursts with coronal jets on the disk show that it may be a general situation for nonthermal electrons to propagate in dense coronal structures. Further, type III electrons must propagate along open magnetic field lines in order to account for their association with electrons detected in the interplanetary medium (e.g. Lin, 1985).

The association of type III's with jets establishes that the acceleration of electrons to speeds of $\sim c/3$ (energies some tens of keV) coincides with the plasma flows. Further, the type III evidence suggests that the jets, which appear to emanate from closed loops, must have open field lines along which the nonthermal electrons propagate. The location of type III bursts at the lower frequency ($164 MHz$) on the invisible or poorly visible part of the jet suggests that the electron density in that part of the jet is adequate to produce plasma radiation, but not high enough for the jet to be visible in soft X-rays. This is consistent with extrapolation of the decreasing density away from the region associated with the jet. Closer to the jet, the density ($6.9 \times 10^8 cm^{-3}$) derived from $236.6 MHz$ type III burst observation (on the plasma radiation interpretation) is only slightly smaller than or even comparable with the X-ray derived density ($6 - 10 \times 10^8 cm^{-3}$) at the top of the jet. The slight difference of density may result from two possibilities: either the $236.6 MHz$ source is higher than the top of the jet; or the jet is thicker than any value within the assumed range (Kundu et al. 1995).

The flare-like brightenings associated with jets arising in XBP's or active regions are typically too weak to be detectable in hard X-rays, so the radio phenomena must serve to identify any nonthermal effects in these events. The type III emission shows clearly that particle acceleration is taking place, as is normal in ordinary flares. The type III emission is extremely short-lived relative to the jet process. The location of the low frequency type III bursts on the extension of the soft X-ray jets, argues that the coronal jets can be much longer than inferred from soft X-ray measurements. The soft X-ray emission and the radio observing frequency provide two independent means of estimating the electron density at the point of observation.

Acknowledgments

This research at the University of Maryland was supported by NASA grants NAGW-1541 and NAGW-4569, and NSF grant ATM 93-16972.

References

Kundu, M. R., Gergely, T. E, Turner, P. J., and Howard, R. A. 1983, *Ap.J.*, **269**, L67.

Kundu, M. R., Raulin, J. P., Nitta, N., Hudson, H. S., Shimojo, M., Shibata, K., and Raoult, A., 1995, *Ap.J.*, **447**, L135.

Lin, R. P., 1985, *Solar Phys.*, **100**, 537.

Shibata, K., et al., 1992, *PASJ*, **44**, L173.

Wild, J. P., Sheridan, K. V., and Neylan, A. A., 1959, *Austr. J.Phys.*, **12**, 369.

STATISTICAL STUDY OF SOLAR X-RAY JETS OBSERVED WITH THE *YOHKOH* SOFT X-RAY TELESCOPE

M. SHIMOJO AND T. HASHIMOTO
Department of Physics, Tokai University
Hiratsuka, Kanagawa 259-12, Japan

K. SHIBATA AND T. HIRAYAMA
National Astronomical Observatory
Mitaka, Tokyo 181, Japan

AND

K. L. HARVEY
Solar Physics Research Corporation
Calle Desecada, Tucson, AZ85718, USA

From a statistical study of 100 X-ray jets, found from the database of full Sun images taken with the Soft X-ray Telescope (SXT) aboard *Yohkoh* during the period between November 1991 and April 1992, we found the following characteristics: 1) Most of the jets are associated with small flares (microflare – subflare) at their footpoints. 2) The length of jet ranges from a few $\times 10^4$ km to a few $\times 10^5$ km. 3) The apparent velocity ranges from $\sim$ 10 km/s to 1000 km/s and the well observed apparent velocity of jet is about 200 km/s. 4) The lifetime ranges from a few min to $\sim$ 10 hours and the distribution of lifetimes is a power law. 5) We found that the shape of the X-ray jet can be classified into about 5 distinct types ; 43% of jet have constant widths, and 33% of the jet appear to be converging. 6) About two thirds (64%) of the jets appear in or near active regions (AR). When a jet is ejected from a bright point like feature in an AR, it occurs to the west (86%). 7) A clear gap ($> 10^4$ km) is seen between the exact footpoint of the jet and the brightest part of the associated flare in 27% of the jets. 8) The X-ray intensity along an X-ray jet often shows an exponential decrease with distance from the footpoint. 9) This exponential intensity distribution holds from the early phase to the decay phase. 10) From co-alignment between SXT images and magnetogram images, we have

Y. Uchida et al. (eds.),
Magnetodynamic Phenomena in the Solar Atmosphere – Prototypes of Stellar Magnetic Activity, 449–450.

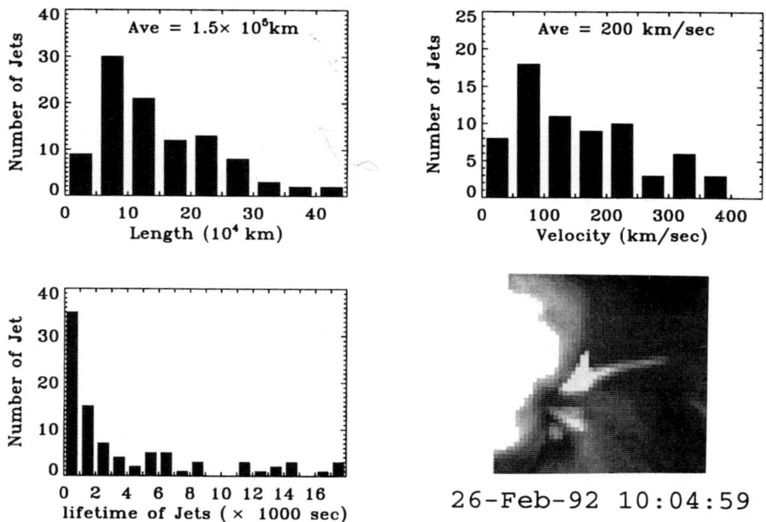

Figure 1. Upper right : Histogram of the projected length of X-ray jet. Upper left : Histogram of the apparent velocity of X-ray jet. Lower right : Histogram of lifetime of jet. Lower left : A example of X-ray jet.

found that jet-producing regions correspond to mixed - polarity regions i.e. regions of satellite polarity.

For more details, we have analyzed partial frame images (PFIs), which were taken at a higher cadence than full frame images (FFIs), and found that the temperatures of the jet and footpoint are almost equally high, $4 \sim 6$ MK, while the emission measures are $\sim 10^{27}$ cm^{-5} and $\sim 10^{28}$ cm^{-5} for the jet and its footpoint, respectively.

On the assumption that the radius of the jet is the same as its width, we estimated density, thermal energy and kinetic energy. The density of jet is $\sim 10^9$ cm^{-3}, the density of footpoint is $10^9 \sim 10^{10}$ cm^{-3}, the thermal energy of jet is $10^{26} \sim 10^{27}$ erg, the kinetic energy of jet is $10^{25} \sim 10^{26}$ erg and the thermal energy of footpoint is $10^{27} \sim 10^{28}$ erg. It is also found that the density of jet increases by more than 2 times over the pre-jet value. This result is suggestive that X-ray jet is the real flow of hot plasma.

References

Shimojo, M. *et. al.*, 1995 *Publ. Astro. Soc. Japan* submitted.

SIMULTANEOUS OBSERVATION OF SOLAR SURGES IN Hα AND X-RAY

MAKI AKIOKA
Hiriaso Solar Terresrial Research Center, Communications Resea
Laboratory
3601 Isozaki, Hitachinaka-shi, Ibaraki 311-12, JAPAN

AND

GIANNA CAUZZI
Institute for Astronomy, Uviversity of Hawaii
Hawaii, USA

Some fundamental questions on the X-ray counterparts of Hα surges remain still controversial although several efforts were made based on the SMM observations (e.g. Švestka, Fárník, and Tang (1990)). Here, some important characteristics of the X-ray counterparts of Hα surges are discussed based on the 26 surges simultaneously observed with *Yohkoh*/SXT and the Hα telescope at Hiraiso. Typically, the spatial sampling of *Yohkoh*/SXT is mainly 2.5"/pixel and the Hiraiso Hα, 0.9 or 1.2 "/pixel, while the observation cadence of *Yohkoh*/SXT for Quiet Mode is 1 or 4 min and the Hα, 1 or 2 min. The combination of instruments is therefore appropriate for comprehensive analyses of surges.

We found the following properties:

- In many cases (20 out of 26), no X-ray emission was seen on the trajectories of the Hα dark surges.
- Where the X ray emission of the surge ejecta was detected, the Hα counterpart was in emission on the filtergram (3 cases).
- In every case, foot-point brightenings in X rays coincides in with Hα.
- In 5 cases, a neighboring loop is brightened in association with the surges.
- In some cases, Hα bright points were slightly displaced from the launch site of the surge .

Many Hα dark surges do not have X-ray counterparts along their trajectories. On the contrary, Hα bright surges accompany X-ray jets along their

Y. Uchida et al. (eds.),
Magnetodynamic Phenomena in the Solar Atmosphere – Prototypes of Stellar Magnetic Activity, 451–452.

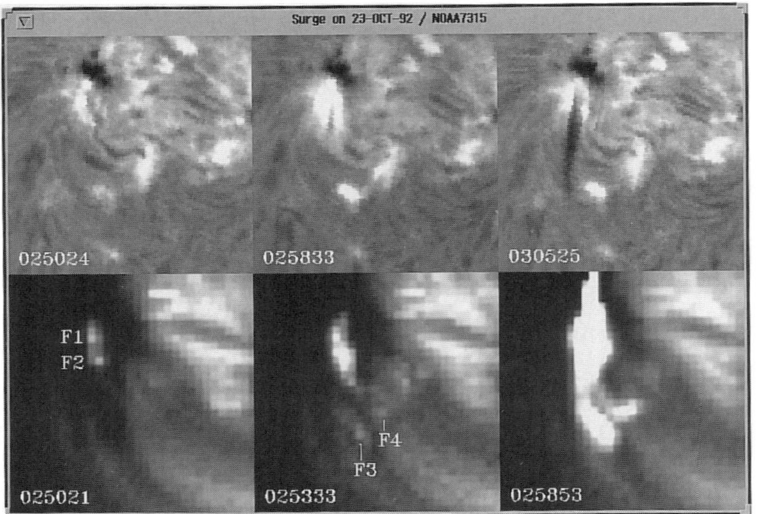

Figure 1. Hα and X-ray images of 23-Oct-92 surge (see text).

trajectories. The SXT intensity is proportional to the emission measure of the hot ($10^6 K$) plasma : Hα clouds with high pressure larger than 1 dyn/cm^2 are observed as emission against the disk with Hα center filtergram (Heinzel and Karlický, 1987). These indicate that Hα bright surges with X-ray ejecta have higher pressure than Hα dark surges.

As described in the previous paragraph, some surges are accompanied by neighboring loop brightenings. Here, we gives the description of the event on 23 October,1992 simultaneously observed with *Yohkoh*/SXT, Mees Polarimeter of University of Hawaii, and Hiraiso Hα. The temporal evolution of the surge observed in Hα and X-rays is shown in Figure 1. An Hα footpoint brightening started at 0248 UT and a dark absorption feature appeared above the bright patch on 0254 UT. The length of the surge reached its maximum of 50 Mm in projection at 0305 UT. In X-ray images, the footpoint brightening started at 0248 UT. Two separate points (F1 and F2) are brightened simultaneously and evolved into a small loop. At 0253 UT, isolated bright points (F3 and F4) appeared and the 0258 UT image shows that F3 and F4 were footpoints of two loops. Later, the whole loops filled with hot plasma. The evolutionary characteristics of this loop brightening show that not the propagation of a conduction front but a plasma evaporation due to high-energy particles plays an essential role in the energy transport in surges.

References

Heinzel, P. and Karlický (1987), *Solar Phys.*, **110**, pp.343–357

Švestka, Fárník, and Tang (1990), *Solar Phys.*, **127**, pp.149–163

STATISTICAL STUDIES OF SOLAR Hα BRIGHTENING EVENTS AND THEIR RELATION TO SOFT X-RAY EVENTS

C.Y. YATINI AND Y. SUEMATSU
National Astronomical Observatory
Mitaka, Tokyo 181, Japan

Extended Abstract

We present some results of studies of solar Hα brightening events whose data were obtained with the Automatic Solar Hα Flare Detection System at NAOJ (Suematsu et al., 1993). The events include not only flares but also small-scale brightenings which are not usually regarded as flares. About 663 events were selected from the data of Sept. 1991 to Dec. 1993. The Hα brightening events were characterized by some quantities such as the peak intensity ratio, duration, rise and fall time, and peak Hα flux, and were compared with soft X-ray events from GOES data. Main results are:

1. The distribution of Hα events as a function of duration or peak intensity ratio shows an exponential-like decrease with increasing parameter (cf. Smith and Smith 1963; Yeung and Pearce 1990).
2. The most frequent events are characterized by short durations and small spatial dimensions.
3. The events with rise ratio (rise time/duration) less than 0.4 comprise about 70 % of all events studied.
4. There are poor correlations between the area and duration or between area and peak Hα flux.
5. Most of Hα brightening events are associated with soft X-rays bursts. Their time profiles are similar to each other, although the peak in soft X-rays tend to come later by about one minute than that in Hα (cf. Thomas and Teske 1971; Teske 1971) .
6. The soft X-rays flux tend to increase with Hα flux, although the correlation is not good (Fig. 1, cf. Thomas and Teske 1971; Teske 1971).

These results seem to indicate that the brightening events have a wide variety of physical conditions in the energy release and transportation.

Y. Uchida et al. (eds.),
Magnetodynamic Phenomena in the Solar Atmosphere – Prototypes of Stellar Magnetic Activity, 453–454.

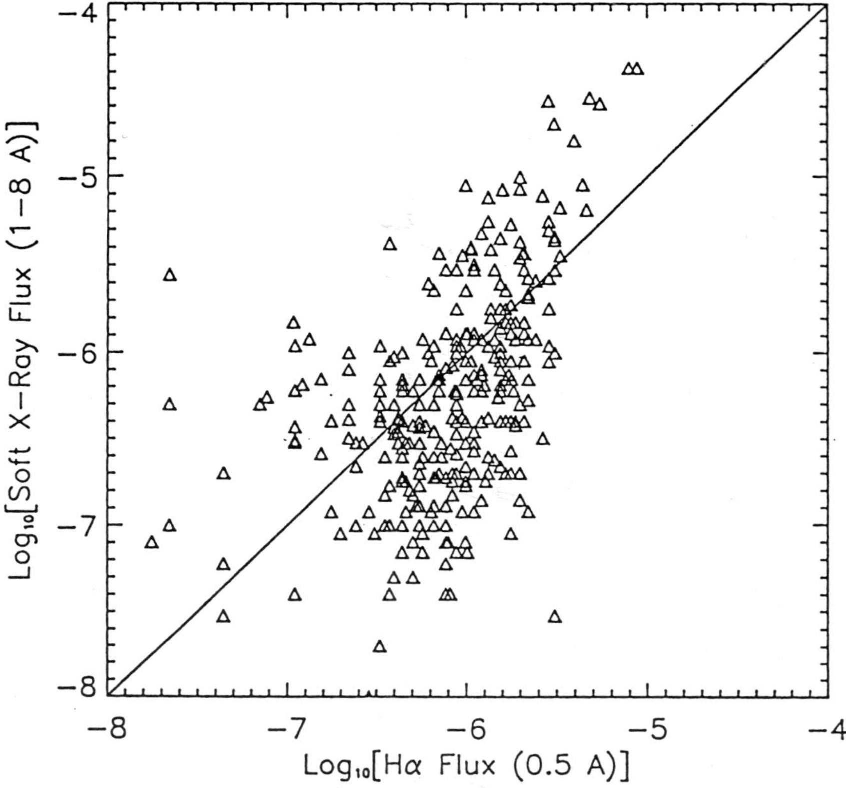

Figure 1. Scatter plots of peak soft X-ray flux (1–8Å band) vs. Hα (0.5Å band) in a logarithmic scale for 301 events

References

Smith, H.J. and Smith, E.V.P., 1963, *Solar Flares*, The Mcmillan Company, New York.

Suematsu, Y., Shimizu, T., and Tanaka, N., 1993, in *X-Ray Solar Physics from Yohkoh*, (eds.) Y.Uchida et al., p.39, Universal Academy Press, Tokyo.

Teske, R.G., 1971, *Solar Phys.*, **21**, 146.

Thomas, R.J. and Teske, R.G., 1971, *Solar Phys.*, **16**, 348.

Yeung, J. and Pearce, G., 1990, it Astr. Ap. Suppl. Ser. **82**, 543.

THE THREE-DIMENSIONAL STRUCTURES OF ELEMENTARY CORONAL HEATING EVENTS

C.E. PARNELL
School of Mathematical and Computational Sciences, University of St Andrews, North Haugh, St Andrews, Fife, Scotland, KY16 9SS.

The poster I presented at this meeting contained work from three papers: Priest, Parnell and Martin (1994); Parnell, Priest and Golub (1994, 1996). The intention was to present the converging flux model, detailed in Priest *et al.* (1994), as an elementary heating mechanism for small-scale coronal phenomena. The model explains in a natural way how these phenomena, in particular bright points, might occur. It satisfies many of the observations of the photosphere, chromosphere and corona relating to bright points. Bright points are observed above pairs of opposite polarity magnetic fragments which converge then mutually lose flux. The model therefore comprises an ambient background field in which a positive and a negative magnetic fragment are situated initially unconnected; observationally, magnetic sources are not linked by chromospheric fibrils. They move together until an x-point forms at the photosphere and rises into the corona, releasing energy in the form of a bright point. The x-point then drops back down to the photosphere and the bright point fades as the magnetic fragments touch. Now reconnection takes place in the photosphere forming a cancelling magnetic feature as the bright point ends, in accordance with observations. The fragments continue to converge and mutually lose flux until they completely cancel and only the ambient field remains.

The work from the paper by Parnell *et al.* (1994) used the philosophy of the converging flux model in an attempt to explain the three-dimensional looped structure of bright points. A comparison was made between soft x-ray images of particular bright points and magnetograms of the underlying photosphere. The magnetic fragments relating to the bright points were identified and then modelled in a simple manner as poles of the appropriate strength – the potential field due to these sources was then derived. As the sources moved in the observed manner newly reconnected field lines, which were assumed to have been injected with hot dense plasma causing them

Y. Uchida et al. (eds.),
Magnetodynamic Phenomena in the Solar Atmosphere – Prototypes of Stellar Magnetic Activity, 455–456.

to brighten, formed and produced the observed bright point structure (see also B. Schmieder in these proceedings).

The final section of the poster considers satellite points, small bright single-loop structures seen in soft x-rays by NIXT (see Golub *et al.*, 1994 for details). The proposed model in the poster suggested that moving magnetic features radiating out from a decaying sunspot interact with surrounding network fields to produce satellite points. From the NIXT images the lifetime of these features is unclear; however we can see that from one end of the bright loops there is a rapidly varying bright spot-like structure. It is possible that these spot brightenings are related to active region transient brightenings since they vary rapidly during the five minutes of the NIXT flight – this suggestion cannot be conclusively proved. However, the single loop structures themselves maintain a steady intensity throughout and are therefore not transient brightenings, though they clearly have an important relationship with the flaring spots. One scenario to explain the observed satellite points is that moving magnetic fragments interact with the network fragments around the sunspot to give rise to reconnection in the corona, thus producing the bright loop structures. The reconnection here may be of an impulsive bursty regime with fluctuations in the intensity of the newly reconnected loops likely to be observed. It is possible that these variations may be intense enough to produce a flaring event as seen in one in ten bright points. This is not an unreasonable assumption since bright point flares and active region transient brightenings release approximately the same amount of energy and have similar structures, sizes and lifetimes. However below active region transient brightenings little evidence has been found of converging fragments and only a few cases have been found of emerging flux, so it is not yet clear what is causing active region transient brightenings (see also Shimizu in these proceedings).

In conclusion the converging flux model is found to model in a natural way elementary heating events such as bright points. The philosophy of the model may also be used to explain the three-dimensional structure of both bright points and satellite points and possibly even active region transient brightenings, thus indicating that small-scale reconnection events are important in the heating of the solar corona.

1. References

Golub, L., Zirin, H. and Wang, H. (1994) *Solar Phys.* **153**, 179.
Parnell, C.E., Priest, E.R. and Golub, L. (1994) *Solar Phys.* **151**, 57.
Parnell, C.E., Priest, E.R. and Golub, L. (1996). *In preparation.*
Priest, E.R., Parnell, C.E., and Martin. S.F. (1994) *Astrophys. J.* **427**, 459.

CAUSAL RELATION BETWEEN Hα ARCH FILAMENT LOOPS AND SOFT X-RAY CORONAL LOOPS

K. YOSHIMURA, H. KUROKAWA AND S. SANO
Kwasan and Hida Obs., Faculty of Science, Kyoto University
Sakyo-ku, Kyoto, 606-01, Japan

AND

H. HUDSON
Inst. for Astronomy, Univ. of Hawaii
Honolulu, HI 96822, USA

We have been studying the spatial and temporal relations between Hα arch filament system(AFS) loops and soft X-ray(SXR) coronal loops, comparing high resolution Hα images of emerging flux regions (EFRs) observed with the Domeless Solar Telescope (DST) at Hida Obs. with SXR images taken by the Soft X-ray Telescope (SXT) aboard *Yohkoh*. The first results of this study were published in Kawai et al. (1992, 1993), showing good spatial correspondence between Hα AFS loops and SXR bright loops especially in the youngest parts of EFRs.

To study their causal relation in more detail, we made coordinated observations of EFRs with Hida DST and *Yohkoh* SXT in June and August, 1994, these gave, for the first time, a good example showing an apparent causal relationship between an AFS loop and a SXR loop, which is demonstrated in Figure 1.

Loop $h1$ and loop $s1$ indicated in Fig. 1 are an emerging Hα loop and a brightening SXR loop, respectively. The loop $h1$ was first found to darken in Hα blue wing images at around 00:48 UT, and became visible in Hα center images at around 00:53 UT,when loop $s1$ started to brighten in SXR. In the Hα red wing images, both legs of loop $h1$ were found to darken even after 00:58 UT. We overlaid the Hα image on the SXR image, and found the Hα emerging loop and the SXR brightening loop to be located very close but tilted with respect to each other. It must be noted, however, that appearances of AFS loops are not always followed by SXR brightenings, and that it is rare to find such a clear causal relation as found here between

Y. Uchida et al. (eds.),
Magnetodynamic Phenomena in the Solar Atmosphere – Prototypes of Stellar Magnetic Activity, 457–458.

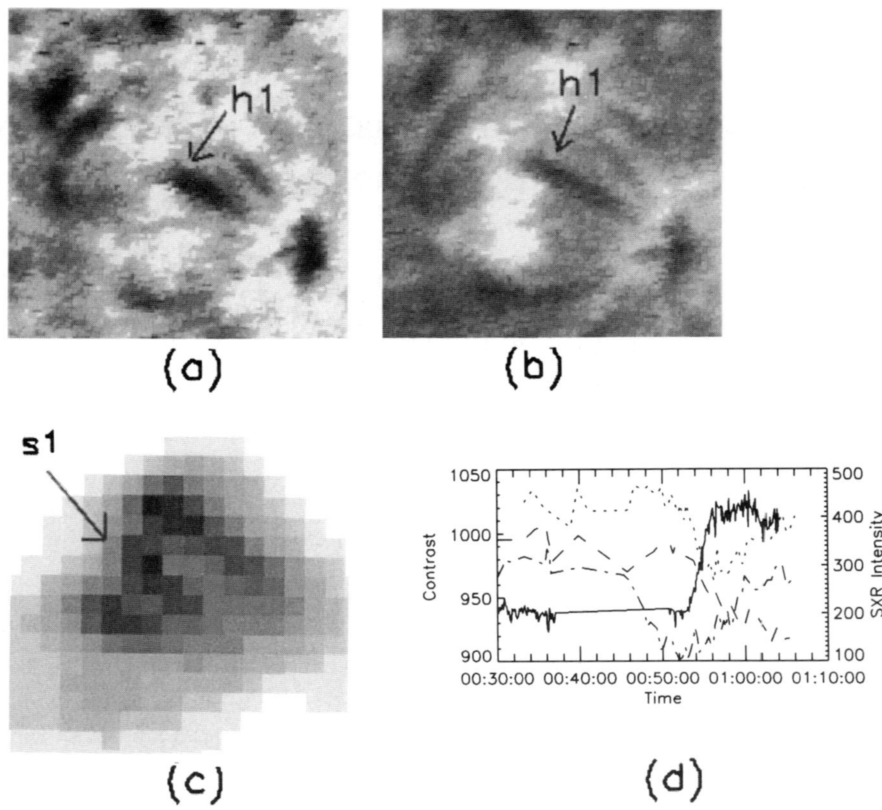

Figure 1. (a)Hα -0.6Åimage showing the emerging Hα AFS loop h1 at 00:54:22UT, 11th June, 1994. (b)Hα line-center image at 00:54:26 UT. (c) SXR image at 00:55:58 UT. Loop s1 was very close to the Hα AFS loop $h1$. (d) Temporal variations of the darkness of the loop $h1$ in Hα line-center (dotted line), in Hα -0.6 Å(dot–dashed), in Hα +0.6 Å(dashed) and the brightness of the $s1$ (solid). For the three Hα curves, the mean intensity in the quiet region is set to 1000 contrast units. Notice the first darkening of the loop $h1$ in Hα -0.6 Å, and the later darkening and brightening in Hα line-center and in SXR, respectively.

loops $h1$ and $s1$. Loop $h1$ showed a prominent upward motion in the Hα blue wing. This may be a necessary condition with which an emergence of an Hα AFS loop produces a brightening of an SXR loop.

References

Kawai, G., Kurokawa, H., Tsuneta, S., Shimizu, T., Shibata, K., Acton, L., Strong, K. and Nitta, N.(1992) *Publ. Astron. Soc. Japan*, **44**, pp.L193–L198

Kawai, G., Kurokawa, H., Akioka, M., Shibata, K., Tsuneta, S., Shimizu, T., Acton, L., Strong, K. and Nitta, N.(1993) *The Proceeding of International Symposium on the Yohkoh Scientific Results. (at Sagamihara, Feb.,1993)* p16, (written in Japanese)

EMERGING FLUX, RECONNECTION, AND XBP

Observation Meets Theory

L. VAN DRIEL-GESZTELYI, B. SCHMIEDER AND P. DEMOULIN
Observatoire de Paris, D.A.S.O.P., 92195 Meudon, France

C. MANDRINI
Instituto de Astronomía y Física del Espacio, IAFE-CONICET, CC.67, Suc.28, 1428 Buenos Aires, Argentina

G. CAUZZI
Institute for Astronomy, University of Hawaii, Honolulu, HI 96822, U.S.A.

A. HOFMANN
Astrophysikalisches Institut Potsdam, Sonnenobservatorium Einsteinturm, D-14473 Potsdam, Germany

N. NITTA
Lockheed Palo Alto Research Laboratory, Palo Alto, CA 94304-1911, U.S.A

H. KUROKAWA
Kwasan and Hida Observatories, Kyoto University, Kamitakara, Gifu 506-13, Japan

AND

N. MEIN AND P. MEIN
Observatoire de Paris, D.A.S.O.P., 92195 Meudon, France

Ground-based optical observations coordinated with *Yohkoh/SXT* of an old, disintegrating bipolar active region AR NOAA 7493 (May 1, 1993) provided a multiwavelength (magnetic fields, Hα and X-ray) data base for the study of a flaring "active region" X-ray bright point (XBP) of about 16 hours lifetime, and of the activity related to it in different layers of the solar atmosphere (van Driel-Gesztelyi *et al.*, 1995).

We find that
(1) This XBP, which appeared in a decaying active region related to a new minor bipole of about 10^{20} Mx, was very similar to XBPs observed in quiet solar regions.

Y. Uchida et al. (eds.),
Magnetodynamic Phenomena in the Solar Atmosphere – Prototypes of Stellar Magnetic Activity, 459–460.

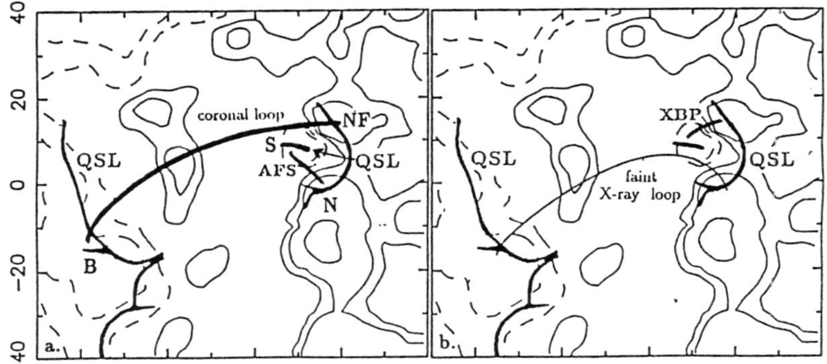

Figure 1. Extrapolated coronal magnetic field lines before reconnection (a) and after reconnection (b).

(2) The XBP flares were due to macroscopic reconnection between a small (emerging) and a big (pre-existing) loop.
Evidence: (i) The trigger of the reconnection was the motion of the new negative polarity pore (S) towards the pre-existing positive polarity plage (NF) with a velocity of 0.2 km s^{-1}. (ii) the XBP was located between the new pore and the pre-existing plage. (iii) The location of the XBP shifted together with the moving new pore in such a way that it always remained between the new pore and the old plage. (iv) The XBP was not point-like, but had an elongated, tiny loop structure connecting the new pore and the old plage. (v) There was another longer loop (NF–B) involved in the XBP flares: brightness enhancements of its remote footpoint suggest that it was heated by both fast particles and a conduction front (v $\approx$ 700 km s^{-1}). (vi) 3D reconnection model calculations are in good agreement with the observations. Extrapolating the observed photospheric magnetic fields we found field lines corresponding to the observed arch filament system (AFS) and X-ray loops on both sides of the quasi separatrice layers (QSL), as expected in a 3D reconnection process (Figure 1). Since the two reconnected loops were very different in volume, the energy released in the reconnection region made the short loop (S–NF; XBP) much brighter than the long loop (N–B).

(3) The XBP flares *looked like* compact simple-loop flares, like the majority of flares observed by *Yohkoh.* Although in general we can not exclude the possibility that flaring of a single loop is due to the relaxation of internal magnetic stresses, these observations suggest that even apparent simple-loop flares can be attributed to macroscopic magnetic reconnection.

References

L. van Driel-Gesztelyi, B. Schmieder, G. Cauzzi, N. Mein, A. Hofmann, N. Nitta, H. Kurokawa, P. Mein and J. Staiger: 1995, *Solar Phys.*, in press.

COLLISION OF PLASMA FLOWS IN A CURRENT LOOP

J. ZHAO, J.I. SAKAI
Faculty of Engineering, Toyama Univ., Toyama 930 Japan

AND

K.I. NISHIKAWA
Space Physics Department, Rice Univ., Houston, TX 77005

Extended Abstract

Chromospheric evaporation is one of the important mechanisms for energy transports at the rise phase of solar flares. With its high resolution, the *Yohkoh* SXT has ability to detect moving evaporation fronts, provided that the instrument takes images since the very beginning of a flare. This anticipation has been proven for the first time by Hudson *et al.* (1993), who show that the *Yohkoh* SXT can observe the upward-moving fronts along loop legs as it happens. In his study of the *Yohkoh* HXT, Sakao (1994) found that double sources are a common feature in the footpoints of a single loop. Since the HXT emissions from the footpoints vary simultaneously, it implies that accelerated electrons stream down along the loop towards both ends. By the subsequent evaporation of the lower solar atmosphere from the ends is possible to be created. The existence of such chromospheric evaporation has been modeled as one of the four phases of a flare on the basis of synthesized SXT images (Reale and Peres 1995).

In this study, we investigate a collision of upward flows from the footpoints of a current loop with a 3-D particle code (Buneman 1994). In the initial state, upward flows formed by both electrons and ions are set separately inside a current loop, which are pushed to move together along the ambient magnetic field with the same bulk velocity, on the order of a quarter of the Alfvén velocity. The two flows collide with each other at $\omega_{pe}t = 15$, and then mix. Figure 1 plots the time history of the drift velocities of electrons and ions in the flows. One can see that the drift velocities of the ions slow down little, suggesting that the ions enter the collision region and pass through it as the evolution is going on. However, the electrons are trapped in the collision region as seen that the drift velocities decrease to near zero.

Y. Uchida et al. (eds.),
Magnetodynamic Phenomena in the Solar Atmosphere – Prototypes of Stellar Magnetic Activity, 461–462.

Accumulation of the electrons may be account for the brightenings at the loop apex, a subject to be discussed later. During the collision process, the initial kinetic energies of the flows in the drifting direction are deposited with an accompanying growth of the electron kinetic energies of the loop and flow in other directions. The temperatures of the electrons in the loop and flows are increased 1.5 times their initial ones.

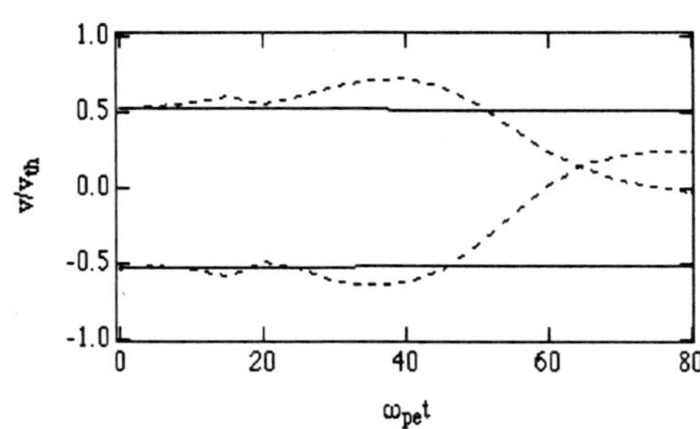

Figure 1. Time history of the drift velocities of electrons and ions in the flows. Dashed lines: electrons; solid lines: the ions.

One of the recent discoveries made with SXT on board *Yohkoh* is that the appearance of bright knots at the tops of loops is a common feature in flares (Acton *et al.* 1992, Doschek *et al.* 1995). As pointed out by Doschek *et al.* (1995), this new phenomenon indicates that the plasma within a flare flux tube may not be in pressure equilibrium for extended time intervals over a flare lifetime; i.e., the plasma at the top of a flare may be both denser and hotter than in the loop legs. The plasma physics of stable overpressure at loop tops still remain unclear. We try to interpret it with the model of a collision of upward fronts in a current loop. The simulation results clearly show that the upward fronts will collide somewhere in the loop, resulting in enhancement of temperature and density of the electrons in the collision region. The more detailed comparison between the simulation and observation is in progress, and will be reported soon.

References

Acton, L.W., et al. (1992), *PASJ*, **44**, L71.

Buneman, O. (1994) in *Computer Space Plasma Physics, Simulation Techniques and Softwares*, ed. H. Matsumoto & Y. Omura (Terra Scientific, Tokyo), p67.

Doschek, G.A., Strong, K.T., and Tsuneta, S. (1995) *ApJ.*,**440**, 370.

Hudson, H.S., Strong, K.T., Dennis, B.R., Zarro, D., Inda, M., Kosugi, T., and Sakao, T. (1993) *ApJ. Letter*, **422**, L25.

Reale, F. and Peres, G. (1995) *A. & A.*, 299,225

Sakao, T. (1994) Ph.D dissertation, University of Tokyo.

OSCILLATIONS IN A QUIESCENT SOLAR PROMINENCE

R. OLIVER AND J. L. BALLESTER
Departament de Física,
Universitat de les Illes Balears,
E-07071 Palma de Mallorca, Spain

Abstract. Many of prominences' physical features still remain enigmatic —questions regarding the formation, long-term equilibrium, dynamic evolution and eventual eruption of quiescent prominences are yet to be answered.

Prominence seismology, i.e., the constructive interaction between the modelling of prominence equilibrium, the theoretical study of the vibrations of the whole structure (normal modes) and direct observation, is a way of improving our knowledge of these objects.

Moreover, prominences are just a piece of a larger puzzle, the corona, in which they are embedded. A better understanding of the processes taking place in the solar atmosphere can be achieved by a deeper inspection of prominence properties; one, for example, would need to explain why the coronal magnetic topology and plasma properties are such that the formation and existence of condensations are allowed in some places while not in some others.

In previous works 1-D equilibrium solutions with a **temperature discontinuity** at the boundary between prominence and corona (e.g. Joarder & Roberts 1992; Oliver *et al.* 1992, 1993) were used. It was found that the coexistence of two media with very different properties gives rise to three types of fast, slow and Alfvén MHD modes: **internal** and **external** modes have properties determined by the isothermal prominence and corona, respectively. **Hybrid** modes are a mixture of internal and external.

In this paper we address the question:

> *Are the modes of oscillation of a system with a* **smooth temperature variation** *from the prominence to the corona different from the modes of a* **two-temperature** *system with a sharp temperature increase at the interface between the prominence and the corona?*

This, of course, may have important implications on the models that should be used in future investigations of this kind.

Y. Uchida et al. (eds.),
Magnetodynamic Phenomena in the Solar Atmosphere – Prototypes of Stellar Magnetic Activity, 463–464.

The model of Low & Wu (1981) is chosen to represent the equilibrium state of the system. This is a solution of the type put forward by Kippenhahn & Schlüter (1957), which incorporates the thermal balance equation with a simplified form of the radiative loss term. The temperature in this model suffers a smooth variation from prominence to corona.

An important feature of internal, external and hybrid modes in a two-temperature system is that their frequencies evolve in a different manner as the half-width of the system (x_c) is varied. **Internal** modes are characterized by a constant ω, **external** modes have $\omega \sim x_c^{-1}$ and **hybrid** modes have $\omega \sim x_c^{-1/2}$.

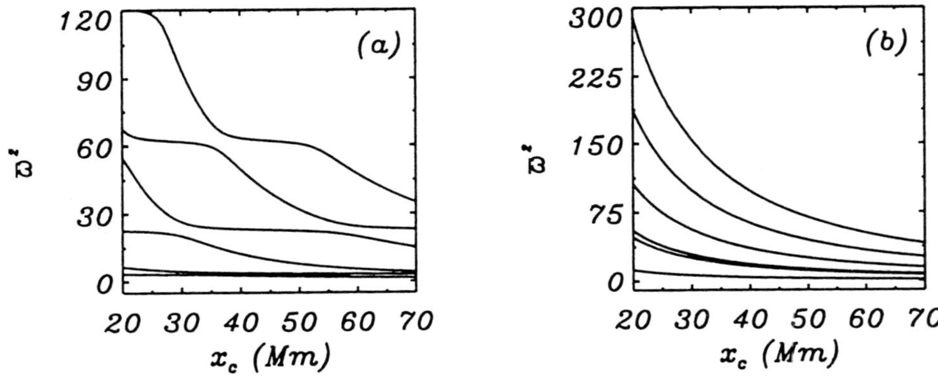

Figure 1. Evolution of the dimensionless frequency, $\bar{\omega}^2$, vs. the half-width of the system, x_c, for **(a)** a two-temperature equilibrium and **(b)** a Low-Wu equilibrium.

This is shown in Figure 1a, where internal and external modes can be seen crossing each other. Modes possess internal features where their frequency remains almost constant, whereas they become external where ω decreases with x_c. The above situation drastically changes in the present equilibrium (Oliver & Ballester 1995), as all modes display a remarkably similar frequency evolution with the parameter x_c (Fig. 1b).

References

Joarder, P. S. & Roberts, B. (1992) *A&A* **261**, 625

Kippenhahn, R. & Schlüter, A. (1957) *Z. Astrophys.* **43**, 36

Low, B. C. & Wu, S. T. (1981) *ApJ* **248**, 335

Oliver, R., Ballester, J. L., Hood, A. W. & Priest, E. R. (1992) *ApJ* **400**, 369

Oliver, R., Ballester, J. L., Hood, A. W. & Priest, E. R. (1993) *ApJ* **409**, 809

Oliver, R. & Ballester, J. L. (1995) *ApJ*, in press

NON-EQUILIBRIUM EFFECTS ON THE OPTICALLY THIN RADIATIVE LOSS FUNCTION

Ø. WIKSTØL AND V. H. HANSTEEN
Institute of Theoretical Astrophysics
P.O. Box 1029, Blindern, 0315 Oslo, Norway
email: oivindw@astro.uio.no, viggoh@astro.uio.no

The optically thin radiative loss function, $f(T) \equiv L_r/n_e n_H$, where L_r represents the radiative losses (erg/s/cm^{-3}) and n_e, n_H the electron and hydrogen number densities is an important quantity in the energetics of coronal and transition region plasmas. We will consider the importance of calculating the ionization state and radiative losses consistently with the dynamic state of the coronal/transition region plasma, and through time dependent calculations we will show the effects of non-equilibrium ionization on the radiative loss function.

Although some authors (e.g. Spadaro et al. 1990, Hansteen 1993) have discussed deviations from ionization equilibrium caused by coronal and transition region dynamics, no systematic study of the effects of this on the radiative loss function have yet been carried out. Wikstøl & Hansteen (1995) present calculations of a loop being heated to and cooling from coronal temperatures. In order to illustrate the magnitude of non-equilibrium effects we will here show calculations of a coronal loop cooling to chromospheric temperatures. The total length of the loop is l = 20 000km and the initial maximum temperature is $T = 2.44 \times 10^6$K.

In the left panel of figure 1 the time evolution of the loop temperature is shown. As the coronal heating is switched off at t = 0s, the loop cools by heat conduction and radiation. The dynamic response of the loop, not shown here, to this cooling is large scale downflow towards the chromospheric footpoints as the coronal material can no longer be sustained against gravity.

The right panel of figure 1 shows the radiative loss function, $f(T)$. The radiative loss function is significantly reduced compared to its equilibrium value ($t = 0s$) as time progresses at temperatures below $T \approx 5 \times 10^5$K. This

Y. Uchida et al. (eds.),
Magnetodynamic Phenomena in the Solar Atmosphere – Prototypes of Stellar Magnetic Activity, 465–466.

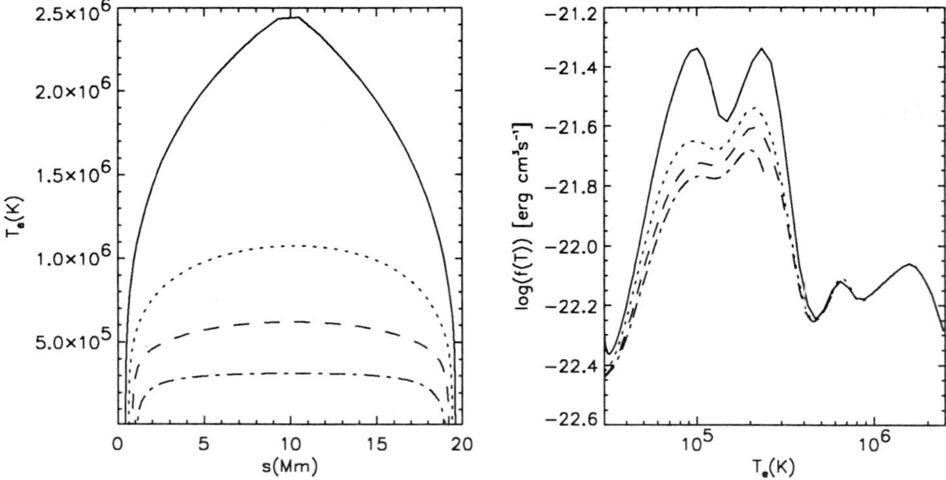

Figure 1. Temperature profile and radiative loss function. solid line: t=0s, dotted: t=400s, dashed: t=750s, dash-dotted: t=1150s.

behavior is a result of deviations from ionization equilibrium which may be explained as follows:

The principal contributions to the radiative loss function in the temperature regime of $T_e < 5 \times 10^5$K stem from the resonance lines of several times ionized carbon and oxygen ions. As the material flows down the steep temperature gradient in the transition region, the timescales for recombination eventually becomes long compared to the dynamic timescale of plasma flowing through a temperature scale height. The slowest processes are recombination from C V to C IV and O VII to O VI. The lower ionization stages, C IV,III and O VI,V are therefore under-abundant at their equilibrium temperatures; these are the ionization stages that contain the lines most important to the radiative losses and their under-abundance leads to the reduction in the radiative loss term shown in figure 1.

We have performed the same calculations using radiative losses calculated assuming ionization equilibrium and find significant differences in the energetics of the cooling. When time dependent radiative losses are used the atmosphere cools from a maximum temperature of $T = 4 \times 10^5$K to $T = 1 \times 10^5$K in 310s. The same cooling takes 175s in the case where (the larger) equilibrium radiative losses are used.

References

Hansteen, V.H. 1993, *ApJ*, **402**, pp. 741–755

Spadaro, D., Zappala, R.A., Antiochos, S.K., Lanzafame, G., Noci, G. 1990, *ApJ* **362**, pp. 370–378

Wikstøl, Ø., Hansteen, V.H. 1995, *in preparation*

NONLINEAR DISSIPATION OF ALFVÉN WAVES

G. CHRISTOPHER BOYNTON[1,2] AND ULF TORKELSSON[1]
[1] *Sterrekundig Instituut*
Postbus 80000, 3508 TA Utrecht, The Netherlands

AND

[2] *Physics Department*
Univ. of Miami, P.O. Box 248046, Coral Gables, FL 33124, USA

Alfvén waves are incompressible to first order and therefore believed to propagate without dissipating. A linearly polarised Alfvén wave is however compressible to second order due to the variations in magnetic pressure, $B^2/(2\mu_0)$, (e.g. Alfvén & Fälthammar 1963). The wave can therefore steepen to discontinuities (Cohen & Kulsrud 1974).

High-amplitude Alfvén waves can be of importance in driving outflows from the Sun and other stars (e.g. Rosner 1995), but current models are hampered by the insufficient understanding of the damping of high-amplitude Alfvén waves, which is the problem we address.

We simulate the propagation of high-amplitude Alfvén waves in homogeneous and stratified media assuming isothermal magnetohydrodynamics (MHD). We restrict ourselves to one-dimensional models, where the waves are propagating along a background magnetic field $\mathbf{B_0}$, on which we impose oscillations $\mathbf{B_1}$ and $\mathbf{v_1} = -\mathbf{B_1}/\sqrt{\mu_0\rho}$ perpendicular to $\mathbf{B_0}$. The numerical calculations are based on the method of flux-corrected transport (Boris & Book 1973).

Figure 1 demonstrates the difference between the propagation of a circularly polarised Alfvén wave and a linearly polarised wave of the same amplitude in a homogeneous medium. The circularly polarised wave propagates with very little distortion, whereas the linearly polarised wave steepens to form a number of current sheets, in which dissipation occurs. The reason is that the magnetic pressure of the circularly polarised wave is independent of phase, whereas it varies with the phase of the linearly polarised wave. Physically the magnetic pressure gradients in the linearly polarised wave push together magnetic field lines of opposite directions in the current sheets.

Y. Uchida et al. (eds.),
Magnetodynamic Phenomena in the Solar Atmosphere – Prototypes of Stellar Magnetic Activity, 467–468.

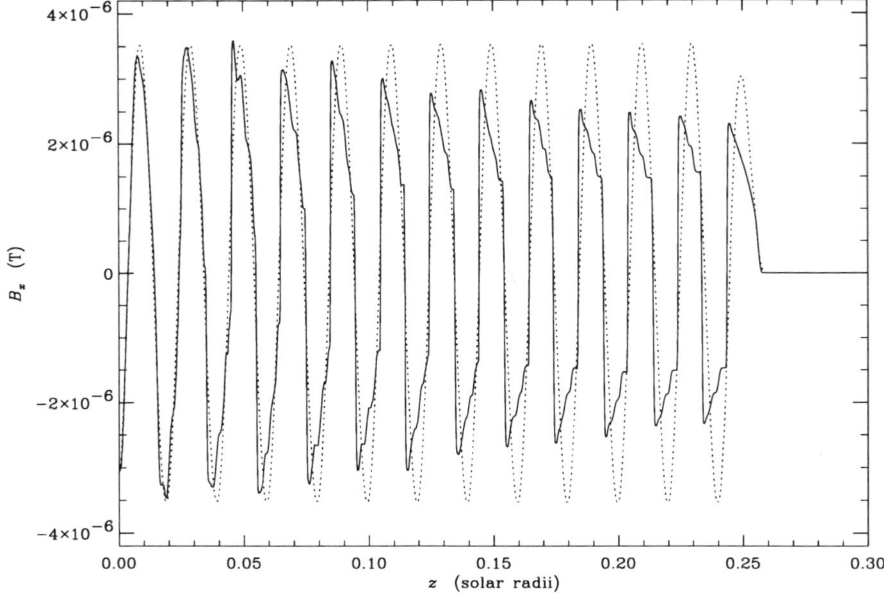

Figure 1. The B_x-components of linearly (solid line) and circularly (dashed line) polarised Alfvén waves propagating along $\hat{\mathbf{z}}$. The plane of polarisation is inclined with $\pi/4$ with respect to $\hat{\mathbf{x}}$. The background medium is homogeneous with a density of 10^{-14} $\mathrm{kg\,m^{-3}}$, a temperature of 10^6 K, the period of the waves is 300 s and $B_1/B_0 = 1.0$

In a stratified medium the amplitude of $\mathbf{B_1}$ decreases as the wave propagates upward. For that reason the wave dissipates only below a certain height, but it keeps its square-shape even at higher altitudes as there is no mechanism that can transform it back to its sinusoidal shape. Even though the amplitude decreases, the relative importance of the magnetic pressure, $B_1^2/(2\mu_0)$, compared with the gas pressure increases. Therefore the wave loses energy by doing work on the background medium even when it cannot dissipate (Boynton & Torkelsson 1995).

References

Alfvén, H. and Fälthammar, C.-G., (1963) *Cosmical electrodynamics. Fundamental principles.* Clarendon Press, Oxford

Boris, J. P. and Book, D. L. (1973) Flux-corrected transport. I. SHASTA, a fluid transport algorithm that works, *J. Comp. Phys.*, **11**, 38–69

Boynton, G. C. and Torkelsson, U. (1995), Dissipation of non-linear Alfvén waves, *A&A*, accepted

Cohen, R. H. and Kulsrud, R. M. (1974) Nonlinear evolution of parallel-propagating hydromagnetic waves, *Phys. Fluids*, **17**, 2215–2225

Rosner, B. (1995) Mechanisms of solar mass-loss, *these proceedings*

ON THE STABILITY OF SOLAR PHOTOSPHERIC MAGNETIC FLUX TUBES

Akitsugu TAKEUCHI
Yonago National College of Technology
4448 Hikona, Yonago, Tottori 683, Japan

In the quiet solar photosphere, most of the magnetic flux is in the form of small-scale intense magnetic flux tubes. It is believed that the intense flux tubes are formed by the convective collapse, which is a kind of the convective instability in a magnetic flux tube. Thus, the nature of the instability has been investigated by many authors. From linear theory it was found that the stability is determined by β (the ratio of gas pressure to magnetic pressure in the flux tube); if $\beta < \beta_c$ the tube is convectively stable, otherwise it is unstable (Spruit, Zweibel 1979). Moreover, Takeuchi (1993) showed that a weak flux tube ($\beta > \beta_c$) evolves into an intense flux tube ($\beta < \beta_c$) in a static equilibrium, due to the convective collapse, carrying out the nonlinear calculations. However, he adopted the adiabatic approximation which is not valid in the photospheric region. Thus, we investigated the nonlinear and the long term evolution of the convective instability in a weak flux tube, adopting thin flux-tube approximation (Roberts, Webb 1978). The radiative heat exchange was taken into account adopting Newton's law of cooling.

In this study, the same set of equations as those used in Takeuchi (1995) were used. The convective instability was initiated by a small downflow added to an initial atmosphere. The level $z = 0$ km corresponded to $\tau_{5000} = 1$ in the external medium, and the closed and open (flow-through) boundary conditions were adopted at the upper ($z = 1000$ km) and lower ($z = -2000$ km) boundaries, respectively. Since the weak flux tubes are formed by the flux expulsion due to the granulation, $\beta = 10$ and a_0 (the radius of the flux tube at the photosphere) $= 200$ km were chosen as the standard initial parameters of a weak flux tube. In figure 1, we showed time variations of the velocity v, the temperature difference $\Delta T (= T(z,t) - T(z,0))$, and the magnetic flux density difference $\Delta B (= B(z,t) - B(z,0))$, at $z = -500$ km. It is apparent that a weak flux tube

Y. Uchida et al. (eds.),
Magnetodynamic Phenomena in the Solar Atmosphere – Prototypes of Stellar Magnetic Activity, 469–470.

evolves into an intense flux tube as reported by Takeuchi (1993). After the collapse ($t > 1000$ s), however, the temperature and the magnetic flux density differences decrease slowly with time, due to the radiative heat exchange. It means that an intense flux tube, formed by the collapse, evolves slowly into a weaker tube. However, Schüssler (1990) estimated the life time of an intense flux tube 30 ~ 60 min, which is much shorter than the time scale of the subsequent evolution. Thus, the intense flux tube, formed by the collapse, does not change so much in the life time. Therefore, our results suggest that an actual photospheric flux tube is probably created by the convective instability.

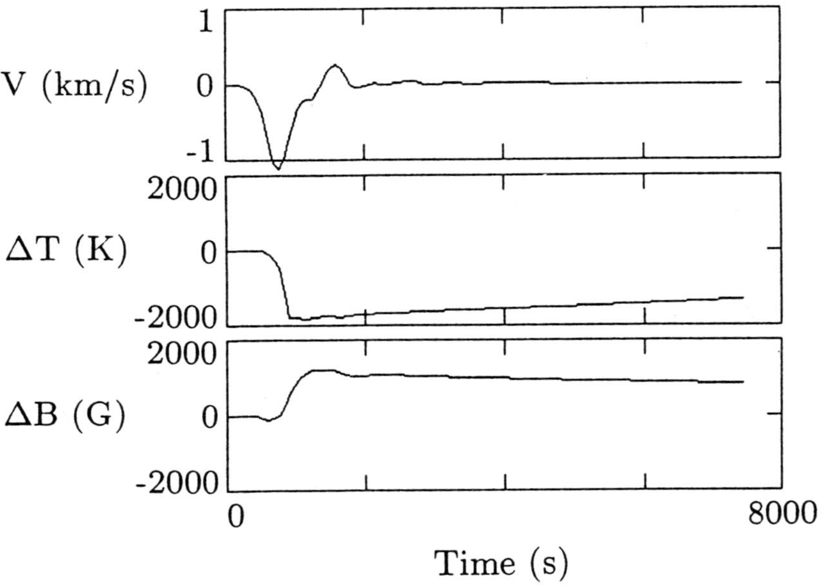

Fig.1. Time variations of the velocity v, the temperature difference ΔT, and the magnetic flux density difference ΔB at $z = -500$ km.

References

Roberts B., Webb A.R. 1978, Sol. Phys. 56, 5

Schüssler M. 1990, in Solar Photosphere: Structure, Convection and Magnetic Fields, IAU Symp No.138 ed J.O.Stenflo (Kluwer Academic Publishers, Dordrecht) p161

Spruit H.C., Zweibel E.G. 1979, Sol. Phys. 62, 15

Takeuchi A. 1993, PASJ 45, 811

Takeuchi A. 1995, PASJ 47, 331

A GENERAL CODE FOR MODELING MHD FLOWS ON PARALLEL COMPUTERS: VERSATILE ADVECTION CODE

GÁBOR TÓTH
Sterrenkundig Instituut, Utrecht University
Postbus 80000, 3508 TA Utrecht, The Netherlands
E-mail: toth@fys.ruu.nl

Abstract. A general tool for solving MHD and hydrodynamical problems typical of astrophysical applications is designed and implemented. The code allows the user to solve a hyperbolic system of partial differential equations with a variety of modern high resolution numerical schemes on 1, 2 or 3D non-uniform Cartesian grids with slab or axial symmetry. The equations may contain fluxes and source terms. The boundary conditions can be chosen from a number of implemented types, – fixed, periodic, reflective or "free" – but the code is structured in a manner that allows the user to write his/her own subroutines for a new type of boundary condition, or even to implement a new set of equations. The choice of algorithms allows the user to find the optimal one for the application at hand and/or to compare results of different schemes thus gaining insight into the numerical effects. An emphasis is put on independence of computer platforms. Efficiency of the code is also a high priority, the options available for the user should not compromise the performance.

The Versatile Advection Code is under development, but it has already reached a stage when it can be used for scientific research. Since the same source code is used for 1, 2 or 3D simulations, a special purpose preprocessor translates the dimension-independent notation into standard Fortran 90. I chose Fortran 90 because it is becoming *the language* of scientific computation, it expresses parallelism clearly, and it is fully compatible with Fortran 77. At the moment, however, Fortran 90 compilers are not yet available for many computers, thus the code can be translated to Fortran 77. For parallel computers ADAPTOR, a freely distributed software by Brandes, can also produce the necessary message passing. It is also possible to translate the Fortran 90 code to CMFortran, the data-parallel language of the

Y. Uchida et al. (eds.),
Magnetodynamic Phenomena in the Solar Atmosphere – Prototypes of Stellar Magnetic Activity, 471–472.

Connection Machine. Simple Perl scripts and a makefile take care of the precompilation procedure.

The modular structure of the code allows for a choice of the equations to be solved, at the moment four systems of equations are available: the Euler equations of compressible hydrodynamics with adiabatic or full energy equation, and the ideal/resistive MHD equations with isothermal or full energy equation. The number of spatial dimensions and the number of components for the vector variables (e.g. velocity and magnetic field) may have 6 different combinations for each system of equations.

At present three algorithms are available for solving the differential equations. The Flux Corrected Transport (FCT) scheme by Boris and Book and its later variation by Odstrčil, a Total Variation Diminishing (TVD) scheme with a Roe-type Riemann solver, and a Lax-Friedrich type TVD scheme with no Riemann solver. Both TVD schemes can use different limiters. The TVD schemes were extensively discussed and developed by Yee. All schemes are explicit, they are able to follow shocks and other discontinuities as well as smooth flows accurately. The methods differ in their abilities to resolve contact discontinuities, to suppress unphysical oscillations, and in their robustness to handle extreme situations. For MHD calculations the divergence of the magnetic field may be forced to remain zero by a projection scheme. The user can select among the available options by setting switches in a parameter file. The initial conditions are created by a separate program, VACINI. The user is not limited by the existing options, new subroutines may be added, or the initial file can be created by other means.

The code has been extensively tested (Tóth and Odstrčil 1995) on standard numerical tests. In spite of its general design, VAC is faster than or as fast as other less general codes, the comparisons were made on a DEC Alpha work station. VAC has successfully run on DEC 5000 and Cray C90 computers compiled in Fortran 77, and on Connection Machine 5 using CMFortran. Tests on other parallel machines with message passing are planned for the near future. The algorithms will be generalized and developed further. Non-Cartesian grids and implicit methods are the two major extensions that are planned.

For more information on this project please contact the author.

Reference

G. Tóth and D. Odstrčil, *Comparison of some Flux Corrected Transport and Total Variation Diminishing Numerical Schemes for Hydrodynamic and Magnetohydrodynamic Problems* , submitted to J. Comp. Phys., 1995

A REGRIDDING ALGORITHM FOR HIGH RESOLUTION HYDRODYNAMICS OF FLARING CORONAL LOOPS

R. M. BETTA, G. PERES, S. SERIO AND F. REALE
Istituto ed Osservatorio Astronomico di Palermo,
Piazza del Parlamento 1, 90134 Palermo, Italy

As part of the upgrade of the Palermo-Harvard hydrodynamic code (PHHC) of coronal loops (Peres et al. 1982), we have devised a regridding algorithm which provides high resolution in the whole loop and, in particular, in the very thin transition region. Most hydrodynamic flare loop models used so far lack this feature which, however, is important for the accurate determination of mass, momentum and energy balance between the chromosphere and the corona, as well as the correct calculation of EUV line emission. The latter is mostly emitted in the transition region and yields crucial diagnostics of loop plasma structure and dynamics.

The PHHC solves the equations of mass, momentum and energy conservation of the plasma from the chromosphere to the corona confined inside a coronal loop, and includes crucial physical effects, such as thermal conduction, radiative losses, the ionization of hydrogen, the presence of gravity and viscosity, and a transient heating term, triggering the flare. The initial conditions are hydrostatic; for more details see Peres et al. (1982).

Our regridding algorithm grants that: i) the fractional jump of temperature, density and pressure between two adjacent points is less than a prescribed small value (typically 10%); ii) the grid is regular, since numerical errors arise when cell sizes vary abruptly along the grid (Betta 1995 and references therein). The spatial grid is continuosly monitored and, when nedeed, modified so as to fulfill the above requirements and therefore to solve numerically the hydrodynamic equations with significant accuracy.

In the grid the cell size decreases according to a geometric progression from the loop base (in chromosphere) to the transition region, where we adopt a small uniform grid; the cell size increases again according to a geometric progression from the base of the transition region to the apex of the loop, in corona. The code automatically chooses the smallest cell size and the ratios of the geometric progressions in order to satisfy conditions

Y. Uchida et al. (eds.),
Magnetodynamic Phenomena in the Solar Atmosphere – Prototypes of Stellar Magnetic Activity, 473–474.

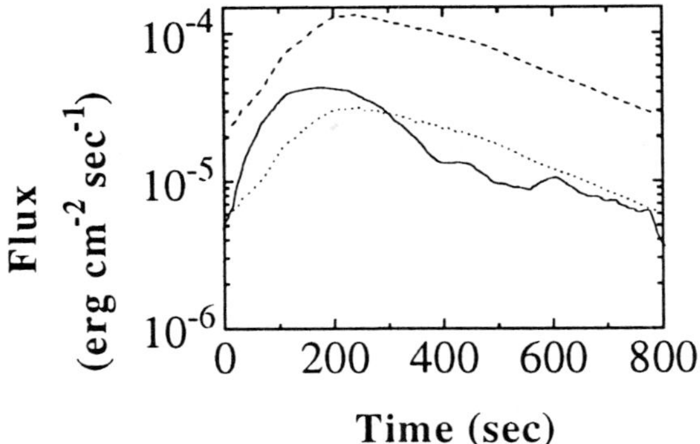

Figure 1. Comparison among the light curve of the flare on Nov 12, 1980 at 17:00 UT observed by the UVSP/SMM (solid line) in the line O V at 1371 Å(Chen & Pallavicini 1988) and the light curves synthesized with the results of the PHHC from the spectral models of Raymond & Smith (1977) (dotted line) and Landini & Monsignori Fossi (1990) (dashed line).

i) and ii). A check is performed at every time-step to ascertain whether the condition i) is satisfied, else the grid is modified. Thus we grant a high resolution even of the transition region in spite of its rapid motion and steepenig during a flare.

We have chosen a well-studied flare, the one observed by SMM on Nov. 12 1980 at 17:00 UT, to show how the better grid resolution can improve the predictions on hydrodynamics and on spectral emission.

The fitting of the X-ray lines light curves in practice confirms the results of Peres et al. (1987) although with a much lower numerical noise. No similar analysis had been done for EUV lines, because only the present version of the code allows to resolve the temperature range of line formation necessary to synthesize the emission from the plasma in the transition region. We report in Figure 1 the flux observed by the Ultraviolet Spectrometer on SMM in the line O V, at 1371 Å(maximum temperature formation $\sim 2\ 10^5 K$, i.e. in the transition region), compared with that predicted by our simulation with two different spectral synthesis models and the model flare parameters for the top-heated flare as in Peres et al. (1987), Figure 4.

References

Betta, R., 1995, Thesis, University of Palermo, Italy

Chen, C.C., Pallavicini, R., 1988, The Astrophysical Journal, **324**, 1138

Landini, M., Monsignori Fossi, B.C., 1990,*Astron. Astrophys. Suppl.Ser.* , **82**, 229

Peres, G., Rosner, R., Serio, S., Vaiana, G.S., 1982,*The Astrophysical Journal*,**252**, 791

Peres, G., Reale, F., Serio, S., Pallavicini, R., 1987, *The Astrophysical Journal*,**312**, 895

Raymond, J.C., Smith, B.W., 1977, *The Astrophysical Journal Suppl. Ser.*, **35**, 419

ROTATIONAL MODULATION OF Hα EMISSION IN RS CVN SYSTEMS

A. FRASCA, S. CATALANO, E. MARILLI
Osservatorio Astrofisico di Catania
Viale A. Doria 6, I-95125 Catania, Italy

The spectroscopic Hα observations of five RS CVn systems (UX Ari, HR 1099, Z Her, HK Lac, RT Lac) and of the contact system VW Cep here discussed were obtained in 1989, 1990, 1991, 1993 and 1994 with the 91-cm telescope at the Serra La Nave station (Mt. Etna, 1750 m a. s. l.) of the Catania Astrophysical Observatory. The spectra recorded on a CCD camera (Bonanno and Di Benedetto, 1990) were obtained with the REOSC echelle spectrograph in the single-mode configuration and in the echelle-mode configuration yielding a resolution of 1.7 Å and 0.46 Å respectively.

The Hα emission in the program star spectra has been evaluated by making a careful subtraction of a synthetic spectrum, built up by a weighted sum of inactive-star spectra (Frasca and Catalano, 1994). The synthetic spectra were defined by taking into account the relative Doppler shifts due to the orbital motion of the two components and the rotational broadening of the Hα profiles whenever it was larger than the instrument spectral resolution.

The net Hα equivalent width ($W_{\mathrm{H}\alpha}$), obtained integrating the flux in the difference spectra, displays, in most cases, noticeable variations correlated with the orbital phase. Amplitude and mean level changes in the Hα "light curves" are frequently observed.

A rough reconstruction of the chromospheric surface structures has been made analyzing the Hα "light curves" by means of the BINARY MAKER program (Bradstreet, 1993), adopting a hot-spot model. A constant emission flux ratio of plages and quiet chromosphere ($\mathrm{F}_{plage}/\mathrm{F}_{chrom}$) for a given star in the different years, allowing the quiet chromospheric level to vary, has been assumed. This allows us to look for a geometrical solution where the size of the "bright spots" is only indicative and may give only some evidence on the changes of configuration from year to year. The solutions essentially locate the plages, with a reasonable constraint on the longitudes

Y. Uchida et al. (eds.),
Magnetodynamic Phenomena in the Solar Atmosphere – Prototypes of Stellar Magnetic Activity, 475–476.

of the major active regions. Results of the solution for the better observed systems are summarized in Table 1.

TABLE 1. Parameters of the Hα plages

System	Plage	Radius	Lat.	Long.	$\frac{F_{plage}}{F_{chrom}}$
UX Ari (1994)	1	30°	-20°	275°	6.8
" "	2	25°	60°	350°	6.8
HR 1099 (1989)	1	23°	10°	20°	6.0
HR 1099 (1994)	1	20°	-10°	140°	6.0
" "	2	25°	40°	70°	6.0
HK Lac (1989)	1	30°	30°	· 35°	6.8
" "	2	30°	30°	165°	6.8
HK Lac (1990/91)	1	30°	30°	35°	6.8
HK Lac (1993)	1	15°	70°	60°	6.8
HK Lac (1994)	1	25°	-15°	30°	6.8

The evolution of the active regions is clearly seen on HR 1099 and HK Lac. This latter system, in particular, displays in 1993 a minimum mean level and a smaller amplitude. It is worthy to note that only one of the two plages needed to fit the wide and flat 1989 curve seems to survive, at the same position, in 1990.

As a progress report on a wider program at the Catania Observatory, we have shown that continuous Hα monitoring of active systems is a very useful tool for studying the activity evolution at chromospheric level. Although Hα emission model calculation would provide physical parameters of chromospheric plages, the simple geometrical solution is able to give some idea of the complex structure of chromospheric active regions.

Acknowledgements This work has been supported by the Italian *Ministero dell' Università e della Ricerca Scientifica e Tecnologica*, the *Gruppo Nazionale di Astronomia of the CNR* and by the *Regione Sicilia*, which are gratefully acknowledged.

References

Bonanno G., Di Benedetto R. (1990) *Publ. of Astron. Society of Pacific* **102**, p. 835
Bradstreet D.H. (1993) *Binary Maker 2.0 User Manual*, Contact Software
Frasca A., Catalano S. (1994) *Astronomy and Astrophysics* **284**, p. 883

RADIO AND X-RAY EMISSION IN STELLAR MAGNETIC LOOPS

E. FRANCIOSINI AND F. CHIUDERI DRAGO
Department of Astronomy and Space Science,
University of Florence, Italy

In this paper we present a model of a magnetic loop in an RS CVn active star, in which a distribution of nonthermal particles is episodically injected in the local thermal plasma and then evolves in time due to the effect of collision and radiative losses. Preliminary calculations in the simplifying assumption of a uniform source with constant magnetic field have shown that this model can explain the observed characteristics of the flaring and quiescent radio emission of RS CVn stars (Chiuderi Drago & Franciosini 1993). Here we consider a more realistic magnetic configuration, assuming that the source is a dipolar magnetic loop connecting two starspots on the stellar surface. In the evolution of the particle population also the electrons escaping into the loss-cone are taken into account, and in the calculation of the radio spectrum the free-free emission and absorption by the background thermal plasma and the Razin effect are considered. The main parameters of the loop, sunspot dimension and maximum photospheric field, are derived from optical observations, while the thermal plasma density is left as a free parameter. A more detailed description of the model can be found in Franciosini & Chiuderi Drago (1995).

The computed radio emission, at different times after the injection, reproduces very well the observed common characteristics of the radio spectra of the flaring and quiescent emission observed at different epochs. A comparison of the results with the observed luminosity-spectral index correlation found by Mutel et al. (1987), however, shows that the agreement with the observations can be achieved only if the thermal plasma density is lower than $\sim 10^7$ cm^{-3}. The calculations also show that the structure of the radio source evolves with time from a core-halo soon after the flare to a halo after a few days (Fig. 1) in agreement with VLBI observations (Mutel et al. 1985, Massi et al. 1988).

Y. Uchida et al. (eds.),
Magnetodynamic Phenomena in the Solar Atmosphere – Prototypes of Stellar Magnetic Activity, 477–478.

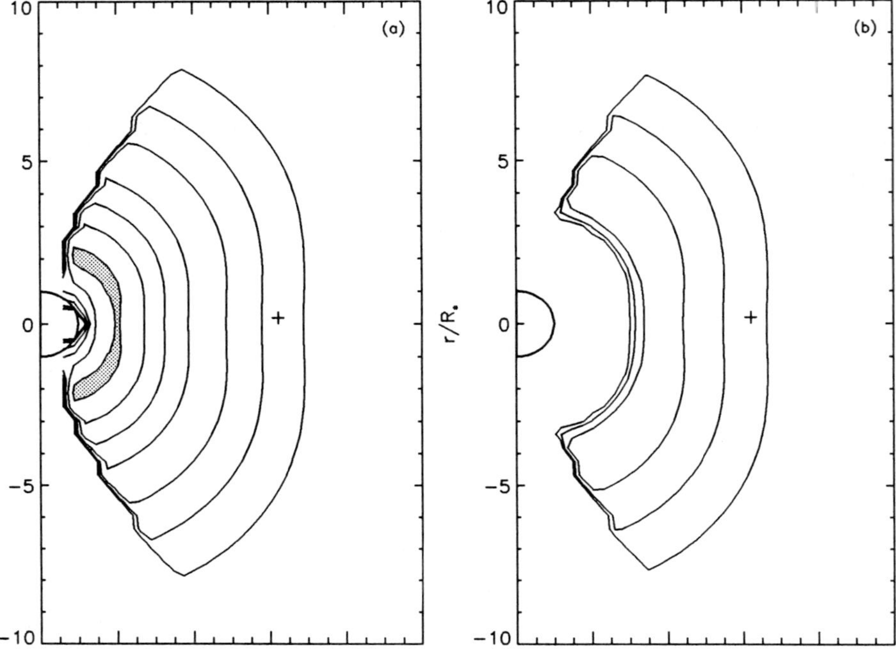

Figure 1. Brightness distribution of the source at $\nu = 5$ GHz, computed at the flare time (a) and after 6 days (b), for a loop with spot radius of 20° and spot separation of 90°. Isocontours are at 5×10^{-10}, 10^{-9}, 2×10^{-9}, 5×10^{-9}, 10^{-8}, 2×10^{-8} and 4×10^{-8} erg cm^{-2} s^{-1} Hz^{-1} sr^{-1}; the highest brightness contour has been shadowed for the sake of clarity. The cross indicates the distance of the companion when the system is in quadrature. Note that the ratio between the highest and lowest intensity contour is a factor of 80 at t=0 and only 4 at t=6^d.

The emission measure, derived from the loop's volume and the above mentioned upper limit for the thermal plasma density, is in agreement with those inferred from the X-ray observations of these type of stars, indicating a possible co-spatiality of the two emissions.

References

Chiuderi Drago, F., & Franciosini, E. (1993), *Astrophys. J.* **410**, 301

Franciosini, E., & Chiuderi Drago, F. (1995), *Astron. Astrophys.* **297**, 535

Massi, M., Felli, M., Pallavicini, R., Tofani, G., Palagi, F., & Catarzi, M. (1988), *Astron. Astrophys.* **197**, 200

Mutel, R.L., Lestrade, J.-F., Preston, R.A., & Phillips, R.B. (1985), *Astrophys. J.* **289**, 262

Mutel, R.L., Morris, D.H., Doiron, D.J., & Lestrade, J.-F. (1987), *Astron. J.* **93**, 1220

CORONAL PROPERTIES OF NEARBY OLD DISK AND HALO DM STARS

G. MICELA AND S. SCIORTINO
Istituto e Osservatorio Astronomico di Palermo, 90134, Italy

AND

J. PYE
Department of Physics and Astronomy, University of Leicester, Leicester, LE1 7RH, UK

We present preliminary results of the analysis of PSPC pointed observation program and archive data of nearby old disk and halo M stars. Our sample consists of 7 M stars classified as Halo (H) or Old-Halo (O/H) by Leggett (1992) plus 4 Old Disk (OD) stars. We have included in our sample also the OD K4 star GJ845.

Since all stars are pointed targets, X-ray source counts are evaluated in a 2' radius circle, while background counts are measured in an annulus centered on the source with internal and external radii of 2.5' and 3', respectively. For GJ 595 we have masked source and background regions to exclude the contribution of two nearby intense sources.

For each of the detected stars we have obtained light curves with different time binning. All stars observed with high counts statistics show variations with amplitude up to a factor 2-3 on time scales larger than a few thousand of seconds. In general, the temporal gaps in the data flow within a single observation prevent us from constraining the characteristics of the variability.

We have evaluated the hardness ratio defined as $HR = \frac{H-S}{H+S}$ where S and H are the number of photons collected in SASS channels 3-10 and 11-30, respectively. With this definition, following Schmitt et al. (1995) we have computed the conversion factor from counts to flux to evaluate the X-ray luminosity. Resulting flux and luminosity, evaluated in the (0.16-2.4) keV bandpass, are reported in table 1. Upper limits for undetected stars have been evaluated assuming a conversion factor equal to the mean of the values adopted for the other stars in the sample. A comparison of our and Schmitt

Y. Uchida et al. (eds.),
Magnetodynamic Phenomena in the Solar Atmosphere – Prototypes of Stellar Magnetic Activity, 479–480.

TABLE 1. X-ray Luminosity for sample stars

GJ	HR	Conversion factor	Count Rate	f_x $erg\ s^{-1}\ cm^{-2}$	$\log L_x$ $erg\ s^{-1}$
1	-0.658	4.8e-12	0.01822	8.79e-14	26.33
191	-0.607	5.1e-12	0.05402	2.75e-13	26.69
191	-0.712	4.5e-12	0.02557	1.16e-13	26.32
213	-0.886	3.6e-12	0.00646	2.33e-14	26.01
299	...	(5.3e-12)	<0.00611	<3.22e-14	<26.25
398	-0.066	8.0e-12	0.20882	1.66e-12	28.57
406	-0.173	7.4e-12	0.23823	1.76e-12	27.08
595	...	(5.3e-12)	<0.00733	<3.87e-14	<26.58
699	-1.000	3.0e-12	0.02122	6.39e-14	25.41
821	-0.705	4.6e-12	0.00616	2.82e-14	26.60
845	-0.591	5.2e-12	0.51985	2.69e-12	27.58
866	-0.335	6.5e-12	0.29667	1.94e-12	27.43
887	-0.565	5.3e-12	0.20675	1.10e-12	27.21

et al. (1995) data points on the L_x-HR plane (cf. their fig.5) shows that our Population II stars tend to occupy the lower part of the envelope defined by nearby stars. It is worth noting that the hardest star (GJ 398) is also the most X-ray luminous. Fleming et al. (1995) do not find any difference between the behavior of kinematically young and old disk stars in their sample. Our old disk stars fall in the body of their distribution, while halo stars seem to cluster in the lower part of the envelope of their diagram. We have performed spectral fits for all the program stars with enough count statistics using the xspectral package inside PROS. In all cases we obtain acceptable fits and the residuals do not show systematic behavior. Hence double temperature fits for these stars are not required. Only for GJ 821 the residual distribution seems to require a high temperature component, however the limited count statistics does not warrant to perform a two-temperature fit. The derived coronal temperatures are of the order of 1-2 10^6 K, indicating that the coronal emission of these stars is softer than that of active dM stars reported by Schmitt et al. (1990).

References

Fleming, T. A., Schmitt, J. H. M. M., & Giampapa, M. S., 1995, ApJ, in press

Leggett, S. K., 1992, ApJS, 82, 351

Schmitt, J. H. M. M., Collura A., Sciortino, S., Vaiana, G. S., Harnden, F. R., Jr., & Rosner, R. 1990, ApJ, 365, 704.

Schmitt, J. H. M. M., Fleming, T. A. & Giampapa, M., S. 1995, ApJ, in press

SPECTRAL AND SPATIAL STRUCTURE OF η CARINAE

YOHKO TSUBOI, KATSUJI KOYAMA
Department of Physics, Faculty of Science, Kyoto University
Kitashirakawa-Oiwake-Cho, Sakyo-ku, Kyoto 606-01, JAPAN

AND

ROBERT PETRE
Laboratory for High Energy Astrophysics, NASA/GSFC
Greenbelt, MD 20771, USA

1. Introduction

We present the ASCA results on the peculiar star η Carinae (here η Car) in the center of the Carina Nebula. The observation was done for 40 ksec on 1993 August 24-25 with two Solid-state Imaging Spectrometers (SIS0, SIS1) and two Gas Imaging Spectrometers (GIS2, GIS3). In this region, three early type stars and diffuse emission from the Carina Nebula are visible as well as η Car, but, we concentrate on η Car here.

2. Spectral Analysis

The background subtracted spectrum of η Car is given in Figure 1. The strong emission line at $\sim$6.7 keV, the energy of the K shell transitions from He-like iron, indicates the presence of a low density plasma at a temperature of several keV. The local flux minimum at $\sim$1.5 keV is indicative of the presence of two spectral components, one hot and heavily absorbed, the other with lower temperature and less absorption (Chlebowski et al. 1984; Koyama et al. 1990). We therefore fitted the spectrum using a two-temperature coronal plasma model (Raymond & Smith 1977), but this left a substantial excess residual at $\sim$0.5 keV, while the model provided a reasonable fit above $\sim$0.7 keV. This structure could either be due to: (1) an additional, cooler plasma component; (2) an absorption edge due to neutral oxygen; or (3) line emission at $\sim$0.5 keV. As for (3), the most probable line is Ly α of nitrogen at 500 eV. After fitting the data with each model, we found that the model with an extreme over-abundance of nitrogen (larger

Y. Uchida et al. (eds.),
Magnetodynamic Phenomena in the Solar Atmosphere – Prototypes of Stellar Magnetic Activity, 481–482.

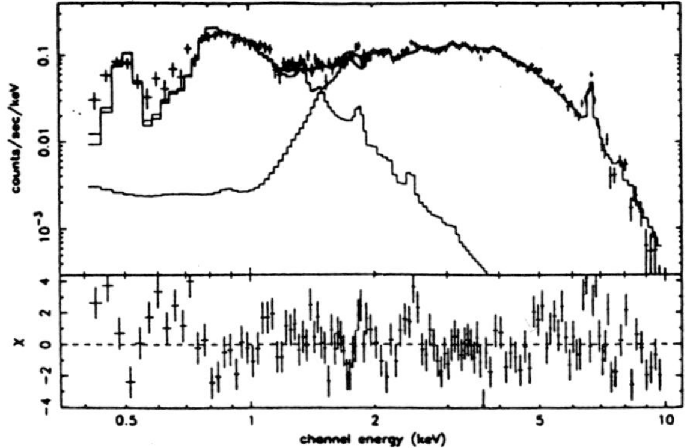

Figure 1. X-ray spectrum of η Car

than the solar by more than 100 times) is the best description of the data (the abundances of other elements were fixed to solar values). This is the first X-ray evidence of the enrichment of CNO-processed material in an evolved massive star.

3. Spatial Analysis

In order to investigate the spatial distribution of the high and low temperature components and the 0.5 keV line, we constructed azimuthally-averaged radial surface brightness distributions in the 0.4-0.6 keV, 0.7-1.4 keV and 2-10 keV bands. After subtracting the diffuse emissions from the Carina Nebula as a background, we fitted the radial profile with a model of the point spread function (PSF), made from the point source 3C 273. The 2-10 keV data agree well with the PSF model; hence we infer that the high temperature component is consistent with a point source; probably η Car itself. The 0.7-1.4 keV data, on the other hand, are more extended than the PSF and therefore the high temperature component as well. This implies that the emission likely arises from the ejecta of past outbursts. The limited statistics in the 0.4-0.6 keV band prevents us from being able to constrain its extent, but its profile appears more similar to the soft component than the hard.

References

Chlebowski, T., Seward, F., Swank, J., Szymkowiak, A. (1984) Hard X-ray emission from the carina Nebula, *Astrophysical Jornal*, **281**, 665-672

Koyama, K., Asaoka, I., Ushimaru, N., Yamauchi, S. (1990) Hard X-ray emission from the carina Nebula, *Astrophysical Jornal*, **362**, 215-218

Raymond, J.C., Smith, B.W. (1977) Soft X-ray spectrum of a hot plasma, *Astrophysical Jornal Supplement Series*, **35**, 419-439

COMPARISON OF HIGH RESOLUTION OPTICAL AND SOFT X-RAY IMAGES OF SOLAR CORONA

A. TAKEDA, H. KUROKAWA, R. KITAI AND K. ISHIURA
Kwasan and Hida Observatories, Kyoto University, Kamitakara, Gifu 506-13, Japan

AND

L. GOLUB
Harvard-Smithsonian Center, 60 Garden Street, Cambridge, MA02138, U.S.A.

We report a brief result of comparing high spatial resolution images of the innermost corona observed at several visible wavelengths and in soft X-rays. The optical images were obtained during a total solar eclipse, at around 19:00 UT on 11 July 1991, by a Kyoto university team. The observed wavelengths are three coronal forbidden lines (FeXIV:530.3nm, FeX:637.4nm, and CaXV:569.4nm), HI:656.3nm, and continuum arround 610.0nm. (Kurokawa *et al.*, 1992, Takeda *et al.*, 1994) Soft X-ray images were taken by the Harvard-Smithsonian group utilizing the Normal Incidence X-ray Telescope' (NIXT), from 17:25 to 17:30 UT. It is sensitive to the temperature from 1M to 3MK, due major contribution from MgX and FeXVI coronal lines. (Golub *et al.*, 1989)

East and west limb images at different wavelengths are shown in Fig.1. Fig.2 illustrates the intensity variation of the east limb images. Comparisons among these images and the plots shows:

(1) The NIXT image most resembles the green line (530.3nm) image. The distribution of brightness as well as the prominent structures are quite similar in the two images.

(2) A general coincidence between the NIXT and the 637.4nm images is seen only along the innermost part of the corona. The red line image is

Y. Uchida et al. (eds.),
Magnetodynamic Phenomena in the Solar Atmosphere – Prototypes of Stellar Magnetic Activity, 483–485.

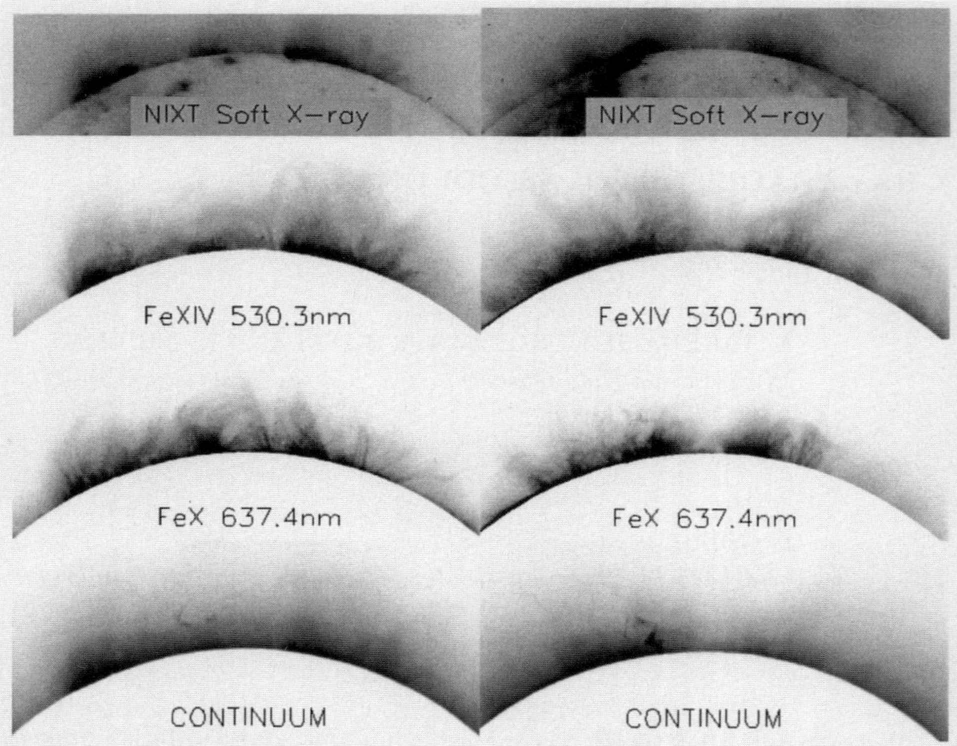

Figure 1. The images of the 11 July, 1991 solar corona on the east limb (left column) and the west limb (right column), observed at soft X-ray, the green line, the red line, and continuum (from top to bottom). Photographic film densities are transformed into relative intensities, and spatial scale between the NIXT and optical images are fitted using the data of the solar and lunar radii of the eclipse day, reported by Japanese Hydrographic Department.

low background, and is composed of many thin loops most of which are invisible in the NIXT image.

(3) The global features of the continuum images are similar to those of the green line images. The NIXT image, therefore, resembles the continuum image as well. But the continuum image is more diffuse than the other images except in some prominences. The yellow line images show no prominent structure in the 11 July, 1991 corona.

NIXT has peak temperature response between 1M–3MK, while 530.3nm, 637.4nm, and 569.4nm images represent 1MK, 2MK, and 3.5MK plasma, respectively. Good coincidence between the NIXT and the green line image indicates that plasma of 2MK is dominant in the observed active regions.

According to *Yohkoh* data analysis, typical quiescent coronal loops observed with SXT are reported to have temperatures of more than 3MK

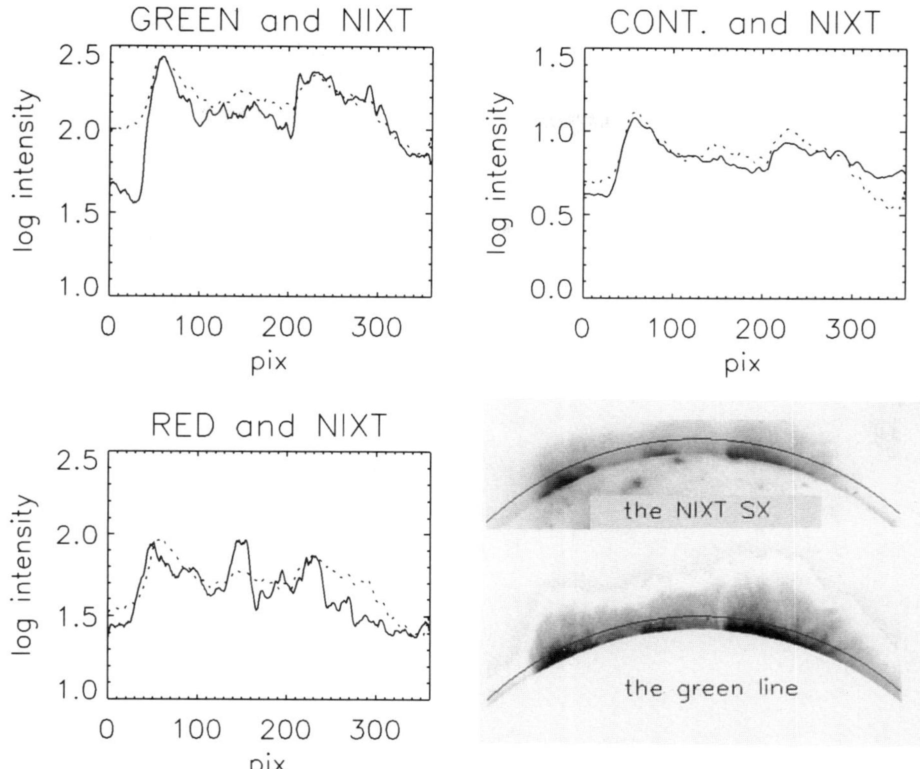

Figure 2. Intensity variation of the east limb images parallel to the limb at height about 40000km (bottom right panel). Solid lines of top left, bottom left, and top right panels show plots of absolute intensities, in logarithmic scale, of the green line, the red line, and continuum images, respectively. Units are in (erg $s^{-1}str^{-1}cm^{-2}$) for the green and the red lines, and in (erg $s^{-1}str^{-1}cm^{-2}Å^{-1}$) for continuum. Dotted lines in these panels are relative intensities of the NIXT image, and vertically shifted to be easily compared with the other curves.

(Kano *et al.*, 1995). The lack of yellow line emission and the dominance of 2MK plasma component in our observation, however, show the general existence of coronal structures not above but below 3MK.

References

Golub, L., and Herant M. (1989), *Proc. SPIE*, **Vol. 629.**, pp.1160

Kano R., and Tsuneta S. (1995), *Astrophys. J.*, in press

Kurokawa H., Kitai R. and Ishiura K. (1992), in Yamashita T. (ed.), *Observations of Solar Corona at the Mexico Total Eclipse of July 11, 1991*, pp.10.

Takeda A., Kurokawa H. Kitai R. and Ishiura K., in Enome S. (eds.), *Proceedings of Kofu Symposium*, pp.381

Poster Papers

Posters for Session II

LARGE-SCALE ARCADE FORMATION ON MAY 15, 1992 AND ITS INTERPLANETARY CONSEQUENCE

Y. NAKAGAWA AND T. WATANABE
Department of Earth Sciences, Ibaraki Univ., Mito 310 Japan
H. HUDSON
Institute for Astronomy, Univ. of Hawaii, Honolulu, Hawaii 96822, U.S.A.
AND
M. KOJIMA
STELAB,Nagoya Univ., Toyokawa 442, Japan

A high-latitude SXR arcade was formed in the northern hemisphare of the Sun on May 15, 1992, from 03h to 22h UT (Fig. 1). In the cource of arcade formation, the area of a coronal hole located immediately to the south of the arcade was quickly decreaced in 23 hours. Decreaçing of the area of the coronal hole is seen both in Kitt Peak He I 10830Å images (Fig. 2) and in YOHKOH SXR images. This means that the decreacing of the area of coronal hale was not caused by a projection effect of three-demensional coronal strucrures. The upper frame of Fig. 3 shows the temporal variation of the average brightness in an area which coveres the brightest portion of the arcade (shown by the square in Fig. 1). The lower frame of the same figure shows that of the coronal hole's area. The <u>increace</u> of the brightness of the arcade and <u>decreace</u> of the coronal hole's area started nearly at the same instance. It is suggested that the large-scale change in coronal hole geometry triggered the arcade formation. In interplanetary space, a transient increase in the solar wind up to 707 km/sec was observed by IPS of 0221+07 at 0.31AU from the Sun, at 17 UT of May 16, 1992, and an SC of geomagnetic storm was observed at 20:21 UT on May 18, 1992. The average speed of the disturbance between the Sun and the Earth is about 530 km/sec. It is suggested that the interplanetary disterbance was generated by grobal change in large-scale magnetic structures including the decreacing of coronal hole's area and associated formation of the arcade.

Y. Uchida et al. (eds.),
Magnetodynamic Phenomena in the Solar Atmosphere – Prototypes of Stellar Magnetic Activity, 489–490.

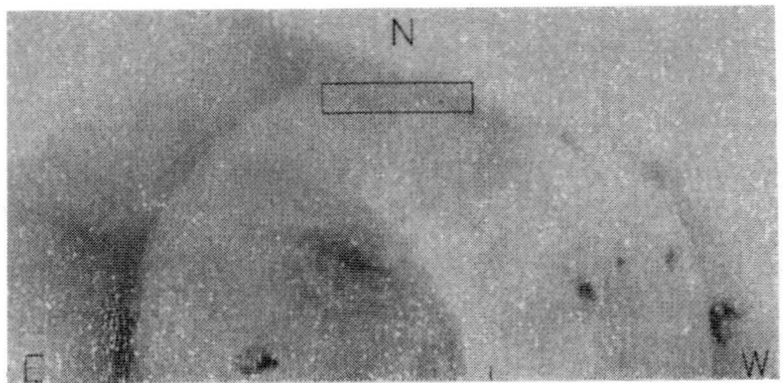

Figure 1. An example of *Yohkoh* SXT image (negative) of the large-scale arcade near the north pole of the Sun taken at 08:41:23 UT on 15 May, 1992. The Square shows the area in which the averaged SXR brightness of the arcade is estimated.

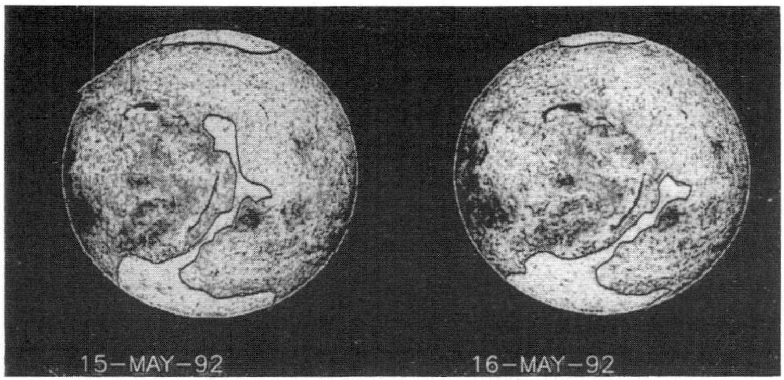

Figure 2. Kitt Peak He I 10830 Åimages taken at 15:21:UT on 15 May, 1992(left) and 15:24 UT on 16 May, 1992 (right), respectively. The coronal hole bounderlies are shown by thick lines.

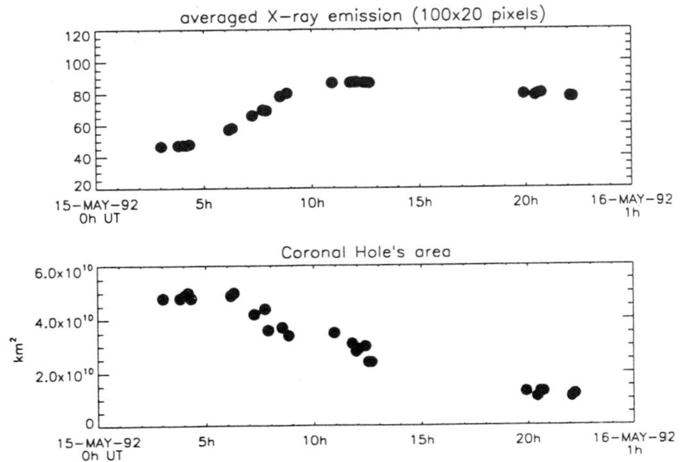

Figure 3. Temporal variations of averaged SXR brightness of the arcade (upper) and that of the coronal hole's area(lower). The averaged brightness is estimated in the area which shown in Fig. 1.

Observations of a Quiescent Prominence Straddling the Solar Limb during the Total Eclipse of 11 July 1991

V. GAIZAUSKAS[1], E. DE LUCA[2], L. GOLUB[2], H. P. JONES[3], AND L. NOVEMBER[4]
[1]*Herzberg Institute of Astrophysics, National Research Council of Canada, Ottawa, Canada K1A 0R6*
[2]*Smithsonian Astrophysical Observatory, Harvard-Smithsonian Center for Astrophysics, Cambridge, MA 02138, USA*
[3]*NASA/GSFC, SW Station, Box 26732, Tucson, AZ 85726-6732, USA*
[4]*NSO/Sacramento Peak, Box 62, Sunspot, NM 88349, USA*

The following instruments viewed a prominence straddling the SW limb during the total eclipse of 11 July 1991: the Normal Incidence X-ray Telescope (NIXT) launched on a NASA sounding rocket at White Sands, New Mexico; a 70 mm cine camera with broad-band red filter at the prime focus of the Canada-France-Hawaii Telescope (CFHT); the NASA/NSO spectromagnetograph at Kitt Peak; and the Hα-scanning photoheliograph at the Ottawa River Solar Observatory (ORSO). Only the CFHT was in the path of totality; the prominence and overlying corona filled the field of view of its 70 mm camera (Koutchmy et al., 1994). The other instruments viewed the entire uneclipsed surface by scanning (NSO, ORSO) or in single exposures (NIXT).

The digitized photographs (CFHT, NIXT, ORSO), the NSO magnetogram, and the NSO He10830 spectroheliogram were scaled (1.3"/pixel) and oriented to the NIXT image as the common base. Widely spaced (> 8') bright and compact features in a composite Hα image of the SW quadrant served to align the Hα and X-ray images to ± 1 pixel. A single notch in the prominence standing above the limb fixed the CFHT and ORSO images to ± 2 pixels.

Figure 1 portrays the prominence in aligned Hα, NIXT, and CFHT images. A sharp dark lane in the background of low-level X-ray emission coincides with the 'spine' of the Hα filament - a narrow dark ridge down its S side. Is the dimming due to reduced heating and/or a smaller effective density scale height in a filament channel where transverse magnetic fields might suppress dynamic effects (e.g., Bastian et al., 1993)? We see a channel at Hα±0.6 Å persisting for a week before the eclipse as an elongated void in the network formed by bushes of spicules and fibrils beside this filament. But the S edge of the void is 9" N of the Hα spine, well beyond our range of error.

Y. Uchida et al. (eds.),
Magnetodynamic Phenomena in the Solar Atmosphere – Prototypes of Stellar Magnetic Activity, 491–492.

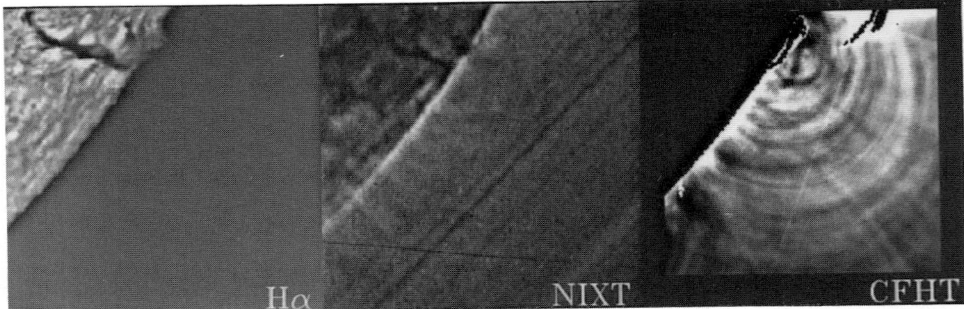

Fig. 1: - Digitally enhanced Hα, soft X-ray and white-light images (left to right) of a low prominence straddling the SW limb during the eclipse of 11 July 1991. The white-light prominence has been blacked out so that details will stand out in and beneath the overlying arcade of loops. Each field of view is 314"x314".

Also, a narrow dark arch in the X-ray image overlaps the portion of the Hα filament above the limb (Figure 1). X-ray emission just above and below the arch indicates obscuration of a source beyond the limb. We conclude that strands within the prominence/filament are absorbing soft X-rays.

From the well-resolved change of X-ray flux across the darkest part of the dark lane we estimate of a minimum column density of 6×10^{19} neutral hydrogen atoms in the line of sight. From the geometrical properties of the filament, derived during its disk passage and at both limb transits, we calculate a volume density of (a) 3×10^{10} if the absorbing path extends from the top of the filament down to the top of the chromosphere, or (b) ~ 10^{11} if the path extends no deeper than the width of the spine.

The bright southern tail of the white-light prominence (Figure 1) has no counterpart in either Hα or soft X-rays. We speculate that the plasma in that feature is either too hot or too cold to be visible in Hα or X-rays. The dark void above the blacked-out white-light prominence, presumably a 'filament cavity', bears no simple relationship to either the prominence below or to the arcade of loops above.

A full version of this paper is being submitted to The Astrophysical Journal.

The NIXT sounding rocket program is supported by NASA Grant NAGW-4081 to the Smithsonian Institute.

References

Bastian, T. S., Ewell. M. W., Jr., and Zirin, H.: 1993, *Astrophys. J.* **418**, 510
Koutchmy, S. and 16 co-authors: 1994, *Astron. Astrophys.* **281**, 249

YOHKOH SXT OBSERVATIONS OF PROMINENCE ERUPTION AND DISAPPEARANCE

H. TONOOKA, R. MATSUMOTO AND S. MIYAJI
Dept. of Physics, Chiba Univ., Chiba 263, Japan

S. F. MARTIN
Helio Research, La Crescenta, CA 91214, USA

R. C. CANFIELD AND K. REARDON
Univ. of Hawaii, Honolulu, HI 96822, USA

A. MCALLISTER
NOAA Space Enviroment Lab., Boulder, CO 80303, USA

AND

K. SHIBATA
National Astronomical Observatory , Mitaka, Tokyo 181, Japan

The aim of this paper is to study the mechanism of the disparition brusque, i.e., the sudden disappearance of a prominence (e.g., Tandberg-Hanssen 1995) observed with an Hα limb filtergraph and compare *Yohkoh* SXT and Hα observations.

Although there have been some studies of eruptive prominences using the *Yohkoh* SXT data (see, e.g., McAlister *et al.* 1992), no systematic study has yet been done. Specifically, this is the first study examining the category known as *quasi-eruptive prominence.*

Using prominence data from the Hα limb filtergraph of the Mees Solar Observatory, it has been verified that there exist at least three categories of prominence eruptions or disappearances. The most common category is the *eruptive prominence* in which some of the filament mass is injected into the high corona and is thought to escape the sun while the remainder of the mass falls back to the chromosphere. The second category is the *quasi-eruptive prominence* in which the prominence begins to rise bodily, similar to a complete eruption, but all of its observable mass in Hα flows to the chromosphere near or beyond the apparent ends of the prominence. The third is the *disappearing prominence* in which the mass drains out of the

Y. Uchida et al. (eds.),
Magnetodynamic Phenomena in the Solar Atmosphere – Prototypes of Stellar Magnetic Activity, 493–494.

prominence without any evidence of either slow ascent or eruption. The latter two categories are thought to be relatively rare.

We picked 12 disparition brusque cases using observations in Hα from the data of Mees Solar Observatory. We categorize 8 of them as *eruptive prominences*, 2 as *quasi-eruptive prominences*, and 2 as *disappearing prominences*. Comparing the Hα images with the data observed by *Yohkoh* SXT, we found the following. (1) Some structural changes are found in soft X-ray images in all of the *eruptive* and *quasi-eruptive prominences*. On the other hand no structural changes in the soft X-ray images were found in the cases of *disappearing prominences*. (2) In the case of *quasi-eruptive prominences*, structural changes in the soft X-rays above the prominence were found even after the prominence had stopped rising. (3) At the beginning of one of the prominence eruptions, we found that the region of the prominence had weaker soft X-ray emission than its surroundings.

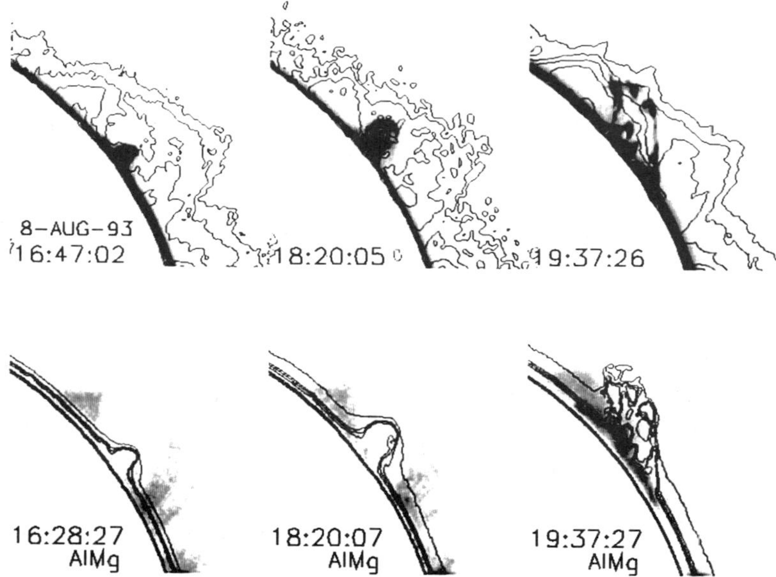

Figure 1. Coaligned SXT AlMg and Mees Hα coronagraph images. Top images are the Hα intensity with soft X-ray contours, bottom images are the opposite. This is a typical example of a *quasi-eruptive promince.* The sequence shows a system of overlying soft X-ray loops (16:28:27) that disappear as the Hα prominence erupts into them. In the final image (19:37:27) two new system of loops are seen on either side of the prominence.

References

Tandberg-Hanssen, E., 1995, *The Nature of Solar Prominences*, Kluwer Acad. Publ.
McAlister, A., *et al.*, 1992, PASJ, 44, L205

POLAR X - RAY ARCADE FORMATION AND GIANT CUSP

K. FUJISAKI, H. OKUBO, Y. UCHIDA, S. HIROSE AND S. CABLE
Department of Physics, Science University of Tokyo,
Shinjuku-ku, Tokyo 162, Japan

AND

S. TSUNETA
Institute of Astronomy, University of Tokyo,
Mitaka 181, Tokyo, Japan

The pre-event structure and its change in the polar arcade formation event of February 24–25, 1993 are examined. The pre-event coronal structure was a double bipolar array (or a quadruple array) type, with the middle region having a mixed photospheric magnetic polarity. A bundle of longitudinal threads ("spine") in soft X-rays rose with a tooth-brush type foot connections, and the arcade formation progressed westwards in time. The westmost part of the "spine" ballooned up, and the shape finally took a form like the giant cusp of Jan 25, 1992. Alerted by this, we examined the latter event in detail. We found that the giant cusp, against the expectation from the classical model, showed its southern foot remaining on the visible disk, whereas the main body of the cusp took place near or beyond the limb. These suggest that, contrary to the conventional interpretation, it may just have formed from a swollen "spine" structure similar to that of the Feb 25, 1993 event.

A circum polar arcade formation is considered to be of the same kind of process as energetic arcade flare, but occurring in much weaker magnetic field having much larger scale in the circum polar region. Scruting of faint events of this type, and the fainter pre-event structures and their changes have been made possible for the first time by *Yohkoh.* We take up one typical event of this type, February 24–25, 1993, observed with *Yohkoh*-SXT, and discuss the detail of the process from the pre-event structure to the final product of the process. This polar arcade formation occurred near the south pole at 17:50 UT, Feb 24, 1993 - 11:40 UT, Feb 25.

Y. Uchida et al. (eds.),
Magnetodynamic Phenomena in the Solar Atmosphere – Prototypes of Stellar Magnetic Activity, 495–496.

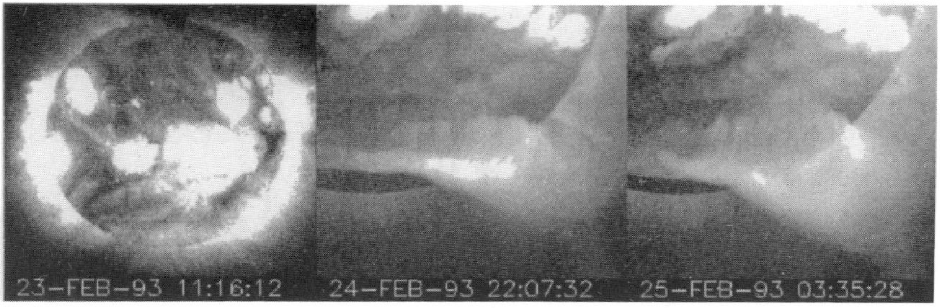

Figure 1. (a) Pre-event coronal structure showing intermingled two arcades. (b),(c) Time development of the formed polar arcade.

The faint background loop structure existed in a pre-event area is shown at around 75° south in Fig.1 (a), together with the later time development in Fig.1 (b) and (c). Kitt Peak magnetogram and the Big Bear Hα picture, rotated precisely to the time of the soft X-ray loops, show that the dark filament coincided in projection with the region where the X-ray loops in the pre-event showed some twin arcades side-by-side, having a crossed intermingled feet in the middle. The magnetogram rotated to the exact time suggested that the region in the middle has a mixed polarity on both sides of the polarity-reversal line in the averaged field. The structure in the middle region is apparently not an arcade whose top part sags as a result of the weight of the dark filament mass loaded at the top of the otherwise convex loops, but seems to be loops connecting the outer sources to the opposite polarity in that region of mixed polarity in the neighborhood of the polarity-reversal line.

In Fig.1 (b) and (c), there appeared a bundle of bright threads along (somewhat shifted because it was high above the region) the polarity-reversal line, showing a "tooth brush" type connections to the photosphere. The "spine", a bundle of such threads, rose, and the brightening progressed towards the west. The "spine" progressively ballooned up at its west end, and finally formed a large protruding arch than opened at its top, just like the giant cusp observed on Jan. 25, 1992. It is clear in this event that the cusp-like feature was not a reconnecting arch.

We therefore re-examined the Jan. 25, 1992 giant cusp formation (Hiei et al. 1994) in comparison with simlar event or the Feb. 24–25, 1993. One conspicuous point was that the giant cusp showed its southern foot on the visible disk while it was forming. This is conspicuous because the main body of s cusp was rooted in a region that hail already rotated beyond the west limb. This suggests that the giant cusp we saw on Jan. 25, 1992 may have been a phenomenon of the same type as the Feb. 24–25, 1993 event which show the remnant of the "spine" (Fig.1(c)), rotated to the limb.

THE SKEW OF X-RAY CORONAL LOOPS OVERLYING Hα FILAMENTS

S.F. MARTIN[1] AND A.H. MCALLISTER
High Altitude Observatory
PO Box 3000, Boulder, CO 80307.

The basic magnetic structure surrounding photospheric magnetic inversion lines, where filament channels have developed, is comprised of axial fields along the inversion zone and overlying coronal arcades [Martin, 1990]. The overlying arcade is thought to be a necessary condition for the development of the axial fields of filament channels [Martin, 1990]. The formation of filament channel axial fields is evidenced by the alignment of the chromospheric fibrils and is known to be a condition for the formation of filaments [Martin, 1990]. Observations have recently led to a general empirical model of Hα filament channels [Martin, Bilimoria and Tracadas, 1994] and filaments [Martin and Echols, 1994]. One of the significant features of this model is that filaments and filament channels fall into two categories, having "dextral" and "sinistral" chirality, recognizable by the systematic orientation of the "barbs" along their two sides. Viewed along the axis, and away from the observer, a filament is called dextral or sinistral when the barbs veer to the right or the left, respectively, away from the filament axis.

The coronal arcades over the axial fields of the filament channels are often observable in soft x-rays, especially following filament eruptions. One of us (AHM) observed that the changing orientation of the loops during dynamic events has either a clockwise or counterclockwise sense as successive loops develop. This pattern of change is interpretable as a change in magnetic shear in altitude indicative of a smooth transition from the lower-lying axial fields to the overlying arcades.

We started our investigation with a survey of the rotational sense of the dynamic arcades. We define the term 'skew' to represent the angle between a tangent at the top of any coronal loop (in the plane of the loop) relative to the long axis of a filament, or the long axis of a filament channel or polarity inversion. Coronal loops, collectively or individually, can be 'right-skewed'

[1] Home Institution: Helio Research, 5212 Maryland Ave., La Crescenta, CA 91214

Y. Uchida et al. (eds.),
Magnetodynamic Phenomena in the Solar Atmosphere – Prototypes of Stellar Magnetic Activity, 497–498.

or 'left-skewed'. In an initial sample of 11 dynamic events right-skewed coronal loops, displaying a clockwise evolution, were found to corresponded to sinistral filaments, and left-skewed coronal loops, displaying counterclockwise evolution, correspond to dextral filaments. All four cases in the northern hemisphere exhibited skew increasing in the counterclockwise sense while all seven cases in the southern hemisphere displayed skew increasing in the clockwise sense.

We then chose a six month period, January through June 1992, and using Hα images from HAO's Mauna Loa Observatory and NOAA, selected all filaments at least 20 degrees in length, except those along the polar crown. Searching the *Yohkoh* SXT images for corresponding arcades we found that not all clear filaments had clear arcades, and visa versa, which reduced our sample from 70 to 33 events. We performed a double blind study in which one of us (SFM) determined the filaments' chirality and the other (AHM) determined the skew of the arcades, each working independently.

All of the dextral filaments were associated with left-skewed coronal loop systems and all of the sinistral filaments were related to right-skewed coronal loops. Thus, we have found evidence for a 1:1 relationship between the skew of the coronal arcades over a filament and the chirality of the filament. Thirteen (13) of the sixteen (16) dextral filaments were in the northern hemisphere while fourteen (14) of the seventeen (17) sinistral filaments were in the southern hemisphere, in agreement with the hemispheric pattern found by Martin, Bilimoria and Tracadas [1994]. The association between the skew of the coronal loops and the dextral/sinistral filament categories is stronger than the association with hemisphere. These results suggest that: 1) that there is a coherence to the large-scale fields in the filament channels and the surrounding corona; and 2) some of the processes which form filament channels and filaments are likely to be related to the generation of the coronal field in general.

We present our results, including detailed discussion of their implications for filament structure and long term coronal evolution, in a full paper submitted to the Astropysical Journal.

References

S. Martin. Conditions for the Formation of Prominences as Inferred from Optical Observations. In V. Ruzdjak and E. Tandberg-Hanssen, editors, *Dynamics of Quiescent Prominences*, pages 1–48. Springer-Verlag, 1990.

S. Martin, R. Bilimoria, and P. Tracadas. Magnetic Field Configurations Basic to Filament Channels and Filaments. In R. Rutten and C. Schrijver, editors, *Solar Surface Magnetism*, pages 303–338. Kluwer Academic Pub., 1994.

S. Martin and C. Echols. An Observational and Conceptual Model of the Magnetic Field of a Filament. In R. Rutten and C. Schrijver, editors, *Solar Surface Magnetism*, pages 339–346. Kluwer Academic Pub., 1994.

STUDIES OF SOLAR WIND ONSET IN CORONAL HOLES PLANNED FOR SOHO

A. H. GABRIEL[1], F. BELY-DUBAU[2] AND C. DAVID[1]

[1] *Institut d'Astrophysique Spatiale, CNRS/Université Paris XI, 91405 Orsay Cedex, France*
[2] *Observatoire de la Côte d'Azur, Lab. Cassini CNRS URA1362, BP 229, 06304 Nice Cedex 04, France*

Since it is now widely assumed that the main source of the high-speed solar wind arises from coronal holes, it is important to reconcile models for the solar wind with the physical conditions observed in holes. Unfortunately such observations are very limited, due to the low luminosity of holes and the consequent instrumental difficulties. Attempts to study conditions in holes have been made from Skylab and more recently from Yohkoh. Measurements made against the disk are dominated by the emission coming from the first scale-height at the base of the corona, and are likely to be contaminated by stray light coming from the brighter regions. With a combination of instruments on SOHO, we plan to measure the temperature in coronal holes above the limb as a function of height in the atmosphere. The resultant temperature gradients will help to understand the mechanism of wind acceleration.

Theoretical models of the wind are based upon two principle processes, the Parker thermally driven wind and momentum transfer to the plasma from MHD waves. The second process arises due to the increase in velocity of the waves as they enter a rarer medium, implying a reduction in the wave momentum. This loss is balanced by a transfer of momentum to the local plasma. Such a process is "reversible" in the thermodynamique sense. It does not require dissipation of wave energy, although this may also occur. For the Parker process to dominate, it is necessary to have high temperatures, of the order 3 to 4 10^6 deg K near the sun. Such values are not excluded by presently available data, although they have never been observed in the low corona in open-field regions. We propose to measure the temperature gradient above the limb in a coronal hole, using the spec-

Y. Uchida et al. (eds.),
Magnetodynamic Phenomena in the Solar Atmosphere – Prototypes of Stellar Magnetic Activity, 499–500.

tral line ratio 173Å/1038Å in oxygen VI. This ratio has been selected for its sensitivity to temperature in the range of interest and for the relatively high abundance of oxygen. These two lines will be observed by the CDS and SUMER instruments respectively. If the observed temperature is found to fall as the height increases, this will eliminate the possibility of a pure thermally driven wind.

Eclipse studies show that the density in holes can be lower than non-hole regions by a factor of 10 (Koutchmy 1975). This would imply that the intensity in UV lines is lower by a factor of 100. Such a high contrast can lead to contamination, either by non-hole material in the line-of-sight or by scattering in the instrument optics. Observations of hole shapes show that tangential viewing at the poles near solar minimum will normally avoid non-hole material in the line-of-sight. For the SOHO observing sequence proposed, the spacecraft must be rolled through 90 deg in order to orientate the spectrometer slits east-west at the pole. The slits are then stepped progressively out to 5 arcmin into the corona, in a sequence lasting a total of 13 hours.

Scattering from the surface of the spectrometer telescope mirrors will have the effect of contaminating the measurements made in the hole regions by light coming from the brighter and extended disk. For this effect, the important property is scattering at angles in the range 2 to 30 arcmins, rather than the smaller angles which effect the spatial resolution directly. For a given surface, the situation improves rapidly with increasing wavelength. We can readily show that for the grazing-incidence telescope of Skylab (Vaiana *et al* 1977), almost all of the signal from holes was due to scattering from non-hole regions. Observations have been reported (Haro, Tsuneta and Acton 1994) of temperature measurements in coronal holes from SXT on Yohkoh, but our estimates show that these should also be dominated by scattering from the telescope. This would lead to the predictable result that the deduced temperature is the same as in non-hole regions. Due to their higher brightness, it is however possible to make such measurements in non-hole regions (Foley, Acton and Culhane, 1996). The CDS telescope, with a surface smoothness no better than that of SXT, will have a lower scattering due to the longer wavelength used. Our modelling shows that for CDS and SUMER, scattering will be sufficiently low to make valid coronal hole diagnostic measurements in the UV region for the first time.

References

Foley C. A., Acton L. W. and Culhane J. L., 1996, (*this volume*)
Hara H., Tsuneta S. and Acton L. W., 1994, Publ. Astron. Soc. Japan, **46**, 493.
Koutchmy, S., 1975, Solar Phys., **51**, 374.
Viana, G. S. *et al*, 1977, Space Sci. Instr.,**3**, 19.

LOW-SPEED SOLAR WINDS FROM ACTIVE REGIONS

H. WATANABE, M. KOJIMA, AND Y. KOZUKA
Solar-Terrestrial Environment Laboratory, Nagoya University, Toyokawa 442, Japan

AND

H. MISAWA
Upper Atmosphere & Space Research Laboratory, Tohoku University, Sendai, 980, Japan

It is well known that the solar wind speed near the heliospheric current sheet is low. However, the speed in the low-speed belt does not show a uniform speed distribution. We sometimes detect a very low speed wind, less than 300 km/sec, from a given region on the Sun (Kojima *et al.*, 1991). Moreover, low-speed wind is distributed not only along the neutral line but also from low magnetic intensity regions on the source surface and/or coronal bright regions which do not correspond to the neutral line (Kojima and Kakinuma, 1987). It has not yet been established from which regions the very low speed wind originates.

Three Carrington rotations in late 1991 have been analyzed to study the relationship between the low-speed solar wind and active regions on the Sun. Figure 1 shows a synoptic chart of solar wind speed on the source surface derived from interplanetary scintillation (IPS) observations at distances of 0.2–0.4 AU. From this figure, we can find some features as follows: (1) The solar wind speed along the neutral line is generally low, less than 400 km/sec, except for longitudes 210°–270°; this might be caused by the effect of the south polar stream extending toward the equator. (2) The low-speed solar wind is distributed not only along the neutral lines, but also in other low latitude regions. (3) Very low speed regions, less than 300 km/sec, are observed at longitudes 350°–110°, 160°–190°, 210°–250°, 290°–300°; the first two regions are near the neutral line, but the last two regions are not located near the neutral line. These four regions are more likely related to the existence of sunspots rather than the neutral line. Figure 2 shows a superposition of a *Yohkoh* SXT image (grey code) and coronal open

Y. Uchida et al. (eds.),
Magnetodynamic Phenomena in the Solar Atmosphere – Prototypes of Stellar Magnetic Activity, 501–502.

magnetic field lines. Magnetic field lines near the longitudes 240° and 300° in the low latitude show converging structures toward the source surface. These regions coincide with the very low speed wind regions in Figure 1.

These results suggest the possibility that some active regions supply the very low speed wind (Watanabe *et al.*, 1995). Uchida *et al.* (1992) pointed out from *Yohkoh* observations that the corona above active regions expands occasionally, contrary to the commonly accepted ideas. It is possible that such an expanding active region could be one of the origins of the very low speed wind. One interesting point is that even the very low speed wind not originating near the neutral line does not disappear near the Sun, surviving up to at least 0.2–0.4 AU. Nakagawa (1993) reported that the sources of peculiar interplanetary clouds observed at 1AU often coincide with active regions, some of which are not located under the neutral line at 2.5 Rs. The flow speed of the clouds is generally low, so this is consistent with our results.

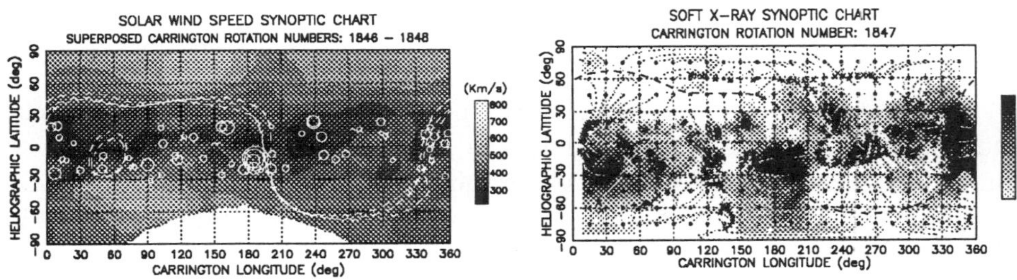

Figure 1 (Left panel) Synoptic chart of solar wind speed on the source surface at 2.5 Rs. This is derived with IPS observations at distances of 0.2–0.4 AU. Dashed lines are Stanford neutral lines, and circles are sunspots.

Figure 2 (Right panel) Superposition of Yohkoh SXT image (grey code) and coronal open magnetic field lines (grey lines) and the neutral line at 2.5 Rs (dashed line), which were inferred from photospheric magnetic field observations (Kitt Peak Observatory) using a potential-field model.

References

Kojima, M., and Kakinuma, T. (1987) *J. Geophys. Res.*, **92**, 7269–7279.

Kojima, M., Washimi, H., and Misawa, H. (1991) *Proc. Pacific regional STEP Conference*, Taipei, Taiwan, pp. 23–26.

Nakagawa, T. (1993) *Solar Phys.*, **47**, 169–197.

Uchida, Y., McAllister, A., Strong, K. T., Ogawara, Y., Shimizu, T., Matsumoto, R., and Hudson, H. S. (1992) *Publ. Astron. Soc. Japan*, **44**, L155–L160.

Watanabe, H., Kojima, M., Kozuka, Y., Yamauchi, Y., and Misawa, H. (1995) *Solar Wind Eight*, in press.

DETECTION OF ACTIVE REGION EXPANSION IN INTERPLANETARY SPACE BY SAKIGAKE SPACECRAFT AT 1 AU

T. NAKAGAWA
Tohoku Institute of Technology
35-1 Yagiyama kasumi-cho, Taihaku, Sendai 982, Japan

AND

Y. UCHIDA
Science University of Tokyo
1-3 Kagurazaka, Shinjuku, Tokyo 162, Japan

1. Introduction

The Japanese interplanetary spacecraft *Sakigake* found peculiar interplanetary objects whose magnetic fields deviate largely from the Archimedian field in the Parker wind. In each of the objects, the magnetic field was highly variable but was restricted in a certain plane at least locally in the vicinity of the spacecraft; thus they were called "planar magnetic structures" (Nakagawa et al., 1989). A planar field variation suggests that the spacecraft encountered part of a large scale structure of sheared magnetic field. The events had higher plasma density and slower bulk speed than those of the average solar wind.

The "planar magnetic structures" were enigmatic because they had no relationship with prominent phenomena on the solar surface such as flares, prominences nor dark filament disappearances (Nakagawa, 1993). Sometimes they were observed recurrently at an interval of 27 days. Since the discovery of active region expansions by *Yohkoh* (Uchida et al., 1992), the question has arisen whether the "planar magnetic structures" might be interplanetary extensions of the active region expansions. A systematic search of possible relationships was made by using data obtained by *Sakigake* and *Yohkoh* spacecraft during the period from January 6 to November 11, 1993, when there were not too many active regions on the visible solar hemisphere.

Y. Uchida et al. (eds.),
***Magnetodynamic Phenomena in the Solar Atmosphere – Prototypes of Stellar Magnetic Activity,* 503–504.**

2. Detection of Interplanetary Planar Magnetic Structures

A "planar magnetic structure" is detected in interplanetary magnetic field data by using two criteria for the field component normal to the plane $B_n < 0.1B$ and for the standard deviation of the magnetic field vector $\sigma_{\mathbf{B}} > 0.7B$, set to pick up planar and variable magnetic field vectors. Here B is the magnitude of the magnetic field throughout each event.

According to the criteria, five typical "planar magnetic structures" were found on May 18, June 3, June 11, July 1, and July 15, 1993.

3. Solar Sources of Planar Magnetic Structures

The structures were traced back onto the Sun on the assumption that the solar wind speed increased linearly within the first 3 solar radii, from 10 km/s up to the real value measured at 1 AU. The solar wind plasma was assumed to be launched radially outward from the Sun, although this is not always the case for real solar wind. We found that the sources of 4 events coincided with NOAA low-latitude active regions which appeared within 5 degrees from the equator. The 80% coincidence is extremely high with respect to the rate of accidental coincidence, 7%, of Sakigake windows of solar wind observation with active regions.

In the *Yohkoh* Soft X-ray images at or near the calculated time of the launch of each object, we found expansion activities at the active regions near the central part of the Sun's disk from which "planar magnetic structures" were supposed to be launched. Very faint loop-like structures were often observed, and the loops, especially their outer portions, seemed to expand repeatedly into interplanetary space. For the interplanetary object which was not traced back to any NOAA active region, we found expansion activity of faint loops. No flares were observed in any of the sources. These observations support the idea that active-region expansions can extend into interplanetary space to be observed as "planar magnetic structures" at 1AU.

References

Nakagawa, T. (1993) Solar source of interplanetary planar magnetic structures, *Solar Physics*, **147**, pp. 169–197.

Nakagawa, T., Nishida, A., and Saito, T. (1989) Planar magnetic structures in the solar wind, *J. Geophys. Res.*, **94**, pp. 11761–11775.

Uchida, Y., et al. (1992) Continual expansion of the active-region corona observed by the Yohkoh soft X-ray telescope, *Publ. Astron. Soc. Japan*, **44**, pp. L155–L160.

KELVIN-HELMHOLTZ INSTABILITY IN CORONAL RAYS

O. T. MATSUURA AND E. PICAZZIO
Instituto Astronômico e Geofísico, Universidade de São Paulo
Caixa Postal 9638, São Paulo, SP, Brasil, CEP 01065-970

AND

H. SHIGUEOKA
Instituto de Física, Universidade Federal Fluminense
Caixa Postal 100294, Niterói, RJ, Brasil, CEP 24020-290

We carried out a morphological analysis on fine structures recorded in large-scale color photographs of white-light corona obtained in the eclipse of July 11, 1991, in Hawaii, Mexico and Brasil by MICE (see Zirker *et al.*, 1992). Images taken through a standard V filter were digitized with a pixel size corresponding to 21,000 km on the Sun. The MadMax algorithm (Koutchmy and Koutchmy, 1989) was applied for enhancing the fine and faint structures. The best results were obtained by reapplication of the algorithm twice. Among several phenomena disclosed by the comparative analysis of the three images, the most striking one was the presence, in several rays (*long, and almost radially oriented structures*), of undulations with an average wavelength of $\simeq 150,000$ km. The wavenumber, k, multiplied by the halfwidth, a, of the structure, was $ka \simeq 1$. In every case, the wave amplitude relative to a was $\simeq 1/3$ (non-linear). In Fig. 1, the most remarkable undulations in the Western sector are depicted over the sketch of Fig. 1b from Zirker *et al.* (1992). The undulation (a) of the neutral sheet above the SW helmet streamer is the highest ($2.5R_o < h < 3.3R_o$). All others (b-f) were found lower ($0,9R_o < h < 1.6R_o$). It is noteworthy that the undulations persisted in the three pictures.

Tests with MadMax on artificially generated gradients, ruled out that the undulations could be artifacts. The fact that, in a given picture, some fine structures appeared wavy, while others did not, also favors the undulations' reality. Only for lack of space, references are omitted here of other eclipse images that also seem to present wavy structures.

We modeled each fine structure as a magnetic neutral layer (Koutchmy *et al.*, 1994) of thickness $2a$, confined between surfaces of antiparallel magnetic tangential discontinuities. At equilibrium, the vector quantities (velocity and magnetic field) have only a longitudinal component along the solar radial direction. For simplicity, the mass density ρ, the temperature T, the longitudinal velocity V and the magnetic field B are constant inside the sheet, as well as outside, with a finite jump at

Y. Uchida et al. (eds.),
Magnetodynamic Phenomena in the Solar Atmosphere – Prototypes of Stellar Magnetic Activity, 505–506.

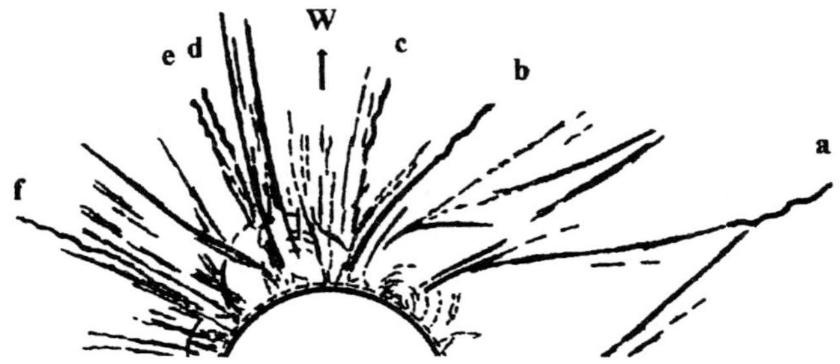

Figure 1. Undulations (a-f) in coronal rays of July 11, 1991 eclipse

the boundary surfaces. The system was linearly analysed for the Kelvin-Helmholtz instability.

For undulation (a) we adopted the following parameters (the indices i and e refer to the plasma internal or external to the neutral layer): $\rho_i = 1.67 \times 10^{-17}\text{gcm}^{-3}; \rho_e = 1.67 \times 10^{-20}\text{gcm}^{-3}; V_i = 4.0 \times 10^7\text{cms}^{-1}; V_e = 2.0 \times 10^7\text{cms}^{-1}$. Yet for other undulations we adopted: $\rho_i = 1.67 \times 10^{-16}\text{gcm}^{-3}; \rho_e = 5.01 \times 10^{-19}\text{gcm}^{-3}; V_i = 1.5 \times 10^7\text{cms}^{-1}; V_e = 5.0 \times 10^7\text{cms}^{-1}$. Always $B_i = 0$, and B_e is such as to guarantee the total pressure equilibrium in the transverse direction, for $T_i = 2 \times 10^6$ K and $T_e = 1.2 \times 10^6$ K.

For the observed wavelength, among the six distinct roots ω [s^{-1}] of the dispersion relation, three were complex. The relevant unstable mode grows in a time scale $\simeq 3.5$ min, which is shorter than the oscillation period: 6 min for (a), and 17 min for undulations (b-f). But, as the amplitude of the unstable mode grows, the magnetic tension grows too, and starts to act as a restoring force. As a consequence, the mode self-stabilizes in the non-linear regime. For incompressible plasmas in a cylinder of radius a, the largest relative self-stabilized wave amplitude is 1/3 when $ka = 1$ (Ershkovich and Chernikov, 1973). These conditions were met in our observations. If our interpretation proves to be right, the observed undulations are non-linear hydromagnetic waves excited by Kelvin-Helmholtz instability. This investigation may offer a new diagnostics of shearing flows and plasma accelerations near the magnetic neutral sheets.
Acknowledgement. OTM obtained support from FAPESP (Proc. 95/0389-5) to attend the IAU Colloq. No.153.

References

Ershkovich, A.I. and Chernikov, A.A. (1973) Non-linear waves in Type-I comet tails *Planet. Spa. Sci.*, **Vol. no. 21**, pp. 663–670

Koutchmy, O. and Koutchmy, S. (1989) Optimum Filter and Frame Integration. Application to Granulation Pictures, in O. von der Luhe (ed.), *Proceedings of the 10th. NSO/SPO Workshop on High Spatial Resolution Solar Observations*, pp. 217–

Koutchmy, S., Molodensky, M.M. and Vial, J.-C (1994) On the 3D solar corona structure, in V. Rusin, P. Heinzel and J.-C. Vial (eds.), *Solar Coronal Structures, Proceedings of the IAU Colloq. 144*, pp. 585–588

Zirker, J.B, Koutchmy, S., Nitschelm, C.,Stellmacher, G., Zimmermann, J.P., Martinez, P., Kim, I., Dzubenko, N., Kurochka, L., Makarov, V., Fatianov,M.,Rusin, V., Klocok, I. and Matsuura, O.T. (1992) Structural changes in the solar corona at the July 1991 eclipse, *Astron. Astrophys.*, **Vol. no. 258**, pp. L1–L4

FORMATION OF CONTACT BINARIES VIA A MAGNETIZED WIND, A PARAMETER-FREE MODEL

K. STĘPIEŃ
Warsaw University Observatory
Al. Ujazdowskie 4, 00-478 Warszawa, Poland
e-mail: kst@astrouw.edu.pl

Based on the model of angular momentum loss (AML) via a magnetized wind (Mestel, 1984) an empirical formula was derived and calibrated for the AML rate of single, solar type stars (Stępień, 1988,1991):

$$\frac{dP_{rot}}{dt} = (6 \pm 3) \times 10^{-9} P_{rot}\, e^{-0.2 P_{rot}} , \tag{1}$$

where time t is in years and P_{rot} in days. Note that, apart from empirical uncertainty, the above formula contains no adjustable free parameters. It reproduces well the Skumanich law $P_{rot} \sim t^{1/2}$ for ages 0.1 Gyr $\lesssim t \lesssim 10$ Gyr and predicts correctly the presently observed spin-down rate of the Sun.

Let us consider a detached close binary with components losing AM according to the formula (1). Its orbital period evolves according to the following relation (Stępień, 1995):

$$\frac{dP_{orb}}{dt} = -(6 \pm 3) \times 10^{-10} P_{orb}{}^{-1/3} e^{-0.2 P_{orb}} , \tag{2}$$

where we adopted masses of the components $M_1 = M_2 = 0.6 M_\odot$ and their radii $R_1 = R_2 = 0.6 R_\odot$. The value of the numerical coefficient in (2) depends on masses of the components and will decrease about twice for $M_1 = M_2 = 1 M_\odot$.

The evolution of orbital period of a detached binary was computed for several different initial values of the period and different masses of the both components (Stępień, 1995). A minimum value of the orbital period of a zero age main sequence (MS) binary was adopted at 2 days. When an orbital period decreases down to a value of 0.3 day it is assumed that a contact system is formed.

The results show that it takes more than 2 Gyr to form a contact system, even under the most favorable conditions (a low mass binary with an

Y. Uchida et al. (eds.),
Magnetodynamic Phenomena in the Solar Atmosphere – Prototypes of Stellar Magnetic Activity, 507–508.

assumed minimum value of the orbital period and a maximum AML rate allowed by Eq. [2]). The presence of contact systems in clusters younger than 2 Gyr indicates that some binaries undergo an exceptionally large AML prior to MS. For ages longer than about 4 Gyr a substantial increase of contact systems is expected among a sample of coeval stars (e. g. a stellar cluster). This result is in a good agreement with the observational data on stellar clusters: the average number of W UMa type stars detected in clusters with age $\leq$ 4 Gyr is 1.3 whereas it is 9 for ages > 4 Gyr (Kaluzny and Rucinski, 1993). Yet, even after 10-12 Gyr. i.e. the age of the galactic disk, only binaries with initial orbital periods not longer than about 4-5 days can form contact systems.

A low rate of formation of contact systems has important implications for a duration of a contact phase. Assuming a uniform star formation rate over the galactic disk age we can estimate the fraction of all binaries that have reached contact during the past 12 Gyr. The result is close to the observed frequency of contact binaries among photometrically surveyed stars (Rucinski, 1994), which indicates that a large fraction of all contact systems formed in the galactic disk stays still in a contact phase. This requires several Gyr for a duration of the contact phase. A similar time scale is obtained if one ignores any possible differences between AML mechanism of a detached binary and a contact system and applies Eq. (2) to the latter. This similarity argues for a uniform AML mechanism operating in single active stars, cool detached binaries and late type contact systems.

Combining time of approach to contact and duration of a contact phase we get a value which is very likely *shorter* than MS life time of either component in case of a low mass binary. After coalescence such binaries should form rapidly rotating MS stars (blue stragglers?). For massive binaries, with components masses of the order of one solar mass, this combined time is very likely *longer* than the life time of either component. After a coalescence, such binaries are expected to form rapidly rotating single giants of FK Com type.

This research was partly supported by a grant from Committee of Scientific Research No. 2P 304.004.07.

References

Kaluzny, J. and Rucinski, S.M., 1993. In 'Blue Stragglers', ed. R.A. Saffer, ASP Conf. Series vol. **53**, ASP, San Francisco, p.164.

Mestel, L., 1984. In 'Cool Stars, Stellar Systems, and the Sun', eds. S.L. Baliunas and L. Hartmann, Springer-Verlag, Berlin, p.49.

Rucinski, S.M., 1994. *Publ. Astr. Soc. Pacific*, **106**, 462.

Stępień, K., 1988. *Astrophys. J.*,**335**, 907.

Stępień, K., 1991. *Acta Astron.*, **41**, 1.

Stępień, K., 1995. *Mon. Not. R. astron. Soc.*, in press.

FORMATION OF RECONNECTION-DRIVEN JETS WITH DISK ACCRETION ONTO MAGNETIZED YOUNG STARS

S. HIROSE AND Y. UCHIDA
Department of Physics, Science University of Tokyo
1-3 Kagurazaka, Shinjuku-ku, Tokyo 162, Japan
K. SHIBATA
National Astronomical Observatory of Japan
Mitaka, Tokyo 181, Japan
AND
R. MATSUMOTO
Department of Physics, Chiba University
1-33 Yayoi-cho, Inage-ku Chiba 263, Japan

High-velocity outflows (optical jets or T Tauri winds) are very common phenomena with young stars in the stage of disk accretion. Thus these outflows are considered to be powered by disk accretion, and to originate from the vicinity of young stars from the energetical point of view (Cabrit *et al.* 1990, Edwards *et al.* 1993a). On the other hand recent observations suggest that these young stars have magnetospheres, which truncate the accretion disk at several stellar radii (Montmerle *et al.*, 1993). This means that the interaction between the magnetosphere and the accretion disk will have relation to the formation of high-velocity outflows. We took a numerical approach (MHD simulation) to this problem.

In our model, the central star has its own magnetosphere and also the accretion disk is penetrated by the large-scale magnetic field, both of the magnetic fields originating from the same interstellar magnetic field (Uchida and Shibata, 1984). Thus there is X-type magnetic neutral point at the interface between the magnetosphere and the accretion disk, which will play an important role on the mass transfer from the accretion disk to the star and the formation of jets.

Numerical results are summerized as follows. In this first numerical attempt, we treat only the case in which the star does not rotate because of numerical limitations. The accretion disk loses its angular momentum by

Y. Uchida et al. (eds.),
Magnetodynamic Phenomena in the Solar Atmosphere – Prototypes of Stellar Magnetic Activity, 509–510.

magnetic braking and then the disk accretion drives the magnetic reconnection between the closed magnetospheric fields and the large-scale magnetic fields threading the accretion disk. We assumed that the enhancement of the diamagnetic current by the disk accretion induces anomalous resistivity around the neutral point (Sato and Hayashi 1979). Most of the mass of the accretion disk is transfered to the magnetosphere through the diffusion region around the neutral point and magnetically braked by the non-rotating star and finally fall towards the polar crown along the stellar fields. On the other hand, the mass at the surface of the accretion disk is accelerated by Lorentz force associated with the magnetic reconnection (sling-shot mechanism) and forms high-velocity bipolar jets. The magnetic reconnection changes the topology of the magnetic fields and the jet is ejected along the opened magnetic fields of the star. Thus the inflow and the outflow coexist along the same line of forces. Note that the above process is not steady but it will continue as long as the disk accretion drives the magnetic reconnection.

The main effect of the stellar rotations which was not taken into account in this simulations is the magnetic braking of the star as the reaction of the magneto-centrifugal acceleration of the reconnection-driven jet along the stellar open magnetic fields. There is a possibility that this magnetic braking by the jet cancels the angular mometum transfer from the accretion disk to the star, which may explain the observed rather slow rotation of young stars (Edwards *et al.*, 1993b).

References

Cabrit,S., Edwards,S., Strom,S.E., and Strom,K.M. 1990, *Astronphys. J.*, **354**, 687.

Edwards,S., Ray,T., and Mundt,R. 1993a, in *Protostars and Planets III*, eds.E.H.Levy and J.I.Lunine (Tucson:University of Arizona Press), p567.

Edwards,S., Strom,S.E., Hartigan,P., Strom,K.M., Hillenbrand,L.A., Herbst,W., Attridge,J., Merrill,K.M., Probst,R., and Gatley,I. 1993b, *Astron. J.*, **106**, 372.

Hirose,S., Uchida,Y., Shibata,K., and Matsumoto,R. 1995, *in preperation.*

Montmerle,T., Feigelson,E.D., Bouvier., and Andre,P. 1993, in *Protostars and Planets III*, eds.E.H.Levy and J.I.Lunine (Tucson:University of Arizona Press), p689.

Sato,T., and Hayashi,T. 1979, *Phys. Fluids*, **22**, 1189.

Uchida,Y., and Shibata,K. 1984, *Publ. Astron. Soc. Japan*, **36**, 105.

MAGNETICALLY DRIVEN JETS FROM ACCRETION DISKS: NONSTEADY AND STEADY SOLUTIONS

T. KUDOH AND K. SHIBATA
National Astronomical Observatory of Japan
2-21-1, Osawa, Mitaka, Tokyo, 181, Japan

We have performed time-dependent 1.5D MHD simulations of astrophysical jets which are magnetically driven from Keplerian disks, in order to study the origin and structure of optical jets ejected from protostars. We clarify the physical properties of the nonsteady jets that had been studied by 2.5D MHD simulations like Shibata & Uchida (1986) (hereafter SU). The 1.5D simulation has the merit that we can perform simulations over many disk orbital periods with large computational regions in wide range of parameters. The formulation of the 1.5D MHD is the same as that of Hollweg et al. (1982), who studied the propagation of nonlinear Alfvén wave in the solar atmosphere.

Two types of nonsteady jets, which are the same types of jets found by SU, appear in the simulation (Figure). We find that the jet ejected from the disk (*disk-jet*) have the same properties of the steady magnetically driven jet that has been investigated by many authors (e.g. Pudritz & Norman 1986, Shu et al. 1994). The results are as follows (Kudoh & Shibata 1995b).

- The mass loss rate strongly depends on the inclination angle between the disk surface and the magnetic field line (Blandford & Payne 1982, Cao & Spruit 1994).
- The scaling law that is known as Michel's minimum energy solution is also satisfied.
- The magnetic energy dependence of the mass loss rate of the nonsteady jet is consistent with that of the steady one (Kudoh & Shibata 1995a). The mass loss rate decreases with decreasing the magnetic energy at the disk surface,

$$\dot{M} \propto E_{\rm mg}^{\alpha}, \tag{1}$$

Y. Uchida et al. (eds.),
Magnetodynamic Phenomena in the Solar Atmosphere – Prototypes of Stellar Magnetic Activity, 511–512.

where E_{mg} is the poloidal magnetic energy normalized by the gravitational energy at the disk. The parameter α ranges from 0 to 0.5 ($\alpha \to 0$ as $E_{mg} \to \infty$, $\alpha \to 0.5$ as $E_{mg} \to 0$). By using the above relation, we can estimate the temperature (T_0) and the strength of the poloidal magnetic field (B_{p0}) at the disk. The mass loss rates of the optical jets are observed as $\dot{M} \sim 10^{-9} - 10^{-7} M_{\odot}$/year. It shows that $T_0 > 10^3$K and $B_{p0} > 1$G.

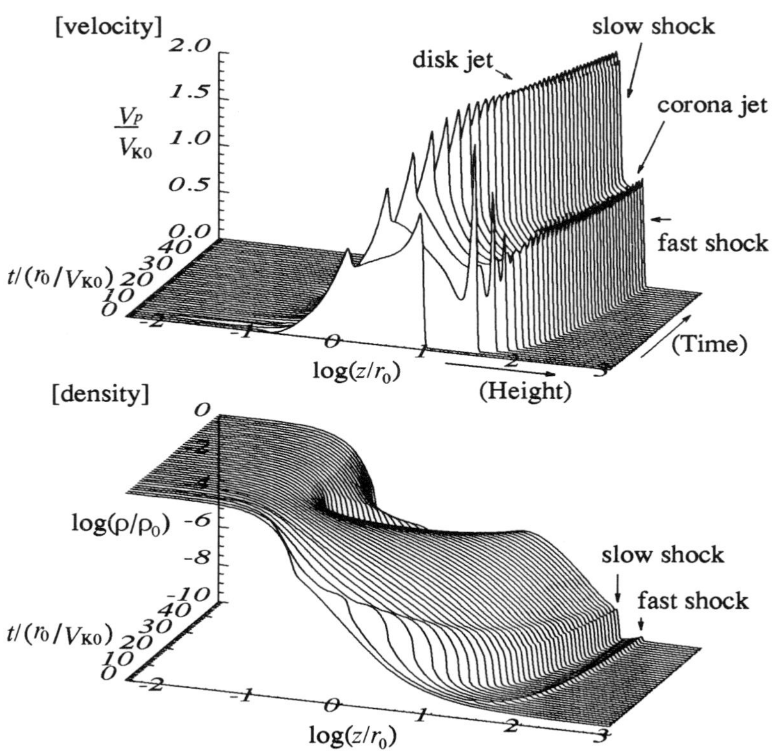

References

Blandford, R.D., & Payne, D.G., 1982, MNRAS, 199, 883
Cao, X., & Spruit, H.C., 1994, A&A, 287, 80
Hollweg, J.V., Jackson, S., & Galloway, D., 1982, Solar Phys., 75, 35
Kudoh, T., & Shibata, K., 1995a, ApJ Let., 452, L41
Kudoh, T., & Shibata, K., 1995b, in preparation
Pudritz, R.E., & Norman, C.A., 1986, ApJ, 301, 571
Shibata, K., & Uchida, Y., 1986, PASJ, 38, 631 (SU)
Shu, F., Najita, J., Ostriker, E., Wilkin, F., Ruden, S., & Lizano, S., 1994, ApJ, 429, 781

Poster Papers

Posters for Session III

ACTIVE REGION EVOLUTION AND FLARE ACTIVITY – A CASE STUDY OF NOAA 7260

N. NITTA

Lockheed Palo Alto Research Laboratory, O/91-30, B/252, 3251 Hanover Street, Palo Alto, CA 94304, U.S.A. and Applied Research Corporation, 8201 Corporate Drive, Landover, MD 20785, U.S.A.

L. VAN DRIEL-GESZTELYI

Observatoire de Paris, Section de Meudon, DASOP, 92 125 Meudon, Principal Cedex, France

K. D. LEKA

High Altitude Observatory/NCAR, P.O.Box 3000, Boulder, CO, 80302, U.S.A.

AND

H. S. HUDSON

Institute for Astronomy, University of Hawaii, 2680 Woodlawn Drive, Honolulu, HI 96822, U.S.A. and Institute of Space and Astronautical Science, 3-1-1 Yoshinodai, Sagamihara, Kanagawa 229, Japan.

Extended Abstract

Taking advantage of good coverage of X-ray and groundbased data for NOAA active region 7260 (AR 7260), we have studied flare activity and the characteristics of individual flares in the context of the region's evolution. AR 7260 consisted of a large preceding spot and an emerging flux region (EFR) which showed remarkable growth. As expected, the EFR produced many flares as it became complicated due to repeated flux emergence. For each flare, we have identified a pair of bipoles that were likely to interact with each other. These flares show a good correlation between the temperature of the brightest core and the X-ray flux from the entire flaring

Y. Uchida et al. (eds.),
Magnetodynamic Phenomena in the Solar Atmosphere – Prototypes of Stellar Magnetic Activity, 515–516.

area, both of which probably peaked when the magnetic shear of the EFR became maximum.

The large preceding spot was already in a decaying stage, as it was surrounded by outward moving magnetic elements similar to moving magnetic features (but distributed anisotropically). Like these features, different sectors around the preceding spot produced flares at different times. The spot itself consisted of two components separated by a light bridge. Most flares around the spot seemed to be connected to the southern component.

Including the ones in the emerging flux region, the flares in AR 7260 generally appeared to be confined, rather than eruptive, except for those that were connected to the northern component of the preceding spot. Both X-ray and Hα images show that this area had field lines that extended longer. It is likely that many eruptions in AR 7260 in the SXT time sequence images are attributable to larger scale shear, i.e. shear between the large spot and the EFR, which may not have directly caused the apparently confined flares.

References

Leka, K. D., van Driel-Gestelyi, L., Nitta, N., Canfield, R. C., Mickey, D. L., Sakurai, T., and Ichimoto, K.: 1994 *Solar Phys.* **155**, 301.

Nitta, N., van Driel-Gestelyi, L., Leka, K. D., Strong, K. T., Lemen, J. R., Freeland, S. L., Kosugi, T., Sakurai, T., Ichimoto, K.,: 1995, *Solar Phys.*, submitted.

RELATIONSHIP BETWEEN COLD AND HOT POST-FLARE LOOPS AND THE IMPACT ON THE RECONNECTION FLARE MODEL

J. F. SEELY AND U. FELDMAN
Naval Research Laboratory
Washington DC 20375 USA

The relationship between cold and hot post-flare loops was studied using the 7 September 1973 and 15 January 1974 flare images recorded by the Skylab SO82 spectroheliograph. The images were of intense emission lines in the extreme ultraviolet region that spanned the 1×10^4 K (He I) to 3.2×10^6 K (Ni XVIII) temperature range. The images of the 15 January 1974 flare at 1240 UT are shown in the Figure. The detailed analysis of the images does not support the widely held notion that cold loops are *always* smaller than hot loops, lie below hot loops, and are similar in shape. It was found that the coldest and hottest loops often differ significantly in size and shape. Based on the analysis of a time sequence of Skylab images of cold and hot loops, there is no evidence that loop systems expand in a discontinuous manner, as would be the case if higher loops were sequentially formed and activated by reconnection of the magnetic field. It was found that the individual coronal loops expand in a gradual manner. These conclusions are consistent with images of post-flare loop systems in the 10^7 K range that were recorded by the soft x-ray telescope on the *Yohkoh* spacecraft. Previous observations and interpretations of post-flare loop systems, which led to the formulation of the reconnection flare model, were re-examined. In light of the recent high-quality imagery that spans the temperature range from 10^4 to 10^7 K, it is concluded that the observations do not support all aspects of the reconnection flare model as presently articulated. Elements of the reconnection flare model that are inconsistent with the modern observations should be reconsidered. A detailed description of this study of Skylab and *Yohkoh* post-flare loops has been accepted for publication in the Astrophysical Journal.

Y. Uchida et al. (eds.),
Magnetodynamic Phenomena in the Solar Atmosphere – Prototypes of Stellar Magnetic Activity, 517–518.

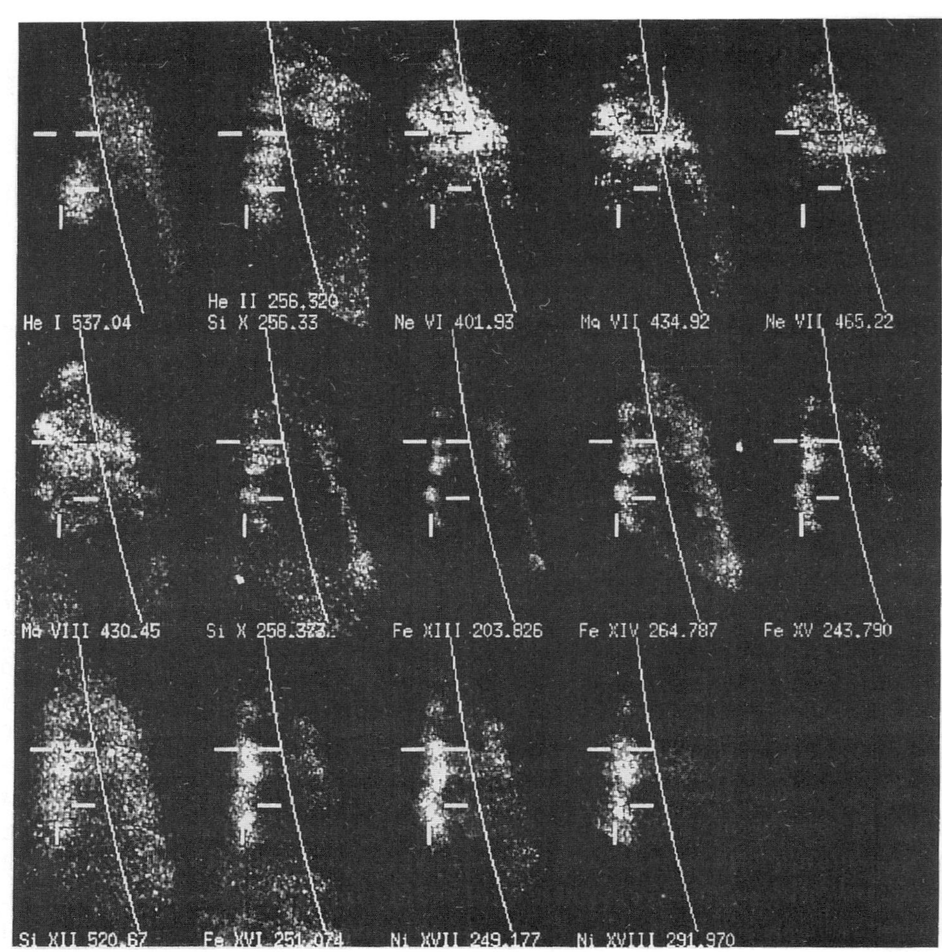
He I 537.04
He II 256.320
Si X 256.33
Ne VI 401.93
Mg VII 434.92
Ne VII 465.22
Mg VIII 430.45
Si X 258.373
Fe XIII 203.826
Fe XIV 264.787
Fe XV 243.790
Si XII 520.67
Fe XVI 251.074
Ni XVII 249.177
Ni XVIII 291.970

PSEUDO-TWO-DIMENSIONAL HYDRODYNAMIC MODELING OF FLARE LOOPS

K. HORI*, T. YOKOYAMA, T. KOSUGI, AND K. SHIBATA
National Astronomical Observatory, Mitaka, Tokyo 181, Japan
**also Tohoku University, Sendai, Miyagi 980-77, Japan*

It has been argued that the solar flare is a phenomenon in which magnetic energy is released through reconnection in the top portions of coronal loops. This view has been recently strengthened from observations with the X-ray telescopes aboard *Yohkoh*, owing to their high spatial and time resolution. Flaring loop models with chromospheric evaporation have been developed for many years (e.g., Nagai 1980), and have succeeded in explaining some of the observed features. However, so far these models were based on 1D hyrdodynamic simulation along a single loop, and are thus not consistent with the reconnection model. In the reconnection model, magnetic field lines reconnect successively to form 2D (or 3D) loop configurations. It is now necessary to extend the previous 1D evaporation model to a 2D or 3D model, to allow for the effects of magnetic reconnection. In this study, we will try to extend the previous 1D evaporation flare model to 2D, using a *"pseudo-two-dimensional"* approach.

Let's assume a system of loops which are formed successively from the inside to the outside through a continuous reconnection. Our flare system consists of (i) thermally isolated, and (ii) dynamically rigid multiple loops. First, we solve one-dimensional hydrodynamic equations including both thermal conduction and radiative cooling effects for each loop. Next, we reproduce a two-dimensional flaring loop system by putting them together. In this sense, our model is a *"pseudo-two-dimensional"* one. We assume that the energy release takes place at each loop top successively from the innermost to the outermost loop with some time delay, which reproduces the successive formation of closed flaring loops due to the continuous reconnection. In order to compare results with the observations, we also derived the surface brightness distribution from the simulation results, which would be detected by *Yohkoh/SXT* with Be119μm filter.

The results are as follows (see Figure 1):

Y. Uchida et al. (eds.),
Magnetodynamic Phenomena in the Solar Atmosphere – Prototypes of Stellar Magnetic Activity, 519–520.

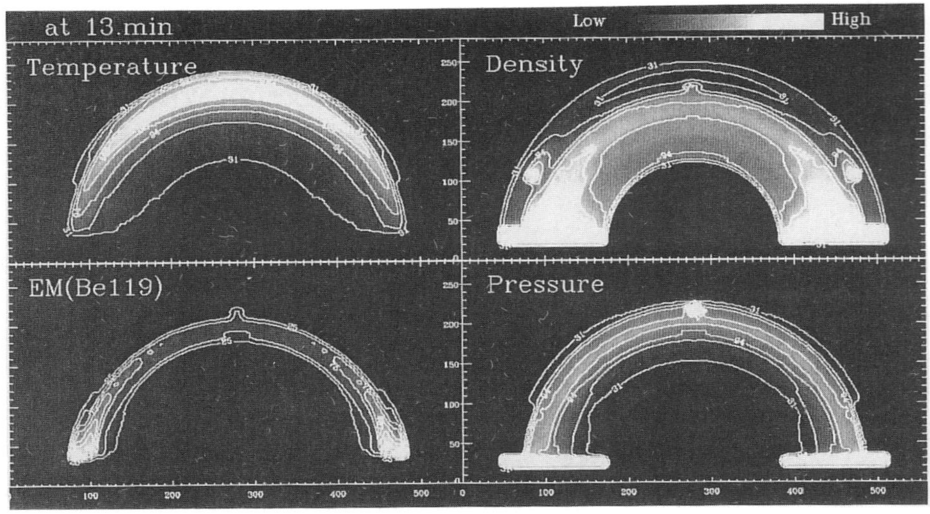

Figure 1. Snapshot at 13minutes after flare heat input. To hide the rather artificial effect which comes from the insufficient resolution of discrete loops, smoothing method is used weakly on each contour map.

(1) The appropriate timing of energy release in each loop causes the successive chromospheric evaporation, which make the highest pressure region continue to be located at the growing loop top.

(2) Iso-temperature contours always lie in the upper part of the high density loop, which would correspond to the bright loop seen through the Be119μ filter.

(3) Contour maps of both temperature and density are not necessarily along the fixed magnetic field lines.

(4) Through the Be119μm filter, the loop brightening begins from footpoints which continue to be the brightest throughout the evolution, with a weaker brightening at the loop top.

References

F.Nagai, (1980) A model of hot loops associated with solar flares , *Solar Phys.*, **68**, pp. 351–379

K.Hori, T.Yokoyama, T.Kosugi, and K.Shibata (1995) Pseudo-Two-Dimensional Hydrodynamic Modeling of Flare Loops (*in preparation*)

INITIAL DYNAMICS IN A LOOP FLARE OF APRIL 22, 1993

S. YASHIRO, N. KOHARA, Y. UCHIDA, S. HIROSE AND S. CABLE
Department of Physics, Science University of Tokyo
Shinjuku-ku, Tokyo 162, Japan

AND

T. WATANABE AND T. KOSUGI
National Astronomical Observatory
Mitaka, Tokyo 181, Japan

The initial dynamical behavior in a loop flare of Apr 22, 1993 is examined by using the *Yohkoh*/BCS after separating the effect of another small flare, and the results show that there existed a clear upflow before any trace of the spiky non-thermal hard X-ray burst. This confirms the long-standing controversy about the electron bombardment picture. We scrutinized the period prior to the hard X-ray impulsive burst by HXT, and found that there was a hard X-ray looptop source with significantly smaller hardness ratio, before the start of the steep rise in the intensities of hard X-rays. This was suspected to be identifiable with the source postulated in the heat conduction-evaporation models, but such interpretation encounters a difficulty because that looptop source was *preceded* by a pair of weaker footpoint sources about 20 seconds prior to it, when the hard X-ray profile had a smooth thermal-like profile with distinctly softer hardness ratio.

It has been reported that blue wings of remarkable width appear in the X-ray emission lines *before* any hard X-ray spiky bursts started, from several high energy flares measured by Bragg Crystal Spectrometers on SMM and *Hinotori* satellites (Doschek *et al.* 1983, Tanaka *et al.* 1983). We noted that this could be a very vital clue for deciding which of the proposed models for the loop flares was viable (Uchida and Shibata 1988, 1990), but this effect tended to be ignored by the researchers working along the electron bombardment-evaporation, or heat conduction-evaporation models. *Yohkoh*, however, reconfirmed that this is a common feature with high energy loop flares, and we have to investigate this more seriously.

Y. Uchida et al. (eds.),
Magnetodynamic Phenomena in the Solar Atmosphere – Prototypes of Stellar Magnetic Activity, 521–522.

We have looked into this effect in a typical on-disk loop flare, April 22, 1993 (GOES class M1.5, location (N11,E04), occurrence time 14:08UT). The BCS data of this flare have not previously been analysed because another small flare occurred almost simultaneously and contaminated the spectra. We, however, found by making a so-called "dynamic spectrogram" that allows us to see the time behavior of the spectra more clearly, that the effects of these two flares can be separated. This procedure makes use of the dependence of the wavelength on the heliographic latitude of the sources, and also requires interpolating the temporal behavior of the "contaminator" reasonably.

We could derive the temperature and velocity of the upflow, as well as the emission measure from the subtracted spectrum. Those provide us with basic information about the rising gas *prior* to the hard X-ray impulsive burst phase. We indeed see the peaks after the hard X-ray bursts, but note that the curves started to rise before them. This confirms the above-mentioned statements of Doschek and Tanaka, and requires specific explanations for this early part (*cf.* Uchida 1996).

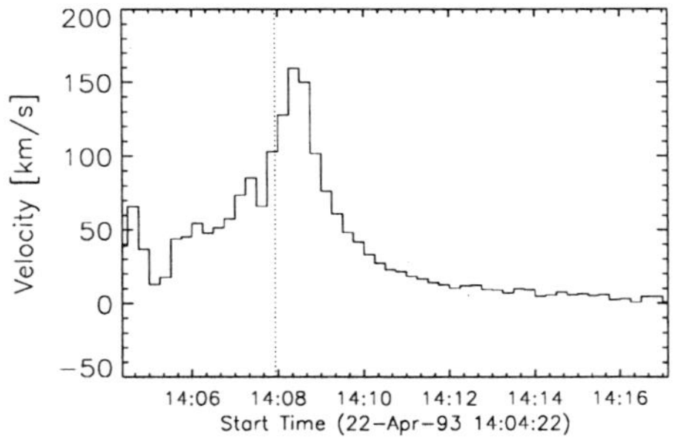

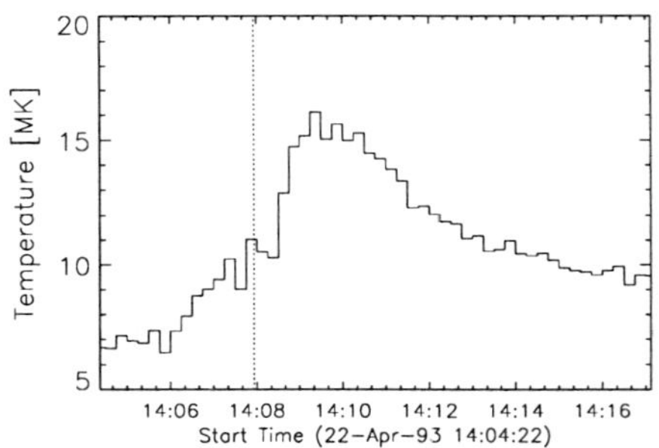

Figure 1. Velocity and Temperature from the cleaned profile of S XV. The dotted line marks the start of the HXR impulsive burst, and the rise started before then.

References

Doschek, G., Cheng, C.C., Oran, E., Boris, J., and Mariska, J., 1983, *Astrophys.J.*, **265**, 1103.

Tanaka, K., Nitta, N., Akita, K., and Watanabe, T., 1983, *Solar Phys.*, **86**, 91.

Uchida, Y., and Shibata, K., 1988, *Solar Phys.*, **116**, 291.

Uchida, Y., and Shibata, K., 1990 : in Y. Uchida *et al.* (eds.) *"Flare Physics in Solar Activity Maximum 22"* (Springer-Verlag), p. 230.

Uchida, Y., 1996 : these Proceedings.

HARD AND SOFT X-RAY OBSERVATIONS OF SOLAR LIMB FLARES

JOHN T. MARISKA
Naval Research Laboratory, Washington, DC 20375 USA
T. SAKAO
National Astronomical Observatory, Mitaka 181, Japan
AND
R. D. BENTLEY
Mullard Space Science Laboratory, University College London
Holmbury St. Mary Dorking, Surrey RH5 6NT
United Kingdom

Many flares show looptop emission in both hard and soft X-rays. By looking at occulted and nonocculted limb flares, it is possible to examine the physical conditions in the various emitting regions in flares. Occulted limb flares have their footpoints behind the solar limb and thus the *Yohkoh* Bragg Crystal Spectrometer (BCS) only observes emission from the looptop source. Nonocculted limb flares have their footpoints exposed. The signal from those footpoints usually dominates the emission observed with the BCS. Thus the nonocculted flares reveal the characteristics of the footpoint sources.

We have examined four nonocculted and four partially to fully occulted limb flares observed by the *Yohkoh* BCS, Hard X-Ray Telescope (HXT), and Soft X-Ray Telescope. In Table 1 we list the flares examined in our study, the spectral index for each flare determined using the HXT, and the peak temperatures and nonthermal broadening determined for each flare using data from the BCS Fe XXV and Ca XIX channels.

For both the nonocculted and occulted flares, we find that the nonthermal broadening measured in both Fe XXV and Ca XIX is large at the earliest time that a useful spectrum can be obtained. In Ca XIX, the nonthermal broadening sometimes increases after the hard X-ray burst begins. The temperatures measured in both Fe XXV and Ca XIX increase with time, peaking

Y. Uchida et al. (eds.),
***Magnetodynamic Phenomena in the Solar Atmosphere – Prototypes of Stellar Magnetic Activity*, 523–524.**

TABLE 1. Limb Flare Characteristics

Date	γ	T (MK)		ξ km s^{-1}	
		Fe XXV	Ca XIX	Fe XXV	Ca XIX
Nonocculted:					
1992 Jun 28	5.3	19.1	14.2	440	250
1992 Aug 11	4.1	20.0	14.7	460	340
1993 Sep 27	4.6	22.0	14.0	320	250
1993 Nov 30	2.6	20.0	16.0	370	200
Occulted:					
1991 Oct 21	8.0	24.5	21.0	250	180
1992 Oct 5	5.2	20.0	14.5	290	160
1993 Feb 1	5.9	21.5	15.0	130	190
1994 Jan 29	6.9	23.0	13.5	270	160

well before peak emission in the respective BCS channel. The temperature behavior appears to be the same in both flare types.

For occulted limb flares, we find that there appears to be a tendency for the nonthermal broadening to be suppressed relative to that seen in nonocculted limb flares. HXT data indicate the the hard X-ray spectra are also softer in the nonocculted flares. These softer hard X-ray spectra are consistent with the hard X-ray emission from the occulted flares coming from thin-target emission high in the flaring loops and the hard X-ray emission from the nonocculted flares coming from thick-target emission originating near the loop footpoints. We have no obvious explanation for the decreased nonthermal broadening in occulted flares. Most explanations for the nonthermal broadening suggest that it should be larger near the looptops.

Our results, which are published in more detail in Mariska, Sakao, and Bentley (1995), differ from those of Khan et al. (1995), who found no differences between occulted and nonocculted limb flares.

References

Khan, J. I., Harra-Murnion, L. K., Hudson, H. S., Lemen, J. R., & Sterling, A. C. 1995, ApJ (Letters), in press.

Mariska, J. T., Sakao, T., & Bentley, R. D. 1995, ApJ, submitted.

X-RAY PLASMA EJECTION IN AN ERUPTIVE FLARE

M. OHYAMA
Solar Terrestrial Environment Laboratory, Nagoya University, Toyokawa 442, Japan

AND

K. SHIBATA
National Astronomical Observatory, Mitaka, Tokyo 181, Japan

In this study we have analyzed an X-ray plasma ejection (or plasmoid) in an eruptive flare ($\leq$ M2.0) on 1992 October 5 which occurred behind the west limb. The plasma ejection is seen as a blob-like feature (i.e., it has the local intensity maximum) as seen in Figure 2. The boundary between the ejction and the flare region is assumed to be the vertical line crossing the intensity saddle point between them. Figure 1 shows the height vs. time for the ejection. From Figure 1, it is found that the plasma ejection was ejected at an apparent velocity of $200 - 450\,\mathrm{km\ s^{-1}}$, and that the acceleration occurred during the impulsive phase.

We derived the temperature and emission measure of the plasma ejection assuming isothermality (Figure 2), then obtained also the other phys-

Table 1. Physical parameters of plasma ejection and flare region. Temperature and emission measure are derived from a beryllium (Be119) and thick aluminum (Al12) filter combination, then the other physical parameters are obtained assuming the line-of-sight thickness $\sim 10^4$ km. (Ejection) 09:25:16 and 09:25:18 UT. (Flare region) 09:24:58 and 09:25:00 UT. The kinetic energy is calculated using the velocity $\sim 200 - 300$ km s^{-1}.

Physical parameter	Plasma Ejection	Flare Region
Temperature (10^6 K)	6 – 13	8 – 17
Emission measure (10^{28}cm^{-5})	7 – 20	60 – 800
Electron density (10^9cm^{-3})	8 – 15	25 – 90
Mass (10^{13}g)	$\simeq 2$	9 – 10
Gas pressure (dyn cm^{-2})	20 – 40	70 – 280
Thermal energy content (10^{28}erg)	4 – 5	20 – 30
Kinetic Energy (10^{27} erg)	3 – 10	—

Y. Uchida et al. (eds.),
Magnetodynamic Phenomena in the Solar Atmosphere – Prototypes of Stellar Magnetic Activity, 525–526.

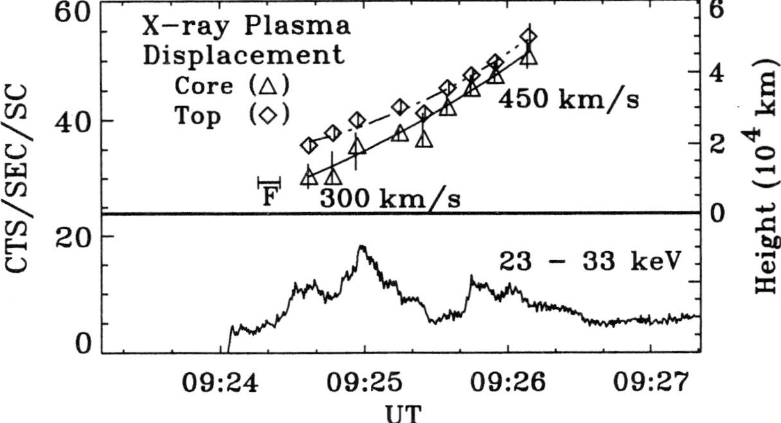

Figure 1. Plasma displacements in thin Al filter images and the hard X-ray (23–33 keV) counting rates. Core corresponds to the position of the maximum intensity of the ejection, and the error bars indicate the region where the intensity is within 90 percent of the maximum value of the ejection. Top corresponds to the height where the intensity is $\sim 1/e$ of the maximum value of the ejection. 'F' indicates the height of flare loops.

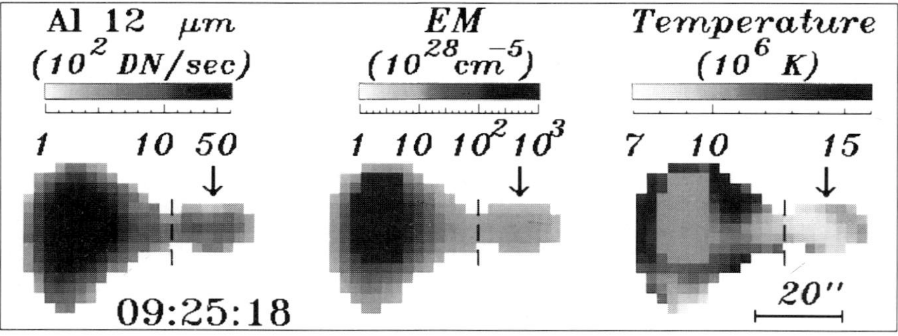

Figure 2. (Left) Soft X-ray intensity map (with Al12 filter). (Middle) Emission measure, and (right) temperature maps obtained using Al12 and Be119 (at 09:25:16 UT) filter images. Temperatures and emission measures are derived after taking 3×3-pixel sums of the intensity in order to improve statistics. The arrows indicate the ejection. The dashed lines show the boundary between the ejection and the flare region. A signal of X-ray scattering of each pixel is less than 15 percent of the observed signal.

ical parameters of both the plasma ejection and the flare region (Table 1). (1) Temperature of the plasma ejection is $6 - 13$ MK which is lower than that of the flare region. (2) Mass of the plasma ejection is about 2×10^{13}g. (3) Kinetic energy of the plasma ejection is smaller than thermal energy content of the plasma ejection. (4) Thermal energy content of the plasma ejection is smaller than that of the flaring region.

ISOLATING THE FOOTPOINT CHARACTERISTICS OF A SOLAR FLARE LOOP

L.K. HARRA–MURNION
School of Physics and Space Research, University of Birmingham, Edgbaston,Birmingham, B15 2TT, UK.

J. L. CULHANE
Mullard Space Science Laboratory, Holmbury St. Mary, Surrey, UK.

T. FUJIWARA
Dept. of Education, Aichi University of Education, Igaya cho, Kariya, 448, Japan.

H. S. HUDSON
Institute of Astronomy, University of Hawaii, Honolulu, HI 96822.

T. KATO
National Institute of Fusion Plasma Science, Nagoya, Japan.

AND

A. C. STERLING
Computational Physics Inc., 2750 Properity Avenue, Suite 600, Fairfax, VA 22031.

1. Introduction

Results from Yohkoh have shown that when loop footpoints are seen in flares they brighten very early and the emission is fainter than emission from the rest of the loop ((Acton *et al.*, 1992), (Hudson *et al.*, 1994)).

BCS (the Bragg Crystal Spectrometer) on Yohkoh has no spatial resolution so it cannot normally observe isolated flare footpoints. It does however have higher spectral and time resolution than seen on earlier missions.

The flare to be discussed occurred on 27th September 1993 at 10:45 U.T., and had a GOES classification of C5.7 and optical importance of

Y. Uchida et al. (eds.),
Magnetodynamic Phenomena in the Solar Atmosphere – Prototypes of Stellar Magnetic Activity, 527–529.

SF. It occurred at N11E80 in NOAA region 7590 with a total duration of approximately 20 minutes.

Both (Khan *et al.*, 1995) and (Mariska *et al.*, 1996) have used SXT (the Soft X-ray Telescope) and BCS to extract spectroscopic information on the bright loop tops (using the limb to isolate the loop tops) seen in solar flaring loops with Yohkoh.

We are most interested, in this case, in the Non-thermal Velocity (V_{nt}), where this is defined as the excess broadening once the thermal broadening has been removed. It has been shown that there is no substantial difference between the V_{nt} in the loop top of flares and the whole loop, (Khan *et al.*, 1995), whereas (Mariska *et al.*, 1996) have shown that the V_{nt} is higher when the footpoints are present. In this paper we attempt to extract BCS spectral information on footpoints.

The SXT images (Figure 1) for this event show the morphology of the flare to be a simple loop-like structure with the bulk of the emission for the first 2 minutes of the event coming from one of the footpoints.

Figure 1. SXT images at 3 times during the flare, illustrating the dominant footpoint.

During the 2 minutes the HXR source comes only from the footpoint, and we assume that the results obtained from BCS provide footpoint information.

2. Results and Discussion

The results are summarized below;

1) Electron temperatures derived from the BCS channels do not show any unusual behaviour. The temperatures at the time of the dominance of the footpoints are not the highest temperatures reached.

2) The SXT temperatures of the looptop reaches $\approx$ 18 MK, which is nearly a factor of 2 larger than the temperature of the footpoints (10MK). Comparing these temperatures with those from BCS would indicate that the hotter Fe XXV emission is coming from the looptop, and the cooler Sxv

and Ca xix temperatures is coming from the footpoints and the legs of the loop ((Doschek *et al.*, 1987)).

3) The HXR emission is coming from only the brightest footpoint at the time of the first peak in the non-thermal velocity. We suggest that the first peak in non-thermal velocity originates from this footpoint. The values of the V_{nt} (Figure 2) at this time are higher than the second peak by 32 km/s, 106 km/s and 184 km/s for Sxv, Ca xix and Fe xxv respectively.

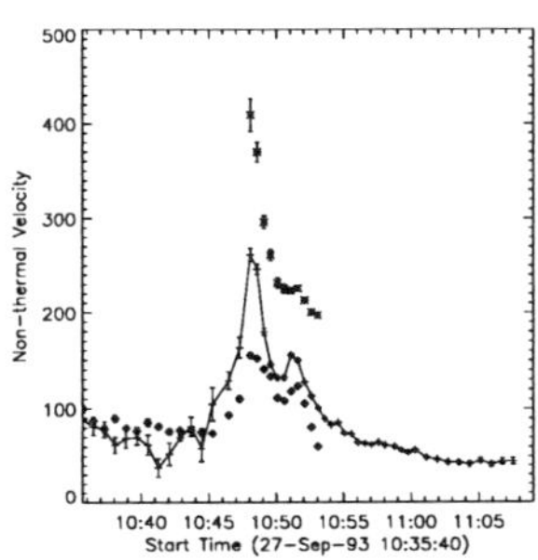

Figure 2. A plot showing the non-thermal velocity derived from BCS Sxv (diamonds), Ca xix (solid) and Fe xxv (crosses) channels. There are no Fe xxv data before 10:48 U.T. due to the low statistics

4) Some evidence exists that the increase in V_{nt} commences before the onset of the HXR emission. However, caution is used as there are no statistically usable Fe xxv spectra at these early times, and there are only a few valid points for Ca xix and Sxv. In summary, these results have a strong implication on the mechanism which produces the broadening of the spectral lines (ie V_{nt}). There is a suggestion that the mechanism begins before the HXR burst and that the mechanism is strongest at the footpoints of the flaring region.

References

Acton, L.W., Feldman, U., Bruner, M.E., Doschek, G.A., Hirayama, T., Hudson, H.S., Lemen, J.R., Ogawara, Y., Strong, K.T. and Tsuneta S., (1992), The Morphology of 20×10^6 K plasma in Large Non-Impulsive Solar Flares *PASJ*, **44**, L71

Doschek, G.A., and Feldman, U, (1987), The temperature of solar flares determined from X-ray spectral line ratios, *ApJ*, **313**, 883.

Hudson, H.S, Strong, K.T., Dennis, B.R., Zarro, D., Inda, M., Kosugi, T., and Sakao, T., (1994), Impulsive Behavior in Solar Soft X-Radiation, *ApJ*, **422**, L25-L27

Khan, J.I., Harra-Murnion, L.K., Hudson, H.S., Lemen, J.R., and Sterling, A.C., (1995), Yohkoh Soft X-ray Spectroscopic Observations of the Bright Loop-Top Kernels of Solar Flares, *ApJ Letters*, in press.

Mariska, J., Sakao, T., and Bentley, R.D., (1996), Hard and Soft X-Ray Observations of Solar Limb Flares, *Kluwer Academic Publishers*, these proceedings.

CORONAL LOOP INTERACTIONS OBSERVED IN OPTICAL EMISSION LINES

R.N. SMARTT
National Solar Observatory/Sacramento Peak †
P.O. Box 62, Sunspot, NM 88349, USA

V.S. AIRAPETIAN
CSC/NASA/GSFC
Code 681, Greenbelt, MD 20771, USA

AND

Z. ZHANG
University of Nanjing
Department of Astronomy, Nanjing, China

Optical coronagraph observations reveal coronal loop interactions in the form of transient localized brightenings (Smartt and Zhang, 1987; Zhang and Smartt, 1991; Smartt et al, 1993). These appear sequentially in time in the three coronagraph channels, showing first in Fe XIV green-line (530.3 nm; T$\sim$ 2 x 10^6K) emission, then in Fe X red-line (637.45 nm; T$\sim$ 10^6K) emission, and finally in Hα. This sequence ($\sim$ 20 minutes) reveals cooling of the plasma at the interaction site following initiation of the interaction. Since an enhancement is not normally observed in a red-line image prior to a corresponding maximum appearing in a green-line image, it is inferred that the plasma is heated rapidly during initial coalescence. The maximum value for the temperature immediately following the initiation of energy release in a particular event was estimated as, $T_{max} \leq 5$ x 10^6K (Airapetian and Smartt, 1995). This indicates that for this, and similar, events, the plasma is already cooling when the maximum in the green-line image occurs.

These brightenings are interpreted as localized coronal loop interactions (CLI) involving two (or more) loops, and described by current loop coalescence involving partial magnetic reconnection (Sakai 1993). Similar events

†Operated by the Association of Universities for Research in Astronomy, Inc. (AURA) under cooperative agreement with the National Science Foundation

Y. Uchida et al. (eds.),
Magnetodynamic Phenomena in the Solar Atmosphere – Prototypes of Stellar Magnetic Activity, 531–532.

with x-type configurations have been observed with the Soft X-ray Telescope of Yohkoh (see, for example, Shimizu et al, 1992; de Jager et al, 1993). Possible exceptions to a two-loop type of interaction are where an apparent single, low-lying loop brightens along most of its length (see, for example, Fig. 1a-f, enhancement B, in Li et al, 1994). However, in some cases traces of a second, faint loop are evident. Alternatively, a kink-like instability in a single twisted loop might produce a CLI-like event.

Commonly Hα becomes concentrated at the interaction site (following a maximum in the green line), then gradually extends down to the solar surface, and fades. Alternatively, in some cases Hα gradually extends in a column from the surface to the site, eventually becoming concentrated there, the emission then gradually fading. In general, these observations, reveal that Hα material is drawn to the site due to some mechanism associated with the interaction. In this paper we have considered the possible role of a thermal instability to account for the Hα morphology, with the following scenario. Initially (T $\sim$ 5 x 10^6K), the characteristic conductive time, t_{cond}, is much shorter than the characteristic radiative time t_{rad} and the plasma cools predominantly by thermal conduction. This is consistent with the observed plasma expansion along the legs of the loops in the vicinity of an interaction site as recorded in the green-line images. At lower temperatures (T $\sim$ 10^6K), radiative cooling of the dense condensation is then more efficient than conductive cooling ($t_{rad} < t_{cond}$), which is the condition for the onset of the radiative (thermal) instability in the solar corona. With further decrease in T, plasma inflow to the interaction site commences – to maintain constant pressure (isobaric regime of thermal instability) ($N_e \propto T_e^{-1}$). For T$< 10^5$K, recombination of hydrogen ions becomes more efficient, producing Hα emission. The plasma becomes optically thick, radiative cooling is then less efficient, and the thermal instability ceases. The "confined" plasma volume then tends to an overpressured state, and loses energy most efficiently by plasma flows along field lines with v_s (initially) $\sim$ 10 km s^{-1}, in agreement with the observations.

References

Airapetian, V.S. & Smartt, R.N. 1995, *ApJ*, **445**, 489.

De Jager, C., India-Koide, M., Koide, S., & Sakai, J.-I. 1993, in Symposium on Current Loop Interaction in Solar Flares, ed. J.-I. Sakai (Toyama: Toyama Univ.), 149.

Finn, J.M., Guzdar, P.N., & Usinov, D. 1994, *ApJ*, **427**, 475.

Li, X-Q., Zhang, Z. & Smartt, R.N. 1994, *A & A*, **290**, 963.

Sakai, J.-I. 1993, in Symposium on Current Loop Interaction in Solar Flares, ed. J.-I. Sakai (Toyama: Toyama Univ.), 13.

Shimizu, T., Tsuneta, S., Acton, L.W., Lemen, J. & Uchida, Y. 1992, *PASJ*, **44**, L147.

Smartt, R.N. & Zhang, Z. 1987, in Theoretical Problems in High-Resolution Solar Physics II (NASA CP-2483), 129.

Smartt, R.N., Zhang, Z. & Smutko, M.F. 1993, *Sol. Phys.*, **148**, 139.

Zhang, Z. & Smartt, R.N. 1991, *Acta Astron. Sinica*, **32**, 233.

ELECTRIC CURRENTS, AND MAGNETIC SHEAR VARIATIONS DURING SOME FLARES OF M- AND X-CLASS

ASHOK AMBASTHA[1], J.M. FONTENLA[2], and MONA J. HAGYARD[3]
1: Udaipur Solar Observatory(PRL), Udaipur-313 001 (India); 2: HAO, NCAR, Boulder, CO 80307 (USA); 3: SSL, NASA/MSFC, Huntsville, AL 35812 (USA)

1. Abstract:

We study the magnetic field evolution during flares of M and X-class, using overlays of cotemporal Hα-filtergrams and magnetograms. Significant decrease in the area-averaged magnetic shear was found in some cases around the flare onset-time, increasing subsequently as the flare progressed. However, the changes were less pronounced for relatively smaller M-class flares. Strong Lorentz forces were found to exist at the photosphere, acting against the observed motions of sunspots. Extrapolated potential field indicated presence of a magnetic null at the upper chromosphere/lower coronal height, directly above a large Hα flare.

2. Introduction:

Strong transverse magnetic fields, gradients, electric currents, and shear are usually found to be associated with locations of large flares (Hagyard 1992). Since flares are believed to derive their energy from magnetic fields, it is expected that these parameters should undergo detectable changes at least during the course of major flares. Previous studies have reported a variety of changes before, during and after some flares (Hagyard et al. 1990, Ambastha et al. 1994, Wang et al. 1994). The main problem is due to uncertainties and errors in measuring the weak linear polarization signal corresponding to the transverse component of magnetic field -- which in turns, is used to derive nonpotentiality or shear. Graphical presentation of shear is usually made only along the neutral lines, which ignores the overall (areal) magnetic field development in the active region. Using a quantitative, area-averaged shear parameter, we have studied some flares of M and X-class where cotemporal and cospatial Hα-filtergrams and magnetograms were available at the NASA/MSFC solar facility. In order to understand the mechanism of the magnetic energy storage, we have also estimated the electrical currents, and Lorentz forces in relation to the observed photospheric motions.

3. Results and Conclusions:

Variation of area-averaged shear for a large X4.7/4B flare in AR6555 is given in Figure 1 (Ambastha et al. 1994). Marked decrease in shear is evident from its preflare value at the flare onset-time, while, no change was observed in other locations. Another important feature, namely a magnetic null, was found to exist at the upper chromosphere/lower coronal height above the flaring location (Fontenla et al. 1995). As the plasma-β would be near or above unity near the null location, large electric currents may easily be produced by externally driven motions. Flaring activity, including three X-class flares during March 25-26, 1991, occurred mostly at this location. Significant horizontal Lorentz forces, with peak magnitude of about 2×10^{-3} N m^{-3} were found to exist very close to the polarity inversion line, opposing the observed sunspot motions. The buildup of free-magnetic energy in the upper atmospheric layers, stored in the form of field-aligned currents can be studied by computing the work done by Lorentz forces at the

Y. Uchida et al. (eds.),
Magnetodynamic Phenomena in the Solar Atmosphere – Prototypes of Stellar Magnetic Activity, 533–534.

photosphere. These Lorentz forces may correspond to the pressure-driven horizontal plasma motions that produce magnetic free-energy storage. The direction and magnitude of the Lorentz forces agree well with the shear buildup and flaring in the active region. It is estimated that the energy buildup over a day is marginally enough to supply the energy released in the X-class flares of AR6555 (viz., ~10^{31} ergs).

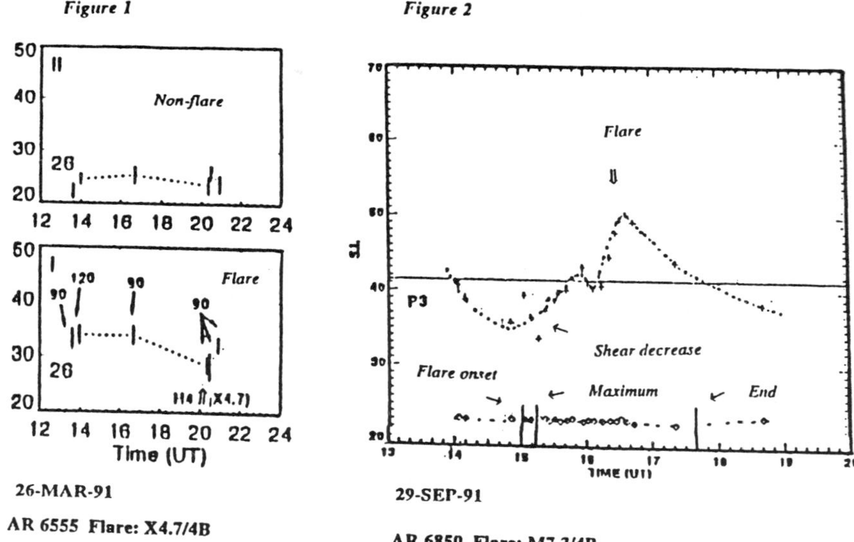

Figure 2 shows another interesting and good case of shear variation obtained for the M7.3/4B class flare of September 29, 1991/15:13UT. This particular flare occurred at a site of weak magnetic field and low shear, and was associated with a massive filament eruption. Significant shear variation was found at the flare-location, before, during and after the flare as compared to other neighboring subareas. However, changes in shear in both these events were not as impulsive or large as reported by Wang et al. (1994). For three other smaller M-class flares, similar shear plots showed only marginal variations in their shear parameters.

Existence of large shear alone was not found sufficient for flaring. Strong shear existed in some subareas of AR6555 which did not produce significant flares, while, large flares occurred in some locations having weak shear, e.g., X4.7 flare of March 26, 1991, and September 29, 1991. Additional factors such as large sunspot motions, δ-spots, and rapid magnetic field evolution appear to be important.

Acknowledgments: *This study was partly carried out while A.A. held a NRC-NASA Research Associateship at the Marshall Space Flight Center, Huntsville. A.A. would also like to thank A. Bhatnagar for his encouragement during this work. Thanks are also due to G. Allen Gary for useful discussions, and to Mitzi Adams for providing some VMGs.*

References:

Ambastha A., Hagyard M.J., and West, E.A.: 1994, Solar Phys., ***148***, 277.
Fontenla J.M., Ambastha A., Kalman B., and Csepura Gy.: 1995, Astrophys. J., ***440***, 894.
Hagyard M.J.: 1992, Mem. Soc. Astron. Ital. ***61***, 337.
Hagyard M.J., Venkatakrishnan P., and Smith J.B., Jr.: 1990, Astrophys. J. Suppl. Series 73, 159.
Wang H., Ewell M.W., Jr., Zirin H., and Ai G.: 1994, Astrophys.J., ***424***, 436.

RELATIONS BETWEEN FILAMENT CURRENT SYSTEMS AND SOLAR FLARES

YU-HUA TANG
Department of Astronomy, Nanjing University, Nanjing, China
LIAN-SHU CUI
Nanjing Artillery Academy, Nanjing, China
CHUN-LIN YIN
Purple Mountain Observatory, Nanjing, China.
AND
AO-AO XU
Department of Astronomy, Nanjing University, Nanjing, China

The flares of May 13 and May 16, 1981 are very notable flares during solar cycle 21. The flare on 4 Feb., 1986 occured in the solar minimum between the solar cycle 21 and cycle 22. It was followed by the strongest magnetic storm ever observed since 1960. For all above three flares, there are complete observational data which include radio, optical and high-energy emission. The occurrence of the each flare was related closely to the movement of active filaments in active regions.

The two-ribbon flares of May 13 and May 16, 1981 took place in active region Hale 17644 . There was a filament running through the active region. Both 3B flares of May 13 and of May 16 happened near the east segment of the large filament. Another major flare on 4 Feb., 1986 occurred in Boulder 4711(Tang *et al.*, 1995). Hα observations show that a number of arch filaments appeared in AR4711. After 0325UT, the activated filaments in the north-east in AR4711 began to rise. The filaments kept rising untill the large flare started(0736UT), when the filaments disappeared completely.

Based on the Kuperus-Raadu model, using the observational data of the three flares, we first measured the rising movement of the active filaments and obtained the h(t) curves. Then we solved the momentum equations and energy equation, obtaining the filament cuurent intensity I(t) and temperature T(t) curves . Finally we analysed and studied the main current systems

Y. Uchida et al. (eds.),
Magnetodynamic Phenomena in the Solar Atmosphere – Prototypes of Stellar Magnetic Activity, 535–536.

of the three flares and different magnetic reconnections. We have obtained the results as follows:

(1) The flare-filament current systems have some varieties. For the flare of May 13, 1981, the current system might be considered as a simple current loop(Xu *et al.*, 1994). If the Lundquist field is adopted as the current distribution within the filament, increase of the filament current implies an enhancement of the magnetic twist within the filament. Calculation showed that the flare of May 13, 1981 was triggered by the tearing mode spontaneous reconnection caused by the magnetic twist within the filament. The flare May 16, it occurred in the same active region as the flare of May 13, and both eruptions were due to the evolution and the motion of the same filament, but the release of the magnetic energy of the May 16 flare might take place instead in the current sheet beneath the rising filament (Martens *et al.*, 1989). This would mean that the flare of May 16, 1981 resulted from sheet reconnection. Furthermore the flare of Feb. 4, 1986 may be result from the coalescence of two filament current loops (Tang *et al.*, 1995). Thus a variety of flares might be explained by different magnetic reconnections caused by different filament rising movements and different filament current systems.

(2) The resistive tearing instability could be regarded as the heating mechanism of preflare phase (for the flare of Feb. 4, 1986) and sometimes may develop into hot phase of the flare (for the flare of May 13, 1981).

(3) When we solve the momentum equations by using different background fields of the active .region at the same conditions for the flare of Feb. 4, 1986, which descending with height exponantially (model I; Wu *et al.*, 1992) and smoothly (model II; Low, 1977), only the force free field which changes quickly with height (model I) could fit the observed rising movement of filament and causes flare eruption.

This work is supported by the National Natural Science Foundation of P.R.China.

References

Low B.C.(1977)Evolving force-free magnetic fields,*Astrophys.J.*,**212**, 234-342.

Martens,P. C. H. and Kuin,N. P. M.(1989)A circuit model for filament eruptions and two-ribbon flares,*Solar Phys.*, **122**, 263-302.

S.T.Wu,C.L.Yin and W.H.Yang(1992)Magnetic diffusion and flare energy build up, *Solar Phys.*,**142**, 313-325.

Tang Yu-hua,Yin chun-lin and Qiu Jiong(1995)The 4 February, 1986 flare and flare-filament current model, *Solar Phys.*, (in press).

Xu Ao-ao,Wu Gui-ping,Tang Yu-hua,Jiang Zhi-bo and Li Qiong-ying(1994)A phenomenological model of the solar flare based on the evolution of filament current, *Chinese Science Bulletin*,**39**,759-765.

MAGNETIC FIELD STRUCTURE AND ACCELERATED ELECTRONS IN A GRADUAL MICROWAVE/HARD X-RAY FLARE

H. NAKAJIMA
Nobeyama Radio Observatory, NAOJ
Nobeyama, Minamisaku, Nagano 384-13, Japan

AND

R. A. SCHWARTZ
Laboratory for Astronomy and Solar Physics, NASA/GSFC
Greenbelt, MD20771 and Hughes STX, Lanham, MD20706, USA

We have studied a flare (GOES class: M2.3) on the west limb (H- alpha: N18, W78) on 1993 March 23 using microwave (17GHz) and Soft X-ray imaging data and also hard X-ray spectrum data. This flare has typical properties of "gradual-hard flares".

Overlays of microwave and soft X-ray images suggest that accelerated high-energy electrons partly precipitate into the chromosphere and are partly trapped in the soft X-ray loop.

Figure 1 shows time profiles of the brightness temperature at various positions in the microwave source during 0130 UT and 0145 UT in the main phase. It can be clearly seen that the time profile at each position along the loop shows increasing peak-delay and decay time with increasing height. This is interpreted as follows. The gyrosynchrotron emission is extremely sensitive to the magnetic field strength, and the magnetic field is probably weaker at the loop top than at the loop footpoint. Therefore, higher-energy electrons contribute to the microwave emission at the loop top, while lower-energy electrons to that at the footpoint. Higher-energy electrons are more easily trapped than the lower-energy electrons.

The microwave time profile at the loop top (position 3) can be obtained by applying an e-folding decay time of 230 s (+/-30 s) to the time profile of low-energy hard X-rays (33-42 keV) and shifting the obtained time profile by 30 s. This suggests that time variations of high-energy electrons in the flaring loop are caused by continuous acceleration with trapping and pre-

Y. Uchida et al. (eds.),
Magnetodynamic Phenomena in the Solar Atmosphere – Prototypes of Stellar Magnetic Activity, 537–538.

cipitation, and that the acceleration of high-energy electrons at the loop top is delayed by 30 s as compared with low energy electrons.

The peak delay time of the 17 GHz emission at the loop footpoint with respect to the low-energy hard X-ray peak time is 47 s which is almost the same as that (43 s) of the hard X-ray emissions of 166 - 1099 keV. This implies that electrons of almost the same energy contribute to both the 17 GHz emission and 166 - 1099 keV hard X-rays. The effective energy of electrons which contribute to the 166 - 1099 hard X-ray emissions is calculated to be 540 keV.

From the gyrosynchrotron emissivity for a source with a uniform magnetic field and an isotropic distribution of electrons whose power-law spectrum index of 2.9, we can estimate the magnetic field at the loop footpoint. The electrons with effective energy of 540 keV mostly contribute to the gyrosynchrotron emission around the 10th harmonic which corresponds to 610 Gauss at 17 GHz. The magnetic field strength at the loop top can be estimated to be 190 Gauss using the observed mirror ratio (= 3.2) of the soft X-ray loop. Then, we find that electrons with effective energy around 1 MeV contribute to most of the microwave emission at the loop top.

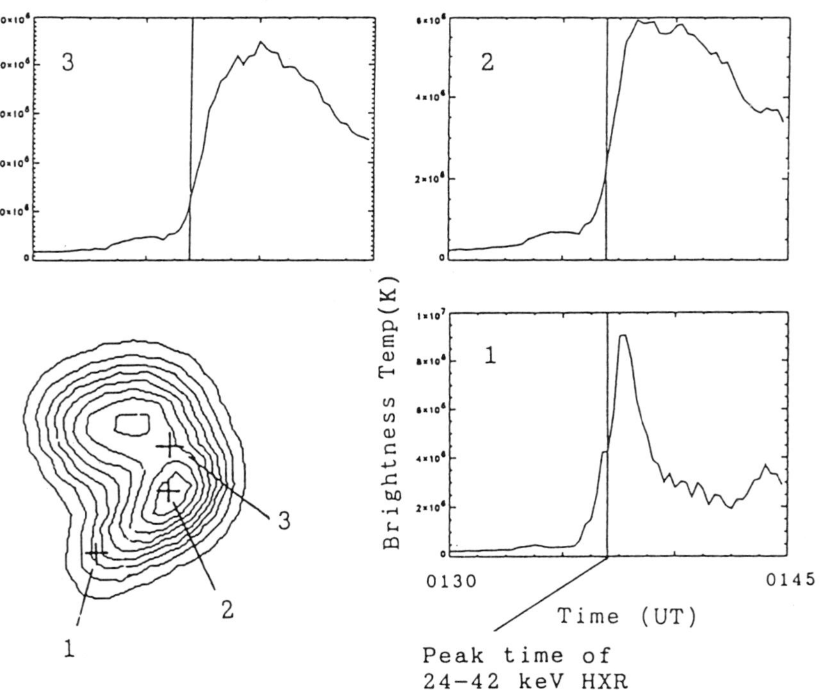

Figure 1. Time Profiles of the brightness temperature at various positions indicated in the microwave images (left-lower panel), during 0130 UT and 0145 UT in the flash phase

Nakajima, H. and Schwarts, R. A. 1995, Ap. J., to be submitted.

HARD X-RAY AND MICROWAVE OBSERVATIONS OF AN IMPULSIVE BURST ON 1994 JANUARY 6

M. NISHIO, K. YAJI AND T. KOSUGI
National Astronomical Observatory, Japan

We describe an impulsive event on 1994 Jan. 6 simultaneously observed by the *Yohkoh* Hard X-ray Telescope (HXT; Kosugi et al. 1991) and Soft X-ray Telescope (SXT; Tsuneta et al. 1991), and the Nobeyama Radioheliograph (Nakajima et al. 1994) to examine if loop-loop interaction is an origin of the impulsive burst. In hard X-rays, onset of the burst is at 04:05:20 UT and duration is about 1 min, during which three spikes, each of about 5-s duration, are involved. The first two spikes are also detected in microwaves, showing a similar time profile to that in hard X-rays. This burst is characterized by a simple, quite hard spectrum of hard X-rays. The time profile is quite similar in HXT four energy bands, namely L (13.9–22.7 keV), M1 (22.7–32.7 keV), M2 (32.7–52.7 keV) and H (52.7–92.8 keV). We believe that this burst straightforwardly represents the behavior of energetic electrons both in hard X-rays and microwaves without serious contamination from the thermal emission of flare hot plasma.

Figure 1 shows overlaid 17-GHz, hard X-ray and soft X-ray images in the second spike and a soft X-ray image 3 min after the onset of the burst. In the 17-GHz images, two compact sources can be seen with a separation of about $60''$ in the east-west direction. They appear bipolar in circular polarization images. The two are quite similar to each other in intensity time profile. In figure 1(b), a loop connecting the microwaves can be seen. In hard X-ray images, a compact single-source structure dominated in the first and third spikes, but the location of the source seen in the third spike shifted to the west by about $10''$. In the second spike, an elongated source structure appeared, which connected the first and third spike sources. Soft X-ray images show a bright patch cospatial with this elongated source. On the other hand, only the western microwave source is cospatial with the hard X-rays. Thus, it seems that the hard X-rays are included in a single loop different from a large loop connecting the microwaves.

Y. Uchida et al. (eds.),
Magnetodynamic Phenomena in the Solar Atmosphere – Prototypes of Stellar Magnetic Activity, 539–540.

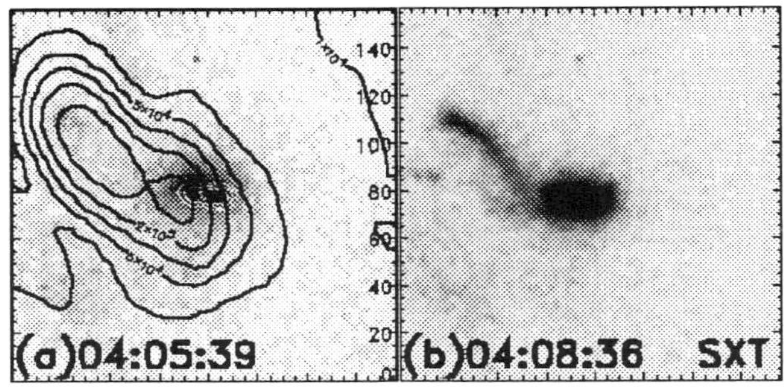

Figure 1. (a) Overlay of 17-GHz images (black contours), HXT M1-band images (white contours) and soft X-ray images (gray scales) taken in the second spike. The 17-GHz images are shown in brightness temperature with contour levels of 12,500 K $\times$ 2^n (n=0–8). Hard X-ray contours are drawn at 50%, 25% and 12.5% levels of the peak intensity. Frame size is $2'.6 \times 2'.6$. (b) soft X-ray image taken 3 min after the onset of the event.

Assuming the gyrosynchrotron mechanism for the microwave radiation and the thick-target model for the hard X-ray radiation, we estimated the electron energy spectral indices for the first and second spikes. The indices derived from microwave spectra, which are obtained by radio polarimeters at Nobeyama and Toyokawa as well as the radioheliograph, are 3.6 and 4.2 to the trapped electrons. The indices derived from the data of HXT M1 and M2 bands are also 3.6 and 4.2. This result strongly supports the idea that the microwave and hard X-ray radiating electrons were produced in a common acceleration site.

From these results, we speculate that the emerging flux model suggested by Heyvaerts et al. (1977) is one of the most plausible; an emerging small loop collides with a pre-existing large loop and produces accelerated electrons. Hanaoka suggest similar loop configuration for a series of flares observed in NOAA 7360 (1995). As we focused to the event where non-thermal electrons play dominant role, we could describe the relation of the loop-loop interaction and the particle acceleration more clearly.

References

Hanaoka, Y. 1995, in this proceedings
Heyvaerts, J. et al. 1977, ApJ, **216**, 123
Kosugi, T. et al. 1991, Solar Phys., **136**, 17
Nakajima, H. et al. 1994, Proc. IEEE, **82**, 705
Tsuneta S. et al. 1991, Solar Phys., **136**, 37

IMAGING SPECTRA OF HARD X-RAY FROM THE FOOTPOINTS OF SOLAR IMPULSIVE LOOP FLARES

T. TAKAKURA
Dept. of Astronomy, School of Science, the University of Tokyo, Tokyo 113

T. KOSUGI AND T. SAKAO
National Astronomical Observatory, Mitaka, Tokyo 181

K.MAKISHIMA
Dept. of Physics, School of Science, the University of Tokyo, Tokyo 113

M.INDA-KOIDE
Laboratory for Plasma Astrophysics and Fusion Science, Dept. of Electronicsand Information, Faculty of Engineering, Toyama University, 3190 Gofuku, Toyama 930

AND

S.MASUDA
Solar-Terrestrial Environment Laboratory, Nagoya University, 3-13 Honohara, Toyokawa, Aichi, 442

Abstract. The imaging spectra of hard X-rays emitted from one or both footpoints of solar impulsive loop flares seem to frequently show a better fit to an extremely hot thermal spectrum, on the order of 10^8 K, than to a power-law spectrum, in the main phase of the bursts.

In the present study this tendency has been verified by the spectra of nine impulsive bursts having a sufficient flux for reliable imaging in the H-band (52.7-92.8 Kev) of the hard X-ray telescope (HXT) aboard the Yohkoh spacecraft. The spectral indexes at the footpoints were derived from the hard X-ray images in four energy bands of the HXT; their reliability was verified by mapping simulations.

Since the X-ray spectrum depends on the location in the X-ray loop, the spectrum of the total flux is generally different from the spatially resolved

Y. Uchida et al. (eds.),
Magnetodynamic Phenomena in the Solar Atmosphere – Prototypes of Stellar Magnetic Activity, 541–542.

spectra. The imaging spectrum is thus crucial for studying the nature of the electrons emitting hard X-rays.

The heat conduction of an extremely hot quasi-thermal plasma seems to play an essential role in impulsive loop flares.

Full paper has been published in *Pub. Ast. Soc. Japan*, 1995, Vol.**47**, 355.

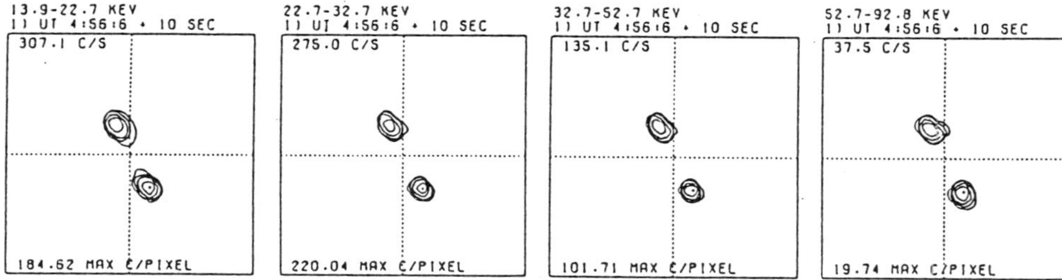

Figure 1. X-ray contour maps of the burst on 1991 December 16 (Data No.5 in Table1). The map size is 91".4 square. Map center is 11'.48 west and 2'.90 south. The minimum contour level is 0.1-times the peak brightness in each map and the contour step is $\sqrt{3}$ -times (i.e., logarithmic step).

Table 1. Physical parameters at the double sources. γ: Spectral index for photon number. T:Equivalent electron temperature in 10^8 K. Th:Thermal spectrum. Pw:Power-law spectrum. n_e:Electron number density in $10^7 cm^{-3}$

	Footpoint A, left (or up)								Footpoint B, right (or down)							
Data No.	γ_1	γ_2	γ_3	T_1	T_2	T_3		n_e	γ_1	γ_2	γ_3	T_1	T_2	T_3		n_e
1		(3.2	3.5)		1.4	2.0	Pw	17.	4.7	(3.4	3.0)	0.60	1.3	2.4	Pw	8.0
2	1.3	3.0	4.0	4.4	(1.5	1.6)	Th	1.0	1.6	2.5	3.6	2.7	(2.0	1.9)	Th	0.26
3-1	5.1	2.9	3.8	0.50	(1.6	1.8)	Th	0.15	1.7	2.0	2.7	2.4	(2.8	2.9)	Th	0.58
3-2	6.0	(3.3	3.0)	0.45	1.4	2.4	Pw	11.	(1.9	2.0	2.2)	2.0	2.9	4.0	Pw	6.4
4	6.0	4.6	3.9	0.45	0.85	1.7	?		1.9	(2.9	2.9)	1.9	1.7	2.5	Pw	19.
5	2.5	3.9	4.6	(1.4	1.1	1.4)	Th	4.1	2.6	3.8	4.3	(1.2	1.1	1.5)	Th?	3.7
6	2.6	3.4	4.5	(1.3	1.3	1.4)	Th	3.2	2.2	3.2	4.6	(1.6	1.4	1.4)	Th	2.0
7	1.7	3.2	4.6	2.4	(1.4	1.4)	Th	2.2	1.9	2.3	3.2	1.9	(2.2	2.2)	Th	0.64
8		(4.1	4.7)		1.0	1.3	Pw?	6.2	1.6	2.2	3.3	2.8	(2.5	2.1)	Th	0.19
9	1.7	2.2	2.9	(2.4	2.4	2.6)	Th	0.25	1.7	(2.3	2.4)	2.4	2.4	3.3	Pw	1.2

NONTHERMAL EMISSIONS IN THE HARD X-RAY AND MICROWAVE RANGES FROM FLARING LOOPS

K. YAJI[1], T. KOSUGI[1,2], AND M. NISHIO[2]
[1] *National Astronomical Observatory, Mitaka 181, Japan*
[2] *Nobeyama Radio Observatory, NAO, Nagano 384-13, Japan*

Hard X-ray and microwave emissions from solar flares are believed to originate from energetic electrons; X-rays are due to Bremsstrahlung while microwaves are due to gyrosynchrotron. Thus the two emissions are expected to provide complementary information on the conditions of flaring loops as well as energization of electrons. With no doubt we need to have simultaneous imaging observations because of complexity of flaring region to derive any definitive conclusions. Since the first hard X-ray imaging from *SMM* and *Hinotori* in early 80s, such attempts have been made. For example, Kundu (1984) and Nakajima (1991) have summarized simultaneous imaging observations for more than several flares. Their conclusions are vague, however; the spatial relationship between microwave and hard X-ray sources is quite complicated. Whether the same population of electrons is responsible for the two types of emission still remains uncertain.

Here we present a report of one more trial study on this context. We have used advanced instruments for both hard X-ray and microwave observations, i.e., the Hard X-ray Telescope (HXT; typical angular resolution $\sim$ 5 arcsec; Kosugi et al. 1991) on board *Yohkoh* and the newly-constructed Nobeyama Radioheliograph operating at 17 GHz (typical angular resolution $\sim$ 15 arcsec; Nakajima et al. 1994). We analyzed 59 flares simultaneously observed with these instruments during the period from June 1992 through December 1994. The two images, at the peak flux in hard X-rays, were precisely coaligned.

A brief and tentative summary of our observations is the following:

1) About 40 % of microwave flares consist of two or more sources (even with the limited spatial resolution). Hard X-ray sources seem to be more compact. This may be partially due to better resolution of HXT, but we have cases where this explanation does not apply, i.e., microwave source(s) does exist even where no hard X-ray source is observed.

Y. Uchida et al. (eds.),
Magnetodynamic Phenomena in the Solar Atmosphere – Prototypes of Stellar Magnetic Activity, 543–544.

2) In the majority ($\sim 85\,\%$) of flares analyzed, the brightest microwave source is accompanied by one or two well-resolved hard X-ray sources. The brightest microwave source is usually weakly polarized; $\sim\ 90\,\%$ of these microwave sources show low degrees of circular polarization less than $30\,\%$.

3) In the remaining minority cases, the brightest microwave source is not associated with any detectable hard X-ray sources; one or more hard X-ray sources appear elsewhere, mostly where a fainter microwave sources is located. Among these cases, about half show relatively high degrees of circular polarization more than $20\,\%$.

These results are consistent with the interpretation (i) that the two types of emission mainly originate from near the footpoints of flaring loops, where the ambient conditions are suitable for producing the emission, i.e., the density is high for hard X-ray production and the magnetic field is strong for microwave production; and (ii) that, since energetic electrons are bounced back in the magnetic loops by magnetic mirroring and since the gyrosynchrotron emissivity strongly depends on the magnetic field strength, microwave sources may appear at the site where no detectable hard X-rays are produced. The dependence of degree of polarization on hard X-ray/microwave flux ratio may be understood in this context.

In-depth analyses are still in progress.

References

Kosugi, T. et al. (1991) *Solar Phys.*, **136**, 17
Kundu, M. et al. (1984) *Adv. Space Res.*, **4**, No.7,157
Nakajima, H. et al. (1991) in *Flare Physics in Solar Activity Maximum*, **22**, 77
Nakajima, H. et al. (1994) *Proc. IEEE*, **82**, 705

HARD AND SOFT X-RAY IMAGES OF AN LDE FLARE

J. SATO, T. KOSUGI, AND T. SAKAO
National Astronomical Observatory, Mitaka, Tokyo 181, Japan

It has long been known that solar flares are classified by their soft X-ray (SXR) time profiles into two classes, i.e., Long Duration Events (LDEs) and impulsive events. Recently *Yohkoh* SXT has revealed that LDEs, characterized by a long duration and gradual decay, usually show a large cusp-shaped structure when seen near the limb (Tsuneta *et al.* 1992, 1996). In contrast, a simple-loop SXR structure is typically seen in impulsive flares with dominant double-footpoint hard X-ray (HXR) sources located at the two ends (Sakao 1994). Sometimes a weak, compact HXR source is also seen in impulsive flares above the apex of the corresponding SXR loop (Masuda *et al.* 1994,1995). Since the HXR emission originates from energetic electrons, it is observationally of crucial importance to examine which portion of the cusp-shaped LDE structure is the brightest in HXRs. With this problem in mind, we present here an imaging observation of an LDE with the Hard and Soft X-ray Telescopes (HXT and SXT, respectively) aboard the *Yohkoh* satellite.

The event occurred on 1993 March 15 at $\sim$ 20:00 UT just behind the west limb. It is a typical LDE; the hard X-ray emission, with a quite soft spectrum even in the rise phase, varies gradually without any remarkable spikes. Soft X-ray (from *Yohkoh*) and microwave (from Nobeyama; at 17 GHz) images in the decay phase are compared by Hanaoka (1994), in which a well-developed cusp-shaped structure, very similar to the famous 1992 February 21 flare (Tsuneta *et al.* 1992, 1996), is seen.

Fig. 1 compares an HXT L-band map (14–23 keV; contours) with the corresponding SXT temperature (right) and emission measure (left) maps for the rise phase of the event. [As one can see, the hard X-ray source is unusually large ($\sim$ 1 arcmin square). This is a common characteristic of LDEs which makes HXT image synthesis difficult. In the case of this event, occultation of footpoint sources by the limb might render this difficulty. In spite of this we incorporated a new technique to stabilize the image, which

Y. Uchida et al. (eds.),
Magnetodynamic Phenomena in the Solar Atmosphere – Prototypes of Stellar Magnetic Activity, 545–546.

is out of scope of this paper.] From Fig. 1, we derive the following:

(1) The brightest core of HXR source is cospatial with the loop-top, high emission-measure (high density) region seen with SXT. The core HXR component seems to extend along the SXR loop structure. Note, however, that the observed amount of HXRs cannot be accounted by the SXR emission measure and temperature; the latter of which is too low (cooler than $\sim 1.0 \times 10^7$ K).

(2) Two bright HXR wings are seen, one extending upward and the other downward. The former may be a manifestation of high-temperature cusp region, while the latter may trace a branch of arcade loops.

A systematic study of HXR/SXR image comparison of LDEs is under way along with improvement of HXT image synthesis tools.

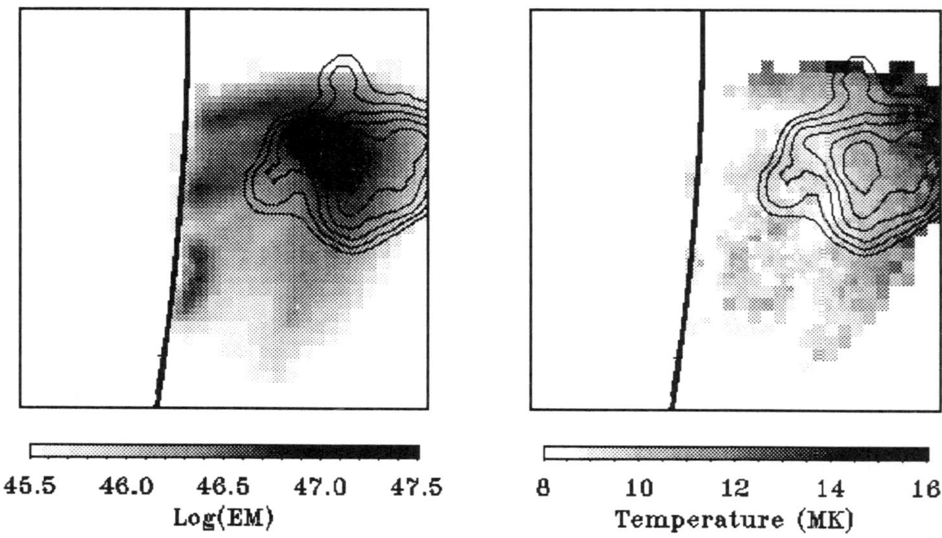

Figure 1. Comparison of HXT map (14–23 keV; contours) with temperature (right) and emission measure (left) maps derived from SXT for the 1993 March 15, LDE flare (157 × 157-arcsec FOV; solar north to the top, west to the right).

References

Masuda, S., Kosugi, T., Hara, H., Tsuneta, S., and Ogawara, Y., (1994), *Nature*, **371**, pp. 495–497

Masuda, S., Kosugi, T., Hara, H., Sakao, T., Shibata, K., and Tsuneta, S., (1995), *Publ. Astron. Soc. Japan*, **47**, pp. 677–689

Sakao, T. (1994) PhD dissertation, University of Tokyo

Tsuneta, S., Hara, H., and Shimizu, T., (1992), *Publ. Astron. Soc. Japan*, **44**, pp. 63–69

Tsuneta, S, (1996), *ApJ*, **456**, pp. 840–849

Y. Hanaoka, (1994), *Proc. of Kofu Symposium*, pp. 181–184

ELECTRON ACCELERATION SITE OF THE 1992 SEP. 6 FLARE

K. FUJIKI
Graduate University for Advanced Studies,
Nobeyama, Minamisaku, Nagano 384-13, Japan

H. NAKAJIMA AND Y. HANAOKA
Nobeyama Radio Observatory, NAOJ
Nobeyama, Minamisaku, Nagano 384-13, Japan

K. YAJI
Graduate University for Advanced Studies,
Mitaka 181, Japan

AND

R. KITAI
Kwasan and Hida Observatory, Kyoto University,
Kamitakara, Gifu 506-13, Japan

In this report, we present results of our analyses of M2.4 -class impulsive flare at 0514 UT on Sep. 6, 1992, based upon microwave, X-ray and optical observations.

In the pre-flare phase (0510 - 0513 UT), the soft X-ray image consists of a bright loop (loop-1) and a faint loop (loop-2). The loop-2 shows no enhancement. Just before the impulsive phase (0514 - 0516 UT), a large loop (loop-3) and a tiny loop (loop-4) begin to be seen above both loop-1 and loop-2 and at the footpoint of the loop-1, respectively. During the impulsive and decay phases, a faint loop (loop-5) develops in the west side of the loop-3 forming a somewhat cusp-shaped structure in this phase. The loop configuration of this flare is drawn in Figure 1 (f).

Hard X-ray images (23 - 33 keV) in the impulsive phase consist of three sources - double footpoint sources located at both footpoints of the bright loop-1 and a loop-top source located around the top of and above the loop-1 (Figure 1 (d), (e)). Loop-5 is not located above loop-1, but is shifted laterally to the west direction. There is a large gap between loop-1 and loop-5. These observational facts indicate that they are separate loop systems with separate evolution.

Y. Uchida et al. (eds.),
Magnetodynamic Phenomena in the Solar Atmosphere – Prototypes of Stellar Magnetic Activity, 547–548.

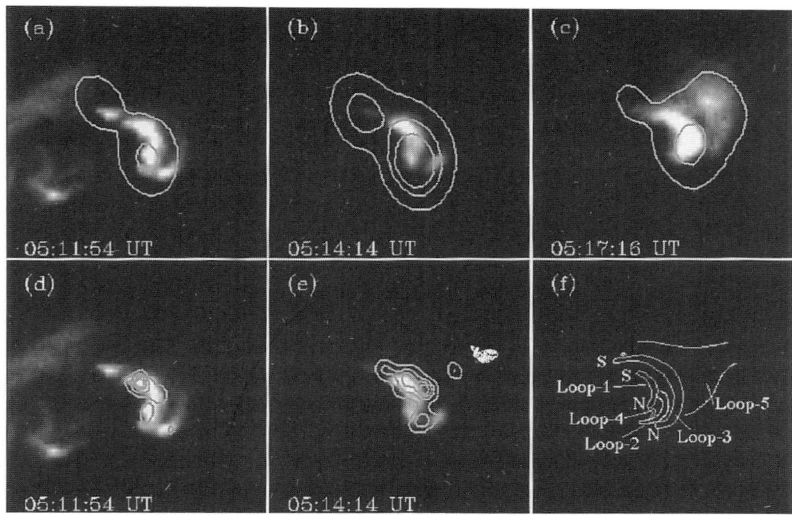

Figure 1. Two dimensional images of soft X-ray (gray scale), microwave (contour of (a)-(c), ($1 \times 10^5, 5 \times 10^5, 1 \times 10^6 K$), hard X-ray(contour of (d)-(e), (20, 40, 60, 80% of peak)) and schematic drawing of magnetic structure (f).

Microwave images at 17 GHz consist of double sources (Figure 1 (a)-(c)). The brighter one (microwave source-A) coincides with the hard X-ray source at the southern footpoint of loop-1, and the weaker one (microwave source-B) is located at the northern footpoint of loop-3. The time profile of the microwave source-A coincides well with that of the hard X-rays. On the other hand, the time profiles of both microwave sources coincide with each other in onset times with accuracy of ± 3 sec but not in fine structures, during 0514 and 0515 UT, and the intensity of the microwave source-B becomes very weak as compared with that of the microwave source-A between 0515 and 0516 UT. Lack of coincidence in fine structures of both microwave time profiles strongly suggests that high energy electrons in both microwave sources are not confined in a single loop but accelerated separately in loop-1 and loop-3, while good coincidence in their onset times suggests that electron acceleration in both sources has close relationship with each other.

One of possible acceleration mechanism to explain the above observational fact is interaction between the loop-1 and loop-3. It is also possible that electrons are accelerated separately in both loops by a high-speed reconnection flow from a magnetic reconnection region above both loops which is not observed in this event.

THE HEIGHT DISTIBUTION OF HARD X-RAYS IN IMPULSIVE SOLAR FLARES

LYNDSAY FLETCHER
Sterrenkundig Instituut, University of Utrecht
Postbus 80.000, 3508 TA Utrecht, The Netherlands

1. Introduction

Matsushita *et al.* (1992) used the Hard X-ray Telescope on board the Yohkoh satellite to image hard X-ray (HXR) emission in 100 impulsive flares, and found that low energy emission comes from higher in the corona than high energy emission. This confirmed observations by Takakura *et al* (1986) with the Hinotori imager. The HXR emission is thought to be bremsstrahlung of non-thermal electrons, but the observed heights are much greater than found in previous theoretical calculations. This led to speculation that a super-hot ($> 10^8$K) plasma was present in impulsive flare loop-tops. However, using a stochastic simulation of electron transport we show that the inferred height distribution can be explained by a non-thermal model.

2. The method

In a stochastic simulation, the progress of numerical test 'particles' is followed in a prescribed field medium by timestepping a set of stochastic differential equations (SDE) for each particle variable (velocity, pitch angle cosine, position), simulating at each timestep the diffusive (scattering) terms by a number chosen at random from a Gaussian distribution, with width depending on $\sqrt{dt}$ scaled to the local diffusion coefficient. The method is based on the equivalence of the set of SDE's and the Fokker-Planck equation, which describes the evolution of the particle distribution function. For further explanation see MacKinnon and Craig (1991). This method allows accurate modelling of the effects of many scattering and loss terms. In the present simulations, Coulomb collisions and magnetic field convergence are included, which limits applicability to regimes dominated by classical colli-

Y. Uchida et al. (eds.),
Magnetodynamic Phenomena in the Solar Atmosphere – Prototypes of Stellar Magnetic Activity, 549–550.

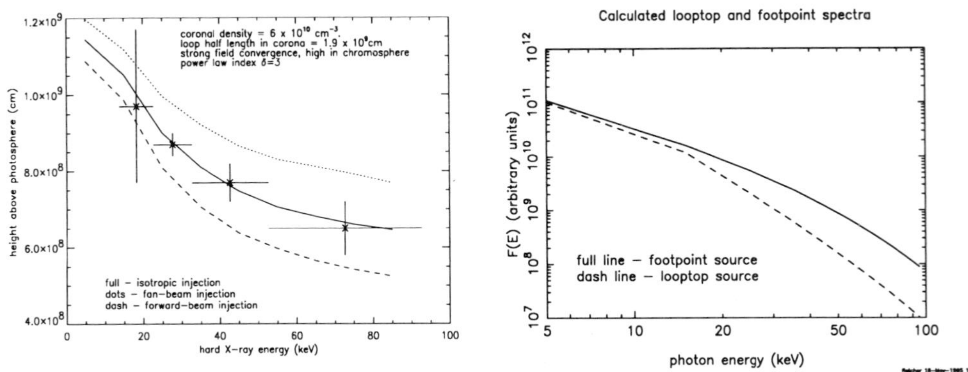

Figure 1. Calculated height distribution and HXR spectra. Loop density is $6\times10^{10}\mathrm{cm}^{-3}$.

sions (so, $T \sim 10^7$ K, $n \sim 10^{10} - 10^{11}$ cm^{-3}, if a beam density of $\sim 10^8\mathrm{cm}^{-3}$ is used - see van den Oord 1990). In the corona the loop model is semi-circular, with either a constant or exponentially varying field strength, and constant density. In the chromosphere, density varies exponentially and the field converges; the field strength and shape can be varied.

3. Results and Conclusions

In figure 1 we show the results of some simulations at a coronal density of $6 \times 10^{10}\mathrm{cm}^{-3}$, and the observations of Matsushita *et al.* It is evident that a non-thermal model can explain the present observations. The process is repeated for other loop and beam parameters (Fletcher, 1995) but with many variables in the simulation (beam pitch angle and energy distribution, loop length, density and field strength) and few data points, the quantitative predictions made are limited. But it is found that the theoretical lines only pass through the error boxes of all four data points for loop densities of $3 \times 10^{10} - 3 \times 10^{11}\mathrm{cm}^{-3}$ and lengths of $1.2 \times 10^9 - 2.7 \times 10^9$ cm. Because the chromosphere is a thick target, the HXR spectrum produced here will always be harder than at the top of the loop, which is less dense. The stochastic simulation also permits the angular distribution to be studied, and calculations of X-ray polarization fractions could further distinguish models if appropriate observations become available.

References

Fletcher, L.: 1996, Astron.Astrophys., in press

Gardiner, C.W.:1985 in H.Haken (ed.), Handbook of Stochastic Methods.

MacKinnon, A.L. and Craig, I.J.D.:1991, Astron.Astrophys. 251, 693.

Matsushita, K., Masuda, S., Kosugi,T., Inda, M. and Yaji, K.:1992, P.A.S.J/ 44. L89

van den Oord, G.H.J.: 1990, Astron.Astrophys. 234, 496n den Oord

A SOLAR FLARE IN THE FEI 5324 LINE ON 24 JUNE 1993

LIN YUANZHANG, ZHANG HONGQI AND ZHANG WENJIAN
Beijing Astronomical Observatory,
Chinese Academy of Sciences,
Beijing 100080, China

The solar telescope magnetograph of the Beijing Astronomical Observatory uses two wavelengths, FeI 5324 and H_β for the observations of the photosphere and the chromosphere (Ai and Hu, 1986). Among hundreds of H_β flares observed since 1987, only a few were accompanied with FeI 5324 emission. A solar two-ribbon flare of 2B/M9.7 class, with both H_β and FeI 5324 emissions, was observed in the region AR7529 (S13, E65) on 24 June 1993. The H_α emission of the flare started at 07:18.5UT, peaked at 07:30.2UT and ended at 09:12UT (Solar Geophysical Data, 1993).

The emission of the FeI 5324 line is originated from the lower photosphere. Our calculation, which is similar to that of Zhang(1986), shows that for this flare with a heliocentric angle of 65 degree, the emission region of the flare observed on FeI 5324 was located at a height of about 180km above $\tau_{5000}=1$. This is far lower than the excitation height (>273km) for the well-known white-light flare on 7 August 1972 (Machado and Rust, 1974). The FeI 5324 line is also not included in the list of outstanding spectral lines of solar flares complied by Švestka (1976). So it seems that the atmospheric excitation of most flares can not reach such depth of the FeI 5324 emission, and this is likely the reason why FeI 5324 flares are so rare.

It is well accepted that the regions of the initial energy release in solar flares are located in the corona and the optical flares are the secondary phenomena resulted in the energy transferring downward from the corona. Evidently, in a FeI 5324 flare, which is an indicator of the excitation in the lower photosphere, the disturbing energy from the upper atmosphere must be useful for understanding the physical process of flares comprehensively and thoroughly, and may be very tremendous in order to transport the energy downward to the dense atmosphere of the lower photosphere. In other words, a FeI 5324 flare must be ranked as a violent event. Therefore, FeI 5324 flares represent another extreme case in flares, and to study them

Y. Uchida et al. (eds.),
Magnetodynamic Phenomena in the Solar Atmosphere – Prototypes of Stellar Magnetic Activity, 551–552.

will be useful for clarifying some arguments in the study of white light flares.

For this flare, besides optical observations, there are rich data of radio observations (Solar Geophysical Data, 1993). The radio bursts of various frequencies, ranging from 33 MHz to 50000 MHz, were recorded by many observatories, including a set of complete records in 3.1, 5.2, 8.4, 11.8, 35.0, and 50.0 GHz at the Bern Observatory in Switzerland. Bern's records show clearly that both the beginning and peak times of the bursts shift gradually from low to high frequency, indicating a downward movement of the particles. Assuming that the source model for the slowly varying component of radio radiation (SVC) is valid during the flare and adopting an approximation of dipole magnetic field (Zhao, 1993), a velocity of about 1000km/s for the downward moving particles could be derived. In the meantime, from 07:24 UT to 09:00 UT, the meter bursts of types II, IV and IIIG were observed at the Observatories of IZMIRAN (Russia), Learmonth (Australia) and San Vito (Italy), indicating a particle movement outward with a velocity of 2500km/s. These confirm the description on the movements of the energetic particles in the flare model proposed by Sturrock (1968).

Based on the synthetic analyses of the optical and radio data, the possible mechanism of the FeI 5324 emission is suggested as follows: As a result of hitting the chromosphere by the particles, a condensation with a temperature higher than that below it was formed near the transition region($h \approx$ 1400 km). Then the lower photosphere was heated through the backwarming (Gan and Mauas, 1994), and the FeI 5324 flare occurred as an indicator of the excitation in the lower photosphere.

References

Ai, G. and Hu, Y. (1986) *Acta Astronomica Sinica*, **27**, 173

Gan, W. Q. and Mauas, P.J. D. (1994) *Ap. J.*, **430**, 891

Machado, M. E. and Rust, D. M. (1974) *Solar Phys.*, **38**, 499

Solar Geophysical Data (1993), No. 587, Part I; No. 588, Part I; No. 592, Part II, NOAA, Boulder.

Sturrock, P.A. (1968)*Structure and Development of Solar Active Regions* IAU Symp. No. 35, K. O. Kiepenhear eds. 471

Švestka, Z. (1976) *Solar Flares*, D. Reidel Publ. Co., Dordrecht, Holland, pp.51-53

Zhang, H. Q. (1986) *Acta Astronomica Sinica* **6** 249

Zhao, R. Y. (1993) *Astron. Soc. Pacific Conference Series*, **46**, 275

ARE WHITE-LIGHT FLARES RELATED TO HIGH-ENERGY PARTICLES?

C. FANG, M. D. DING, J. HU AND S. Y. YIN
Department of Astronomy, Nanjing University, Nanjing, China

Solar white-light flares (WLFs) are defined as those flares which are visible in optical continuum. Up to now, only less than 100 WLFs have been reported. WLFs are of great importance because they present the most extreme conditions in solar flares and provide a severe challenge to energy transport mechanisms and atmospheric models.

In studying the mechanisms and the atmospheric models of WLFs, it is important to distinguish two extreme types of WLF (Machado *et al.*, 1986): Type I, showing a Balmer jump and strong Balmer lines, and Type II, showing weak Balmer emission and without Balmer jump. Recently, we have analysed the spectra of four WLFs (Fang and Ding, 1995 and references therein). Among them, the flare of 1991 October 24, 1974 September 10 and 1974 November 11 belong to Type I WLF, while the 1979 September 19 event may be classified as a Type II WLF. By comparing the data of the four WLFs, we noticed that there are obvious differences in three respects for the two types of WLF:

(1) For the Type I WLF, there is a good time correlation between the peak of continuum emission (CE) and the associated peaks in hard X-ray (HXR) and the microwave radiation. The mean lag of the CE peaks for the three type I WLFs is less than about 10s. However, for the Type II WLF, the 1979 September 19 event, the maximum of CE appeared 2–3 minutes earlier than that of the radio burst.

(2) For the Type I WLF, there is a strong Balmer jump in the spectrum; while for the Type II WLF, the Balmer jump is very weak or disappears.

(3) For the Type I WLF, the Balmer lines are strong and very broad, with generally strong central reversals at the strongest Balmer lines ($H\alpha$ etc.). For example, in the case of the 1974 September 10 WLF, the Balmer lines as high as H15–H16 can be distinguished. The full widths of H14–H15 lines may attain as broad as 4–5 Å. On the contrary, for the 1979 September 19 event, as a typical Type II WLF, Balmer lines are weak and no Balmer lines higher than H11 appears.

By use of the non-LTE semi-empirical modelling, we pointed out that the atmospheric structures of the two types of WLFs are different: for

Y. Uchida et al. (eds.),
Magnetodynamic Phenomena in the Solar Atmosphere – Prototypes of Stellar Magnetic Activity, 553–554.

the Type I, the chromospheric temperature is greatly enhanced, while the increase of the photospheric temperature is relatively small; for the Type II, the situation is just the contrary. Our analyses also indicated some evidences which show that the Type I WLF being related to electron beam bombardment:

First of all, as indicated above, the maximum of CE in the Type I WLF coincides well with the peaks of HXR and microwave bursts, with a time lag less than about 10s. It is a strong evidence that indicates the Type I WLF being related to non-thermal particle bombardment (see also Neidig and Kane, 1993).

Second, the appearance of a strong Balmer jump implies the great enhancement of H_{bf} emission, which not only is a result of the obvious increase of temperature in the chromosphere, but also mainly comes from the non-thermal excitation and ionization of hydrogen by the electron beam bombardment (Fang *et al.*, 1993). The role of a proton beam is less important.

Third, one of the typical characteristics of the Type I WLF, i.e., strong and very broad Balmer lines with central reversal in the strongest Balmer lines, can be well explained by the electron beam bombardment.

It is obvious that electron beams can not heat the photosphere directly. However, the radiative backwarming proposed by Machado *et al.* (1989) may be a viable mechanism for the transportation of energy from the upper chromosphere to the lower atmosphere.

Our calculations indicated that if the flare occurs in the penumbra of a sunspot, the Balmer jump caused by non-thermal effects would be much stronger than in the photosphere and in the flaring atmosphere. It is also showed that the Balmer jump caused by the electron beam bombardment would be strong just at the beginning of the flare, when the coronal column mass is low. Another important point is that the continuum contrast depends much on the position of the flare at the solar disk.

Thus, the Type I WLF is probably related to electron beam bombardment, while the Type II WLF is not.

References

Fang, C., Hénoux, J. C. and Gan, W. Q. (1993) Diagnostics of non-thermal processes in chromospheric flares I., *Astron. Astrophys.*, **274**, 917

Fang, C. and Ding, M. D. (1995) On the spectral characteristics and atmospheric models of two types of white-light flares, *Astron. Astrophys. Suppl. Ser.*, **110**, 99

Machado, M. E. *et al.* (1986) White-light flares and atmospheric modeling, in D. F. Neidig (ed.), *The Lower Atmosphere of Solar Flares*, National Solar Observatory, Sunspot, NM, pp. 483

Machado, M. E., Emslie, A. G. and Avrett, E. H. (1989) Radiative backwarming in white-light flares, *Solar Phys.*, **124**, 303

Neidig, D. F. and Kane, S. R. (1993) Energetics and timing of the hard and soft X-ray emissions in white light flares, *Solar Phys.*, **143**, 201

X-RAY LINE SHIFT AND PLASMA FLOW IN SOLAR FLARE

M. TAKAHASHI
Graduate University for Advanced Studies
Mitaka, Tokyo 181, Japan

AND

T. WATANABE
National Astronomical Observatory of Japan
Mitaka, Tokyo 181, Japan

The flare of 1992 August 17 23:58 UT, which was studied by Enome et al. (1994) and Takahashi et al. (1995), occurred around the disk center of the sun. Because the flare had a single bright loop, the upflow of hot thermal plasma was able to be observed by Yohkoh with high time and spatial resolutions. Helium-like calcium (CaXIX) spectra observed by BCS give us information on the line-of-sight velocity of the evaporating plasma. By fitting two-component theoretical spectra, the upward velocities were obtained (see Table 1.). The velocity at 23:59:26 UT decreased to about three-fifth of that at the beginning of the flare, 23:58:56 UT.

On the other hand, the temporal changes of SXT intensity distributions (*Figure 1*) give the transverse velocity V_t, if evaporating plasma is regarded as a blob which corresponds to a SXT pixel of intensity peak in the time-differentiated SXT images. Consistency of the BCS vertical velocity and the SXT transverse velocity can be checked, if an upright and semi-circular

TABLE 1. Plasma velocity observed by *Yohkoh*

Time(UT)	BCS(km/s)	SXT(km/s)		
		V_t	V_l	V_s
23:58:56	280	180	460	490
23:59:06	260	350	280	450
23:59:16	190	180	70	190
23:59:26	170	180	20	180

Y. Uchida et al. (eds.),
Magnetodynamic Phenomena in the Solar Atmosphere – Prototypes of Stellar Magnetic Activity, 555–556.

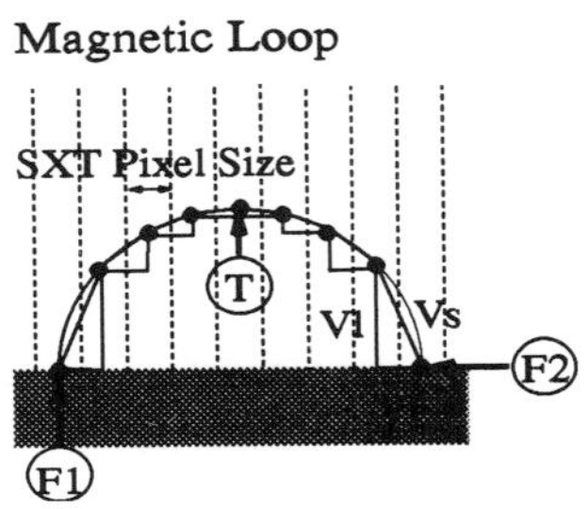

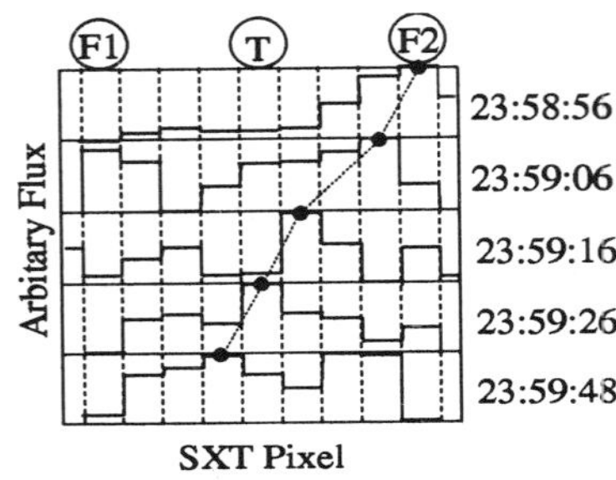

Figure 1. Schematic Diagram (left) and temporal changes of SXT intensity distrubutions (right) of the flaring loop. In right, circles mean the movement of plasma in the loop.

flaring loop is assumed. The vertical motion of the blob is derived from the observed transverse velocity and the blob position along the loop. Then, the blob velocity V_s in the loop can be calculated (Table 1.). The line-of-sight velocity V_l estimated from SXT continues to decrease from the beginning. The decrease rate ($-dV_l/dt$) is larger than that from BCS. The dissimilarity might come from the difference of plasma temperatures observed in the SXT Be-filter and the CaXIX resonance line.

The net increases of the thermal energy and the kinetic energy due to the excess line broadening during the period of 23:58:56 – 23:59:58 UT are 1.3×10^{29} ergs, and 1.8×10^{28} ergs, respectively. The net increase of the total energy is, therefore, 1.5×10^{29} ergs. The net enthalpy increase through the upflow is estimated to 1.8×10^{29} ergs, taking the radiative and conductive losses from the loop into account. The result is consistent with the scenario that thermal energy is injected from footpoints of a flaring loop (Antonucci et al. 1982)

The conductive and radiative loss of the SXT blob is estimated to be about 4.4×10^{27} ergs during its travel along the loop (23:58:46 – 23:59:16 UT), if the plasma parameters of the blob (log Te = 6.86 and log EM = 47.7) at the footpoints (23:58:46 UT) are used. The difference of the thermal energy of the blob between the epochs of 23:58:46 UT (when it was at the footpoints) and 23:59:16 UT (at the loop top) is about 7.7×10^{27} ergs. This implies that the blob was heated only at the footpoints and that no additional heating took place during its travel along the loop in the corona.

References

Antonucci, E. et al. (1982) Impulsive phase of flares in soft X-ray emission, Solar Physics **78**, 107–123.

Enome, S. et al. (1994) Alignment of radio, soft X-ray, hard X-ray images of sources in impulsive and gradual phases of the flare of 1992 August 17–18, PASJ **46**, L27–L31.

Takahashi, M. et al. (1995), The flare of 1992 August 17 23:58 UT, PASJ, submitted

X-RAY OBSERVATIONS OF AN OVER-THE-LIMB SOLAR FLARE WITH LARGE SPECTRAL LINE SHIFTS

A.C. STERLING
Computational Physics Inc.
2750 Prosperity Ave., Suite 600, Fairfax, VA 22031, USA[1]

L.K. HARRA-MURNION
Mullard Space Science Laboratory
U. College London, Holmbury St. Mary, Surrey RH5 6NT, UK

H.S. HUDSON
Institute for Astronomy, University of Hawai'i
Honolulu, HI 96822, USA[1]

AND

J.R. LEMEN AND K.T. STRONG
Lockheed Solar and Astrophysics Laboratory, Dept. 91-30
B/252, 3251 Hanover St., Palo Alto, CA 94304, USA

1. Introduction

Using data from the *Yohkoh* Bragg crystal spectrometer (BCS) and soft X-ray telescope (SXT), we study a solar flare of 15 April 1993 which occurred at about 9 UT. SXT observations indicate that the flare occurred well beyond the western solar limb, meaning that our observations are restricted to the upper-most portions of the flaring structure. Rise-phase BCS spectra show strong bulk lineshifts, that is, shifts of the entire spectral line profiles. Concurrent with these bulk lineshifts, the spectra show "blue wing" asymmetries. Both bulk lineshifts and blue wings are very infrequent characteristics of flares observed near the solar limb. Our studies give us insight into mass motions along flaring loops, and information about the uppermost regions of solar flares. Sterling et al. (1996, ApJ., in press) give a full report on this event.

[1]Current address: ISAS, Yoshinodai 3-1-1, Sagamihara, Kanagawa 229, Japan

Y. Uchida et al. (eds.),
Magnetodynamic Phenomena in the Solar Atmosphere – Prototypes of Stellar Magnetic Activity, 557–558.

2. Observations

We find for the flare of 15 April 1993:

1. The event occurs beyond the solar limb, with at least the lower 80×10^3 km of the flaring structures occulted by the solar limb.
2. BCS spectra from the Fe XXV (1.8298 – 1.8942 Å), Ca XIX (3.1631 – 3.1912 Å), and S XV (5.0160 – 5.1143 Å) channels show strong bulk blue shifts early in the rise phase ($\sim$ 9:02 – $\sim$ 9:04 UT).
3. BCS spectra later in the rise phase show distinct bulk red line shifts ($\sim$ 9:04 – $\sim$ 9:12 UT).
4. Concurrent with the bulk blue and red line shifts, BCS spectra also show blue wing asymmetries throughout the rise phase.
5. Non-thermal velocities during the event are highest at early times, and then decrease toward a constant, non-zero value later in the event.

3. Discussion

Our observations are consistent with assuming that the coronal structure of the 15 April 1993 flare consisted of a single loop oriented edge-on with respect to the Earth, with upward-directed mass flows along the loop. One leg of the loop would be nearer the Earth than the other leg. Most of the time, flows along the near-leg would be red-shifted as the material recedes away from Earth while climbing up the loop leg toward the loop apex. Similarly, flows along the far-leg would be blue-shifted, assuming upflows from each loop footpoint. Seen from Earth, more of the far-leg of the loop would be occulted by the Sun's limb than the near-leg, so that more of the red-shifted material will be visible than the blue-shifted material. That would lead to more of the observed emission measure being red-shifted than blue-shifted. This is just what we see in the BCS spectra: except for the earliest times, red-shifted emission contains most of the emission measure, resulting in bulk red shifts, whereas a minority of the emission measure is blue-shifted, resulting in blue wing asymmetries to the spectral lines rather than complete (bulk) blue shifts of the lines. If the bottom portion of the near-leg is inclined toward Earth, then motions along this lower portion of the loop can explain the early-time bulk blue shifts of the spectral lines.

We find non-thermal line broadenings in spectra of this flare which are qualitatively and quantitatively similar to line broadenings in spectra of non-occulted disk flares. This implies that the mechanism which creates the non-thermal line broadenings must be operative high in the coronal portion of the flaring structure. This is similar to the result found by Khan et al. (ApJ *Letters*, Oct 1995) for flares that were less occulted than the 15 April 1993 event.

DYNAMICAL RESPONSE OF THE SOLAR ATMOSPHERE TO FLARE HEATING

E.K. NEWTON, A.G. EMSLIE
Department of Physics
UAH, Huntsville, AL 35899, USA

AND

JOHN T. MARISKA
E. O. Hulburt Center for Space Research
Naval Research Laboratory, Washington, D.C. 20375

Our ability to test aspects of solar flare energy transport models hinges on our adequately characterizing the SXR–emitting plasma's motions. In particular, to test the electron beam model of energy transport, we must obtain an accurate representation of the plasma's velocity distribution, and thereby momentum, contemporaneously with profiles of the HXR flux and spectrum.

By defining the velocity differential emission measure (VDEM; Newton, Emslie, & Mariska 1995), we can obtain a physical measure of the amount of SXR–emitting plasma flowing at each velocity v, without approximation or parametric fitting. Formally we write the observed line profile $I(\lambda)$ as a *convolution* of this emission measure with a Doppler–displaced, unit line profile $\mathcal{K}(\lambda, v)$:

$$I(\lambda) = \frac{1}{4\pi D^2} \int VDEM(v)\ \mathcal{K}(\lambda, v)\ dv, \tag{1}$$

where

$$VDEM(v) = n_e^2 G(T) \frac{dV}{dv}. \tag{2}$$

VDEM measures the amount of power emitted by a volume element of plasma (with a specified electron number density n_e and emissivity $G(T)$) moving with a given line–of–sight velocity v. Simplistically, $\mathcal{K}(\lambda, v)$ is a

Y. Uchida et al. (eds.),
Magnetodynamic Phenomena in the Solar Atmosphere – Prototypes of Stellar Magnetic Activity, 559–560.

Gaussian centered on $\lambda_o(1 - v/c)$, but in practice the shape of this elementary profile is complicated somewhat by the presence of other spectral features. Spectral line profiles may be synthesized from theoretical hydrodynamic models by first forming VDEM(v) using equation (2) and then substituting the result in equation (1). Alternatively, we may *deconvolve* Ca XIX spectral line profiles observed by the *Yohkoh* BCS using equation (1) to recover VDEM and hence characterize the flare's SXR plasma flows.

We use VDEMs inferred from observed solar flares to compute the momentum M of the emitting plasma and its temporal evolution. We can then correlate this information with the HXR flux observed by the *Yohkoh* Hard X–Ray Telescope. For electron beam heating, we expect, from Newton's law and a more rigorous analysis of the hydrodynamic equations, that the second time derivative of the momentum will be proportional to the intensity of the HXR burst for short–duration events. This $\ddot{M} \sim I_{\rm HXR}$ relationship is born out by an analysis of 18 disk–center impulsive flares of varying durations. Impulsive flares with short–duration HXR bursts ($t \lesssim 20$s) exhibit the anticipated correlation. The longer the rise times and burst durations, the poorer the correlation, as expected. These results lend strong support to the idea that an energetic electron flux is indeed responsible for the SXR atmospheric response observed in solar flares.

VDEM can also be correlated with the *Yohkoh* Soft X–Ray Telescope (SXT) images. While VDEM provides information on the *line–of–sight* velocity component of the emitting plasma, changes in SXT images reflect the *transverse* velocity of the moving plasma. We are therefore able to build a more comprehensive understanding of the flow structure by correlating the appearance of VDEM features with cotemporal brightenings in the SXT images. In addition, SXT images permit a distinction to be made between actual plasma flows and the appearance of new emission in a different area of the active region, both of which are manifested in VDEM as plasma motion. In this way, we are able for the first time to couple spatially–unresolved spectroscopy with spatially–resolved images.

Newton, E.K., A.G. Emslie, & J.T. Mariska 1995, *ApJ*, 447, 915

EFFECTS OF NON-THERMAL PARTICLES ON THE UV AND OPTICAL SPECTRA OF SOLAR FLARES

J.-C. HENOUX
Observatoire de Paris, DASOP URA326, Meudon 92195, France

AND

C. FANG
Department of Astronomy, Nanjing University, China

1. Introduction

In solar flares, energetic electrons or protons can be present with non-maxwellian and non-isotropic velocity distribution functions. The simplest case is the atmospheric bombardment by electrons of more than 10 keV or by protons of more than 1 MeV. Beams of these particles are known to be detectable by their X-ray continuum or γ-ray line emission. However, non-thermal electrons and protons of energy as low as 100 keV can be detected by the use of less energetic photons in the UV and visible wavelength range.

2. Continuum Emission

2.0.1. *White light flares*

The enhanced hydrogen ionization due to electron beam bombardment leads to an increase of the recombination rates and consequently to an enhancement of the bound-free continua. A significant Balmer jump is one of the typical signatures of electron bombardment [1], [2].

2.0.2. *Negative flares*

A two steps process has been suggested to explain the heating of the upper photosphere in white light flares, i.e. a non-thermal ionization leading to an increase of the opacity of the upper photosphere and consequently followed in a few seconds by its radiative heating. Therefore, a small and short duration darkening is expected to precede the white light flares [3].

Y. Uchida et al. (eds.),
Magnetodynamic Phenomena in the Solar Atmosphere – Prototypes of Stellar Magnetic Activity, 561–562.

3. Line Profiles

3.1. ELECTRON BEAMS

As soon as the energy flux in the beam is higher than a few 10^{10} ergs cm^{-2} s^{-1}, the hydrogen Hα line intensity increases by a factor 2 to 5. The line is widely broadened and its central reversal becomes very large. In the UV, significant enhancements of the line wings of the Lyα and Ly β lines are predicted at the beginning of the impulsive phase of flares [4].

3.2. PROTON BEAMS

No strong broadening and no central reversal are expected from proton bombardment [5]. The main effect of protons on atomic lines profiles is the enhancement of the red wing of the profile of some hydrogen lines resulting from proton-hydrogen charge exchange. Negligible in the Hα line wing, the line emission resulting from proton-hydrogen charge exchange is expected to be significant in the red wings of the hydrogen Lyα and Lyβ lines [6].

4. Line Impact Polarization

Another spectroscopic signature of particle bombardment is the impact polarization of atomic lines. Protons beams can generate linear impact polarization in atomic lines in the radial direction [7]. Local anisotropies of the cold chromospheric protons associated to current closure during electron bombardment could also lead to radial linear polarization [8]. Electron beams may produce tangential line linear polarization [9].

References

1. Fang, C., Hénoux, J.C., and Gan, W.Q. (1993), *Astron. Astrophys.* **274**, pp. 917–922
2. Fang, C., Hénoux, J.C., Hu Ju, Xue Yin-zhang, Gao Xiu-fa and Fu Qi-Jun, (1995), *Solar Phys.* **157**, pp. 271–283
3. Hénoux, J.C., Aboudarham, J., Brown, J.C., Van Den Oord, G.H.J., Van Driel-Gesztelyi, L. and Gerlei, O. (1990), *Astron. Astrophys.* **233**, pp. 577–582
4. Hénoux, J.C., Fang, C., and Gan, W.Q. (1995), *Astron. Astrophys.* **297**, pp. 574–578
5. Hénoux, J.C., Fang, C., and Gan, W.Q. (1993), *Astron. Astrophys.* **274**, pp. 923–930
6. Fang, C., Feautrier, N. and Hénoux, J.C., (1995), *Astron. Astrophys.* **297**, pp. 854–860
7. Vogt, E., and Hénoux, J.C. (1995), Pulkovo Workshop on Solar Polarization, in press
8. Emslie, A.G., and Hénoux, J.C. (1995), *ApJ.* **446**, pp. 371–376
9. Firstova, N.M. and Boulatov, A.V. (1995), Pulkovo Workshop on Solar Polarization, in press

A STUDY OF THE THERMAL/NONTHERMAL ELECTRIC FIELD MODEL USING SPECTRAL AND SPATIAL X-RAY DATA FROM CGRO AND YOHKOH

T. A. KUCERA
National Research Council / NASA/Goddard
Code 682, NASA/GSFC, Greenbelt, MD 20771, U.S.A.

We have analyzed the solar flare of 1538 UT, 17 Feb. 1992 in terms of a model in which electric fields produce both thermal emission from Joule heated electrons and nonthermal emission from runaway electrons (Holman 1985, Benka & Holman 1994). The flare was observed by the Compton Gamma Ray Observatory Burst and Transient Source Experiment (BATSE), the Yohkoh satellite, and GOES. The Yohkoh images show a single soft X-ray loop with hard X-ray sources at the loop footpoints and at the looptop. We assume in our model that the electric fields are parallel to the length of the loop, the hard X-ray footpoint emission is nonthermal thick-target emission, and the looptop emission is thermal emission due to Joule heating. The measured BATSE hard X-ray spectra were fit with functions derived from the electric field model. The work is described in greater detail in Kucera et al. 1995.

We find the following:

1. It is not possible to determine unique spectral fits using the thermal/nonthermal model, because a wide range of model parameters result in essentially the same X-ray spectra. This is because the different sets of parameters produce identical electron distribution at high energies, and these high energy electrons dominate the entire nonthermal photon spectrum.
2. We are able to determine the temperature and emission measure in our model source ($T_e \sim 4 - 5 \times 10^7$ K, and $EM \sim 1 \times 10^{46} - 2 \times 10^{47}$ cm^{-3}). These values are similar to those of the hard X-ray loop-top source observed by the Yohkoh/Hard X-ray Telescope (HXT), although the values derived from HXT vary over a large range.

Y. Uchida et al. (eds.),
Magnetodynamic Phenomena in the Solar Atmosphere – Prototypes of Stellar Magnetic Activity, 563–564.

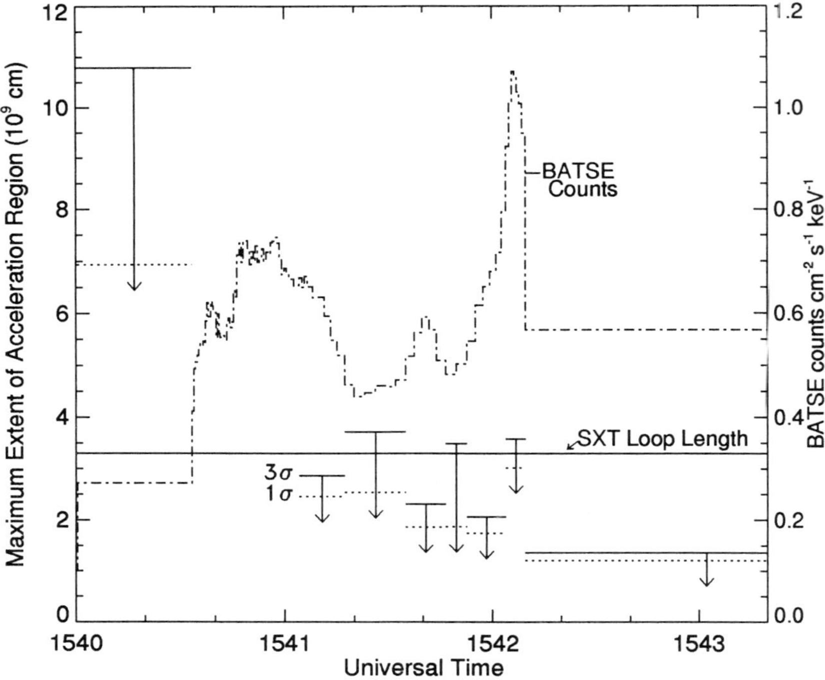

Figure 1. The 1σ (dotted line) and 3σ (solid line) upper limits on the length of the acceleration region at different times during the flare. No upper limits could be found from 1540:33 – 1541:04 UT, because there were no maximum values in the range of parameter space for which our model is valid. We believe that the time variation in the upper limit is due more to changes in the energy range of the flare data (determined by the measured count rate above the background level) than to actual changes in the length of the acceleration region. The dashed line is the BATSE Spectral Detector (SD) time profile (High time resolution SD data is only available from 1540:33 – 1542:10 UT).

3. Limits to nonthermal parameters were also calculated, including the extent of the region in which the electric field causes runaway electron acceleration (see Figure 1). We find that this distance must be shorter than the length of the flaring loop (*i.e.* $< 3 \times 10^9$ cm). Variations in the density and temperature could lead to this result, while still allowing the bulk drift current to reach the chromosphere.

References

Benka, S.G. & Holman, G.D. 1994, *ApJ.*, **435**, 469.

Holman, G.D. 1985, *ApJ.*, **239**, 584.

Kucera, T.A., Love, P.J., Dennis, B.R., Holman, G.D., Schwartz, R.A., & Zarro, D.M. 1995, *ApJ.*, submitted.

ENERGIZATION OF THE 10-100 KEV ELECTRONS IN SOLAR FLARES BY STRONGLY-DRIVEN RECONNECTION

R.L. MOORE
NASA/MSFC, ES82, Huntsville, AL 35812
T.N. LAROSA
Kennesaw State College, Marietta, GA 30061
J.A. MILLER
University of Alabama in Huntsville, Huntsville, AL 35899
AND
S.N. SHORE
Indiana University South Bend, South Bend, IN 46615

Abstract. We have developed a cascading-turbulence model for the energization of the electrons for the impulsive hard X-ray emission from flares (LaRosa & Moore 1993; LaRosa, Moore, & Shore 1994; Moore, LaRosa, & Orwig 1995). In our model, these electrons are energized via strongly-driven reconnection within the flaring magnetic field. The reconnecting fields press together over a large interface ($\gtrsim 10^9$ cm across in large flares). Because the field has fine-scale substructure (close-packed flux tubes, each $\lesssim 10^8$ cm in diameter), the reconnection is very disordered, proceeding in many fine-scale bursts thickly scattered over the interface. Each burst is the reconnection of a pair of opposing flux tubes. Because the reconnection is strongly driven, each reconnection burst lasts for about the Alfven crossing time of a flux tube. The outflows from each of these fine-scale reconnections are Alfvenic and hence MHD turbulent. The largest eddies in this turbulence have Alfvenic circulation and have diameters no larger than those of the reconnecting flux tubes. Each large eddy dissipates in one turnover time ($\sim$ its Alfven crossing time) by cascading to smaller scales. The compressive component of this cascading MHD turbulence dissipates into the electrons via Fermi acceleration. This process dumps energy into the electrons but not the protons because at coronal temperatures the electrons are fast enough to interact with the magnetic compressions in the turbulence but the pro-

Y. Uchida et al. (eds.),
Magnetodynamic Phenomena in the Solar Atmosphere – Prototypes of Stellar Magnetic Activity, 565–566.

tons are much too slow (see Figure 1). This dissipation gives the average electron the kinetic energy of a proton moving at the Alfven speed. Thus, for the Alfven speeds ($\sim 2\times 10^8$ cm/s) and electron densities ($\sim 10^{10}$ cm^{-3}) expected in the reconnecting coronal fields in active-region flares, each reconnection event produces a subsecond burst of $\lesssim 10^{34}$ electrons energized to $\sim$ 20 keV (having a total energy of $\lesssim 10^{27}$ erg), in agreement with the subsecond spikes observed to be typical in the impulsive-phase hard X-ray emission from flares. In large flares, $\sim$ 100 fine-scale reconnection events are in progress at once over the large-scale interface, bulk-energizing electrons to hard X-ray energies at a total rate of $\sim 10^{36}$ electrons/s or $\sim 10^{29}$ erg/s, as required by the observed hard X-ray emission.

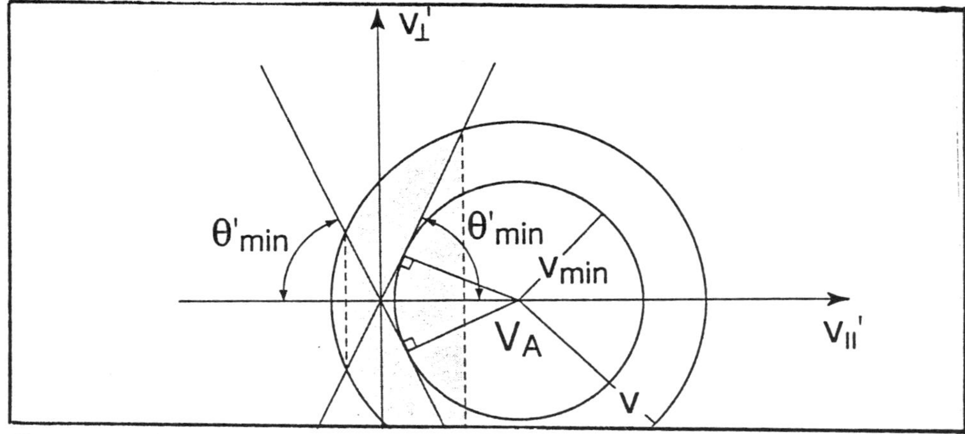

Figure 1. **Diagram of particle velocities $(v'_{\parallel}, v'_{\perp})$ in the frame of a moving magnetic compression.** This diagram shows that there is a minimum speed that particles (electrons and protons) must exceed to be reflected by the compression. Because the magnetic compression (of amplitude δB) propagates along the ambient magnetic field (of strength B_0) at nearly the Alfven speed V_A, the thermal electrons and protons are spherically distributed about a point near V_A on the $v'_{\parallel}$ axis. Only particles with large enough pitch angles Θ' can be reflected and Fermi accelerated by the moving magnetic compression; the minimum pitch angle for reflection is given by $\sin\Theta'_{min} = [B_0/(B_0+\delta B)]^{1/2}$. The diagram shows that particles with speeds less than $v_{min} = V_A \sin\Theta'_{min}$ have pitch angles less than Θ'_{min}. Therefore, particles with speeds less than v_{min} are not Fermi accelerated. For $(\delta B)/B_0 << 1$, $\sin\Theta'_{min} \approx 1$, and $v_{min} \approx V_A$. For coronal conditions in active regions (temperature $\sim 10^6$ K and $V_A \sim 2\times 10^8$ cm/s), most of the electrons have speeds greater than V_A but hardly any protons have speeds greater than V_A. Consequently, the electrons are efficiently Fermi accelerated by the magnetic compressions in the reconnection-driven MHD turbulence in flares but the protons are not.

References

LaRosa, T.N., & Moore, R.L. 1993, Astrophys. J., 418, 912.
LaRosa, T.N., Moore, R.L., & Shore, S.N. 1994, Astrophys. J., 425, 856.
Moore, R.L., LaRosa, T.N., & Orwig, L.E. 1995, Astrophys. J., 438, 985.

OBSERVATIONS OF THE CREATION OF HOT PLASMA IN SOLAR FLARES

S.K. SAVY
Mullard Space Science Laboratory
University College London, Surrey RH5 6NT, UK

Abstract.

During the impulsive phase of many solar flares, expanding soft X-ray sources are observed in high resolution soft X-ray images of the flare loops. Identifying the precise nature of the expansion and its relation to blueshifted emissions is important in resolving the mechanism of energy transfer to the hot plasma in the impulsive phase.

Fourteen flares were observed with the Soft X-Ray Telescope (SXT), Bragg Crystal Spectrometer (BCS), and the Hard X-Ray Telescope (HXT) on the *Yohkoh* satellite. Each of these events consisted mainly of a well resolved single soft X-ray loop with sources that expanded and merged during the impulsive phase. It was found that: *(i)* Most of the flares exhibited asymmetric line broadening due to blueshifted emissions. The intensity of the CaXIX blueshifted emission varied in proportion to the time differential of the stationary emission intensity, and in proportion to the hard X-ray flux. All the flares exhibited the Neupert effect. *(ii)* Soft X-ray footpoint sources expanded along coronal loops during the period of blueshifted emission. The flares eventually evolved into single stationary loop top sources at the end of the impulsive phase. Hard X-ray sources were spatially coincident with soft X-ray footpoints. *(iii)* Flow velocities estimated from soft X-ray source expansion rates in SXT spatial image sequences indicated the presence of high velocity mass motions ($\sim$150$-$550 km s^{-1}), and are consistent with velocities extracted from two-component fits to CaXIX spectra of some of the flares ($\sim$150$-$300 km s^{-1}), (see Table 1).

The results of the analysis show that flares with a simple loop structure and bright flows from footpoints exhibit asymmetric line broadening with bluewing intensities that evolve in proportion to the simultaneous hard X-ray count rate. The evolution of the bluewing in relation to the stationary

Y. Uchida et al. (eds.),
Magnetodynamic Phenomena in the Solar Atmosphere – Prototypes of Stellar Magnetic Activity, 567–568.

component is consistent with generation of both components by the same source of energy.

Spatial evolution of the soft X-ray flare is consistent with the evolution of the bluewing. Dynamically active footpoint sources rapidly evolve into a single stationary source during the impulsive phase, at the same time as the CaXIX resonance line rapidly evolves from a Doppler blueshifted position to a stationary position.

Correlation between the time derivative of the total soft X-ray signal and the hard X-ray light curves (the Neupert effect), and spatial coincidence between the expanding soft X-ray sources and hard X-ray sources, suggests, but does not demonstrate, that the soft X-ray flare is heated by electron bombardment of the chromosphere.

The spatially resolved plasma flows indicate that high velocity mass motions occur in the impulsive phase, as predicted by numerical simulations. Overall the results show that hot plasma is created in a moving state at coronal loop footpoints, however, a more detailed study of the flares in the sample is required to fully reconcile observations of chromospheric evaporation with numerical simulations.

TABLE 1. Flare plasma flow velocities.

Event (UT)		GOES Class	Location	Velocity (km s^{-1}) SXT Be119	Velocity (km s^{-1}) BCS CaXIX
7 Feb 1992	03:42	C6.2	S15W43	241 ± 20	186 ± 15
14 Jul 1992	17:47	C6.0	S09W32	551 ± 114	154 ± 14
5 Aug 1992	21:22	C5.1	S19E32	515 ± 37	-
11 Aug 1992	13:47	C7.2	S11W45	≥265±88	-
21 Aug 1992	10:57	M1.0	N14W40	176 ± 86	196 ± 11
6 Sep 1992	05:14	M2.4	S09W39	529 ± 91	150 ± 19
6 Sep 1992	06:54	M1.3	S13W32	265 ± 87	-
9 Sep 1992	12:44	-	-	191 ± 73	-
10 Sep 1992	22:51	M3.2	N12E41	212 ± 18	-
12 Sep 1992	15:38	C5.2	N18E21	481 ± 40	307 ± 17
8 Feb 1993	02:21	C9.2	S07E34	441 ± 111	245 ± 20
8 Feb 1993	07:00	C3.9	S07E36	529 ± 88	206 ± 22
2 Mar 1993	15:03	C5.0	S07E82	481 ± 40	198 ± 26
22 Apr 1993	14:07	M1.5	N11E04	331 ± 110	311 ± 15

DISCOVERY OF SUB-SECOND BRIGHTENINGS IN SOLAR FLARES WITH THE NOBEYAMA RADIOHELIOGRAPH

T. TAKANO

AND

THE NOBEYAMA RADIOHELIOGRAPH GROUP
Nobeyama Radio Observatory
Nobeyama, Minamisaku, Nagano 384-13, Japan

1. Introduction

The new Nobeyama Radioheliograph was constructed and started routine observations in 1992 June (Nakajima et al. 1994). It provides 17-GHz images of the whole sun with moderate spatial (~18 arcsec) and high temporal (up to 50 ms) resolution. Since the beginning we have observed many flares along with other interesting phenomena (see, e.g., a series of papers on Initial Results from the Nobeyama Radioheliograph, PASJ vol. **46**, No.1). Here we report discovery of what we call sub-second brightenings (SSBs) in solar flares.

2. Observations and Results

We searched rapidly-fluctuating phenomena during flares and found sub-second brightenings (duration of each pulse 50–200 ms) in four flares (1992/ Aug/12, 0232 UT; 1992/Sep/07, 0509 UT; 1992/Oct/08, 0537 UT; and 1992/ Oct/21, 0607 UT) from a dataset between 1992 August and October.

A typical example is shown in Fig. 1, which was observed in association with a GOES M1.3-class flare on 1992/Sep/07 in NOAA7276. Notice that the time profile shown only covers 30-s interval. Individual pulses are as short as 50–200 ms and repeat with intervals of 1–3 s. Though SSB sources are too small to be resolved with the ~18 arcsec beamsize, we estimate them to be less than about 4 arcsec from visibility data. They are located near a sunspot and no positional shift greater than 1 arcsec is observed.

Y. Uchida et al. (eds.),
Magnetodynamic Phenomena in the Solar Atmosphere – Prototypes of Stellar Magnetic Activity, 569–570.

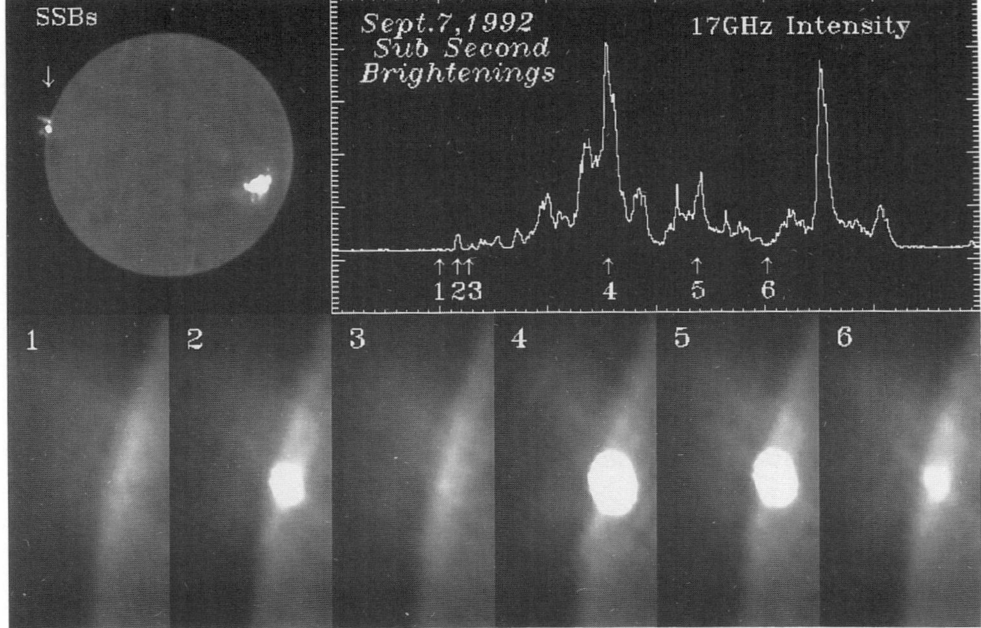

Figure 1. 17-GHz images of SSBs on 1992/Sep/07. Each frame is 79"×157" (solar north is to the top, west to the right). An image of the whole sun (upper left) and the light curve (upper right) are attached. Note that the light curve only covers 30 s from 05:09:09 UT. The images were obtained at the times indicated in the light curve.

3. Discussion and Conclusions

Most plausibly SSB sources are located at a foot point of a coronal loop and compact, less than 4 arcsec ($\sim$ 3000 km). The radio emission, thus, may be interpreted as radiation from high-energy electrons that precipitate down along the coronal loop. SSBs repeat with 1–3 s intervals, which may be interpreted by intermittent acceleration of electrons in a certain region in the coronal loop. The acceleration site need be quite small since the duration of each spike is 50–200 ms; it is estimated 50–500 km if we adopt the Alfven velocity of 1000–10000 km s^{-1} as the controlling speed of the acceleration process.

References

Nakajima H. et al. 1994, The Nobeyama Radioheliograph, *Special Issue of the Proceedings of the IEEE on the Design and Instrumentation of Antennas for Deep Space Telecommunications and Radio astronomy,* **Vol.82, No.5**, pp.705-713

GAMMA-RAY SPECTRAL CHARACTERISTICS OF SOLAR FLARES

K. MORIMOTO, M. YOSHIMORI, K. SUGA AND T. MATSUDA
Rikkyo University, Toshima-ku, Tokyo 171, Japan

AND

K. KAWABATA
Nagoya University, Chikusa-ku, Nagoya 464-01, Japan

1. Abstract

Yohkoh observed three impulsive gamma-ray flares with different spectral characteristics: a narrow line flare on 27 October 1991, a broad line flare on 15 November 1991 and a continuum flare on 3 December 1991. The narrow gamma-ray lines are produced by accelerated proton interactions with heavy nuclei in the solar atmosphere, the broad gamma-ray lines by accelerated heavy nuclei interactions with protons in the solar atmosphere (reverse interactions) and the continuum by relativistic electron bremsstrahlung. The gamma-ray count spectrum of the flare (X6.1 / 3B) on 27 October 1991 is shown in Fig. 1. Several strong narrow lines (neutron capture line and deexcitation lines) and broad lines are apparent. These lines result from accelerated particles of the coronal abundances [1]. Fermi acceleration is most probable, because it is most effective in the early stage of the acceleration and the accelerated particle abundances are similar to the source abundances. The gamma-ray count spectrum of the flare (X1.5 / 3B) on 15 November 1991 is shown in Fig. 2. The spectrum shows broad C and O lines and a complex of broad Fe, Mg, Ne and Si lines without detectable narrow lines. Since the acceleration rate of the shock acceleration is proportional to the ratio of square root of mass to charge, heavy nuclei are accelerated in a different manner than protons. Assuming a two-stage acceleration process, Fermi acceleration (first stage) plus shock acceleration (second stage), we can explain the observed high variabilities of the ratio of heavy nuclei to proton abundances (the ratio varies from 0.3 to 0.01)[3]. The gamma-ray

Y. Uchida et al. (eds.),
Magnetodynamic Phenomena in the Solar Atmosphere – Prototypes of Stellar Magnetic Activity, 571–572.

the electron bremsstrahlung continuum. This type of spectrum is called an electron-dominated event [3]. Since betatron acceleration is more effective for electrons than ions, this spectrum suggests the possibility that betatron acceleration was dominant in this flare.

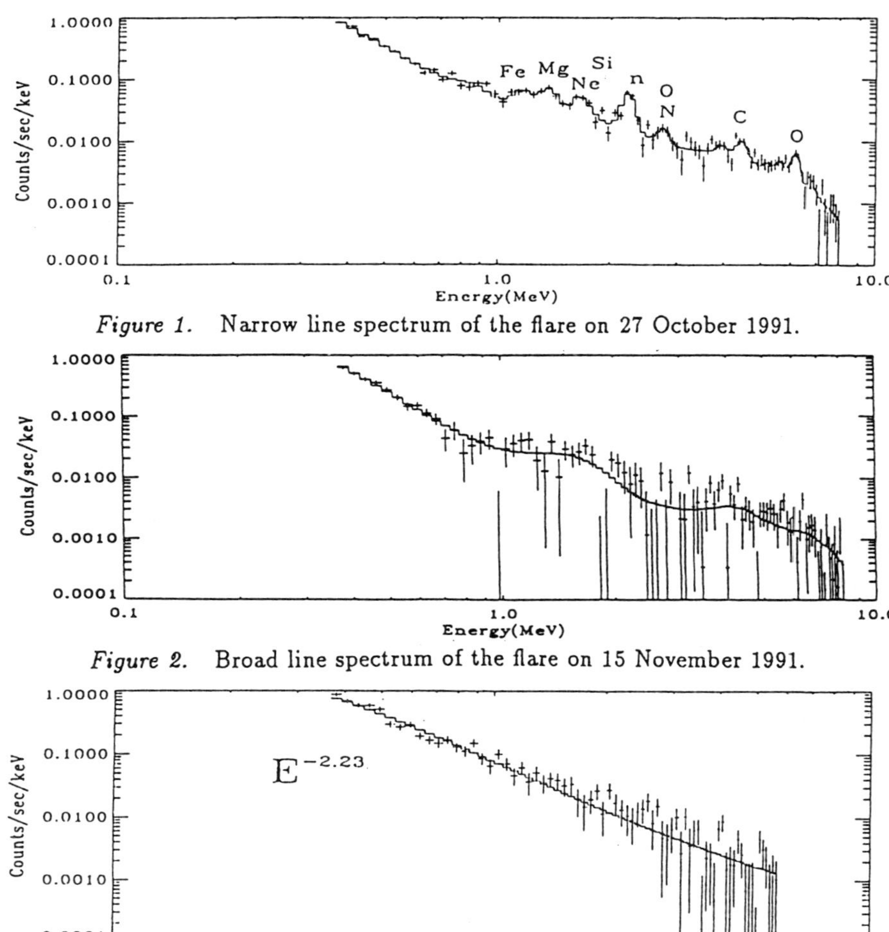

Figure 1. Narrow line spectrum of the flare on 27 October 1991.

Figure 2. Broad line spectrum of the flare on 15 November 1991.

Figure 3. Continuum spectrum of the flare on 3 December 1991.

References

[1]Murphy R.J. (1990) *Astrophys. J.* **358** 298.
[2]Reames D.V. (1990) *Astrophys. J.* **357** 259.
[3]Rieger E and Marschhauser H.Max'91 Workshop No.3: Max'91/SMM Solar Flares: Observations and Theory, p.68.

DOWNWARD BEAMING OF ACCELERATED PARTICLES IN THE 1991 NOVEMBER 15 FLARE

M.YOSHIMORI, K.MORIMOTO, K.SUGA AND T.MATSUDA
Rikkyo University, Toshima-ku, Tokyo 171, Japan

AND

K.KAWABATA
Nagoya University, Chikusa-ku, Nagoya 464-01, Japan

A Yohkoh hard X-ray spectrometer observed gamma-ray line features around 400 keV from a flare (X1.0 / 3B) at 22:35 UT on 15 November 1991. The time profile of hard X-ray count rate consisted of three significant peaks. The line features were detected at the third peak (22:37:46 - 22:37:50 UT) where strong hard X-rays and gamma-rays were emitted. A continuum-subtracted count spectrum of the hard X-ray spectrometer is shown in Fig. 1. The line features are interpreted in terms of the complex of Li (478 keV) and Be (429 keV) nuclear deexcitation lines which are produced by He(α, p)Li and He(α, n)Be reactions, respectively. Since the two line profiles are sensitive to the angular distribution of the interacting α particles, they can provide an essential key to the directivity of accelerated ions [1]. Although the two lines were not resolved due to low energy resolution of the NaI detector, we can derive the best fitting incident Li and Be line profiles using a hypothesis-testing approach. The derived center energies of the Li and Be lines are 431 $\pm$ 32 and 401 $\pm$ 32 keV, respectively, and the derived width of each line (FWHM) is 15 keV. The *Yohkoh* observation suggests that the two lines were redshifted by about 7 %. Murphy et al.[2] calculated the Li and Be line profiles in detail using a transport model. They took account both magnetic mirroring and MHD scattering of accelerated α particles in a flaring loop. Their calculation for a disk-centered flare such as the flare on 15 November indicated that the Li and Be lines are redshifted when α particles stream down to the chromosphere. The present result is consistent with a redshift of 7 %, suggesting the possibility of downward beaming of accelerated α particles. Downward beaming could be caused by strong pitch-angle scattering with MHD turbulence in the flaring loop.

Y. Uchida et al. (eds.),
Magnetodynamic Phenomena in the Solar Atmosphere – Prototypes of Stellar Magnetic Activity, 573–574.

Strong MHD turbulence is expected in impulsive flares. Sakao [3] studied the temporal evolution of the X-ray images in the flare on 15 November. The X-ray image in the 33 - 53 keV band showed a clear double source structure at the third peak. We analyzed the hard X-ray image in the 53 - 93 keV band. It is similar to that in the 33 - 53 keV band. The X-ray image in the 53 - 93 keV band is shown in Fig. 2. The double X-ray sources emitted hard X-rays simultaneously within a fraction of a second. The magnetogram observation showed that the hard X-ray double sources are located on both sides of the magnetic neutral line [4]. This suggests that the hard X-rays were emitted near the footpoints of a single flaring loop. It indicates the possibility that 100 keV electrons suffered strong pitch-angle scattering from MHD turbulence and efficiently precipitated to both footpoints of the single flaring loop. We conclude that the angular distribution of both the accelerated ions and electrons peaked in the downward direction in the flare on 15 November 1991.

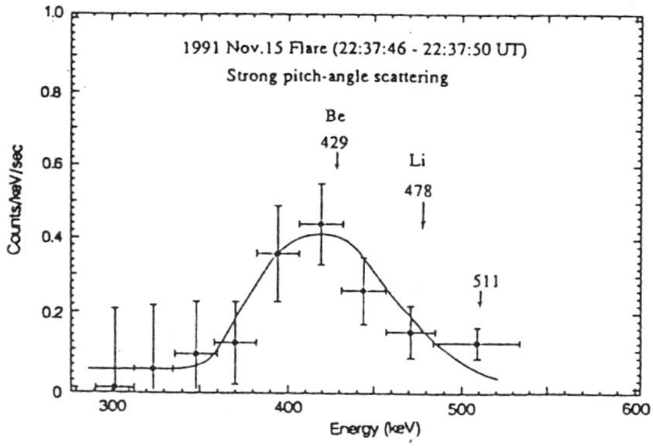

Figure 1. Continuum-subtracted count spectrum.

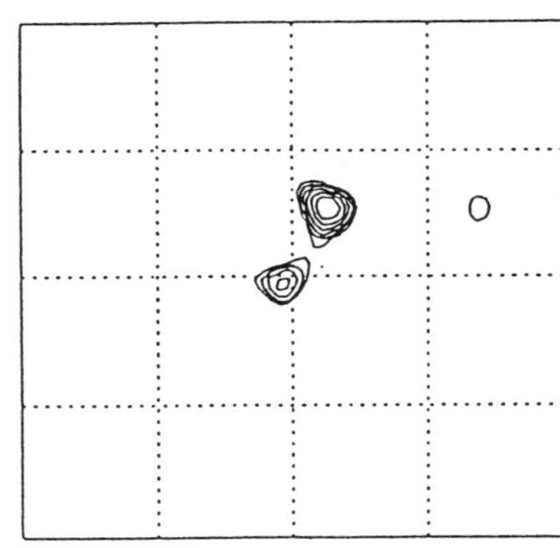

Figure 2. Hard X-ray image in the 53 - 93 keV band.

References

[1]Kozlovsky B. and Ramaty R. Astrophys. Letters Comm. **19** (1977) 19.
[2]Murphy R.J. et al. Astrophys. J. **351** (1990) 299.
[3]Sakao T. PhD Dissertation, University of Tokyo (1994).
[4]Canfield R.C. Publ. Astron. Soc. Japan **44** (1992) L111.

HIGH ENERGY LIMIT OF PARTICLE ACCELERATION IN SOLAR FLARES

K. OHKI
National Astronomical Observatory
Mitaka, Tokyo 181, Japan

In several intense γ-ray flares, a broad line-shaped spectral component peaking at about 70 MeV has been observed. This component is interpreted as π^0 decay emission produced with GeV protons. In the case of the largest flares like June 3, 1982 event, we have to assume a power-law spectrum for the accelerated protons. It would be important to determine the highest cut-off energy E_{cut} in the proton spectrum in order to decide the depth of the acceleration region if the acceleration is due to a shock. On the other hand, the p-p collision cross-section for π^0 production has a sharp low energy cutoff at around 300 MeV. Considering both high and low energy cutoffs in the energy distribution for protons, virtually a mono-energetic distribution of high energy protons might be responsible for the π^0 production. In this paper, we have calculated the line width of π^0 emission with various E_{cut} of the incident protons. In this calculation of γ-ray spectrum, the energy distribution of the π^0 is needed. The most important finding in this calculation is that the calculated 70-MeV line width varies very sensitively with the E_{cut} in the energy range 300MeV - 3GeV. Within this energy range, the line width varies more than a factor of two. We can thus determine the unique energy (the peak energy in a narrow distribution between the high and low energy cutoffs) of the accelerated protons by comparing the line width for each energy to the observed line width. Consequently we can determine the actual E_{cut} in the energy spectrum of accelerated protons by using the above obtained unique energy. The high energy cutoff thus obtained is about 1 GeV for the June 3, 1982 flare.

The high energy cutoff may vary from event to event. So, we have checked another event of June 11, 1991 which was observed by CGRO satellite, because the high energy γ-ray emission from this event extends to the higher energy above 1 GeV. The same kind of comparison of the observed width to the calculated γ-ray line width has been done, indicating that 2 GeV is the most probable unique energy and that the high energy limit of the accelerated protons should exceed several GeV in the case of this event.

Y. Uchida et al. (eds.),
Magnetodynamic Phenomena in the Solar Atmosphere – Prototypes of Stellar Magnetic Activity, 575.

CHARACTERISTICS OF STELLAR FLARES ON EV LAC FROM A LONG-TERM PHOTOMETRIC PROGRAM

G. LETO AND C. BUEMI
Institute of Radioastronomy, VLBI Station
C.P. 141, I-96017 Noto (SR), Italy

AND

M. RODONÒ[1,2] AND I. PAGANO[2]
[1] *Institute of Astronomy of Catania University,*
[2] *Catania Astrophysical Observatory*
V.le A. Doria 6, I-95125 Catania, Italy

Abstract. We present a statistical analysis of a large sample of homogeneous photoelectric monitoring of the red dwarf EV Lac. The data were collected in 10 years at Catania Astrophysical Observatory on Mt. Etna and consist of 175, 90 and 27 flares detected in the U, B and V bands during a total monitoring times of 861, 947, and 826 hours respectively.

The peak luminosity, the total energy emitted, the rise and decay times were determined for each flare. Full report about this work will be found in a forthcoming paper.

1. The Data

EV Lac (BD +43°4305; d=5 pc; Sp=dM4.5e) is a well known flare star, also classified as a BY Dra-type variable with a period $P=4^d.373\pm0^d.002$ (Pettersen 1980). In BY Dra stars the photometric light curve periodic modulation is usually interpreted as evidence of large photospheric spotted regions (see, e.g., Rodonò et al. 1986).

The observations were made from 1967 to 1977 using four telescopes at the Serra La Nave (SLN) mountain station of Catania Observatory: a 91-cm and two 30-cm Cassegrain, and a 61-cm quasi-Cassegrain reflectors, equipped with similar photometers and filters in order to match the standard U B V Johnson magnitude system.

Y. Uchida et al. (eds.),
Magnetodynamic Phenomena in the Solar Atmosphere – Prototypes of Stellar Magnetic Activity, 577–578.

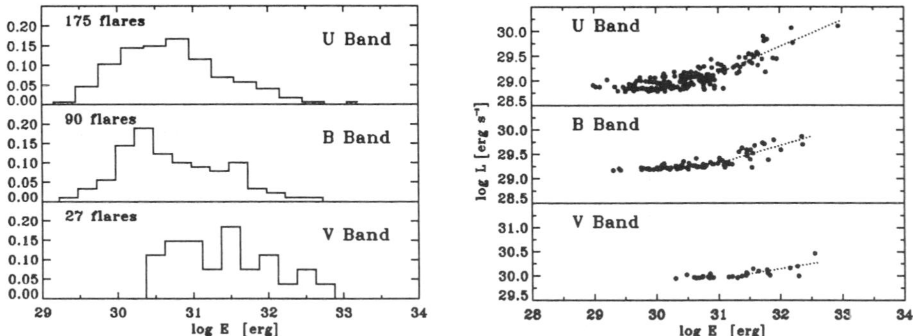

Figure 1. *Left panel:* The distributions of the U, B, and V flare number versus the total flare energy log E. *Right panel:* The peak luminosity (log L) versus the total energy emitted during the flare (log E).

2. Some results

We found mean flare occurrence rates of 0.20 h^{-1}, 0.09 h^{-1}, 0.03 h^{-1} for the U, B, and V bands, respectively.

In Fig. 1 (*left panel*) the distribution of the flare energies is given. At low energies the flare number decreases dramatically because of instrumental sensitivity limit. The coverage time of our sample is sufficiently extended to render significant a statistical analysis: in 861 h of patrol only 1 flare with $E > 10^{33}$ erg is expected (Shakhovskaya 1989), consistently with our data. Therefore the observed decrease of the number of flares at high energies is indicative of the maximum energy limit that can be released through the operating flare mechanism on EV Lac.

The distribution of ΔU at flare maximum can be represented by the empirical law: $N = 38\,e^{-(\Delta U - 0.4)}$ for $\Delta U \geq 0.4$. Similar behaviors are suggested by the ΔB and ΔV distributions.

In Fig. 1 (*right panel*) $\log L$ versus $\log E$ for the U, B and V bands are shown. These plots are suggestive of two different regimes operating at flare energies respectively fewer and higher than about $10^{31} erg$. For $E \geq 10^{31}\, erg$ the slope of the correlation between log L and log E is clearly larger than for $E \leq 10^{31}\, erg$. Actually, for $E \leq 10^{31}\, erg$ the peak luminosity appears to be independent of E.

More details can be found in a forthcoming paper (Leto et al. 1995).

References

Gershberg R.E., Grinin V.P., Ilyin I.V. et al. (1991) Flares on the red dwarf star EV Lac in 1986-1987, *Sov. Astron.*, **Vol. no. 35(3)**, pp. 269–277.

Leto G. et al., 1995, *in preparation.*

Pettersen B.R. (1980) Starspot and the rotation of the flare star EV Lac, *AJ*, **Vol. no. 85**, pp. 871–874.

Rodonò M., Cutispoto G., Pazzani V., et al (1986) Rotational modulation and flares in RS CVn and BY Dra type stars, *A&A* , **Vol. no. 165**, pp. 135–156.

Shakhovskaya N.I. (1989) Stellar flare statistics-physical consequences, *IAU Coll. (104), Solar Phys.*, **Vol. no. 121**, pp. 375–386.

MAXIMUM ENERGY OF FORCE-FREE MAGNETIC FIELDS

T. SAKURAI
National Astronomical Observatory
Mitaka, Tokyo, Japan

AND

J.C. CHAE
Department of Astronomy, Seoul National University
Seoul, Korea

Aly (1984) showed that the energy of any force-free field is bounded from above by the energy of the open magnetic field having the same distribution of magnetic flux on the boundary. His argument is based on integral relationships for force-free magnetic fields. The aim of this paper is to demonstrate this property directly by using a simple force-free field model formulated by Low and Lou (1990).

The model by Low and Lou (1990) is an axisymmetric solution for the force-free field equation

$$(\nabla \times \boldsymbol{B}) \times \boldsymbol{B} = 0,$$

where the field $\boldsymbol{B}$ is assumed to have the form

$$\boldsymbol{B} = \frac{1}{r \sin\theta} \left[\frac{1}{r} \frac{\partial A}{\partial \theta} \hat{\boldsymbol{r}} - \frac{\partial A}{\partial r} \hat{\boldsymbol{\theta}} + Q(A) \hat{\boldsymbol{\phi}} \right],$$

in spherical polar coordinates. Here $(\hat{\boldsymbol{r}}, \hat{\boldsymbol{\theta}}, \hat{\boldsymbol{\phi}})$ represent unit vectors in the (r, θ, ϕ)-directions, respectively. We assume a particular form for A and Q:

$$A = A_0 \frac{P(\mu)}{(r/a)^n}, \quad Q = k \frac{A_0}{a} \left(\frac{A}{A_0} \right)^{1+\frac{1}{n}},$$

where $\mu = \cos\theta$, and a, k, n and A_0 are constants. Then the equation reduces to

$$(1 - \mu^2) \frac{d^2 P}{d\mu^2} + n(n+1)P + \frac{n+1}{n} k^2 P^{1+\frac{1}{n}} = 0.$$

Y. Uchida et al. (eds.),
Magnetodynamic Phenomena in the Solar Atmosphere – Prototypes of Stellar Magnetic Activity, 579–580.

For the solutions to be regular at $\mu = \pm 1$, P should satisfy the boundary condition $P(\pm 1) = 0$. We adopted the normalization for P as $dP/d\mu = 10$ at $\mu = -1$, following Low and Lou (1990).

The solutions characterized by $k = 0$ and $n = 1, 2, 3, \ldots$ are current-free fields in the form of multipoles (dipole: $n = 1$, quadrupole: $n = 2$, hexapole: $n = 3$). Figure 1 shows the eigenvalues in the (k, n)-plane. As the field deviates from a current-free multipole solution, first k grows (i.e. the current density grows) to a certain value and then decreases asymptotically to zero. The value of n monotonically decreases to zero along the equilibrium sequence. Since the magnetic field decays as $B \sim r^{-n-2}$, small values of n indicate that the field decays very slowly in radial distance, namely a spatially extended field geometry.

We also evaluated the magnetic energy as a function of footpoint shear. We found that, as the shear increases, the energy of the force-free field approaches (but never exceeds) the energy of the open magnetic field.

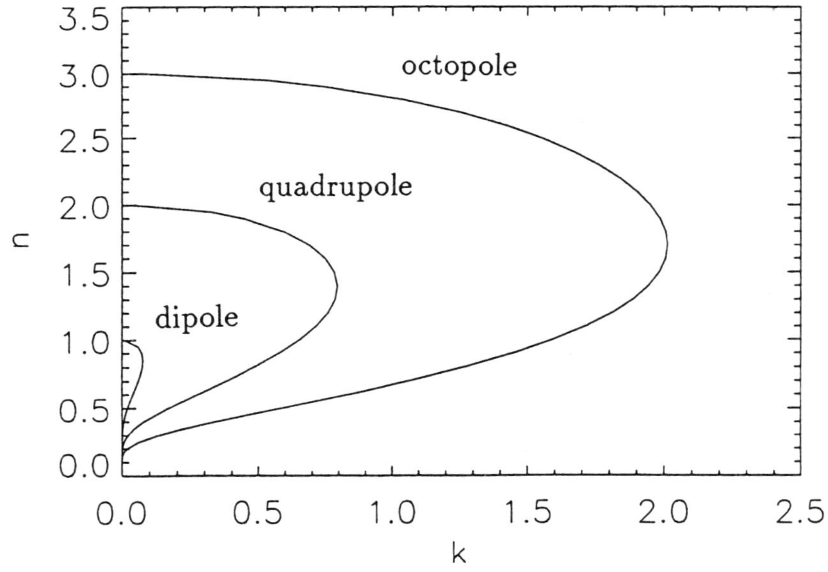

Figure 1. Eigenvalues k of Low and Lou's solution as a function of n.

References

Low, B.C. and Lou, Y.Q.(1990) Modeling of Solar Force-Free Magnetic Fields, *Astrophys. J.* **352**, 343–352.

Aly, J.J. (1984), On Some Properties of Force-Free Magnetic Fields in Infinite Regions of Space, *Astrophys. J.* **283**, 349–362.

A KINETIC MODEL OF MAGNETIC FIELD RECONNECTION

S.TAKEUCHI
Electrical Engineering and Computer Science
Yamanashi University, Kofu 400, Japan

Recently, various kinds of high-energy phenomena caused by solar activity which has been observed in the *Yohkoh* mission have been reported (PASJ, 1994). It has been suggested that magnetic field reconnection plays an important role in solar flares, in which high energy emmisions from the soft X-rays to the γ-rays are generated. The purpose of this article is to report fundamental features of the acceleration mechanism and the conversion process of energy in the magnetic field reconnection through physical pictures between electromagnetic fields and individual plasma particles. What we present here is not an MHD model but a non-steady kinetic model. The kinetic model is a plausible candidate to promise clear understanding of the reconnection process and particle dynamics from microscopic points of view.

We consider a configuration of electromagnetic fields in moving plasma in which two magnetized plasmas having anti-parallel magnetic field approach and collide each other (Takeuchi, 1994). A schematic model is shown in Fig.1(a). It is well known that particles are trapped around a magnetic neutral sheet and stay there with meandering motion. For the same reason, the particles in the present case are also trapped in the neutral sheet, and further it is accelerated by a transverse electric field. This implies that the particle will be continuously accelerated as long as it stays within the neutral sheet. The mechanism presented here is similar to the magnetic trapping acceleration (MTA) mechanism investigated by the author (Takeuchi *et al.*, 1987; Takeuchi *et al.*, 1990). Such efficient acceleration, however, can never continue limitlessly. Since the magnetic field exists across the neutral sheet, a particle is deffected by the Lorentz force during the acceleration and its energy gain will saturate sooner or later.

Numerical calculations have been carried out to verify the theoretical consideration. In the initial stage, many particles are randomly located within a box centered in the reconnection region. The moving velocity of

Y. Uchida et al. (eds.),
Magnetodynamic Phenomena in the Solar Atmosphere – Prototypes of Stellar Magnetic Activity, 581–582.

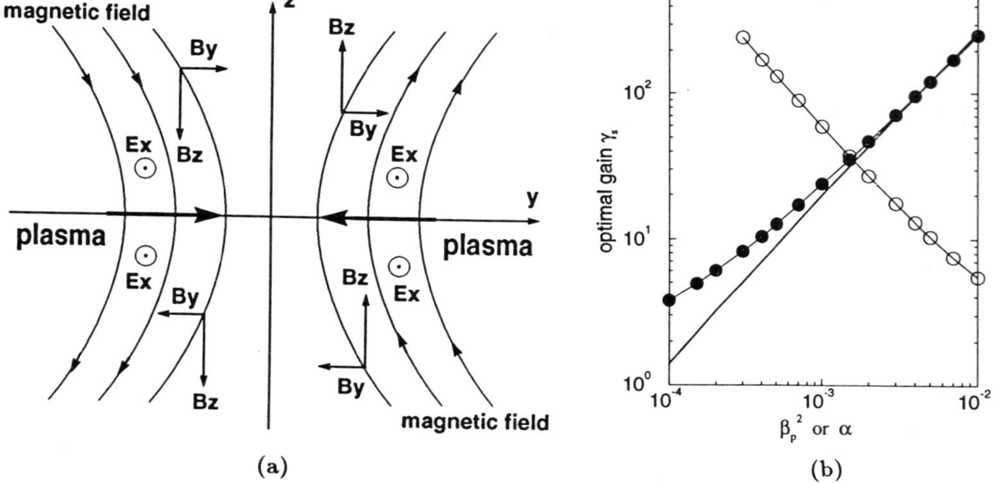

Figure 1. (a) Model of magnetic field reconnection; (b) Optimal gains: black circles for β_p, white circles for α, the solid line for eq.(1) with $\alpha = 10^{-3}$ and $\zeta_0 = 0.1$.

the two magnetized plasmas is β_p, which is given as $\beta_p \equiv |\mathbf{B} \times \mathbf{E}/B^2|$. In addition, the initial velocity of each test particle is also given by β_p, so that the particle moves together with the magnetized plasma. The optimal gains γ_s are plotted in Fig.1(b) for various values of β_p^2 and α, here the radius of curvature of magnetic field line is also an important parameter and its inverse value is given by α. In the figure, the optimal gains linearly increase proportional to β_p^2 and decreases proportional to α. As a result, a simple formula for the optimal gain is approximately given by

$$\gamma_s = A\beta_p^2/\alpha\zeta_0^{1/4}, \tag{1}$$

where $A \approx 15.0$ is a non-dimensional constant determined by numerical calculations and ζ_0 is the initial position of accelerated particle. The solid line in the figure stands for this relation.

References

Special Feature, Initial Results from Yohkoh, (1992) *Publ. Astron. Soc. Japan*,**44**, No.5.

Takeuchi, S. (1994) Particle Dynamics in Two Colliding Plasmas, *ed. S.Enome and T.Hirayama, Proceedings of Kofu Symposium, NRO Report No.360* , pp. 251-254.

Takeuchi, S., Sakai, K., Matsumoto, M. and Sugihara, R. (1987) Unlimited Acceleration of a Charged Particle by an Electromagnetic Wave with a Purely Transverse Electric Field, *Phys. Lett. A* **122** , pp. 257-261.

Takeuchi, S., Sakai, K., Matsumoto, M. and Kawata, S. (1990) Electric and Magnetic Fields Generated by Particles Trapped by an Electromagnetic Wave in a Transverse Static Magnetic Field, *J. Phys. Soc. Jpn.* **59**, pp. 152-158.

COMPUTER SIMULATIONS ON THE SPONTANEOUS FAST RECONNECTION MECHANISM

M. UGAI, T. SHIMIZU and T. OBAYASHI

Department of Computer Science, Faculty of Engineering
Ehime University, Matsuyama 790, Japan

It may now be clear that magnetic reconnection plays a crucial role in large dissipative events in space plasmas. Strong plasma heating just outside the loop top has recently been observed in compact flares and cited as evidence of magnetic reconnection (Masuda et al., 1994); for geomagnetic substorms too, large-scale plasmoids are observed to propagate and believed to result from sudden onset of fast reconnection. Thus, a very essential question may arise concerning how fast reconnection can develop so drastically as to cause solar flares where the magnetic Reynolds number is extremely large.

Historically, two theoretical models have been proposed with respect to the evolutionary mechanism of fast reconnection. One is the so-called externally driven model, where the fast reconnection mechanism involving slow shocks, first proposed by Petschek, is basically determined by external boundary conditions (Petschek, 1964; Forbes and Priest, 1987). On the other hand, we have proposed the spontaneous fast reconnection model, where the fast reconnection mechanism develops by the self-consistent coupling between (macroscopic) reconnection flows (MHD waves) and (microscopic) anomalous resistivities (Ugai, 1984, 1986): The fast reconnection mechanism develops when a current-driven anomalous resistivity is coupled to the global reconnection flow and locally enhanced near a neutral point. In this paper, we argue how the spontaneous fast reconnection model is applied to solar flares on the basis of our recent computer simulations. Here, the following two topics will be considered. The first is concerned with a large-scale plasmoid, which is one of the most distinct features of the spontaneous fast reconnection mechanism. The second is concerned with formation of large-scale magnetic loops and strong loop-top heating, which is of high current interest in view of recent *Yohkoh* observations.

(1) A large-scale plasmoid is the direct outcome of the spontaneous fast reconnection mechanism. If the system is open to free space, the plasmoid grows and propagates along the ambient magnetic field lines. The temporal dynamics of plasmoid was studied in detail by Ugai (1989, 1995). The plasmoid propagation speed is about $0.8V_{A0}$, where V_{A0} is the Alfven velocity measured ahead of the plasmoid. The plasmoid features obtained are found to be quite consistent with recent GEOTAIL sattelite observations.

(2) If plasma cannot flow across one end of the system, the reconnected field lines are accumulated near the boundary, giving rise to formation of a large-scale magnetic loop. The spontaneous fast reconnection.model was applied to this problem by Ugai (1987). It is demonstrated that a cusp-like field structure is formed near the boundary, consistent with Tsuneta (1995); when the magnetic loop has just begun to expand outwards, the reconnection jet collides with the loop most drastically, when a strong fast shock is caused. Figure 1 shows the result of Ugai (1987), where

Y. Uchida et al. (eds.),
Magnetodynamic Phenomena in the Solar Atmosphere – Prototypes of Stellar Magnetic Activity, 583–584.

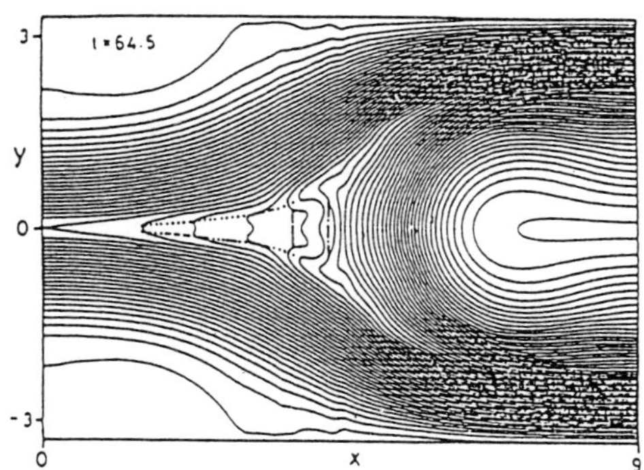

Figure 1. Development of a large magnetic loop, where dotted lines and dashed lines indicate the slow and fast shocks, respectively [after Ugai (1987)]. The right end boundary may correspond to the solar surface.

the fast shock causes strong plasma heating just ahead of the magnetic loop. This is in good agreement with the picture recently suggested from *Yohkoh* observations (Masuda et al., 1994). The formation of a fast shock had been suggested by other authors, but such a definite and strong fast shock as shown in Fig. 1 was first demonstrated by the spontaneous fast reconnection model.

We have argued that the spontaneous fast reconnection model should be quite applicable to solar flares. Tsuneta (1995) convincingly showed that a large flare occured without any definite external agnecy, in good agreement with this model. In the future, it will be important to study the three-dimensional processes as well as the microscopic plasma processes near a neutral point.

Refereces

Forbes, T. G., Malherbe, J. M. and Priest, E. R. (1989) The formation of flare loops by magnetic reconnection and chromospheric ablation, *Solar Phys.*, **120**, 285-307.

Forbes, T. G. and Priest, E. R. (1987) A comparison of analytical and numerical models for steadily driven magnetic reconnection, *Rev. Geophys.* **25**, 1583-1607.

Masuda, S., Kosugi, T., Hara, H., Tsuneta, S. and Ogawara, Y. (1994) A loop-top hard X-ray source in a compact solar flare as evidence for magnetic reconnection, *Nature* **371**, 495-497.

Petschek, H. E. (1964) Magnetic field annihilation, *NASA SP-50*,425-439.

Tsuneta, S. (1995) Structure and dynamics of magnetic reconnection in a solar flare, *Astrophys. J.* (in press).

Ugai, M. (1984) Self-consistent development of fast magnetic reconnection with anomalous plasma resistivity, *Plasma Phys. Controlled Fusion* **26**, 1549-1563.

Ugai, M. (1986) Global dynamics and rapid collapse of an isolated current-sheet system enclosed by free boundaries, *Phys. Fluids* **29**,3659-3667.

Ugai, M. (1987) Strong loop heating by fast reconnection in a closed system, *Geophys. Res. Lett.* **14**, 103-106.

Ugai, M. (1989) Computer studies of a large-scale plasmoid driven by spontaneous fast reconnection, *Phys. Fluids B* **1**, 942-948.

Ugai, M. (1995) Computer studies on plasmoid dynamics associated with the spontaneous fast reconnection mechanism, *Phys. Plasmas* **2**, 3320-3328.

NUMERICAL SIMULATION OF MAGNETIC RECONNECTION IN ERUPTIVE FLARES

T. MAGARA AND S. MINESHIGE
Dept. of Astronomy, Faculty of Science, Kyoto University Sakyo-ku, Kyoto, 606-01, Japan

AND

T. YOKOYAMA AND K. SHIBATA
National Astronomical Observatory, Mitaka, Tokyo 181, Japan

Prompted by the *Yohkoh* observations of solar flares, which have established the essential role of magnetic reconnection in the release of energy, we have studied the evolution of eruptive flares in some detail based on the reconnection model by means of two-dimensional magnetohydrodynamic (MHD) simulations. We are interested in what factor affects the time evolution of solar flares and how the related phenomena, particularly observed loop top source and plasmoid eruption, can be explained by this model. We have studied the dependence of the structure and evolution of the system on plasma β (ratio of gas pressure to magnetic pressure), adiabatic index, γ, and ρ_c (initial density in the current sheet). If the time scale and velocity are normalized by Alfvén time and Alfvén speed, respectively, we find that the main results (e.g., reconnection rate, plasmoid velocity, etc.) are rather insensitive to the plasma beta. The ρ_c-value, on the other hand, crucially affects the motion of a plasmoid. When the adiabatic index γ is small, corresponding to the case of efficient thermal conduction, plasma heating will be generally suppressed but the compression effect can be rather enhanced, which plays an important role of forming high-density loop-top sources. We discuss loop top sources, plasmoid eruption, and the rise motion of a loop in comparison with the observations. Our simulations can well account for the existence of the loop-top hard X-ray sources discovered in the impulsive flares. We concluded that both the impulsive flares and the LDE (long duration event) flares can be generally understood by the reconnection model for the *cusp-type* flares. More details will be presented elsewhere (e.g. Magara et al).

Y. Uchida et al. (eds.),
Magnetodynamic Phenomena in the Solar Atmosphere – Prototypes of Stellar Magnetic Activity, 585–586.

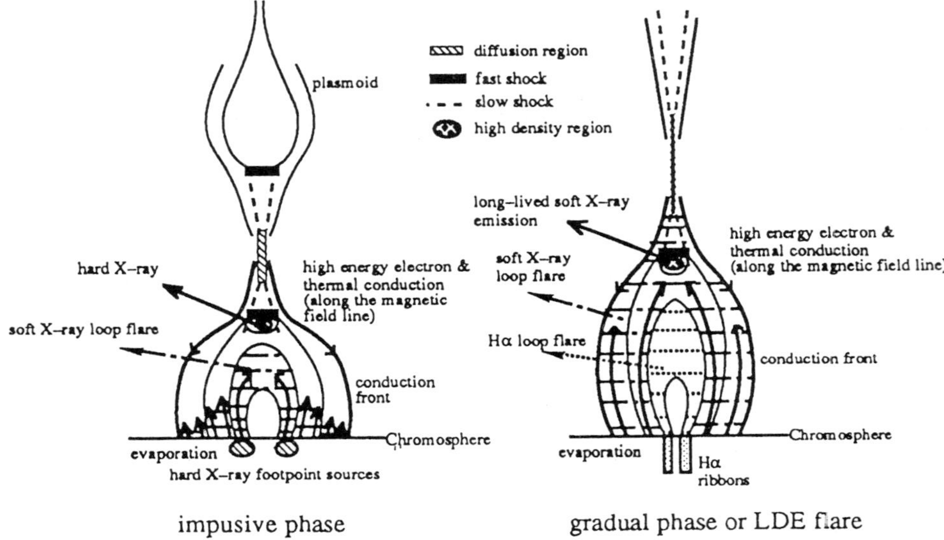

impusive phase gradual phase or LDE flare

In the impulsive phase, there are downward flows of high–energy electrons and heat conduction originating from the neutral point and directing along the magnetic fields toward the chromosphere. These downflows form the conduction front (cf. Forbes and Malherbe 1986) and heat the neighboring plasmas. The chromospheric gas is heated and evaporates into the corona. Since the emissivity is more sensitive to density than temperature, very hot regions will not always be observed, unless density is very high. We thus conjecture that the soft X–ray bright loop could be a dense region filled with evaporated gas. The reconnection jet collides with the loop top, forming the hot and dense core which will emit hard X–rays. If the evaporated flow has not yet reached this energetic region, the hard X-ray source above the soft X-ray flaring loop will be observed.

In the gradual phase or LDE flare, thermal conduction would be efficient and the reconnected loop under the neutral point has sufficiently filled with the evaporated flow. The conduction front will heat up the material along the boundary of the loop, which will be observed as a bright cusp-typed structure in soft X-rays. The loop top source is observed to be embedded by these evaporated gases. As for the inside loops, heating is no longer effective so that the loops can cool via radiative cooling process, thereby forming Hα post flare loops.

References

Forbes, T. G., and Malherbe, J. M. (1986) A SHOCK CONDENSATION MECHANISM FOR LOOP PROMINENCES, *ApJ*, **Vol. 302**, pp. L67–L70

Magara, T., Mineshige, S., Yokoyama, T., and Shibata, K. NUMERICAL SIMULATION OF MAGNETIC RECONNECTION IN ERUPTIVE FLARES submitted to *ApJ*

THREE DIMENSIONAL SIMULATION STUDY OF THE MAGNETOHYDRODYNAMIC RELAXATION PROCESS IN THE SOLAR CORONA

Y. SUZUKI, K. KUSANO, AND K. NISHIKAWA
Department of Materials Science, Faculty of Science, Hiroshima University, Higashi-hiroshima 739, Japan

1. Introduction

The magnetohydrodynamic (MHD) relaxation process in the solar coronal magnetic field is investigated by using a three-dimensional MHD numerical simulation. Our main objective is to explain the physical mechanisms of solar flares based on bifurcation theory, arguing that the solar flare is a transition process between two different solutions (the coupled solution (CS) and the mixed solution (MS)) of the linear force free field (LFFF) (Kusano *et al.*, 1995). To achieve this, the dynamics of a system of model magnetic loops, periodically aligned on the photosphere, is numerically considered. Magnetic helicity and magnetic energy are assumed to be injected into the coronal region by the rotational motion of the photosphere.

2. 3D Dynamics in Magnetic Loops

The simulation model is fundamentally the same as that in Kusano *et al.* 1994, but the simulation box is extended upward. We carry out five runs with different values for the electric resistivity η and the photospheric rotation period T_r.

Simulation results show that isolated flux tubes (plasmoids) are generated on the top of the magnetic loops through magnetic reconnection. It is also observed that during the reconnection process the magnetic energy of the component coupled with the boundary condition is partially transferred into the decoupled component. This causes the transition of the magnetic system between the CS and the MS. Figure 1 shows the time evolution of the magnetic helicity and the magnetic energy for five different runs. The

Y. Uchida et al. (eds.),
Magnetodynamic Phenomena in the Solar Atmosphere – Prototypes of Stellar Magnetic Activity, 587–588.

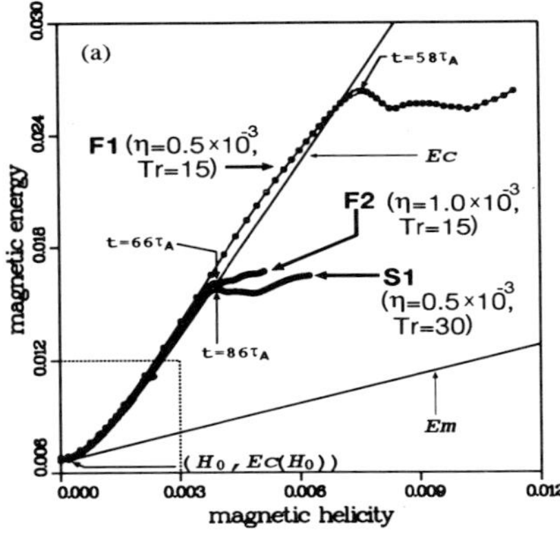

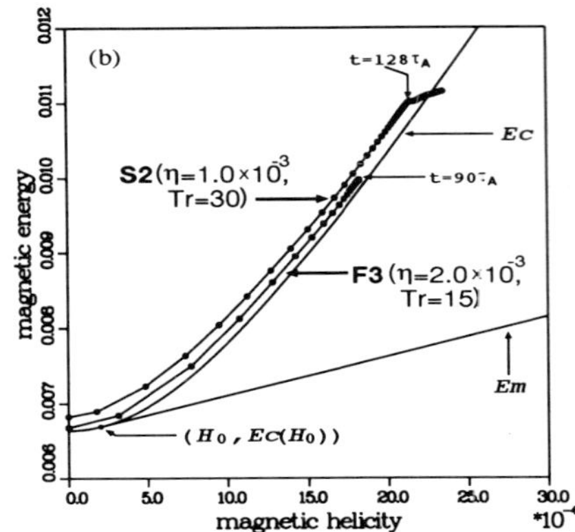

Figure 1. The time evolution of the magnetic energy and the magnetic helicity for five different simulation runs where the time is normalized by the Alfvén time. The small solid circles show the simulation results and the times when the transition takes place are represented. The curve E_c and the line E_m represent the energy for the CS and for the MS, respectively and H_0 is the magnetic helicity at the bifurcation point. Fig. 1(b) magnifies the small square region bounded by the dotted line in Fig. 1(a).

trajectories enter the lower region of the curve for the CS denoted by E_c. This means that a transition into the MS takes place. Furthermore we show that this dynamics is related with an instability which is destabilized when the magnetic loops rise above a critical height. These simulation results support our theoretical prediction (Kusano *et al.*, 1995).

Based on the results above, we propose the physical scenario of the solar flare as following: (1) The bifurcation of the LFFF, which is caused by the rising of the magnetic loops, destabilizes the magnetic configuration. (2) Instabilities grow exponentially. (In this linear phase the configuration of the magnetic loops is gradually transformed and a current sheet is generated.) (3) When it reaches the nonlinear phase, plasmoids are formed through magnetic reconnection. (4) If the resistivity is small enough, intensive heating is caused in a very narrow region and the transition takes place impulsively. A solar flare might be triggered.

References

Kusano, K., Suzuki, Y., Kubo, H., Miyoshi, T., and Nishikawa, K. (1994) Three dimensional simulation study of magnetohydrodynamic (MHD) relaxation process in solar corona. I, *The Astrophysical Journal*, **433. no. 1**, pp. 361–378.

Kusano, K., Suzuki, Y. and Nishikawa, K. (1995) A solar flare trigger mechanism based on the minimum energy principle, *The Astrophysical Journal*, **441. no. 2**, pp. 942–951.

COMPUTER SIMULATION AND VISUALIZATION OF SOLAR FLARES

G. A. HANOUN AND K. J. H. PHILLIPS
Space Science Department, Rutherford Appleton Laboratory
Chilton, Didcot, Oxon. OX11 0QX, UK

1. Introduction

It is commonly assumed that, after the impulsive stage of a solar flare, hot ($T_e \approx 10^7$ K) plasma fills a magnetic flux tube and subsequently cools by both conduction to the chromosphere (where the temperature is about 10^4 K) and radiation (see, e.g., Phillips 1992; Dennis and Schwartz 1989). On the "chromospheric evaporation" picture, the hot plasma is formed by the heating of non-thermal particles that are accelerated during the impulsive stage, travel down the loop legs and impact against the chromosphere where the amount of energy released is too large to be radiated away so convection of the heated chromospheric gas occurs. This paper describes a computer simulation of the later, thermal phase of a solar flare in which the heated gas cools by conduction and radiation, allowing for the possibility of further, continuous heating. The resulting computer program, Hyperion 95, combines analysis and visualization capabilities that allow the flare to be viewed as with an X-ray telescope.

2. The Hyperion 95 Program

The hot plasma produced at the flare impulsive stage is assumed to be contained in a static magnetic flux tube of circular cross section and with a radius that varies in a specified way with height. The heat equation,

$$3Nk\frac{\partial T}{\partial t} = \frac{1}{A(s)}\frac{\partial}{\partial s}\left[\kappa_0 T^{5/2} A(s)\frac{\partial T}{\partial s}\right] - \mathcal{R}(T) - \mathcal{Q}(s,t) \qquad (1)$$

Y. Uchida et al. (eds.),
Magnetodynamic Phenomena in the Solar Atmosphere – Prototypes of Stellar Magnetic Activity, 589–590.

was solved assuming electron density N constant with height, thermal conductivity coefficient equal to the Spitzer value, and the radiation loss function ($\mathcal{R}(T)$ given by Raymond & Smith (1976). The distance along the loop s is measured from the loop top, while various forms of the loop cross section area $A(s)$ were assumed. The boundary conditions are $(\partial T/\partial s) = 0$ at the loop top and $T(s = L, t)$ (L = loop semilength) set to the chromosphere's temperature (= 10^4K) respectively. The initial condition describing the form of $T(s, t = 0)$ can be chosen by the user; for most runs $T(s, t = 0)$ was taken to be isothermal apart from the footpoint, set equal to the chromospheric temperature. The energy input term $\mathcal{Q}$ was arbitrarily chosen to simulate the continuing energy input as in a Kopp–Pneuman model (Kopp and Pneuman 1976).

An implicit finite-differences method was used to solve (1) to find the distribution of temperature and radiation in X-ray bands to those on the *Yohkoh* Soft X-ray Telescope (SXT). An IDL program is then used to display these distributions. In practice the presence of bright loop-top sources in SXT flare images (see Acton *et al.* 1992) has proved very difficult to reproduce in the loop models calculated by the Hyperion 95 program. Such bright sources require either very dense or hot localized sources. A dense source in a flux tube with parallel magnetic fields should freely expand in a few seconds, while a temperature distribution with localized hot region should be rapidly (again in a few seconds) be smoothed out by the very high conductivity coefficient.

3. Summary

The Hyperion 95 program allows the simulation of flare loop models with arbitrary input, provided by either models or observational data, and the visualization of X-ray emission for comparison with flare images as from, e.g., the *Yohkoh* SXT instrument. The presence of bright loop-top sources in SXT images is difficult to explain with conductively cooling loop models.

References

Acton, L. W., Feldman, U., Bruner, M. E., Doschek, G. A., Hirayama, T., Hudson, H. S., Lemen, J. R., Ogawara, Y., Strong, K. T., and Tsuneta, S. (1992) *Pub. Astron. Soc. Japan*, **44,** L71.

Dennis, B. R. and R. A. Schwartz (1989) *Solar Phys.*, **121,** 75.

Kopp, R. A. and Pneuman, G. W. (1976) *Solar Phys.*, **50,** 85.

Phillips, K. J. H. (1992) *Guide to the Sun*, Chapter 6. Cambridge University Press.

Raymond, J. C., Cox, D. P., and Smith, B. W. (1976) *Astrophys. J.*, **204,** 290.

PINCH PROCESSES NEAR THE TRANSITION REGION AND ELECTRON BOMBARDMENT OF THE CHROMOSPHERE IN THE PRESENCE OF ELECTRIC CURRENT

MARIAN KARLICKÝ
Astronomical Institute of the Czech Academy of Sciences
251 65 Ondřejov, Czech Republic

1. Introduction

Recently, a 2-4 GHz type II-like radio burst observed during the impulsive phase of the February 27, 1992 flare was interpreted as the radio emission of an MHD (shock) wave generated near the transition region (Karlický and Odstrčil, 1994). Moreover, Hudson *et al.* (1994) show that during the impulsive phase of some flares the soft X-ray radiation comes from chromospheric layers. These phenomena indicate that processes near the transition region can be important for some flares, mainly for their impulsive phases. Analyzing the physical conditions in this region, two interesting processes were recognized:

2. Pinch Processes in Transition Region Induced by Radiative Losses

To demonstrate pinch processes we made a numerical model. In this model the initial temperature was constant throughout the whole current sheet and the value corresponding to the transition region temperature was chosen as $T_e=10^5$ K. The magnetic field out of the current sheet was chosen as 0.001 T, therefore the initial current density in the center of the current sheet is 0.025 A m^{-2}. The initial density was computed from the pressure equilibrium condition; the density outside the current sheet was chosen as $n_e=5\times10^{16}$ m^{-3}, and thus the maximum density in the current sheet center of $n_e=1.22\times10^{17}$ m^{-3} was found. Further we assumed that in the central part of the current sheet the heating is stopped and radiative losses are not compensated by this heating. Due to these radiative losses the cen-

Y. Uchida et al. (eds.),
***Magnetodynamic Phenomena in the Solar Atmosphere – Prototypes of Stellar Magnetic Activity*, 591–592.**

tral part of the current sheet cooled down rapidly, which led to a collapse of the current sheet. In our case, for example, during this process (after about 1 s) the current density in the current sheet increased to a value of 20 times greater than the intial one. Simultaneously, the current sheet was compressed and the MHD wave escaped from the space of the current sheet. We think that this process is important for the triggering of fast magnetic field reconnection and non-thermal processes during the impulsive phase of the solar flare. For example, it can be the case of the flare triggering after a MHD destabilization of the whole pre-flare magnetic field system, in deep layers, where the vertical current sheet is expected to be formed. For details, see Karlický and Jiřička (1995).

3. Electron Bombardment of the Chromosphere in the Presence of Electric Current

Assuming the current filament along the whole flare loop, we computed the Ohmic electric field in such a filament. For electric current densities in the corona $j_c = 6.53$ and 0.653 A m^{-2}, which correspond to the current drift velocities $v_D = v_{Ti}$ and 0.1 v_{Ti}, and for the VAL-C model of the solar atmosphere it was found that Ohmic electric field in the upper chromosphere is greater than that equivalent to collisional losses, mainly for sufficiently high current densities and electron energies. That means that in such a case for an appropriate orientation of electric current and electric field, the downwards propagating superthermal electrons can be accelerated in these layers and stopped in deeper layers. This effect increases the efficiency of hard X-ray emission of electron beams bombarding the chromosphere and thus reduces the electron transport problem known in the interpretation of the flare hard X-ray emission. For details, see Karlický (1995).

Acknowledgements. This work was supported by CSAS Grant 303404.

References

Hudson, H.S., Strong, K.T., Dennis, B.R., Zarro, D., Inda, M., Kosugi, T., and Sakao, T. (1994) Impulsive behavior in solar soft X-radiation, *Astrophys. J.* **422**, L25–L27.

Karlický, M. (1995) Electron beam bombardment of the chromosphere in a flare loop with electric current, *Astron. Astrophys.*, in press.

Karlický, M., Jiřička, K. (1995) 1-4 GHz type II-like radio bursts and pinch processes near the transition region, *Solar Phys.*, in press.

Karlický, M., Odstrčil, D. (1994) The generation of MHD shock waves during the impulsive phase of the February 27, 1992 flare, *Solar Phys.* **155**, 171–184.

ALFVÉN WAVE PROTON INJECTION AND ACCELERATION IN SOLAR FLARES

DEAN F. SMITH

Berkeley Research Associates and Department of Astrophysical, Planetary, and Atmospheric Sciences, University of Colorado, Boulder; also Guest Worker at NOAA Space Environment Laboratory, Boulder, Colorado, U.S.A.

JAMES A. MILLER

Department of Physics, University of Alabama, Huntsville, Alabama, U.S.A.

We consider coronal proton acceleration by a spectrum of parallel-propagating Alfvén waves. Both the nonlinear and linear aspects of this acceleration are included, but the Alfvén wave spectrum and total wave energy density are taken as fixed. The proton distribution is followed in detail numerically. The process of nonlinear Landau damping or beat-wave acceleration acts effectively to pull protons from the bulk to beyond the Alfvén velocity, the threshold for gyroresonant or linear acceleration. The gyroresonant acceleration then accelerates a density of 10^6 cm^{-3} of protons beyond 10 MeV with a Bessel function energy distribution as required by the observations. Results are shown in Figure 1 for a Kolmogorov spectrum of equal quantities of left- and right-hand polarized waves, moving in both directions along the magnetic field of energy density 1.8 ergs cm^{-3} at 3 s.

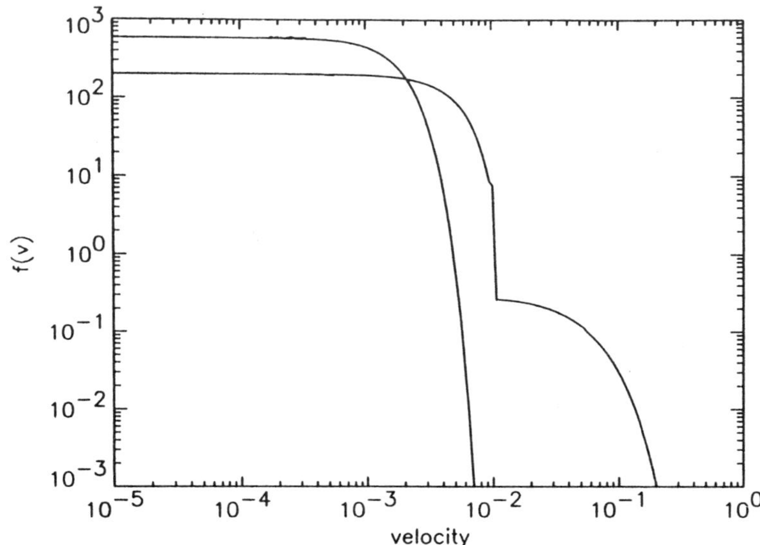

Fig. 1a. The velocity distributions in units of c of protons, initially and at 3 s (with the extended tail).

Y. Uchida et al. (eds.),
Magnetodynamic Phenomena in the Solar Atmosphere – Prototypes of Stellar Magnetic Activity, 593–594.

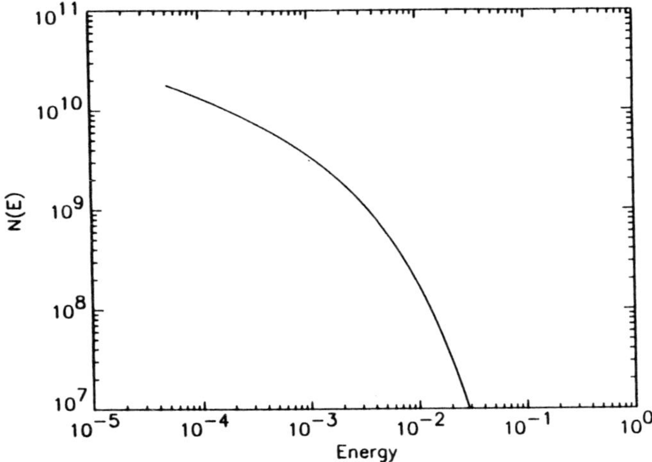

Fig. 1b. The energy distribution of the tail, in units of mc^2 at 3 s.

There is no production of >10 MeV protons in the first second for an initial temperature of 2.0 x 10^7 K and density of 5.2 x 10^9 cm^{-3}. The total energy expended is 100.2 ergs cm^{-3}, and the efficiency is 16.5% to beyond 10 MeV. These results satisfy all existing observations. Within the limitation of a fixed wave spectrum and level, this is the first demonstration that Alfvén wave turbulence alone can, without any auxiliary injection, produce all of the proton acceleration required in impulsive solar flares.

For more details, see Smith, D.F., and Miller, J.A. 1995, *Astrophys. J.*, 446, 390. This work was supported by NSF grant ATM–9214678.

3D MHD SIMULATION OF THE DYNAMIC CURRENT INJECTION MODEL OF LOOP FLARES

S. CABLE AND Y. UCHIDA
Dept. of Physics, Science University of Tokyo
R. MATSUMOTO
College of Arts and Sciences, Chiba University
Yayoi-cho, Chiba 260, JAPAN
AND
S. HIROSE
Dept. of Physics, Science University of Tokyo
1-3 Kagurazaka, Shinjuku-ku, Tokyo 162, JAPAN

Extended Abstract. Models of loop flares based on chromospheric evaporation caused by an *assumed* loop-top source of particles or heat leave unexplained the formation of such a source. Recent Yohkoh observations place certain restrictions on these models (see Doschek et al. (1995)). We suggest here a physical process for producing such a source, namely the dynamic current injection model of Uchida and Shibata (1988), and we report on our 3D MHD simulations of this model.

In this model, the flare loop is assumed to be part of a magnetic flux tube risen out of the photosphere. Twists in the flux tube field lines, generated in the convection zone, relax into the loop and drive chromospheric plasma into the high part of the loop. A single twist packet may produce "an active region transient brightening". If, with lower probability, there is one twist packet rising in each footpoint, the plasma streams collide in the high part of the loop. This, we suppose, is the origin of the simple loop flare.

Uchida and Shibata [1988] studied this model by 2.5D MHD simulations and were able to reproduce many qualitative features of observed flares. However, 2.5D simulations cannot account for some potentially important effects. Among these are the kink instability at the incidence of the twist packets and the effect of fast shocks which, in 3D, can propagate from one loop leg to the other directly through the outer corona.

Y. Uchida et al. (eds.),
Magnetodynamic Phenomena in the Solar Atmosphere – Prototypes of Stellar Magnetic Activity, 595–596.

We have performed full 3D MHD simulations of this model. The simulation results presented at this conference were obtained from the following initial conditions. The plasma density and pressure were set by a simple two temperature model: 3×10^6 K in the corona of our concern and 5000 K in the chromosphere and photosphere. A loop-shaped potential field with a span of about 15,000 km between the footpoints was assumed initially. Toroidal field packets were released from the phtosphere below the footpoints.

As the simulation progresses, plasma is driven into the corona, following the lines of the potential field. The plasma enters the corona in three distinct fronts, which we interpret as a fast shock front followed by a torsional Alfven front and a slow shock front. The first two of these fronts are are shown in the plots of thermal pressure in Figure 1. Following the collision of the fast fronts, the loop top density increases by a factor of about 4, the pressure by 100, and the temperature by 25.

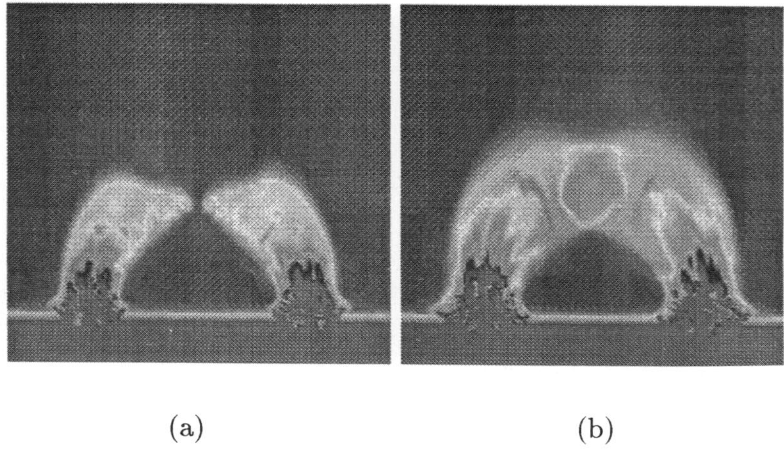

(a) (b)

Figure 1. Thermal pressure. (a) 4.3 seconds after release of magnetic twist. In each column, the front at the loop top represents an enhancement of a factor of 60 and the front about halfway up the loop represents an enhancement of 100 and higher. (b) 6.3 seconds after release of magnetic twist. The region at the loop top has a thermal pressure about 100 times that of the corona.

More extensive parameter studies intended for publication are in progress. One of the authors (S. Cable) thanks the Japan Society for the Promotion of Science Foreign Researcher Post-Doctoral Fellowship program for its support of his part in this research.

References

Doschek, G., Strong, K.T., Tsuneta, T. (1995) The Bright Knots at the Tops of Soft X-Ray Flare Loops: Quantitative Results from *Yohkoh*, *Ap. J.*, **440**, pp. 370–385.

Uchida, Y., Shibata, K. (1988) A Magnetodynamic Mechanism for the Heating of Emerging Magnetic Flux Tubes and Loop Flares, *Solar Phys.*, **116**, pp. 291–307.

NONLINEAR SIMULATION OF TEARING INSTABILITY OF A MAGNETIC ARCADE

SHINJI KOIDE
Toyama University
3190 Gofuku, Toyama 930 JAPAN

1. Introduction

It is an important problem of solar physics whether the coronal heating can be explained by tearing instability in coronal magnetic field. The "line-tied" boundary condition have been used for the ideal magnetohydrodynamical (MHD) evolution of coronal magnetic loops (Mikić, Barnes, and Schnack 1988; Van Hoven, Mok, and Mikić 1995). These boundary conditions were applied to tearing instability and it was found that unless the axial magnetic field possessed a null on some magnetic surfaces, solar coronal loops were completely stable to the tearing instability (Hood and Priest 1979; Mok and Van Hoven 1982). Hassam (1990) showed that the assumption of the line tying is inapplicable for the tearing instability because the growth rate is sub-Alfvénic in the entire loop, including the submerged part. He suggested that we have to consider the whole loop through the solar corona and photosphere. We have investigated the tearing instability of a magnetic arcade through the corona and photosphere by three-dimensional full MHD simulation code. We simplify the model of the tearing instability of the arcade magnetic sheet that the magnetic sheet meanders with artificial gravity and bouyancy and the current flows along the magnetic arcade.

2. Results

Vector plots of Figure 1 show the magnetic field ((a),(b)) and velocity ((c),(d)) on $x-y$ plane ($z=0$) ((a),(c)) and $z-y$ plane ($x=0$) ((b),(d)) at nonlinear stage $t=100\tau_A$. Here "Alfvén transit time" τ_A is defined to be a unit distance (as shown in Fig. 1) divided by Alfvén speed in the initial state. The contour maps of Figs. 1 show density ((a),(c)) and pressure

Y. Uchida et al. (eds.),
Magnetodynamic Phenomena in the Solar Atmosphere – Prototypes of Stellar Magnetic Activity, 597–598.

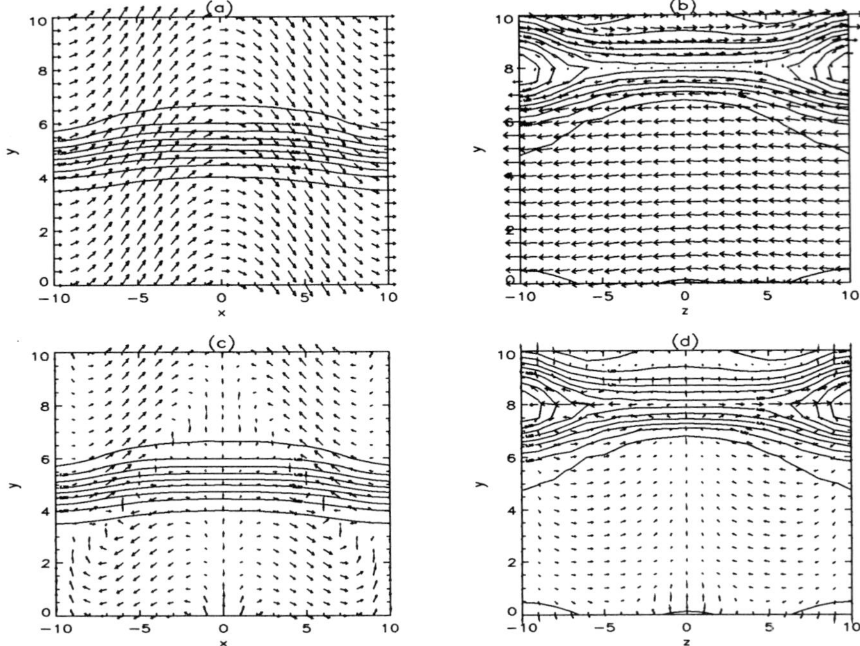

Figure 1. The arcade tearing instability through the corona and photosphere

((b),(d)). The contour maps of the density show the lower, dense part (photosphere) and the upper, rare part (corona). The magnetic sheet meanders through the photosphere part and coronal part (Fig. 1(a)). The current sheet located along one of the magnetic sheet and forms the reversed magnetic field of z-direction (Fig. 1(b)). We can see the magnetic islands at the edges of the current sheet(Fig. 1 (b)). High pressure region are seen in the magnetic island (Fig. 1 (b)). Fig. 1(d) shows that the plasma flows into the the current sheet at $z = 0$, and accelerates into z-direction strongly. This line is identified with the magnetic reconnection point (X-point) of tearing instability. This acceleration of the plasma makes the magnetic energy convert the kinetic energy. This mechanism in the arcade sheet would explain the coronal heating.

References

Hassam, A. B. 1990, *Astrophysical Journal,* **440**, L105.
Hood, A. W., and Priest, E. R. 1979, *Solar Phys.,* **64**, 303.
Mikić, Z., Barnes, D. C., and Schnack, D., D. 1988, *Astrophysical Journal,* **328**, 830-847.
Mok, Y., and Van Hoven, G. 1982, *Phys. Fluids,***25**, 636.
Van Hoven, G., Mok, Y., and Mikić, Z. 1995, *Astrophysical Journal,* **440**, L105-108.

FORMATION AND EMERGENCE OF CURRENT LOOPS IN WEAKLY IONIZED PLASMAS

J.I. SAKAI, M. SUZUKI
Faculty of Engineering, Toyama University, Toyama 930 Japan

AND

T. FUSHIKI
Hokuden Information System Service, 3-1, Sakurabashi, Toyama, 930 Japan

Magnetic field dynamics in weakly ionized plasmas is an important research subject in both astrophysical and laboratory plasmas. There is great interest in understanding the generation of magnetic field and filament structure (current loops) in weakly inonized plasmas (Brandenburg and Zweibel, 1994, Patel and Pudritz, 1994). Recently Suzuki and Sakai (1995) investigated nonlinear dynamics of Alfvén waves in weakly ionized plasmas, and found the formation of current filament structures associated with the Alfvénic perturbation. We show simulation results for the formation of current loops under the external gravity due to rotational motions in weakly ionized plasmas.

We used a 3-D simulation code (Suzuki and Sakai, 1995) describing a weakly ionized plasma with coupling between ions and neutrals caused by ion-neutral collisions. It is assumed that ion inertia and ion pressure terms can be neglected as compared with Lorentz force and friction force in the ion equation of motion. Then the induction equation for magnetic field is modified with a nonlinear term (called the ambipolar diffusion) as follows,

$$\frac{\partial \rho_n}{\partial t} + \nabla \cdot (\rho_n \mathbf{v}_n) = 0 \tag{1}$$

$$\rho_n \left\{ \frac{\partial \mathbf{v}_n}{\partial t} + (\mathbf{v}_n \cdot \nabla)\mathbf{v}_n \right\} = -\nabla p_n + \frac{1}{4\pi}(\nabla \times \mathbf{B}) \times \mathbf{B} \tag{2}$$

$$\frac{\partial \mathbf{B}}{\partial t} = \nabla \times (\mathbf{v}_n \times \mathbf{B}) + \frac{c^2}{4\pi\sigma}\Delta \mathbf{B} + \nabla \times \left\{ \frac{1}{4\pi \rho_i \rho_n \gamma_c} \mathbf{B} \times [\mathbf{B} \times (\nabla \times \mathbf{B})] \right\} \tag{3}$$

Y. Uchida et al. (eds.),
Magnetodynamic Phenomena in the Solar Atmosphere – Prototypes of Stellar Magnetic Activity, 599–600.

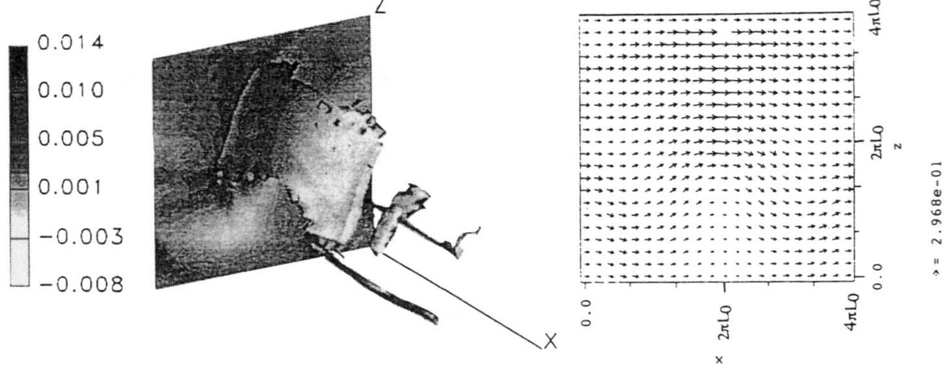

The system size in the z-direction is $4\pi L_O$, while the scale height is about $100L_O$. The initial density and pressure are taken as $\rho = p = e^{-z/100}$, and $B_x = 0.2, B_y = B_z = 0$. The imposed initial velocities are

$$v_x = 0.1(tanh(z - 0.5\pi) - tanh(z - 2\pi))sin(0.5x)$$
$$v_y = 0.5(tanh(z - 0.5\pi) - tanh(z - 2\pi))sin(0.5x)$$
$$v_z = 0.5(tanh(z - 0.5\pi) - tanh(z - 2\pi))(cos(0.5y) - 0.2sin(3x/4))$$

These vortex motions cause filament currents along the magnetic field (B_x). Due to the velocity perturbation (v_x), the current loops formed emerge upwards as seen in the figure (left). This figure shows the isosurface of current intensity $J_z = 0.00447$ at $t = 5\tau_A$. The gray scale shows the intensity of the current J_z on a plane ($x = 0$). The right figure shows the vector plot of magnetic field (B_x-B_z) on a plane ($y = 2\pi L_O$) at $t = 5\tau_A$.

The conclusions are summarized as follows: (1) The convective motions produce current loops, which emerge with upward velocity of about 0.2 v_A. (2) About half of the convective kinetic energy is converted to magnetic energy associated with current loops, which are transfered upwards. (3) The effect of ambipolar diffusion (A_D=(ion-neutral collision time)/τ_A =$10^{-5} \sim 10^{-1}$) is weak for the formation of the current loops.

One of the authors (J. Sakai) is partly supported by a Grant-in-Aid for Scientific Research from the Ministry of Education (07640352) and Densoku Company.

References

Brandenburg, A. and Zweibel, E.G. :1994, Astophys. J. 427, L91.

Patel, K. and Pudritz, R.E. :1994, Astrophys. J. 424, 688.

Suzuki, M. and Sakai, J.I.: 1995, PAFS-33, Research Report on Plasma Astrophysics and Fusion Science, Toyama Univ., appear in Ap. J.

FORMATION OF CURRENT LOOPS DRIVEN BY THE KELVIN-HELMHOLTZ INSTABILITY

Y. HASEGAWA AND J. I. SAKAI
Faculty of Engineering, Toyama University, Toyama, 930 Japan

AND

K. I. NISHIKAWA
Space Physics Department, Rice University, Houston, Texas 77005

1. Introduction

We consider the generation mechanism of current loops which may be related to convective motions near the Sun's surface. Recently Sakai and Fushiki (1995) showed that a shell current loop with a diffuse closure current can be generated due to the twisting motions of the loop footpoints. Among a variety of convective motions the Kelvin-Helmholtz instability triggered in shear flows is also important.

2. Initial Condition

We present results for the generation process of current loops driven by the Kelvin-Helmholtz instability by using a three-dimensional ideal MHD (ZEUS-3D) code. The length is normalized by L_0 and the time is normalized by $\tau_A = L_0/v_A$. The system size are $0 \leq x \leq 2\pi L_0$, $0 \leq y \leq 2\pi L_0$ *and* $0 \leq z \leq 2\pi L_0$ in x, y and z directions. The initial velocity field and plasma density are given as follows:

$$V_x(x,y,z) = V_z(x,y,z) = 0,$$

$$V_y(x,y,z) = (V_0/2)\tanh x(1 - \tanh z),$$

$$B_x(x,y,z) = B_y(x,y,z) = 0,$$

Y. Uchida et al. (eds.),
Magnetodynamic Phenomena in the Solar Atmosphere – Prototypes of Stellar Magnetic Activity, 601–602.

$$B_z(x,y,z) = B_0,$$

$$\rho(x,y,z) = \begin{cases} \rho_0 & (uniform) \\ 1+\rho_0(1-\tanh z)/2 & (non-uniform) \end{cases}$$

where V_y is the shear velocity, (B_x, B_y, B_z) is the magnetic field, $\rho(x,y,z)$ is the density, V_0 is the relative velocity, B_0 is the uniform background magnetic field, The plasma β is 1.0.

3. Simulation Results

In order to develop the KH instability, a perturbation was imposed on the initial configuration. Four cases are studied: the shear velocity V_0 is 1.5 v_A and 0.4 v_A, the density is uniform and not uniform. Figure 1 shows current streamlines, at $5\tau_A$, for the case of $V_0 = 1.5v_A$ and density is not uniform.

4. Conclusion

The Kelvin-Helmholtz instability grows in the shear flow and the kinetic energy of shear flow is converted to the magnetic energy by the instability and the current loops are generated from the Kelvin-Helmholtz vortices. TABLE. 1 shows conversion rate of magnetic energy from the initial kinetic energy associated with the Kelvin-Helmholtz vortices.

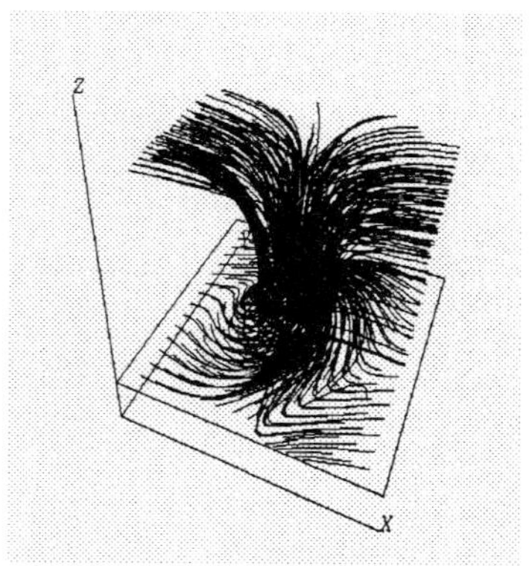

Figure 1. current streamlines.

density	uniform	non-uniform
$V_0 = 1.5v_A$	36	26
$V_0 = 0.4v_A$	37	24

TABLE 1. conversion rate of magnetic energy from the initial kinetic energy (%).

Acknowledgements

We are thankful to Dr.M.Norman for use of ZEUS-3D code.

References

Sakai, J.I. and Fushiki, T. (1995) *Solar Physics*, **156**, 281.

SIMULATION OF A CLOUD-CLOUD COLLISION IN AN ELECTRON-POSITRON PLASMA

T. KITANISHI AND J.-I. SAKAI
Faculty of Engineering, Toyama University, Toyama930, Japan
AND
K.-I. NISHIKAWA
Space Physics Department, Rice University, Houston, Texas 77005, USA

The dynamics of the clouds in magnetized plasmas is one of the basic processes found in many areas ranging from astrophysical to laboratory plasmas. This cloud dynamics in electron–ion and electron–positron plasmas has been investigated theoretically (Peter and Rostoker, 1982) and by using particle simulations (Galvez, 1987; Neubert. *et al.*, 1992; Kitanishi *et al.*, 1994). Results of these simulations have shown that (1) the cloud keeps moving across an ambient magneticfield as $E \times B$ drift, (2) electron and ion (positron) charge sheaths expand along the magnetic field, and (3) electromagnetic waves are emitted from the cloud due to the coherent gyration of particles. For the case of electron–positron plasma in strong magnetic field, the kinetic energy of the cloud is very efficiently transformed into the emitted wave energy (Kitanishi *et al.*, 1994).

In this study, we investigate a collision of clouds moving perpendicular to an ambient magnetic field in an electron–positron plasma by using a 3–Dimentional electromagnetic particle code (Buneman, 1993). The clouds do not mix up each other. Instead, front of the clouds extend to the transverse direction (the y direction). The backgroud plasmas initially between the clouds are compressed in the x direction and driven to the y direction. In addition, the tip of the background plasma tend to move around in the x direction (like the I-shape. see Figures). In the collision process, some of kinetic energy of the cloud is converted to field energy and kinetic energy of the background plasma within $1T_c$. More than 50% of the released kinetic energy of the clouds is converted to that of background plasmas.

Y. Uchida et al. (eds.),
Magnetodynamic Phenomena in the Solar Atmosphere – Prototypes of Stellar Magnetic Activity, 603–604.

TABLE 1. Variation of the kinetic energy of the cloud (K_{cl}), that of background plasmas(K_{bg}), and field energy in the simulation domain (E_{fld}) from $T = 0$ to $T = 1T_c$.

	single cloud	collision cloud
loss of K_{cl}	495	951
increase in K_{bg}	175	526
increase in E_{fld}	207	303

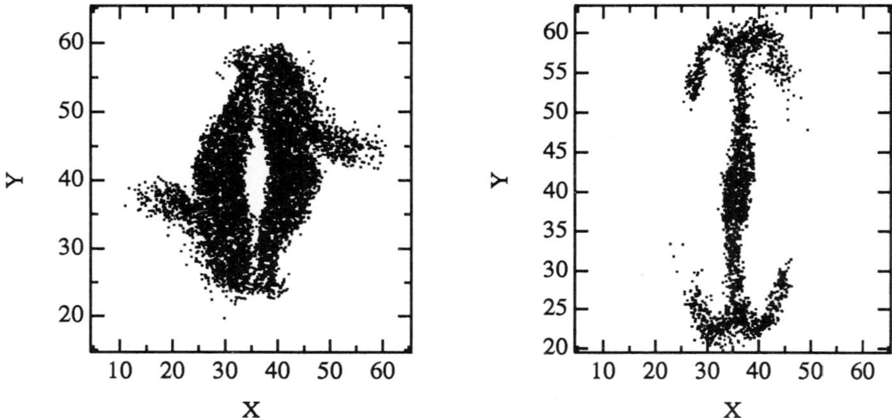

Figure 1. The location of the cloud positrons (left) and background positrons initially between the clouds (right) projected onto $x - y$ plane at $T = 4T_c$.

References

Buneman, O.: in *Computer Space Plasma Physics, Simulation Technique and Softwares*, edited by H. Matumoto and Y. Omura (Terra Scientific, Tokyo, 1993), p. 67.

Galvez, M. (1987) *Phys. Fluids* **30**, 2729.

Kitanishi, T., Sakai, J. I., Nishikawa, K. I., and Zhao, J. (1994) PAFS-28 (preprint).

Neubert, T., Miller, R. H., Buneman, O., and Nishikawa, K. I. (1992) *J. Geophys. Res.* **97**, 12057.

Peter, W. and N. Rostoker, N. (1982) *Phys. Fluids* **25**, 730.

COSMIC RAY CHAOS REQUIRES NEW PHYSICS

Wasaburo UNNO, Takashi KITAMURA and Soji OHARA
Kinki University, Institute for Science and Technology
Higashi-Osaka, Osaka 577, Japan

1. Introduction

Cosmic ray air showers have been considered as random phenomena, since the quantum mechanics of high energy particles is involved intrinsically. Yet, we have detected low dimensional chaos which can be distinguished from colored noise by the surrogate data test. The origin of this chaos is difficult to understand on current astrophysical grounds.

2. Observations and Analysis

Extensive air showers with energies larger than 3×10^{14}eV have been observed at Kinki University for more than 3 years since June 1991. About 4×10^5 air shower events were observed with 4 to 5 min average separations. The sequence of these 0.4 million air shower arrival time intervals (ASATIs) are the primary data of the analysis. Let x_i be the i-th time interval, $t_{i+1} - t_i$, then a m-dimensional vector $V_m^{(i)}$ which defines N points in the m-dimensional embedding space can be constructed by $V_m^{(i)} = [x_i, x_{i+1}, ..., x_{i+m-1}]$ (i=1,2,3,...,N). The correlation dimension D_m is defined by

$$D_m = \mathrm{d}\ \ln C_m(r)/\mathrm{d}\ (\ln r) \tag{1}$$

where $C_m(r)$ denotes number of pairs of vector points, $V_m^{(i)}$ and $V_m^{(j)}$, of mutual distance r_{ij} less than r. $D_m(\ln r)$ can be obtained in sufficient accuracy for N $\geq$ 128 as tested empirically by changing N. If $D_m(\ln r)$ is practically constant over some range in $\ln r$, the statistical property of the data is not random but is either colored noise or chaos. The distinction between the latter two cases can be distinguished by the surrogate data test. The embedding dimension m should be taken to be larger than $2D + 1$.

3. Results

Almost all of about 3,000 sets of 128-ASATI data show random noise characteristics as expected. There are, however, 13 cases showing fractal dimension D_7 less than 4. The uncertainty in the definition of the low dimensionality can be estimated empirically from the range of flatness of the D_m curve compared with the degree of fluctuation of the determination of D_m. The probability of random noise being taken as chaos or colored noise is found

Y. Uchida et al. (eds.),
Magnetodynamic Phenomena in the Solar Atmosphere – Prototypes of Stellar Magnetic Activity, 605–606.

to be 10^{-16}. The same analysis for the Osaka City University data (the observing site is 115 km distant) finds 5 cases of low fractal dimensionality that occurred simultaneously with some of the 13 cases of the Kinki University. This excludes the possibility of the low dimensionality by means of either instrumental or accidental causes by the probability of 10^{-12}. In the surrogate data test, the order of 128-ASATI data showing a low fractal dimension (flat D_m) is completely randomized. The same analysis as before is then repeated. We found that we have actually observed chaos, since the flat part of D_m has practically disappeared with the surrogate data. Colored noise characteristics have also been recognized, however, in a few cases.

4. 1991 Sept.19 Events

A series of surprising events occurred on Sept.19, 1991. The D_7 was first detected as 4.5 on Sept.12.7 (JST). Writing this as (4.5,12.7), we can describe the events as [(4.5,12.7), (3.7,19.1), (2.2,19.2), (1.2,19.4), (1.6,19.6), (2.6,19.8), (4.0,20.2)]. Low fractal dimension is detected very rarely. Yet, we have detected series of low fractal dimensions. The D_7 attains a sharp minimum value of 1.2 within a few hours. The rapid variation of the fractal dimension is not easy to understand. Clustering of events is noticed at the minimum. The average time interval is reduced to about 3 min as compared with the usual average of 4 to 5 min.

5. Concluding Remark

The origin of cosmic ray particles showing extensive air showers is considered to be in active stars such as supernovae, X-ray binaries, pulsars, or in galactic center activities. The origin of chaos in air showers is more difficult to identify. If it is too far in the Galaxy, chaos will disappear during the propagation through interstellar magnetic fields. The energy of the cosmic ray, however, is too high for chaos to originat within the solar system. The chaos should have been formed before air shower formation, since it is rarely observed. Anomalous atmospheric structure causing chaos is also excluded in view of the simultaneous detection by the Osaka City University data and of the Sept.19 event. Our tentative model is the chaos produced by disintegration of a dust cosmic ray having originated from a comet impinging into a magnetic loop outburst in an accretion disk of an active star or in an inner Galactic arm and accelerated therein. Further study is vitally needed. (see, Ohara et al., 1995)

References

Ohara, S., Kitamura, T., Chikawa, M., Unno, W., Tsuji, K., Konishi, T. and Masaki, I.(1995) Chaotic Features of Extensive Cosmic Ray Air Showers, *Science and Technology*, Kinki University, No.7, 39-46.

SOLUTIONS OF THE HYDROMAGNETIC EQUATION WITH LOCALIZED AND UNIFORM RESISTIVITY

F. JAMITZKY AND M. SCHOLER
Max-Planck-Institut für extraterrestrische Physik
D-85740 Garching, Germany

1. Introduction

A new analytical solution of the hydromagnetic equation is presented. The general solution for the flux function is given for general incompressible and irrotational two-dimensional plasma flows. Scaling laws for the width and length of the current sheet are obtained for uniform and localized resistivity. For uniform resistivity the scaling laws predict an infinite growth of the current sheet length for increasing magnetic Reynolds number, while for localized resistivity the current sheet length possesses an upper bound.

2. The Hydromagnetic Equation

The electric field is given by Ohm's law:

$$\mathbf{E} + \mathbf{v} \times \mathbf{b} = \eta \mathbf{j} \tag{1}$$

where $\mathbf{E}$, $\mathbf{b}$, $\mathbf{v}$, η and $\mathbf{j} = \nabla \times \mathbf{b}$ denote the electric field, the magnetic field, plasma velocity, the electric resistivity and the electric current density respectively. For an incompressible two-dimensional flow in the $x - y$ plane one can define a stream function ϕ and a flux function a such that $\mathbf{v} = \nabla\phi \times \hat{\mathbf{z}}$ and $\mathbf{b} = \nabla a \times \hat{\mathbf{z}}$. Ohm's law can be written as:

$$-E + [a, \phi] = \eta \nabla^2 a \tag{2}$$

where the bracket is defined by $[u, v] = \partial_x u \partial_y v - \partial_y u \partial_x v$. This equation is called the stationary hydromagnetic equation.

3. General Solution

For an irrotational flow defined by $\nabla^2\phi = 0$ one can find a complex function $\chi(x + iy)$ and a function $\psi(x, y)$ such that $\chi(x + iy) = \phi(x, y) + i\psi(x, y)$.

Y. Uchida et al. (eds.),
Magnetodynamic Phenomena in the Solar Atmosphere – Prototypes of Stellar Magnetic Activity, 607–608.

The hydromagnetic equation can then be transformed to:

$$\eta\left(\partial_\phi^2+\partial_\psi^2\right)a+\partial_\psi a=-E\Gamma(\phi,\psi) \tag{3}$$

with $\Gamma(\phi,\psi)=|dz/d\chi|^2$. This equation is an inhomogeneous linear partial differential equation with constant coefficients and can be solved by standard methods. One obtains as general solution:

$$\begin{aligned} a(\phi,\psi) &= -\frac{Ee^{\psi/2\eta}}{8\pi\eta}\int\int\Gamma(\phi_1,\psi_1)K_0\left(\frac{\sqrt{(\psi_1-\psi)^2+(\phi_1-\phi)^2}}{2\eta}\right)d\psi_1\,d\phi_1 \\ &+e^{\psi/2\eta}\int_{-\infty}^{\infty}\gamma_1(\alpha)e^{\alpha\phi+\beta\psi}d\alpha+e^{\psi/2\eta}\int_{-\infty}^{\infty}\gamma_2(\alpha)e^{\alpha\phi-\beta\psi}d\alpha \end{aligned} \tag{4}$$

where $4\eta^2\left(\alpha^2+\beta^2\right)=1$, the functions γ_1 and γ_2 are given by the boundary conditions and K_0 denotes the modified Bessel function of second kind and zeroth order.

4. Scaling Laws

The general solution can be applied to magnetic reconnection. A normalisation is given by $b_x(0,L_e)=b_e$ and $v_y(0,L_e)=-v_e$. The magnetic Reynolds number and the reconnection rate are then defined by $R_m=b_eL_e/\eta$ and $M_e=v_e/b_e$ respectively. The current sheet is oriented along the x -axis. From the general solution one can derive scaling laws for the width l and the length L of the current sheet. One obtains:

$$l/L_e\sim R_m^{-1/2}M_e^{-1/2} \tag{5}$$

and

$$L/L_e\sim R_m^{1/2}M_e^{-1/2} \tag{6}$$

In the case when the resistivity is localized to a strip along the y axis of halfwidth x_0 one obtains as upper bound of the current sheet length

$$L\leq x_0 \tag{7}$$

and for the current sheet width:

$$l/L_e\sim(R_mM_e+L_e^2/x_0^2)^{-1/2} \tag{8}$$

Even for large magnetic Reynolds number the diffusion region is confined to a region of size x_0.

A full paper is to appear in the Journal of Plasma Physics.

References

Jamitzky, F. and Scholer, M. (1995) Steady state magnetic reconnection at high magnetic Reynolds number: A boundary layer analysis, *J. G. R.*, **100**, pp. 19,277-19,285

LARGE-SCALE ACTIVE CORONAL PHENOMENA IN YOHKOH SXT IMAGES

Z. ŠVESTKA
CASS, UCSD, La Jolla, CA 92093-0111, U.S.A.

F. FÁRNÍK
Astronomical Institute ASCR, Ondřejoz, Czech Republic

HUGH S. HUDSON
Institute for Astronomy, University of Hawaii, Honolulu, HI 96822, U.S.A.

Y. UCHIDA
Physics Dept., Science Univ. of Tokyo, Tokyo 162, Japan

P. HICK
CASS, UCSD, La Jolla, CA 92093-0111, U.S.A.

AND

J. R. LEMEN
Lockheed Palo Alto Res. Lab., Palo Alto, CA 94304, U.S.A.

We have found several occurrences of slowly expanding giant arches in *Yohkoh* images. These are similar to the giant post-flare arches previously discovered by SMM instruments in the 80s. However, we see them now with 3–5 times better spatial resolution and can recognize well their loop-like structure. As a rule, these arches follow *eruptive flares* with gradual soft X-ray bursts and rise with speeds in the range of 1.1–2.4 km s^{-1} which keep constant for $>$ 5 to 24 hours, reaching altitudes up to 250,000 km above the solar limb. These arches differ from post-flare loop systems by their (much higher) altitudes, (much longer) lifetimes, and (constant) speed of growth. One event appears to be a rise of a transequatorial interconnectiong loop

In the event of 21–22 February 1992 one can see both the loop system, rising with a gradually decreasing speed to an altitude of 120,000 km, and the arch, emerging from behind the loops and continuing to rise with a constant speed for many more hours up to 240,000 km above the solar limb. In the event of 2–3 November 1991 three subsequent rising large-scale

Y. Uchida et al. (eds.),
Magnetodynamic Phenomena in the Solar Atmosphere – Prototypes of Stellar Magnetic Activity, 609–610.

coronal systems can be recognized: first a fast one with speed increasing with altitude and ceasing to be visible at about 300,000 km. This most probably shows the X-ray signature of a coronal mass ejection (CME). A second one, with gradually decreasing speed, might represent very high rising flare loops. A third one continues to rise slowly with a constant speed up to 230,000 km (and up to 285,000 km after the speed begins to decay), and this is the giant arch. This event, including an arch revival on November 4–5, is very similar to rising giant arches observed by the SMM on 6–7 November 1980. Other events of this kind were observed on 27–28 April 1992, 15 March 1993, and 4–6 November 1993, all seen above the solar limb, where it is much easier to identify them.

The temperature in the brightest part of the arch of 2–3 November 1991 was increasing with its altitude, from 2 to 4 $\times$ 10^6 K, which seems to be an effect of slower cooling at lower densities. Under the assumption of a line-of-sight thickness of 50,000 km, the emission measure indicates arch densities form 1.1 $\times$ 10^{10} cm^{-3} at an altitude of 150,000 km to 1.0 $\times$ 10^9 cm^{-3} at 245,000 km 11.5 hours later. It appears that the arch is composed of plasma of widely different temperatures, and that hot plasma rises faster than the cool component. Thus the whole arch expands upward and its density gradient increases with time which explains why *Yohkoh* images show only the lowest and coolest parts of the expanding structure. The whole arch may represent an energy in excess of 10^{31} erg, and more if conduction contributes to the arch cooling.

We suggest that the rise of the arch is initiated by a CME which removes the magnetic field and plasma in the upper corona and the coronal structures remaining below this cavity begin into expand into the "magnetic vacuum" left behind the CME. However, we are unable to explain why the speed of rise stays constant for so many hours.

The complete paper will appear in *Solar Phys.* **160**, issue 2 (November 1995).

Poster Papers

Posters for Session IV

3D MHD SIMULATION OF MAGNETIC FLUX TUBES IN CO-ROTATING COORDINATES

Comparison with Yohkoh Data and Observations of Fast Rotating Stars

W. CHOU, T. TAJIMA,
Department of Physics, University of Texas, Austin, TX 78712
R. MATSUMOTO,
Chiba University, Chiba, Japan, and ASRC, JAERI, Naka, Japan
AND
K. SHIBATA
National Astronomical Observatory of Japan, Mitaka, Japan

The solar soft X-ray pictures taken by the *Yohkoh* satellite show a sequence of regularly spaced, S-shaped active regions at low latitude of the north and the south hemispheres. It appears that these active regions are emergent parts, due to magnetic buoyancy (Parker) instability, of a global helical magnetic flux tube embedded in the convection zone. Also, the sunspots, which are the foot points of the emergent flux tube, are arranged as: the preceding polarity of an active region tends to be near the equator, whereas the following polarity is farther from the equator. Here in this paper, we simulate a 3-D flux tube to compare with these observations.

The observation of some fast rotating stars shows that stellar spots appear near the polar region but not near the equator. We provide an explanation from our linear analysis: the magnetic field in the low latitude has very small growth rate, due to fast rotation, while the field in the polar region has much larger growth rate, so that it can emerge outside the surface of star.

In our simulation, we assume an ideal gas with adiabatic $\gamma = 1.05$. The resistivity, viscosity, and thermal diffusivity are negleted. Cartesian coordinates are used, with z in the vertical direction opposite to the constant gravity and the cylindrical flux tube along the x direction. Magnetic field is imbedded inside the cylinder and parallell to the x-direction with plasma $\beta = 1$, and $\mathbf{B} = 0$ outside. To initiate the dynamics, a small velocity perturbation v_x is added inside the tube. The code uses modified Lax-Wendroff scheme with artificial viscosity. The rotational speed used

Y. Uchida et al. (eds.),
Magnetodynamic Phenomena in the Solar Atmosphere – Prototypes of Stellar Magnetic Activity, 613–614.

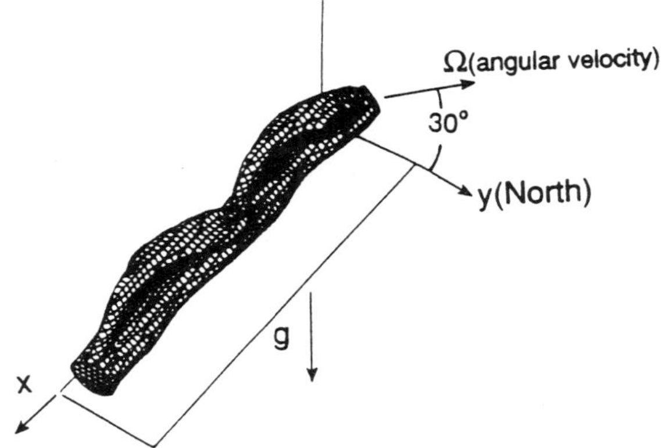

Figure 1. Isosurface of $B^2 = 1[\rho_0 C_s^2]$ of toroidal flux tube in north 30 degree after the system rotates 3/4 turn.

here is $\Omega H/v_A = 0.1$, which is the same order of magnitude as in the real solar case, where $v_A = 2 \times 10^5$cm/sec, and scale height $H = 10^{10}$cm.

We see magnetic buoyancy when there is no rotation, similar to Matsumoto et. al. (1993). When the flux tube is put in the north pole and includes rotation, the tube is twisted by Coriolis force. In this case, it looks like a rubber tube wrung from two ends, and the magnetic field lines wound along the twisted tube. The direction of tilting is the same as the observation. Figure 1 shows the simulation result of a toroidal flux tube in north 30 degree of the Sun. The Coriolis force not only serves as a twisting mechanism, but also causes asymmetry in east-west direction, because the Coriolis force induces the flow in negative x-direction (solar east).

Using the technique similar to Parker (1979) and Horiuchi et al. (1988), we peformed the linear stability analysis of magnetic flux sheet (instead of flux tube) in co-rotating frame, assuming isothermal atmosphere. We found that the linear growth rate of Parker instability in the low latitude region is much smaller than that in the high latitude region. If the rotation speed is greater than some critical value, the toridal magnetic flux sheet in the equator will be stable. We note that the growth (or decay) mode in the north hemisphere travels toward the positive x (direction of magnetic field) and positive y (north) direction.

References

Horiuchi, T., Matsumoto, R., Hanawa, T. and Shibata, K., 1988 Publ. Astron. Soc. Japan, 40:147-169

Matsumoto, R., Tajima, T., Shibata, K. and Kaisig, M., 1993 Astrophy. J., 414:357-371

Parker, E. N., Cosmical magnetic fields, Oxford, 1979, pp. 325-331

DISTRIBUTIONS OF THE MAGNETIC FIELD INCLINATION IN SUNSPOTS

T. SHINKAWA AND M. MAKITA
Kwasan Observatory, Kyoto University
Yamashina, Kyoto 607, Japan

Many authors have observed sunspot magnetic fields. Most of their results agreed that the inclination angle of the magnetic fields to the vertical line (γ) is 0° at the center of sunspots. But at the outer edge of the penumbrae, while the results by Beckers and Schröter (1969), Wittmann (1974), etc. are $\gamma = 90°$, the results by Kawakami (1983) , Skumanich *et al.* (1994), etc. suggest $\gamma = 70° \sim 75°$. The difference is a question to be settled, because the Evershed flow is considered to be horizontal.

There is a simple way to determine the inclination angle of the magnetic field of a nearly circular sunspot. When a sunspot is observed near the Sun's limb the magnetic field in a meridional plane through the center of the spot is nearly normal to the line of sight. When the magnetic field is normal to the line of sight, its inclination angle to the vertical line can be obtained from the azimuth (α) of the linear polarization. We can determine it from the ratio between Q and U. ($\alpha = \frac{1}{2}\tan^{-1}(\frac{U}{Q})$.) The advantage of this method is to be able to determine the inclination angle geometrically, without solving the complicated line-formation problem (Hale and Nicholson, 1938).

In this work, we use the data obtained with the vector magnetograph of the Okayama Astrophysical Observatory (OAO). Two compensators eliminate the instrumental polarization of the telescope.

Among all data which were observed with OAO's magnetograph from December 1982 to December 1990, we look for large axially-symmetric sunspots near the limb (heliocentric angle θ is more than 60°). We selected 10 sunspots from this data base.

Near the center of the spot the error of the inclination angle ($\Delta\gamma$) is about $\pm 1°$, at the outer penumbrae $\Delta\gamma \sim \pm 5°$.

Figure 1 shows the averaged distribution of the magnetic field inclinations of the 10 sunspots.

Y. Uchida et al. (eds.),
Magnetodynamic Phenomena in the Solar Atmosphere – Prototypes of Stellar Magnetic Activity, 615–616.

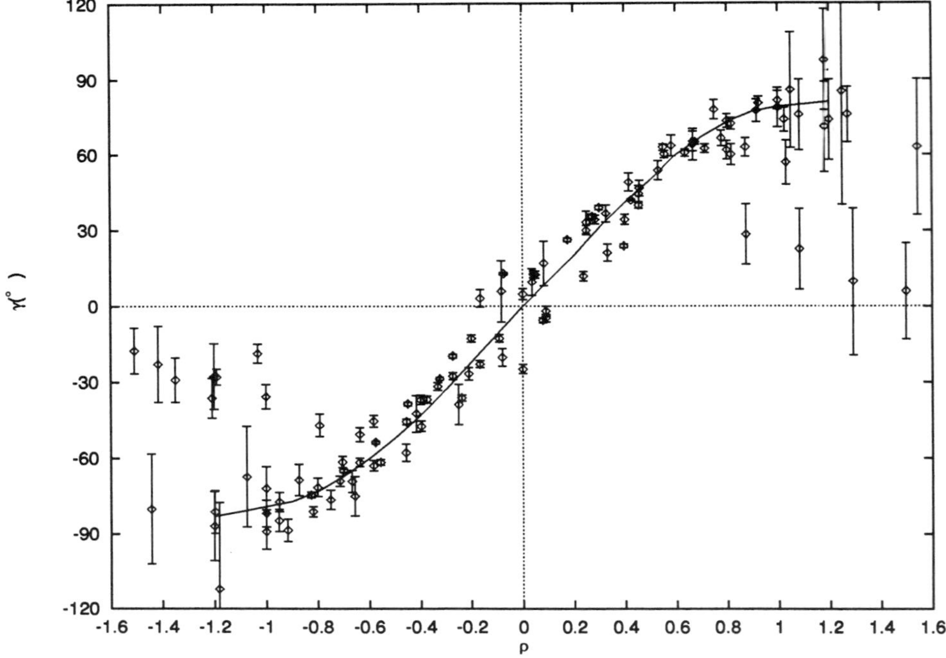

Figure 1. The spatial variation of the inclination angle of the magnetic field of sunspots. The distance (ρ) from the center of the spot is normalized to the sunspot radius determined by the line intensity images. We show error bars. Negative or positive ρ means the direction to the equator or (north or south) pole respectively.

At the center of spot $\gamma \sim 0°$. At the boundary of the penumbrae with the photosphere $\gamma \sim 80°$. Considering errors, at the outside of sunspots γ is definitely smaller than $90°$.

Our result is similar to the distributions of inclination of Skumanich *et al.* (1994), or Kawakami (1983) but is different from those of Beckers and Schröter (1969), or Wittmann (1974).

References

Beckers, J.M. and Schröter, E.H.: 1969, *Solar Physics*, **10**, 384

Hale, G.E. and Nicholson, S.B.:1938, in *Magnetic Observations of Sunspots 1917-1924 part 1*, Carnegie Institution of Washington, Washington,D.C., p24-29

Kawakami, H. : 1983, *Publ. Astron. Soc. Japan*, **35**, 459

Skumanich, A.,Lites, B.W.,Pillet, V.M.:1994, in R.J. Rutten and C.J. Schrijver (eds.), *Solar Surface Magnetism*,Kluwer Academic Publishers,Netherlands, 99

Wittmann, A. : 1974, *Solar Physics*, **36**, 29

DYNAMICS OF THE SOLAR GRANULATION: ITS INTERACTION WITH THE MAGNETIC FIELD

A. NESIS, R. HAMMER AND H. SCHLEICHER
Kiepenheuer-Institut für Sonnenphysik
Schöneckstr. 6, D-79104 Freiburg, Germany

1. Introduction

The dynamics of the granulation manifests itself in the Doppler shift fluctuations of the cores of spectral lines, which are associated with spatially resolved intensity structures (granules). Another part of the dynamical behavior of the granulation, especially the dynamics of the intergranular space, leads to unresolved Doppler velocity fluctuations detectable as line broadening enhancements of magnetically insensitive absorption lines.

Our study is based on high-resolution spectra of Ca network and internetwork regions taken at the VTT on Tenerife. The enhanced magnetic activity of the network could change the behavior of convective flow and turbulence. We analyze the statistical properties of line shifts and broadening of magnetically sensitive and insensitive lines, in particular the intermittency of the turbulent velocity field. We study the variation of these statistical parameters with height and with the level of magnetic activity. The results could be relevant for the heating of the overlying layers.

2. Results and Conclusions

Granulation interacts with the magnetic field mainly via the unresolved velocity field, which is particularly prominent in the intergranular space near granular borders. The unresolved velocity often does not occupy all the available intergranular space; therefore, it is intermittent. We determine the intermittency coefficient to be in the range 0.6–0.7 and find that neither the vertical stratification in the granular overshoot layers nor the magnetic activity appear to affect the character of intermittency of the enhanced line broadening.

Y. Uchida et al. (eds.),
Magnetodynamic Phenomena in the Solar Atmosphere – Prototypes of Stellar Magnetic Activity, 617–618.

Strong velocity gradients of the convective overshoot flow at the borders of granules (cf. Nesis et al. 1993) could be the sources of the unresolved velocity field in the intergranular space. In such a case the velocity field can be considered as shear turbulence.

In the supergranular network we found the magnetic field also located at granular borders, at those places that exhibit an enhanced turbulent unresolved velocity field. Therefore, *magnetic and turbulent fields are not only intermittent, but also co-located.*

Furthermore, we calculated the power spectra of the unresolved velocity field and of the resolved velocity field corresponding to the convective Doppler velocity shift. For granular size scales, the power spectrum of the convective Doppler shift shows a slope of $\approx 1/k^{5/3}$, whereas the unresolved velocity field exhibits a slope of $\approx 1/k^2$, where k is the wave number. These slopes are the same for network and internetwork regions, although the Doppler shift velocity as well as the unresolved velocity field are clearly inhibited by the magnetic field in the network. This confirms results by Kida et al. (1991), who found power spectra with slopes of $\approx 1/k^{5/3}$ and $\approx 1/k^2$ in their theoretical study of the properties of MHD turbulence.

It is conceivable that the heating mechanisms of the chromosphere and corona are related to the presence of magnetic fields and their excitation by turbulent processes in the granular overshoot layers. Our results suggest that the sources of these heating mechanisms are intermittently distributed within the intergranular space and tightly connected with the existence of large gradients in the convective flow field. The co-location of the magnetic and unresolved velocity field supports this possibility.

References

Kida, S., Yanase, S., Mizushima, J. (1991): Statistical properties of MHD turbulence and turbulent dynamo. *Phys. Fluids A,* **3** *(3) pp. 457-465*

Nesis, A., Hanslmeier, A., Hammer, R., Komm, R., Mattig, W., and Staiger, J. (1993): Dynamics of the solar granulation II. A quantitative approach. *Astron. Astrophys.* **279** *pp. 599-609*

A MODEL OF MAGNETIC FIELD ON SUPERGRANULES

SAN JAE LI AND OK RUN KIM
Pyongyang Astronomical Obs., National Academy of Science, Pyongyang, DPR of Korea

AND

SONG HAK KIM
Dept. of Physics, Kim Hyong Jik Univ. of Education, Pyongyang, DPR of Korea

The magnetic fields in supergranules and in the chromospheric network have been studied by Kopp & Kuperus (1968), Gabriel (1976), Delache (1976), Simon, Weiss & Nye (1983) and Anzer & Galloway (1983).

From the previous models, we understand that the magnetic field frozen-in to the plasma flow of a supergranular convective cell is concentrated around the boundary of the supergranule to form imaginary magnetic charges, which determine the distribution of the magnetic field in the photosphere and the chromosphere. By using Frazier (1970)'s observation we obtained the depth of IMC – $\approx$ 5200 km and the quantity of IMC – $\approx 1.3 \times 10^8$ G km^2.

The two-dimensional shapes of magnetic force-lines distributed by this IMC are similar to the results of Kopp & Kuperus (1968) and Gabriel (1976). We assume that the supergranular convective cells form in a regular hexagonal shape and the IMCs are arranged at their boundaries. We are then able to explain Frazier (1970)'s three curves on the line-of-sight component of magnetic field at photospheric level quiet well, as shown in Fig. 1.

From divIB=0, we obtain a formula for the line-of-sight component of magnetic field of supergranules:

$$B_z(Z) = \frac{2\pi}{\sqrt{3}L^2} \sum_{i=1}^{6} Q_m^{(i)} = \text{const},$$

where L(=3.2$\times$ 10^4 km) is the size of the supergranular cell and $Q_m^{(i)}$ is the quantity of IMC. Taking account of const $\approx$ 1G., IMCs in magnetic elements

Y. Uchida et al. (eds.),
Magnetodynamic Phenomena in the Solar Atmosphere – Prototypes of Stellar Magnetic Activity, 619–620.

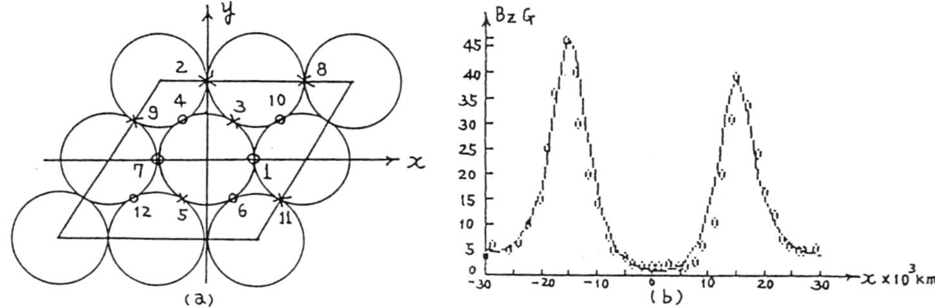

Figure 1. Interpretation of Frazier's magnetic field curve
(a) distribution of IMCs in the fundamental magnetic element
(b) comparison of theoretic curve (solid line) with observational curve (dotted line)

are distributed with bipolar shapes. Therefore, IMCs may be related to the formation of XBPs in the corona.

In fact, forming XBPs by a crossing of bipolar magnetic fields with opposite polarities will generate 2.6×10^3 of XBPs [*Skylab* observation: 1.5×10^3 (Golub et al., 1974); *Yohkoh* observation: 4×10^3 (Harvey et al., 1994)], and the spaces between those bipolar points are $\gtrsim L/2 = 1.6 \times 10^4$ km [*Yohkoh* observation: $\gtrsim 2 \times 10^4$ km (Harvey et al., 1994)].

Also, the concentrating time of IMC is evaluated to be $\approx$ 25 hours, which is consistent with the life time of supergranules 20 $\sim$ 28 hours and the lifetime of ephemeral-region $\approx$ 24 hours.

Acknowledgements. The authors would like to thank Profs. T. Sakurai and K. Shibata for helpful comments. We are specially grateful for Prof. T. Kosugi for his effort to make the Colloquium a successful one for the IAU for support.

References

Anzer, U. and Galloway, D. J. (1983) *Mon. Not. Astr. Soc.* **203**, 637.
Beckers, J. M. (1968) *Solar Phys.*, **3**, 367.
Delache P. (1976) *Phil. Trans. Roy. Soc. Lond.*, A. **281**, 339. Delache P. (1976) in R. M. Bonnet and P. Delache(eds.), *The Energy Balance and Hydrodynamics of the Solar Chromosphere and Corona* G. DE Bussac, pp. 412.
Frazier, E. N. (1970) *Solar Phys.*, **14**, 89.
Gabriel, A. M. (1976) *Phil. Trans. Roy. Soc. Lond.*, A. li.tex**281**, 339
Golub, L., Krieger, A.S., Silk, J., Timothy, A., and Vaiana, G. (1974) *Ap. J.*, **189**, 293.
Harvey, K. L., Nitta, N., Strong, K. T., and Tsuneta, S. (1994) in Y. Uchida, T. Watanabe, K. Shibata, and H. S. Hudson (eds.), *X-Ray Solar Physics from Yohkoh*, Universal Acad. Press. Inc., Tokyo, Japan, pp. 21-24.
Kopp, R. A. and Kuperus, M. (1968) *Solar Phys.*, **4**, 212.
Simon, G. W., Weiss, N. C. and Nye, A. (1983) *Solar Phys.*, **81**, 250.

SOLAR CONVECTION AND MAGNETO-CONVECTION

SHANTANU BASU AND DAVID J. BERCIK
Physics and Astronomy Department, Michigan State University East Lansing, MI 48824, USA

ÅKE NORDLUND
Copenhagen University Observatory, Ostervoldgade 40, DK-1350, Copenhagen, Denmark

AND

ROBERT F. STEIN
Physics and Astronomy Department, Michigan State University East Lansing, MI 48824, USA

1. Discussion

We explore the general properties of convection in a solar surface layer. Our three-dimensional simulations use a tabulated equation of state that includes excitation and ionization of hydrogen and other abundant atoms, and the formation of H_2 molecules. The radiative energy exchange is determined by solving the radiative transfer equation. This is particularly important in the upper cooling layer, where the observed granulation pattern is created. Our hydrodynamic runs simulate a region $6 \times 6 \times 3$ Mm, extending vertically from the temperature minimum at 0.5 Mm above the visible surface, down to 2.5 Mm below it. We have run a new simulation at twice the resolution (25 km horizontally, and 15 - 35 km vertically) of a previous run. The bulk properties of the convection are very similar in both runs. The general picture is one in which most of the gas is found in slowly rising, expanding upflows. The gas that does reach the surface moves horizontally, cools, and descends back into the stratified atmosphere, being compressed into narrow, rapidly descending filaments. This general topology of the convection is described in detail by Stein and Nordlund (1989). The high resolution run reveals more small granules, and has more small scale structure in the granulation pattern and in the downdrafts. Figure 1

Y. Uchida et al. (eds.),
Magnetodynamic Phenomena in the Solar Atmosphere – Prototypes of Stellar Magnetic Activity, 621–622.

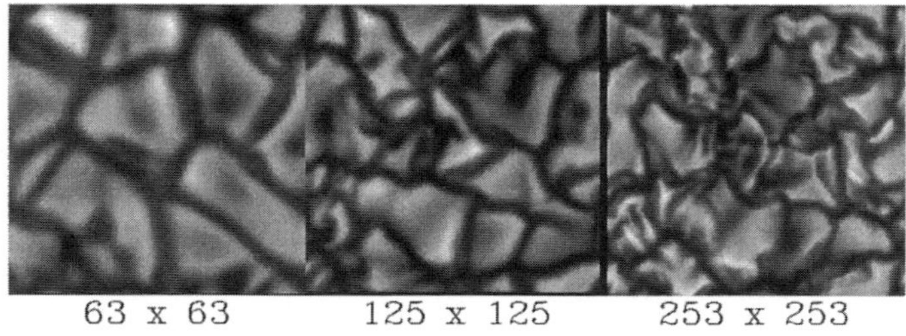

Figure 1. The progressive change in the surface emergent intensity (granulation) pattern as we go from horizontal grids of 63×63 to 125×125 to 253×253.

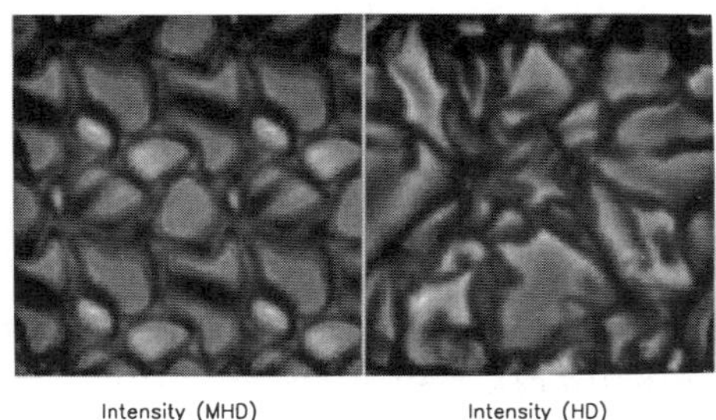

Figure 2. A comparison of the granulation patterns for hydrodynamic and magnetohydrodynamic convection. The runs are made at identical horizontal resolution (50 km).

compares the granulation pattern in three different runs, with horizontal grids of 63×63, 125×125, and 253×253 (new run), respectively.

Figure 2 shows the granulation pattern from a preliminary magnetoconvection run (see also Nordlund and Stein 1990; Stein et al. 1992). An initially uniform vertical magnetic field of 500 G was imposed on a snapshot of hydrodynamic convection. The computed region is $3 \times 3 \times 1.5$ Mm. We plot a 6×6 Mm region of the granulation pattern, assuming periodicity in the horizontal directions, to facilitiate comparison with the hydrodynamic run. It is seen that the granule sizes tend to be smaller in the presence of the imposed magnetic field.

References

Stein, R. F., and Nordlund, A. 1989, ApJ, 342, L95

Nordlund, A., and Stein, R. F. 1990, in IAU Symp. 138, Solar Photosphere: Structure, Convection, Magnetic Fields, ed. J. O. Stenflo, 191

Stein, R. F., Brandenburg, A., and Nordlund, A. 1992, in ASP Conf. Series, 26, ed. M. S. Giampapa and J. A. Bookbinder, 148

TRIPLE–DIPOLE MODEL ON THE LARGE–SCALE MAGNETIC STRUCTURE OF THE SUN

T.SAITO [1] , Y. KOZUKA [2] , M. KOJIMA [2] ,
Y. MORI [3] , M. YASUE [3] , & Y. MATSUURA [3]
[1] Tohoku U.,Emer. Prof., [2] STE Lab.Nagoya Univ., [3] Miyagi Univ. Education

1. Introduction

It is necessary from a viewpoint of solar–terrestrial physics to study complex heliomagnetospheric structure and its solar cycle variation. The purpose of the present paper is to investigate large–scale structure of the solar magnetic fields, especially in relation to high–latitude prominences.

2. Large–scale Magnetic Structure and Its Solar–cycle Variation

Most of the PNL's (magnetic neutral lines at the photosphere with radius of r=1.0Rs) is connected to the SNL (neutral line at the source surface with r=2.5Rs) by one neutral sheet in the coronal region (Saito, et al., 1993). Since the magnetic field distribution at the source–surface is nearly dipolar (Saito, et al., 1995), it is convenient to represent the distribution by a large–scale neutral sheet (LSNS) as shown in Figure 1B. Associating with the reversal of the polar magnetic field of the sun, the LSNS shows clearly a 180° reversal once every one solar cycle. This solar cycle variation is well expressed by the triple–dipole model (Saito, et al.1994) as shown by the LSNS calculated from the three dipoles in Figure 1C.

3. High–latitude Prominence

Solar prominences, which tend to appear along the PNL, are classified into the midlatitude type and the high–latitude type. The observed high–latitude type is at $\sim 40^{\circ}$ heliolatitude in sunspot minimum phase, while $\sim 90^{\circ}$ in maximum phase. Then assuming that the prominences tend to appear under the SNL, the latitude of SNL is read from 1976 to the present as expressed by the dots in Figure 1D. The envelope of the dots mean the highest latitude of the SNL. The simple SNL is connected via LSNS to the meandering PNL, whose latitudes are, therefore, deviated from the latitude of the SNL. Since the solid curve in Figure 1D is drawn by considering the latitudinal deviation from the envelope (broken curve), the solid curve means the highest latitude of the PNL, along which the high–latitude prominences are expected to appear. It is clear that the solid curve expected from the LSNS agrees well with the observed tendency of the high–latitude prominence.

4. CME along the Large–scale Neutral Sheet

The July 31, 1992 event on the outstanding loop prominence were found to burst out

Y. Uchida et al. (eds.),
Magnetodynamic Phenomena in the Solar Atmosphere – Prototypes of Stellar Magnetic Activity, 623–625.

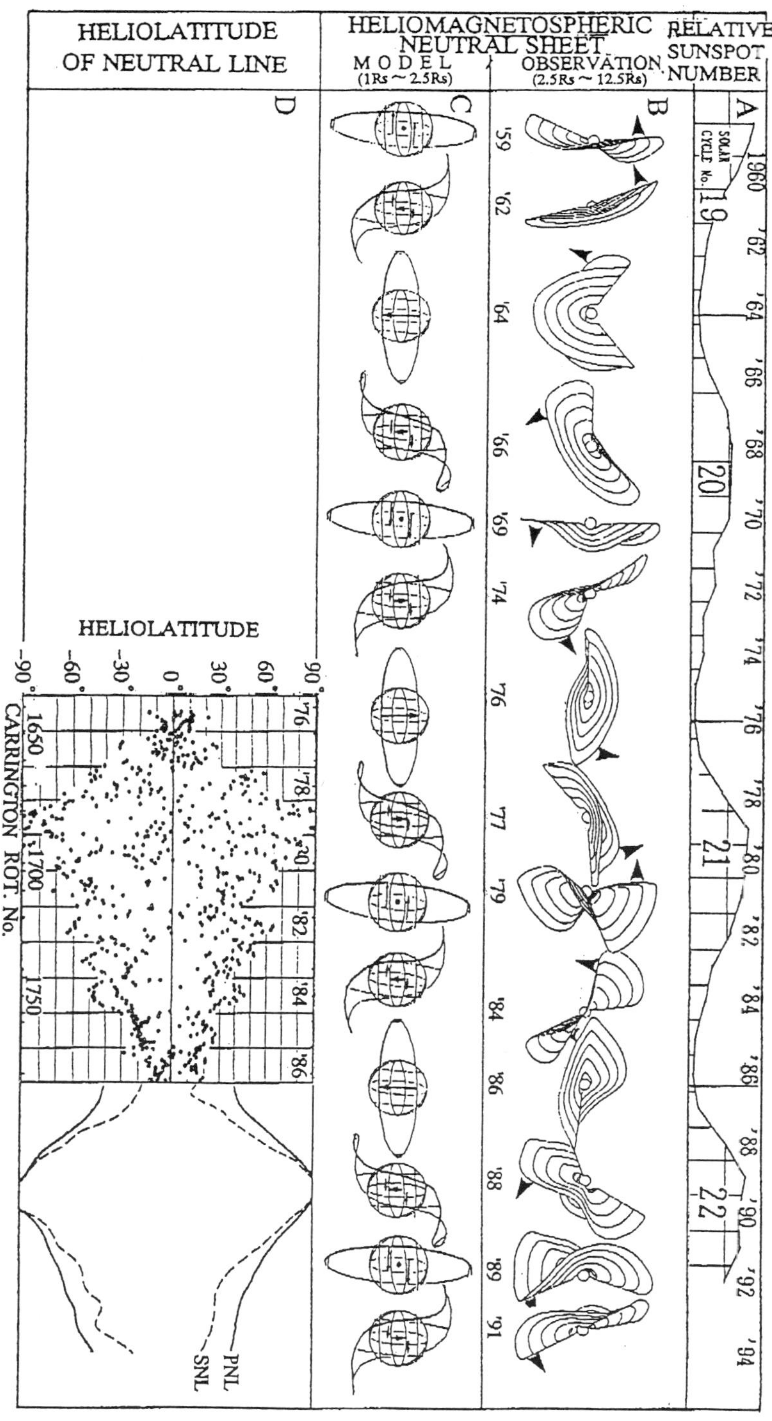
HELIOLATITUDE OF NEUTRAL LINE
HELIOMAGNETOSPHERIC NEUTRAL SHEET
MODEL (1Rs ~ 2.5Rs)
OBSERVATION (2.5Rs ~ 12.5Rs)
RELATIVE SUNSPOT NUMBER
A
B
C
D
SOLAR CYCLE No.
19
20
21
22
HELIOLATITUDE
CARRINGTON ROT. No.
PNL
SNL

along an LSNS. The April 14, 1994 event on a filament disappearance in southern high latitude gave rise to a CME (Hudson, 1995), which is also revealed to expand along the LSNS. Hara (1995) advocated the importance of high-latitude filament disappearance citing the November 22, 1991 event, which is confirmed to be closely related to the LSNS (cf. Figure 3 of Saito, et al., 1994). Hence, it is concluded that the role of the LSNS must be very important for CME originated from filament disappearance.

Reference

Hara, H. (1995) Activity of high-latitude corona associated with low-latitude corona, Abstract of the spring meeting of Japanese Astronomical Society of Japan, S09w.

Hudson, H. S. (1995) The origin of coronal transients and mass ejections observed by Yohkoh, Abstract of IAU Colloquium No. 153, Magnetodynamic Phenomena in the Solar Atmosphere, p.60.

Saito, T., S.-I. Akasofu, Y. Kozuka, T. Takahashi, and S. Numazawa,(1993) The solar coronal sheet during the period of sunspot maximum, J. Geophys. Res., 98: 5639-5644.

Saito, T., Y. Kozuka, S. Tsuneta, and S. Minami (1994) Rotational reversing model and triple dipole model as substantiated by YOHKOH SXT data, X-Ray Solar Physics from YOHKOH, ed. by Uchida, et al., 211-216.

Saito, T., S.-I. Akasofu, Y. Kozuka, S. Minami, and S. Tsuneta (1995) Plasma corona and dust corona of the sun, Proc. International Conference on Dusty Plasmas, Plenum Publ.Co., New York, 449-476.

SOLAR CYCLE VARIATION OF THE ROTATION OF THE LARGE-SCALE MAGNETIC FIELD OF THE SUN

Y. KOZUKA AND M. KOJIMA
Solar-Terrestrial Environment Laboratory, Nagoya University, 3-13, Honohara, Toyokawa 442, Japan

AND

T. SAITO
Taihaku 3-6-29, Taihaku-ku, Sendai 982-02, Japan

Large-scale structures of the solar magnetic field tend to rotate more rigidly than smaller-scale structure. It is known that coronal holes rotate quasi-rigidly while sunspots show a differential rotation. The rotational characteristics of some solar structures have been examined using autocorrelation techniques. Large-scale magnetic field, however, has not been discussed in detail. In the present study, we investigate the rotational characteristics of the large-scale solar magnetic field during about two solar activity cycles. A spherical-harmonic analysis is used for analyzing a distribution of the photospheric magnetic field.

Synoptic charts of the photospheric magnetic field observed at National Solar Observatory/Kitt Peak during the period from 1985 to 1994 are used in this analysis. A multipole expansion of the photospheric magnetic field was performed for every Carrington rotation. The Carrington longitudes of the horizontal magnetic axes of $(n, m) = (1, 1)$ component ('horizontal dipole') and $(n, m) = (2, 2)$ component ('quadrupole') were calculated from spherical harmonic coefficients. The harmonic coefficients during solar cycle 21 have been calculated by Hoeksema and Scherrer (1986) using the Stanford solar magnetic field data. Rotation periods of the horizontal dipole and quadrupole components were given from changes of the longitudes of the magnetic axes. We obtained a long-term behavior of the rotation period of each harmonic component during about two solar cycles.

Figure 1 shows the rotation periods of the positive magnetic axes of the horizontal dipole and quadrupole components during about two solar activity cycles. The rotation period of the horizontal dipole component is

Y. Uchida et al. (eds.),
Magnetodynamic Phenomena in the Solar Atmosphere – Prototypes of Stellar Magnetic Activity, 627–628.

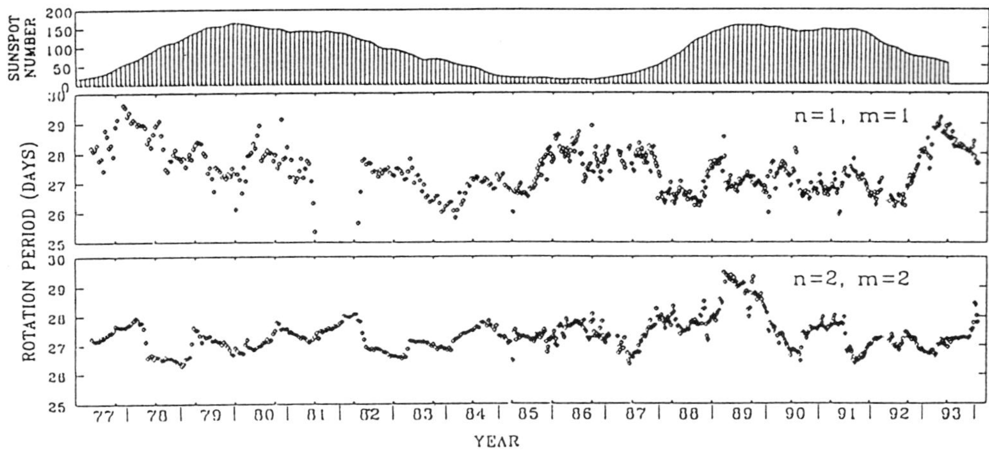

Figure 1. Long-term variation of the rotation periods of the positive magnetic axes of the horizontal dipole and quadrupole components from 1977 to 1993. Top panel shows the relative sunspot numbers.

about 28-29 days in the rising phase of the solar activity cycle and about 27 days in the declining phase. In cycle 22 the 28-days rotation period also appears from the end of 1992 to 1993. On the other hand, the rotation period of the quadrupole component is different from that of the horizontal dipole. The quadrupole shows the rotation period of 27-28 days in cycle 21, while more than 29-days period appears in the rising phase of cycle 22. A rapid change of the rotation period from 29 days to 26.5 days is seen at the middle of 1990 near the solar maximum of cycle 22.

These results show that each mode of the spherical harmonics of the photospheric magnetic field has own rotation period. It is also concluded that the rotation periods of the horizontal dipole and quadrupole exhibit a solar cycle dependence (Kozuka *et al.*, 1995).

References

Hoeksema, J.T., and Scherrer, P.H. (1986) The solar magnetic field — 1976 through 1985, *Rep. UAG-94, World Data Cent. A for Sol. Terr. Phys.*, Boulder, Colo.

Kozuka, Y., Kojima, M., and Saito, T. (1995) Long-term variation of the rotation of the large-scale solar magnetic field, to be submitted.

GRAND MINIMA IN SOLAR MAGNETIC ACTIVITY

S.M. TOBIAS
Department of Applied Mathematics and Theoretical Physics
University of Cambridge, Cambridge CB3 9EW, UK

The basic 11 year solar cycle is aperiodically modulated on a longer timescale of approximately 80–90 years (the Gleissberg cycle) and solar activity is recurrently interrupted by Grand Minima on a timescale of about 200 years (see Weiss in these proceedings). It is widely accepted that the Sun's magnetic field is generated by a dynamo acting at (or near) the base of the convection zone, and it is possible to construct a very simple third-order model (Tobias *et al.* 1995), using physical arguments and normal form theory, that demonstrates that aperiodic modulation of the basic cycle follows immediately from the assumption that dynamo action sets in at a supercritical Hopf bifurcation. In their paper this modulation is associated with the appearance of grand minima, but another interpretation is that the modulation leads naturally to the Gleissberg cycle and that another explanation must be found for the appearance of minima.

One possible mechanism has been proposed by M. Schüssler (Ferriz-Mas *et al* 1994). Here the normal solar cycle is produced by a 'dynamic α-effect' due to the motion of flux tubes rising under the influence of buoyancy forces to induce helicity at the base of the convection zone. When the dynamic α-effect is in operation strong fields are easily generated and the dynamo works 'normally'; i.e. (using the variables and equations of Tobias *et al.*).

$$\left.\begin{array}{rcl} \dot{r} &=& (\lambda + z + cz^2)\,r + ar^3 + bzr^2 \sin\phi, \\ \dot{\phi} &=& \Omega - bzr\cos\phi, \\ \dot{z} &=& -z - r^2, \end{array}\right\} \text{if } |z| > z_0 \text{ or } r > r_0. \quad (1)$$

These strong fields may act back on the flows that produce them and switch off the dynamic α-effect. With this α-effect no longer working the magnetic field has to build up slowly due to the action of the turbulence, so

$$\left.\begin{array}{rcl} \dot{r} &=& \epsilon\,\mathrm{rand}(0,1), \\ \dot{\phi} &=& \mathrm{rand}(0,1) - bzr\cos\phi, \\ \dot{z} &=& -z - r^2, \end{array}\right\} \text{if } |z| < z_0,\ r < r_0 \quad (2)$$

Y. Uchida et al. (eds.),
Magnetodynamic Phenomena in the Solar Atmosphere – Prototypes of Stellar Magnetic Activity, 629–630.

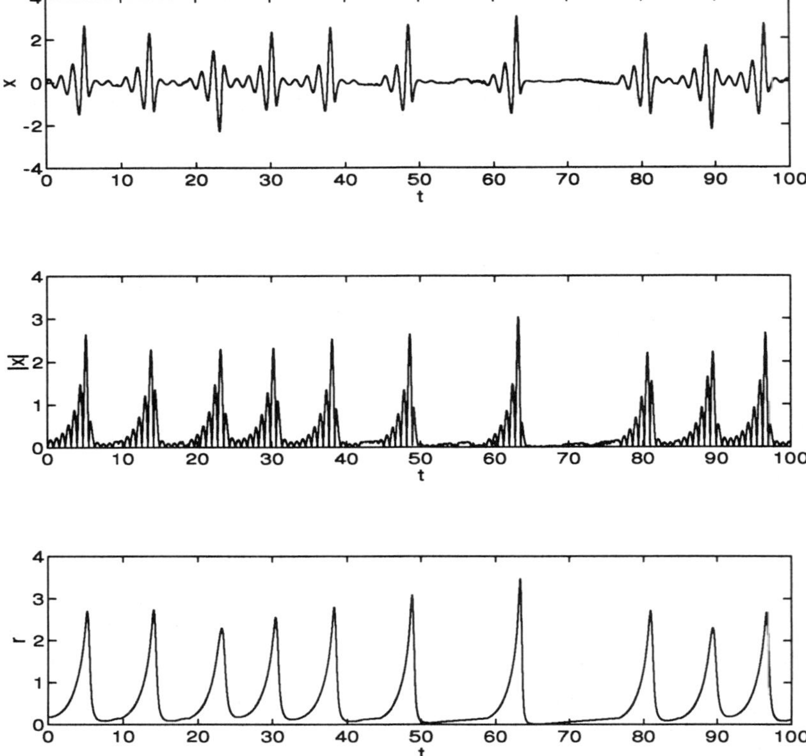

Figure 1. a) Time series for toroidal field $x(t)$. b) Time series for toroidal field strength $|x|(t)$. c) Time series for magnetic field strength $r = (x^2 + y^2)^{1/2}$. The modulation and occurrencce of minima are clearly shown.

where ϵ is small and rand$(0,1)$ is a random number between zero and one.

These equations are integrated numerically and the results are shown in Figure 1. It is clear that both the modulation of the basic cycle (The Gleissberg cycle) and the appearance of minima (due to switching) can be successfully modelled.

References

Ferriz-Mas, A., Schmitt, D., & Schüssler, M. (1994) A dynamo effect due to the instability of magnetic flux tubes *Astron. and Astrophys.* **289**, 949

Tobias, S.M., Weiss, N.O. & Kirk, V. (1995) Chaotically Modulated Stellar Dynamos *Mon. Not. Roy. Ast. Soc.*, **273**, 1150

SPECTRAL IMAGING OF AR LAC FROM 1981 TO 1994

I. PAGANO AND M. RODONÒ
Catania Astrophysical Observatory and Institute of Astronomy of Catania University
F.M. WALTER
New York State University, Stony Brook
AND
J.E. NEFF
Penn State University

Abstract. The eclipsing RS CVn-type system AR Lacertae (K0 IV-G2 IV) has been observed with IUE satellite approximately every two years from 1981 to 1994. The principal purpose of these observations was to obtain systematic series of high-resolution ($\Delta\lambda \simeq 0.24$ Å) Mg II h & k line spectra suitable for Spectral Imaging analysis and long-term activity study (Walter et al. 1987, Neff et al. 1989).

We present here preliminary results of the 1994 observation run, and discuss the long-term chromospheric activity evolution since 1981.

1. Preliminary results of the 1994 campaign

59 SWP low-resolution and 59 LWP high-resolution spectra were obtained by IUE between Oct 02 and 06, with a continuous coverage of $\sim$102 hours. As shown in Figure 1, the Mg II k -line chromospheric emission is mainly modulated by the primary and secondary eclipses (vertical dashed lines), whilst only the primary eclipse is clearly visible in the C IV light curve. This is indicative of a confined transition region on the G star, in agreement with Ottmann (1993), who found a low corona on the G component from the analysis of the X-ray emission. At least three TR flares are visible in the C IV light curve, which were not detected in chromospheric lines.

2. Spectral Imaging results and long term variability

In Figure 2 an illustrative drawing of the AR Lac chromospheric spatial configurations as resolved by our *Spectral Imaging* analysis of data acquired

Y. Uchida et al. (eds.),
Magnetodynamic Phenomena in the Solar Atmosphere – Prototypes of Stellar Magnetic Activity, 631–632.

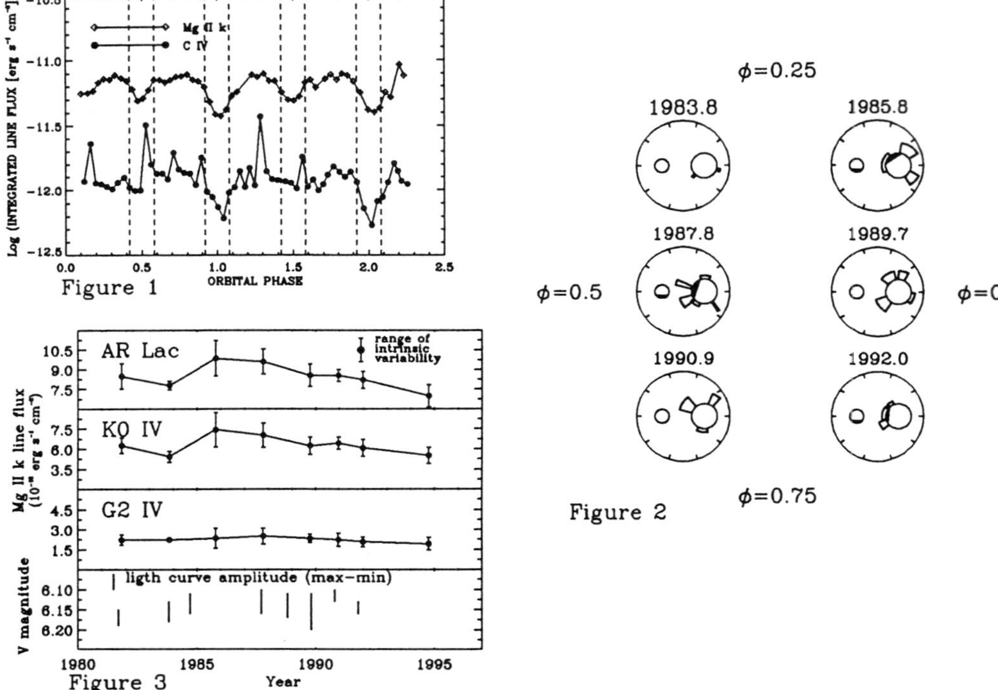

Figure 1. AR Lac Mg II *k* and C IV fluxes versus orbital phase between Oct 02-06,1994.
Figure 2. A drawing of the AR Lac chromospheric spatial configurations as resolved by our *Spectral Imaging* analysis.
Figure 3. AR Lac long-term variability in Mg II lines and in the V magnitude.

between 1983 and 1992, is shown.

Mean out-of-eclipses Mg II *k* line fluxes versus time are shown in Figure 3 for the system (*top panel*) and the separate components (*middle panels*). The vertical bars indicate the range of variability during each observation run. In the years 1986-1988, AR Lac showed a maximum of chromospheric emission. The long-term modulation of the AR Lac Mg II light seems to be mainly due to the K star. In fact, the G star emission was almost constant during the investigated time interval. The amplitudes (max-min) of the AR Lac V light-curves are plotted versus time in Figure 3 (*bottom panel*). The period of maximum chromospheric emission apparently occurs close to the phase of minimum spot coverage. However, no conclusive evidence of correlation or anticorrelation between the chromospheric and photospheric variability can be drawn.

References

Neff J.E., Walter F.M., Rodonò M., Linsky J.L., 1989, Astron. Astrophys.215, 79
Ottmann R., Schmitt J.H.M.M, Kürster M., 1993, ApJ 413, 710
Walter F.M., et al., 1987, Astron. Astrophys.186, 241

TWO MECHANISMS OF A CURRENT SHEET CREATION IN THE SOLAR CORONA.

A.I. PODGORNY
Lebedev Physical Institute
Leninsky Prospect 53, Moscow, 117924, Russia

I.M. PODGORNY
Institute for Astronomy
Pyatnitskaya Str. 48, Moscow,109017, Russia

AND

S. MINAMI
Department of Electrical Engendering, Osaka City University
3 Sugimoto Sumiyoshi Osaka, 558 Japan.

Many papers confirm the possibility of energy storage in a current sheet (CS) and its fast transfer into thermal energy and plasma acceleration. The energy storage in the solar corona may occur by disturbances focusing in the singular magnetic line or at powerful plasma flow interaction with the transverse magnetic field. Both possibilities were numerically simulated.

One-fluid MHD equations for compressible plasma has been numerically solved for the magnetic field $\vec{B}$, the velocity $\vec{V}$, the temperature T, and the plasma density n. The PERESVET program has been used [1, 2] for solving these equations.

Current sheet creation by small disturbance on the photosphere has been investigated by numerically solving the equations in the 3D space for $\beta = 10^{-2}$. The initial magnetic field with a neural line is approximated by 4 dipoles. The disturbance focusing and CS creation has been observed at different types of disturbance on the photosphere. The stable current sheet is appeared in the Alfvenic time. Simultaneously plasma acceleration takes place along the CS, because of a normal magnetic field component in the CS. The $\vec{j}\times\vec{B}$ force produces plasma acceleration in both direction along the CS up to the super Alfvenic velocity.

Y. Uchida et al. (eds.),
Magnetodynamic Phenomena in the Solar Atmosphere – Prototypes of Stellar Magnetic Activity, 633–634.

In the process of the CS evolution the CS transfers in an unstable condition. The energy storage in the magnetic field can release and a flare occurs [1].

Besides the mechanism of long time CS creation at disturbance focusing the alternative possibility can occur at supper Alfvenic plasma jet interaction with the perpendicular magnetic field with the plasma at $\beta = 0.01$ [3]. The numerical solution is carried out for 2D case. The jet velocity is four Alfven velocity. In time order of one Alfvenic time after injection a typical stationary CS is created. No MHD instability has been observed. Apparently, any arising disturbance transfers from the region by supper Alfvenic plasma flow. When the plasma injection is ceased the CS become unstable and fast energy release is observed.

The electrodynamical solar flare model [4] is based on a possibility of the CS creation in the solar corona and possibility of the field-aligned currents (FAC) generation in a CS. Measurements in the Earth magnetosphere and laboratory experiments [5] demonstrate electric field $\vec{j}\times\vec{B}/nec$ creation along a CS. Such electric field produces FAC in the Earth magnetotail. FAC closed in the ionosphere by the Pedersen currents. The similar circuit must exist in the solar corona. FAC should connect in the photosphere.

In the process of CS development, plasma outflow along the CS may exceed plasma inflow because of magnetic reconnection. CS thickness decreases and CS becomes unstable [3]. The thickens decreasing accompanied by sharp increasing of E and FAC and fast plasma acceleration along CS takes place. If FAC achieve some critical value (current velocity $\gg$ electron thermal velocity) the FAC becomes unstable and anomalous resistance or/and electrical double layers arise producing electron acceleration. Fast electrons create soft X-rays and H_α in the chromosphere. Electrons, which are accelerated upward in downward FAC, produce radio bursts. The very important feature of H_α emission is the matter motion downward. Upward motion (in $Ca\ XIX$ lines) is observed in the same place. Usually, the effect is explained by explosive heating in the chromosphere due to the pulse of fast electrons. But usually the momentum of downward motion is bigger than upward one. This effect can be explained by electrodynamical acceleration in the chromosphere between a pare of FAC.

References

1. Podgorny,A.I. (1989) *Solar Phys.*, **Vol. 123. No. 1.**, pp. 285–308
2. Podgorny,A.I. (1995) *Solar Phys.*, **Vol. 156. No. 1.**, pp. 41–64
3. Podgorny,A.I. and Podgorny,I.M. (1995) *Solar Phys.*, In press.
4. Podgorny,A.I. and Podgorny,I.M. (1992) *Solar Phys.*, **Vol. 139. No. 1.**, pp. 125–145
5. Minami,S., Podgorny,A.I. and Podgorny,I.M. (1992) *Geophys. Res. Letts.*, **Vol. 20. No. 1.**, pp. 9–12

List of Participants

Belgium
Erdelyi, Robert

Brazil
Matsuura, O. Toshiaki

Canada
Gaizauskas, Victor

China
Fang, Cheng
Gan, Wei-Qun
Lin, Yuanzhang
Tang, Yu-Hua

Czech Republic
Karlicky, Marian

DPR of Korea
Cha, Gi Ung
Li, Sang Jae

Egypt
Mosalam-Shaltout,M.A.

France
Bely-Dubau, Francoise
Montmerle, Thierry
Schmieder, Brigitte

Germany
Axford, W. I.
Jamitzky, Ferdinand
Kliem, Bernhard
Mundt, Reinhard
Nesis, Anastasios
Schmitt, J. H. M. M.
Schumacher, Joerg

India
Ambastha, Ashok K.

Italy
Catalano, Santo V.
Drago, Franca Chiuderi
Falciani, Roberto
Micela, Giuseppina
Noci, Giancarlo
Pallavicini, Roberto
Peres, Giovanni
Reale, Fabio
Rodono, Marcello
Sciortino, Salvatore
Serio, Salvatore

Japan
Akioka, Maki
Cable, Samuel B.
Enome, Shinzo
Fujiki, Kenichi
Fujisaki, Kozo
Hanaoka, Yoichiro
Hara, Hirohisa
Hasegawa, Youichi
Hayashi, Keiji
Herdiwijaya, Dhani
Hiei, Eijiro
Hirayama, Tadashi
Hirose, Shigenobu
Hori, Kuniko
Huang, Wenhong
Ichimoto, Kiyoshi
Iwata, Fumi
Kano, Ryouhei
Kawabata, Kinaki
Kitai, Reizaburo
Kitamoto, Shunji
Kitamura, Masatoshi
Kitanishi, Tadashi
Kocharov, Grant E.
Kodaira, Keiichi
Koide, Shinji
Kojima, Masayoshi
Koshiishi, Hideki
Kosugi, Takeo
Koyama, Katsuji
Kozuka, Yukio
Kudoh, Takahiro
Kurokawa, Hiroki
Kusano, Kanya
Magara, Tetsuya
Masuda, Satoshi
Matsuda, Takeshi
Matsumoto, Ryoji
Minami, Shigeyuki
Morimoto, Kouji
Morita, Satoshi
Nakagawa, Tomoko
Nakagawa, Yoshimichi
Nakajima, Hiroshi
Nakakubo, Kayoko
Nakamura, Masanori
Nishikawa, Takara
Nishio, Masanori
Ohki, Kenichiro
Ohyama, Masamitsu
Okubo, Akane
Saito, Takao
Sakai, Jun-ichi
Sakao, Taro
Sakurai, Takashi
Sano, Shusaku
Sato, Jun
Shibasaki, Kiyoto
Shibata, Kazunari
Shimizu, Toshifumi
Shimojo, Masumi
Shin, Junho
Shinkawa, Takehiko
Suematsu, Yoshinori
Summers, Danny
Suzuki, Yoshio
Takahashi, Masaaki
Takakura, Tatsuo
Takano, Toshiaki
Takeda, Aki